Liquid Metals

Liquid Metals

Properties, Mechanisms, and Applications

Lei Fu
Mengqi Zeng

WILEY-VCH

Authors

Prof. Lei Fu
Wuhan University
College of Chemistry and Molecular Sciences
No. 299, Bayi Road
Wuchang District
430072 Wuhan
China

Prof. Mengqi Zeng
Wuhan University
College of Chemistry and Molecular Sciences
No. 299, Bayi Road
Wuchang District
430072 Wuhan
China

Cover Image: © Oxygen/Getty images

Library of Congress Card No.: applied for

British Library Cataloguing-in-Publication Data
A catalogue record for this book is available from the British Library.

Bibliographic information published by the Deutsche Nationalbibliothek
The Deutsche Nationalbibliothek lists this publication in the Deutsche Nationalbibliografie; detailed bibliographic data are available on the Internet at <http://dnb.d-nb.de>.

Print ISBN: 978-3-527-82818-0
ePDF ISBN: 978-3-527-82877-7
ePub ISBN: 978-3-527-82879-1
oBook ISBN: 978-3-527-82878-4

Typesetting Straive, Chennai, India

Printed on acid-free paper

C106986_080322

Contents

Preface

Liquid metals can flow freely like water at relatively low temperatures, thus changing the perception of metals as solid. Taking gallium (Ga) as an example, hold it in your hand for a while and, interestingly, it melts and turns into a silvery drop that rolls around in your hand. More interestingly, it can be encapsulated in carbon tubes to form the smallest thermometer in the world. This kind of fascinating materials bring people into a new world. Nowadays, liquid metals are emerging as a new cutting-edge material in the frontiers of science and technology and nurturing a revolution in both fundamental discovery and unconventional applications.

Combining the characteristics of liquid (with deformable and injectable characteristics) with metal (with electrically and thermally conductive properties), liquid metals are favorable for future flexible electronics such as high-resolution conductors, keypads, pressure sensors, triboelectric nanogenerator, and other wearable sensors. For precisely constructed wearable electronics, higher integrated resolution is always in pursuit. Liquid metal inks can enable high-resolution, reconfigurable three-dimensional (3D) printing, as they are intrinsically stretchable and exhibit low toxicity, negligible volatility, and superb electrical conductivity.

Like in the movies of Terminator, liquid metals can deform to any shape and are capable of autonomous healing after suffering damage. They can exhibit more biological-like behaviors, such as breathing, eating, self-moving, and leaping. They could even bestow a soft robot with a sense of touch. Furthermore, to meet the requirement of invasive surgery or drug delivery in human body, microrobots are in urgent need. More recently, the core parts of robots, soft electromagnetic actuators, were formed by replacing solid metal coils with liquid metal channels. Thus, advanced microrobots for medical applications were enabled. In this way, the classic movie scene from *Fantastic Voyage*, a miniaturized submarine destroys a blood clot and saves the patient's life may come true one day. Nowadays, many researchers from different disciplines are entering into the field of liquid metals. It is believed that this kind of interdisciplinary crossing will certainly give liquid metal robots more new life.

On account of high deformability, high conductivity, and superior electrochemical properties, liquid metals are also promising as electrodes and they can endow rechargeable batteries with high theoretical specific capacity, and meanwhile intrinsically avoid uneven dendrites issues. Compared to traditional Na electrode, the emerging room temperature liquid metals are considered as electrode materials with more promising features, because they can keep the liquid advantages without external energy input. Liquid metals also hold enormous potential as routes to catalysts for many industries and applications, owing to the catalytic ability given by their metal nature and the liquid characteristic, offering a low-energy way to reduce atmospheric carbon and help combat climate change.

In addition, unlike solids, liquids can facilitate reactions below the surface, where a sea of ions and electrons can be exploited. Due to their wide melting range, reducing atmosphere, and electron-rich environment, liquid metals can dissolve different metals and are recently emerging as medium for chemical reactions.

The field of liquid metals owns great and limitless potentiality, providing a solid foundation for future discoveries and developments. Over the last half century, a large number of research and review articles on the properties and applications of liquid metals have been reported. Even so, accurate and reliable information of their structures and physicochemical properties are still not plentiful. Therefore, it is urgent to unveil the mystery of liquid metals. And that's what I and Professor Zeng try to do in the book. After several years of accumulating knowledge in the liquid metal field, we would like to share our understanding about this unique material family in a more professional and widespread way. The contributors of this book are all the members of our lab, Laboratory of Advanced Nanomaterials of Wuhan University.

To give a thorough understanding of liquid metals and promote their further researches and rapid applications, this book is divided into two parts, namely, fundamental properties and emergent applications of liquid metals. The first part of this book is the fundamentals of liquid metals. The book first gives a brief introduction of liquid metals in Chapter 1 (contributors: Yao Wang, Linyi Li, Shasha Zhao). Then it describes the structures of liquid metals and discusses their physicochemical properties, as dictated by the structures in Chapter 2 (contributors: Yiran Ding, Chenyang Wang). Based on the understanding of the atomic structure of liquid metals, we move toward describing the properties of liquid metals: rheological, electrical, thermal, and surface properties. It is these amazing properties that have generated the exuberant fascination in liquid metals. Then we have included Chapter 3 (contributors: Linyang Li, Xuejiao Chang) as a summary of the preparation methods of liquid metals. It covers preparation of liquid metals from zero-dimensional to two-dimensional. Then Chapter 4 (contributors: Jiaqian Zhang, Jinxin Liu, Jia Lei) gives a description of external field tuning of liquid metals, such as electrical, magnetic, chemical, force field, and others. The other part including Chapters 5–9 gives an overview of the representative emergent applications of liquid metals: new reaction medium (Chapter 5, contributors: Kena Yang, Yunxu Chen, Guanghui Cao), constructing architectures (Chapter 6, contributors: Jingjing Si, Feifei Cao, Jingjing Liang), stretchable electronics (Chapter 7, contributors: Yao Xiao, Jinglu Liu, Mengyue Zhou), biomedicine (Chapter 8, contributors: Luyang Wang, Yu Ding, Xian Sun), and energy (Chapter 9, contributors: Wenjie Wang, Xiaohui Zhu, Fangyun Lu). Finally, the advancements achieved recently will be summarized, and perspectives for future development in the applications of liquid metals will be outlined. As technology advances and fundamental discoveries are made in this direction, liquid metals are promising to offer more fantastic opportunities as powerful new generation functional materials.

Over the past few years, my group and collaborators have devoted great efforts to shaping the fields of liquid metals. We would like to thank those people for their professional contribution. Meanwhile, we also wish to express our gratitude toward the editors for their invitation and advices for completing this book.

We trust that the book will serve as an introduction to those who want to take part in the exciting developments of this field. This book can help readers understand the technical advances and fundamental discoveries at the leading edge of liquid metals and solve the puzzles of various phenomena about liquid metals. It is believed that the intriguing and inspiring findings presented in this book will lead to more activities in the research of liquid metals and provide further opportunities for advancing liquid metal science.

January 2021 *Lei Fu*

Part I

Fundamentals of Liquid Metal

1 Introduction

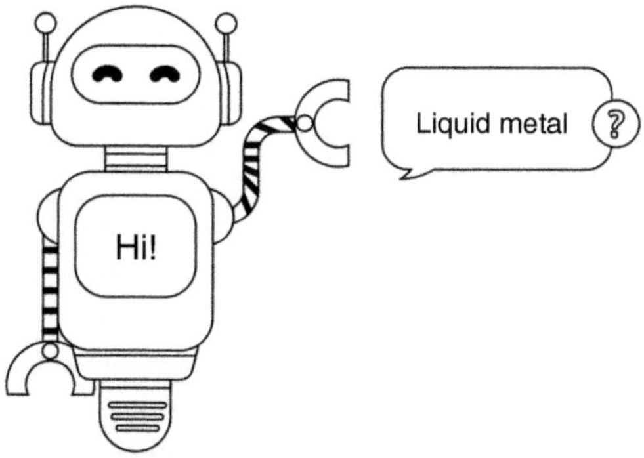

The seeds of science are grown for the harvest of the people.
Source: Dmitri Mendeleev (1834–1907). Public Domain.

1.1 The Discovery and Development History of Liquid Metal

In the world of metals, the general "members" are solid. Generally, only when the metals are heated to a relatively high temperature, the metals will exhibit a reddish liquid state. In fact, even if there are no severe high-temperature conditions, we can observe the flowing liquid metal. The most common liquid metal is mercury, which is found in the thermometers we use every day. In ancient China, mercury was also endowed with mysterious colors. Emperor Qin Shi Huang, the first emperor of China, built a model of a river infused with mercury in his tomb, representing his ruling power over the land and his expectation of immortality [1]. In Europe, it is well-known that the philosophy of alchemy laid the foundation for the principles and rules of modern chemistry, and mercury is the core of alchemy research. European alchemists believed that mercury is the main component of all metals and can be combined with other metals to become gold [2]. Although not as expected by ancient alchemists, mercury can indeed dissolve almost all metals to form a soft alloy, that is, amalgam, which is regarded as a metal solvent and has a

Liquid Metals: Properties, Mechanisms, and Applications, First Edition. Lei Fu and Mengqi Zeng.

wide range of applications. Unlike mercury, which always appears in a liquid state in a general environment, the melting point of gallium is just around room temperature. We can easily switch the state of gallium between solid and liquid. For example, if we put gallium on the palm, we can observe its transition from solid to liquid (different from mercury). This may be the reason why people are fascinated by gallium.

As early as the nineteenth century, the special property of gallium became the subject of a classic prank among chemists. One popular trick is "a disappearing spoon." Since gallium looks like aluminum and molds easily, is to fashion gallium spoons and serve the unsuspecting guests with tea. Upon stirring the tea, the gallium spoon melts rapidly. And watch your guests' surprised looks when their tea "eats" their spoons [3]. More in practical, its low-melting point and high boiling point mean that gallium could exhibit metallic properties in the liquid phase across a range of useful, desirable, and accessible temperature.

Recently, scientists have played a new trick with gallium. They have written the smallest book in the world with the help of focused gallium ion beam. Teeny Ted from Turnip Town (2007), published by Robert Chaplin, is certified by Guinness World Records as the world's smallest reproduction of a printed book. The book, costing an enormous \$15 000 (around £10 000), yet made is a 30 micro-tablet book carved on a pure crystalline silicon page by using a focused ion beam, is measured to be just 0.07 mm by 0.10 mm, according to Simon Fraser University. As we know, gallium is typically used as a liquid metal ion source for a focused ion beam. It is gallium as the liquid metal ion source used in the focused ion beam that created the smallest book in the world. The book even has its own ISBN reference, ISBN-978-1-894897-17-4, though the readers will require a scanning electron microscope to read the story of a turnip contest [4].

Liquid metals can bring us much more than that. In fact, they have become a part of our daily lives. Liquid metal may exist as a dental implant in our teeth, as fusible metal in automatic fire-fighting devices, or it could exist as a central processing unit (CPU) coolant in our computers. In the future, liquid metal could also be found in our foldable smartphones. Now, it's time to enter the world of liquid metals and learn more about this fascinating substance.

1.1.1 A Vindication for Mendeleev's Prediction

Gallium has only a short history. It was first discovered only a little over a 100 years ago. Particularly, gallium was the first chemical element in the history of chemistry that was predicted theoretically and then verified in nature in 1875.

The periodic table is undoubtedly the most commonly used tool in chemistry. In fact, it greatly facilitates scientific research and the memory and analysis of elements. Before Dmitri Mendeleev arranged the elements as the periodic tables that we are familiar with now in 1869, there were various periodic tables. However, in contrast to the previous scientists, Mendeleev's method of dividing the elements is more concise and beautiful, which is widely recognized and has been used ever since. He reasoned that there should be an as-yet-undiscovered element. Based on the position of the element in the table, he predicted some of their characteristics and properties. The main reason that Mendeleev's periodic table is so convincing may come from the fact that he initially left gaps in it to predict the existence of elements and then to be confirmed later.

As early as 1875, the French chemist Lecoq de Boisbaudran discovered gallium and successfully extracted and purified this new metal. The measured properties of this new element were surprisingly in agreement with Mendeleev's prediction and confirmed his placement of the eka-elements in the periodic table. At present, the industry of gallium production and purification technology has been highly mature, and the purity of gallium can reach 99.9999% [3, 5].

1.1.2 Gallium's Applications in the Semiconductor Industry

With the discovery of gallium arsenide (GaAs) compounds as semiconductor materials, in the early 1960s, gallium alloy began to attract the attention of various researchers. Perhaps alloying with various metals easily is the most crucial characteristic of gallium. And the low-melting point property of gallium can be conferred to the final gallium alloy, making the resulting material more easily workable, stable, and cost-effective. In the semiconductor industry, its alloying with other elements and phase transition properties are highly valuable. Thus, a lot of gallium applications are concentrated in the semiconductor industry. GaAs, as gallium's most common alloy and a new type of high-quality semiconductor, has launched a research boom, which is widely used in preamplifiers and high-speed logic chips in mobile phones, while AlGaAs and InGaAs are often applied as the light-emitting materials. Blue light-emitting diodes (LEDs) were successfully developed in the early 1990s, and the development of white LEDs also followed. Since then, a "lighting revolution" have begun. With gallium consumption soaring, coupled by commercial speculation, the price of gallium rose substantially. After nearly 20 years of development, white LED lighting technology has made great achievements. Compared with the traditional lighting technology, LED lighting technology has the advantages of high-efficiency, energy-saving, long life, green environmental protection, and high light efficiency, which has been supported by governments all over the world. At present, the research and production of GaAs have been mostly turned to the LED industry [3].

1.1.3 Tackling Fundamental Problems in Fuel in Energy Science

In addition to its contribution to the semiconductor field, gallium chemistry has also addressed many fundamental questions in energy science. Early studies have shown that the zeolite doped with gallium can effectively catalyze methylcyclohexane [6]. The ring is opened and broken to form short alkanes, which is used to recover gasoline pyrolysis products. Moreover, gallium zeolites can also catalyze the aromatization of *n*-decane [7]. Emerging studies have shown that GaN nanowires can even catalyze the formation of benzene from methane [8]. Breaking the C—H bond of methane is not easy, which will contribute to these processes mentioned above.

The relative inactive methane exhalations and undesired by-products of gasoline production can be converted into useful petrochemicals. These reactions are crucial for the conversion, which are of great significance for fine chemical production and fuel storage.

1.2 Liquid Metal Family

It is common knowledge that the existing form (solid, liquid, and gas) of all elements and chemical compounds rely on conditions of temperature and pressure. But when it comes to metal, the first thing we think about is bulky objects and high melting point. It is hard to believe that some metals exist in liquid form at room temperature. One of the most intriguing properties of gallium is its liquid state at room temperature. When you cut off a piece of solid gallium with a knife and put it in your hand for a closer look, an interesting thing happens: It melts and turns into a silvery drop that rolls around in your hand like a drop of water on a lotus leaf. The reason for this curious phenomenon is that the melting point of gallium is only 29.8 °C, below the human body temperature, so it will melt in the hand. In addition, although gallium's melting point is low, the boiling point can be as high as 2200 °C. In other words, from about 30 °C to about 2000 °C, gallium would remain in liquid state, while mercury would boil at 360 °C. Therefore, gallium can be used to make high-temperature thermometers that can measure temperatures below 1500 °C [3].

Herein, we define metals and alloys with melting point between room temperature and 300 °C as liquid metals. Single-component liquid metals are predominantly composed of post-transition metals (Ga, In, Tl, Sn, Pb, Al, and Bi), zinc-group metals (Zn, Cd, and Hg), and alkali metals (Li and Na). We note that alloys possess much lower melting point than the pure metal. For example, the melting point of eutectic gallium and indium alloy (EGaIn) is only 16 °C [9]. So there are much more liquid metal alloys. Tables 1.1 and 1.2 list the melting point of several liquid metals and alloys. To enrich the content of this book, we will also introduce some molten metals (such as copper, gold, etc.), which have similar properties compared to liquid metals [10] (Figure 1.1).

In recent decades, low-melting point liquid metals have been used in many different fields. Gallium base liquid metal at room temperature has unique surface properties and physicochemical properties. It can achieve a variety of morphological changes such as deformation, movement, separation, and fusion through a variety of energy fields such as electric, magnetic, and concentration gradient fields or surface modification. Liquid metals show a promising application in microfluidics, biomedicine, and robotics, and other fields.

At present, numerous studies devoted to liquid metals are dispersed and need to be unified. All the emerging applications in the fields of catalysis, synthesis, microfluidics, soft electronics, sensor, and therapy exploit the same phenomena that occur in liquids and at metal interfaces. Integrating these seemingly unrelated research directions into a single field of liquid metal is quite challenging. But it will facilitate the development of innovative applications that take full advantage of the unique properties of liquid metals and its alloys [10].

Table 1.1 Atomic number and melting point of liquid metals.

Elements	Atomic number	Name (abb.)	Melting point (°C)
Lithium	3	Li	180.5
Sodium	11	Na	97.8
Aluminum	13	Al	660.3
Cadmium	48	Cd	321.1
Mercury	80	Hg	−38.8
Zinc	30	Zn	419.5
Gallium	31	Ga	29.8
Indium	49	In	156.6
Thallium	81	Tl	304.0
Tin (white)	50	Sn	231.9
Lead	82	Pb	327.5
Bismuth	83	Bi	271.4

Table 1.2 The composition and melting point of several eutectic binary alloys, ternary alloys, quaternary alloys, and quinary alloys.

Alloyed elements	Element A (at%)	Element B (at%)	Element C (at%)	Element D (at%)	Element E (at%)	Melting point (°C)
Ga/In(EGaIn)	85.8	14.2	0	0	0	15.4
Ga/Sn	91.7	8.3	0	0	0	21.0
Ga/Bi	38.3	61.7	0	0	0	222.0
In/Sn	52.7	47.3	0	0	0	116.9
In/Bi	79.2	20.8	0	0	0	72.5
Pb/Au	84.6	15.4	0	0	0	215.0
Sn/Tl	56.6	43.4	0	0	0	170.2
Ga/In/Sn (Galinstan)	78.3	14.9	6.8	0	0	13.2
In/Sn/Bi (Field's alloy)	60.1	18.8	21.1	0	0	62.0
Bi/Pb/Sn (Rose's alloy)	43.1	23.5	33.4	0	0	95.0
Bi/Pb/Sn/Cd (Wood's alloy)	41.5	20.9	18.3	19.3	0	70.0
Bi/Pb/Sn/Cd/In (French's alloy)	31.5	17.1	14.4	11.7	25.3	46.9

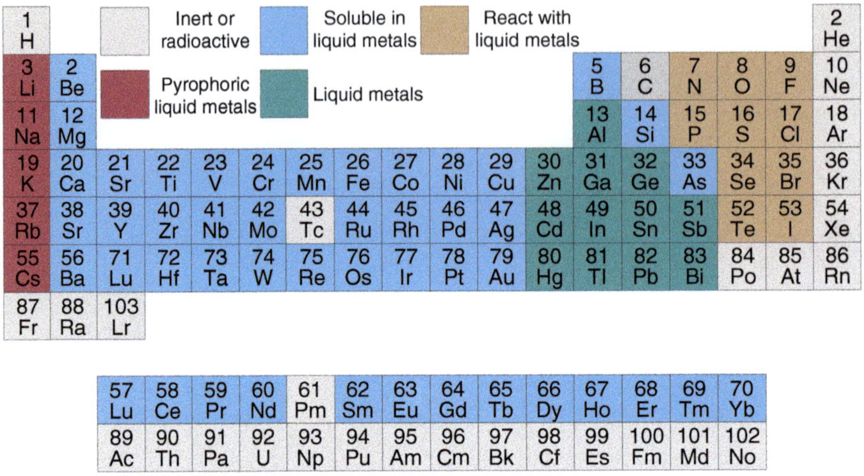

Figure 1.1 Liquid metal base elements. [10]. Source: Daeneke et al. [10]. Reproduced with permission of Royal Society of Chemistry.

1.3 Overview of Liquid Metal

What is well-known is that mercury can dissolve almost all metals to form an amalgam and is regarded as a metal solvent with considerable application. Similarly, liquid metals can also dissolve some metallic elements and nonmetallic elements. And unlike conventional molecular liquid and ionic liquid, liquid metals possess some unique properties (high electrical conductivity, etc.). So liquid metals can serve as a new reaction medium to synthesis different dimensions, compositions, and morphology materials. Liquid metals can spontaneously form smooth, soft, defect-free, stress-free interfaces, which provide an ideal substrate for crystal growth. There are also self-limiting metal oxide films on the liquid metal surface, which are naturally available high-quality two-dimensional films and can serve as precursors to synthesis other two-dimensional materials. Moreover, many two-dimensional atomic crystals have been predicted to grow on liquid metals [11] (Figure 1.2).

In the Terminator series of science fiction movies, the fact that the T-1000 is the most amazing terminator is indisputable. He can transform into anyone and anything, and can heal wounds back to their original state as quickly as a fluid when attacked. It was the presence of liquid metal that endow the killer magical ability to transform and repair himself, making him an unbreakable killer. Thanks to this cool sci-fi fantasy about liquid metal robots, it brought liquid metals into the public eye and aroused the interest of scientists in the study of liquid metals. Compared to conventional rigid robotics, liquid metals hold huge merits, such as low-melting point, high electrical conductivity and thermal conductivity, and liquidity, which broaden their utilization in stretchable and wearable electronics, such as flexible pressure sensors, a stretchable electromagnetic actuator, stretchable wireless power transfer, and stretchable loudspeaker. Recently, a series of magical phenomena have been discovered successively, such as the movement and deformation of liquid metal through

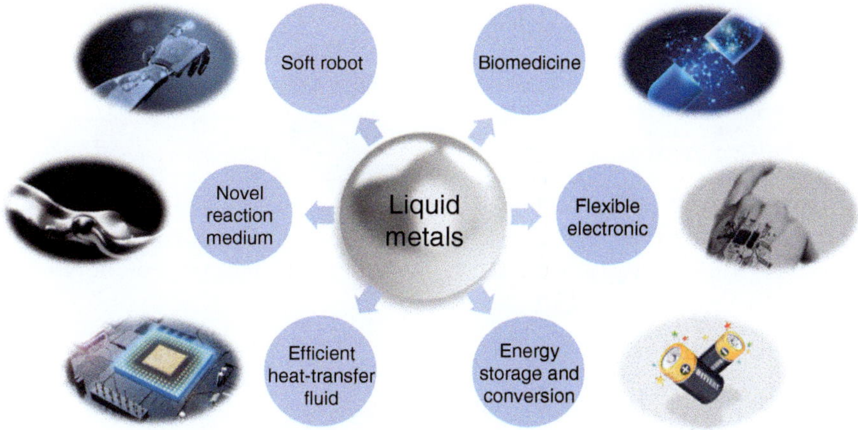

Figure 1.2 The application of liquid metals [11]. Source: Zavabeti et al. [11].

electric field control or magnetic field, the repair of biological fractured nerve with liquid metal, and the autonomous movement of liquid metal without external power after swallowing a small number of substances [12–14]. Liquid metals can also be applied to manufacture a new artificial bioinspired optoelectronic sensorimotor system, which is similar to the neuron system [15]. Besides, the color of liquid metals, which is generally shiny silver-white, can be changed by applying an electrical field, such as pink, purple, and blue. This provides a new route to create kaleidoscopic and colorful liquid metals and will expand their applications in biomimetic intelligent soft robots. These findings lay the theoretical and technical foundation for complex liquid metal robots in the future. So, will robots like Terminator become a reality in the future? Let's wait and see!

Although the pure liquid gallium has no biological activity, gallium ions can influence the reaction in our body, similar to iron ions. Several gallium salts have been used in pharmaceutical and radiopharmaceutical industries. For example, gallium nitrate has been used to treat hypercalcemia associated with tumor metastasis to bones. Besides, owing to the nontoxic and extraordinary flexibility, liquid metals can also be applied in biomedical applications, such as drug delivery, molecular, cancer therapy, and biomedical devices. The nanomachines prepared by liquid metals can be injected into cells to treat some cell diseases [16]. Due to its self-driven ability, the liquid metals robot containing medicine can be automatically delivered into targeted tissues to release medicines to cure diseases [17]. As for the biocompatibility problem of theses liquid metal microrobots, it can be solved by the functionalization of liquid metal (Figure 1.3).

Recently, various technology giants have started the launch boom of foldable smartphones. The emergence of foldable phones (mobile phones with folding screen) has provided a new application opportunity for liquid metals and attracts people's attention to liquid metals again. Going back to the problem of screens, folding screens is not as easy as it seems. How to solve this technical problem? The liquid metal gives the answer. The liquid metal itself has excellent shaping

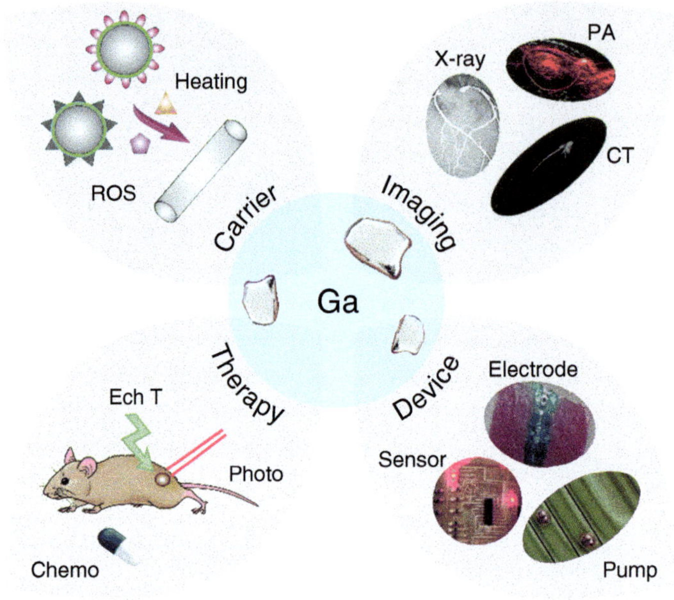

Figure 1.3 Schematic of liquid metals (or alloys) and gallium-based LMs for biomedical applications [18]. Source: Yan et al. [18]. Reproduced with permission of Royal Society of Chemistry.

ability and excellent dimensional stability. Therefore, it can meet the requirements of thickness, strength, and precision of hinge support parts, which makes liquid metal one of the main solutions for support. In this case, the prospects for liquid metals with outstanding performance advantages are clearly limitless. In addition, liquid metals combine the advantages of traditional metal conductivity with the flexibility of nonmetallic materials. At present, liquid metals have been widely used to make flexible electrons, such as stretchable wire, interconnects, and antennas. Liquid metals can also be applied in memory devices, diodes, electrodes, and capacitors. In principle, the deformation ability of liquid metals is not limited. In addition, liquid metals can form self-healing and shape-reconfigurable circuits, and it can be utilized in soft sensors, such as strain sensing, pressure sensor, and touch sensing, because their resistance or capacitance will correspondingly change when deformed. Liquid metals can also be utilized as electronics to sense the change of relative humidity, oxygen, and temperature. Compared with the traditional rigid electron, the flexible electron has great bendability and stretchability, leading to promising applications in the fields of information, energy, medical treatment, and national defense (Figure 1.4).

Energy is the basis of human survival and development. How to use energy efficiently has been extensively studied? Considering the intermittent and nonuniformity of energy, energy storage and conversion is very important. At present, batteries, capacitors, and inductors have been widely used in energy storage and conversion. The battery is widely used in people's life because of high efficiency and

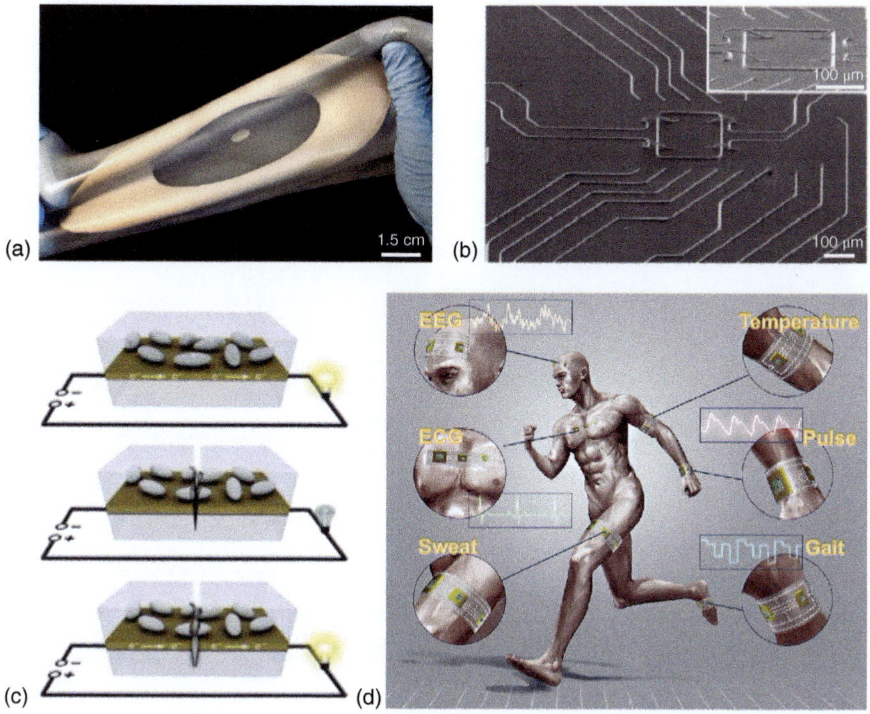

Figure 1.4 (a) Stretchable materials based on liquid metals [19]. Source: Pan et al.[19]. (b) The shape-reconfigurable circuits [20]. Source: Park et al. [20]. (c) The self-healing circuits [21]. Source: Blaiszik et al. [21]. Reproduced with permission of John Wiley and Sons. (d) The sensor based on liquid metals [22]. Source: Rui et al. [22]. Reproduced with permission of Springer Nature.

convenience. But the major challenges of traditional lithium batteries lie in dendrite growth. Liquid metals and alloys, which possess inherent deformability, high electronic conductivity, self-healing, and excellent electrochemical properties, have attracted considerable attention. Utilizing liquid metals can realize dendrite-free, self-healing, low reduction potential, high capacity and energy density, and fast mass/charge transport [23] (Figure 1.5).

Liquid metals usually exhibit outstanding thermal conductivity. For example, the thermal conductivity of water is 0.6 $W (m K)^{-1}$, which is seriously lower than those of gallium (29.6 $W m^{-1} K^{-1}$) at 50 °C [24]. So it is a good coolant. In fact, the application of gallium is particularly prominent in the field of heat dissipation [25]. As a liquid metal, it is inherently more conductive than air, water, and many nonmetallic media. Sapphire Technology, a famous video card manufacturer, displayed a graphics card with liquid metals at the E3 game show and Taipei International Computer Show in 2005. The data showed liquid metal cooled the working temperature of the strongest video card at that time, the Radeon X850 XT PE, to 12 °C, and its heat conduction efficiency was 65 times or faster than that of water at that time. At the beginning of 2019, liquid metals cooling systems appeared in high-end gaming laptops, such

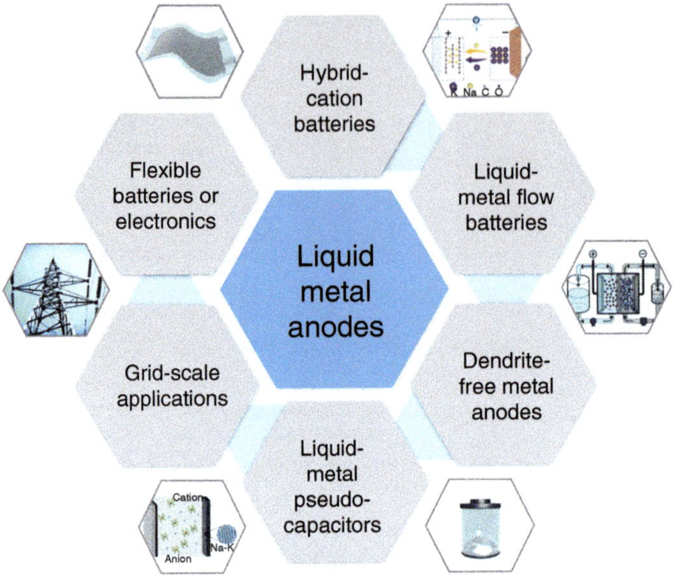

Figure 1.5 Prospective designs and applications of the liquid-metal-based electrodes in various energy storage strategies [23]. Source: Guo et al. [23]. Reproduced with permission of Royal Society of Chemistry.

as ASUS, HP, and MACHENIKE. There is no doubt that the price will increase, but it can give you a very enjoyable experience. Making good use of liquid metals as a radiator of laptops would bring rapid advances in technology.

In recent years, people focus on exploring, developing, and utilizing the universe, so new space materials with a strong performance is essential to overcome the evil environment, such as zero gravity, vacuum, larger temperature, and strong radiation. Up till now, space materials mainly include carbon-based materials, porous metals, polyimide, and fiber. Compared to the above materials, liquid metals possess a lot of distinguished properties, such as high electric conductivity, high thermal conductivity, fluidity, and low volatility, which make them a good choice for utilization in space science and technology. It has also been introduced that liquid metal has very good heat dissipation capacity. For space exploration, a liquid metal cooling system has been proved to be the most effective space nuclear electric heating management method [26]. In addition, liquid metal can be used as a thermal interface material for heat conduction. Solar energy is the only energy source in spacecraft operation, so how to store solar energy is very important. Usually, solar energy is converted into heat energy, and then heat energy is converted into electrical energy for storage. Compared with the common thermal storage materials (fluorine), liquid metal has high thermal conductivity, high latent heat of phase change, and low thermal expansion rate, which can be used as a new type for energy storage material. Three-dimensional printing technology of liquid metals provides a very simple method for rapid manufacturing and maintaining electronic circuits. More importantly, there is a large amount of electromagnetic radiation in the universe, which is

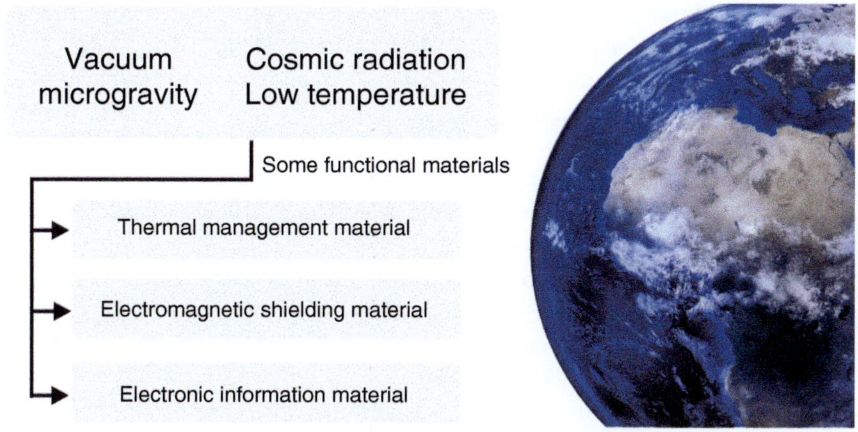

Figure 1.6 Candidate materials needed to satisfy the space physical environment [27]. Source: Zhang and Liu [27]. Reproduced with permission of Springer Nature.

not conducive to the operation of devices. The flexible liquid metal electromagnetic shielding material has good wettability and plasticity, which can meet the requirements of complex shape and effectively protect the equipment from radiation. Moreover, the liquid metals usage in science and technology is still in its infancy and we need to keep exploring and look forward to your joining (Figure 1.6).

References

1 Wright, D.C. (2001). *The History of China*. Santa Barbara: *Greenwood Publishing Group*.
2 Stillman, J.M. (2003). *Story of Alchemy and Early Chemistry*. Santa Barbara: *Kessinger Publishing*.
3 Brennan, M. (2014). Gregarious gallium. *Nat. Chem.* 6: 1108.
4 Simon Fraser University (2007). *Nano Lab Produces World's Smallest book*. Simon Fraser University.
5 Robinson, G.H. (2000). On the organometallic chemistry of gallium and the dynamics of Ga–Ga bond formation. *Chem. Commun.* 2000 (22): 2175.
6 Raichle, A., Moser, S., Traa, Y. et al. (2001). Gallium-containing zeolites – valuable catalysts for the conversion of cycloalkanes into a premium synthetic steam cracker feedstock. *Catal. Commun.* 2: 23.
7 Pradhan, S., Lloyd, R., Bartley, J.K. et al. (2012). Multi-functionality of Ga/ZSM-5 catalysts during anaerobic and aerobic aromatisation of *n*-decane. *Chem. Sci.* 3: 2958.
8 Li, L., Fan, S., Mu, X. et al. (2014). Photoinduced conversion of methane into benzene over gan nanowires. *J. Am. Chem. Soc.* 136: 7793.
9 Liu, T. (2012). Characterization of nontoxic liquid-metal alloy galinstan for applications in microdevices. *J. Microelectromech. Syst.* 21: 443.

10 Daeneke, T., Khoshmanesh, K., Mahmood, N. et al. (2018). Liquid metals: fundamentals and applications in chemistry. *Chem. Soc. Rev.* 47: 4073.
11 Zavabeti, A., Ou, J.Z., Carey, B.J. et al. (2017). A liquid metal reaction environment for the room-temperature synthesis of atomically thin metal oxides. *Science* 358: 332.
12 Zhang, J., Sheng, L., and Liu, J. (2014). Synthetically chemical-electrical mechanism for controlling large scale reversible deformation of liquid metal objects. *Sci. Rep.* 4: 7116.
13 Zhang, J., Sheng, L., Jin, C. et al. (2014).Liquid metal as connecting or functional recovery channel for the transected sciatic nerve. Eprint Arxiv. 1204.
14 Zhang, J., Yao, Y., Sheng, L., and Liu, J. (2015). Self-fueled biomimetic liquid metal mollusk. *Adv. Mater.* 27: 2648–2655.
15 Akbari, M.K. and Zhuiykov, S. (2019). A bioinspired optoelectronically engineered artificial neurorobotics device with sensorimotor functionalities. *Nat. Commun.* 10: 3873.
16 Toumey, C. (2013). Nanobots today. *Nat. Nanotechnol.* 8: 475.
17 Zhang, J., Guo, R., and Liu, J. (2016). Self-propelled liquid metal motors steered by a magnetic or electrical field for drug delivery. *J. Mater. Chem. B* 4: 5349.
18 Yan, J., Lu, Y., and Chen, G. (2018). Advances in liquid metals for biomedical applications. *Chem. Soc. Rev.* 47: 2518.
19 Pan, C., Markvicka, E.J., Malakooti, M.H. et al. (2019). A liquid-metal – elastomer nanocomposite for stretchable dielectric materials. *Adv. Mater.* 31: 1900663.
20 Park, Y.G., An, H.S., Kim, J.Y., and Park, J.U. (2019). High-resolution, reconfigurable printing of liquid metals with three-dimensional structures. *Sci. Adv.* 5: 2844.
21 Blaiszik, B.J., Kramer, S.L.B., Grady, M.E. et al. (2012). Autonomic restoration of electrical conductivity. *Adv. Mater.* 24: 398.
22 Rui, G., Xuelin, W., Wenzhuo, Y.U. et al. (2018). A highly conductive and stretchable wearable liquid metal electronic skin for long-term conformable health monitoring. *Sci. China Technol. Sci.* 61: 1031.
23 Guo, X., Zhang, L., Ding, Y. et al. (2019). Room-temperature liquid metal and alloy systems for energy storage applications. *Energy Environ. Sci.* 12: 2605.
24 Wang, L. and Liu, J. (2013). Liquid metal material genome: initiation of a new research track towards discovery of advanced energy materials. *Front. Energy.* 7: 317.
25 Deng, Y. and Liu, J. (2013). Optimization and evaluation of a high-performance liquid metal CPU cooling product. *IEEE Trans. Compon. Packag. Manuf. Technol.* 3: 1171.
26 Rachkov, V.I., Sorokin, A.P., and Zhukov, A.V. (2018). Thermal hydraulic studies of liquid-metal coolants in nuclear-power facilities. *High Temp.* 56: 124.
27 Zhang, X.D. and Liu, J. (2020). Perspective on liquid metal enabled space science and technology. *Sci. China Technol. Sci.* 63: 1127.

2

Structure and Physical Properties of Liquid Metal

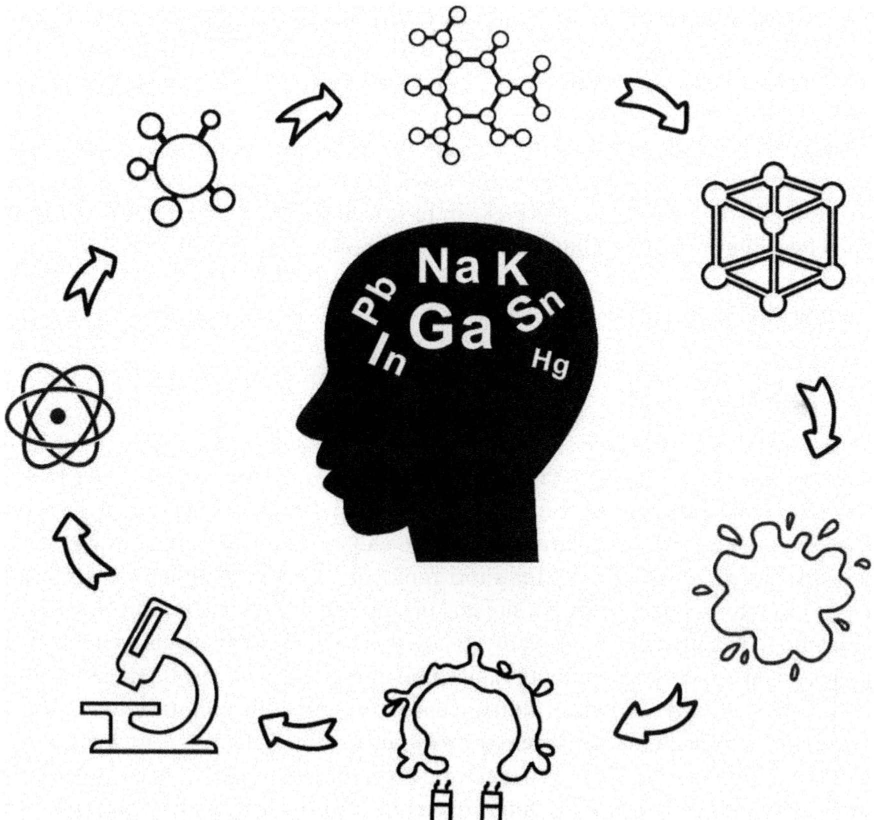

All of the reasons must come from observation and experiments.
Source: Galileo (1564–1642). Public Domain

Liquid Metals: Properties, Mechanisms, and Applications, First Edition. Lei Fu and Mengqi Zeng.

2.1 Structure of Liquid Metal

The structure of liquid metal (LM) is different from periodic crystalline solid and total disorder feature of gas; it features a short-range order and long-range disorder structures. The structure of liquid metal can be characterized by static structure factor (SSF, $S(q)$) and pair distribution function (PDF, $g(r)$), which can be obtained by experiment or theoretical simulation. The structure of liquid metal is also affected by temperature and pressure. Exploring the structure of liquid metal and alloy is of great significance in the field of solid metals and condensed matter physics. We classify liquid metal into single-component liquid metal, monophasic liquid metal, and biphasic liquid metal mix based on their component and phase, and their corresponding liquid structures will be introduced in this section. Moreover, monocomponent liquid metal contains hard-sphere-like liquid metal atoms, merely distorted structure, and strongly anormal structure. The local structure of monophasic liquid metal alloy is more disordered than that of monocomponent liquid metal, and it can be described as the mixture of several types of local structures. As for biphasicliquid metal mix, it contains liquid phase and solid phase at the same time, and can exist stably for a long time if the size of particles is appropriate. Besides, the composition of biphasic liquid metal mix can be affected by the temperature and time. Up to now, the structure of liquid metal is controversial. To analyze the structure of liquid metal, it is necessary to combine the advanced experimental technology and theoretical simulation.

2.1.1 Monocomponent Liquid Metal

Unlike the long-range order structure of crystalline solid and total disorder feature of gas, the atomic arrangement in liquid is a compromise between the two, which is commonly considered as short-range order and long-range disorder. When the metal is in crystal solid phase, its structure can be well explained by three-dimensional (3D) crystallography. However, the short-range-order structure of liquid metal cannot be directly "seen." For one thing, it is invalid to characterize liquid structure by experimental techniques based on crystallography. For another, crystallography theory based on 3D periodicity is not applicable to explain different liquid structures. Generally, SSF ($S(q)$) and PDF ($g(r)$) are used to characterize the structure of liquid metal. SSF is a mathematical description of how a material scatters incident radiation. PDF is a statistic approach to describe the atomic arrangement around a given central atom, generated from the sine Fourier transformation of the normalized scattering function SSF. We can get structure information of liquid metal at statistic average by X-ray diffraction and neutron diffraction. Moreover, taking advantages of computer simulation, we can obtain more detailed information of liquid metal, such as 3D atomic configuration, dynamic evolution of atomic structures with the temperature or time, and even thermodynamic or kinetic quantities and their dynamic evolutions. The combination of X-ray and neutron diffraction in experiments and numerical simulations in theory provides an effective method to study the structure of liquid metal and alloy, and some progresses have been made [1]. Nonetheless, there is still a lack of comprehensive theory and model to describe the structure of liquid metal. Understanding the local structure of liquid metal plays an important role in the specialized alloy design and multiple property analysis.

Based on Waseda's classification, the structure of single-component liquid metal can be essentially classified into three types, which is mainly dependent on the relative relation of positions of peaks in $S(q)$ and $g(r)$. Q_1 and Q_2 represent the first-peak and second-peak positions of $S(q)$, and R_1 and R_2 are the first-peak and second-peak positions of $g(r)$, respectively. These three types of structures have the following characteristics.

(1) The first type of liquid structure (hard-sphere-like structure) has the following features:
  (i) The first peaks of both the $S(q)$ and $g(r)$ are symmetrical.
  (ii) $Q_2/Q_1 = 1.86$ and $R_2/R_1 = 1.91$.
  (iii) The coordination number (Nc) is 9.0–11.0.

Liquid IA and IIA elements, Al and Pb, are classified into this class.

(2) The second type of liquid structure (merely distorted structure) is characterized by the following features:
  (i) The first peaks in $S(q)$ and $g(r)$ are asymmetric.
  (ii) The ratios Q_2/Q_1 and R_2/R_1 are slightly nonideal.
  (iii) The coordination numbers are essentially metallic, that is $Nc = 10.0$–11.0.

Liquid IIB elements, In and Tl, belong to these "distorted structures."

(3) The third type of liquid structure (strong anomaly structure) has the following characteristics:
  (i) There is a shoulder on the high-q side of the first peak in $S(q)$, and subsidiary maxima shoulder in $g(r)$ located in between the two maxima.
  (ii) The peak positions are strongly nonideal, in which Q_2/Q_1 is 1.96–2.1 and R_2/R_1 is 2.0–2.3.
  (iii) The coordination number is exceptionally low ($Nc = 6$–7), suggesting a relatively open local structure.

The liquid structures of many polyvalent liquid metals, such as Ga, Sn, Ge, Si, Ab, Sb, and Bi, are considered to exhibit these "anomalous structures" [1–4].

This classification can be explained by Hafner' theory, in which the liquid structure is mostly affected by the compatibility of the geometric and the electronic factors [5]. The first type of liquid structure, which can be described as hard-sphere-like structure, is compatible with low electron densities and large core radii. And IA and IIA elements belong to this class. Light alkali metals have a highly symmetric structure and can be described as free-electron-like models at ambient conditions. In terms of liquid Pb, the interatomic potential is purely repulsive around the close-neighbor distance, the Friedel oscillations damp, and only the geometric factors dominant [5], which can be explained by the relativistic effects in heavy elements [6]. In these simple liquids, the interatomic or ion–ion interaction can be described by spherical potential [7].

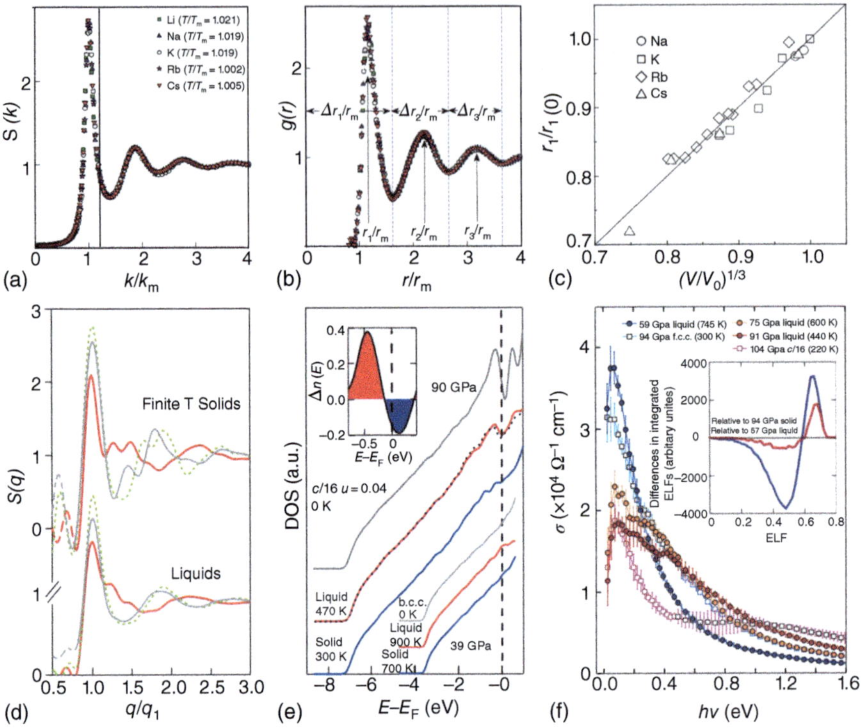

Figure 2.1 The structure of liquid alkali metals. (a) $S(q)$ and (b) $g(r)$ near melting point. (a, b) Source: Mokshin et al. [8]. Reproduced with permission of Springer Nature. (c) Volume dependence of nearest-neighbor distance (r_1). (c) Source: Katayama and Tsuji [9]. Reproduced with permission of IOP Publishing, Ltd. (d) $S(q)$ and (e) electronic density of states in solid and liquid Na under different temperatures and pressures. (f) Electrical conductivity of liquid and solid Na as a function of pressure. (d, e, f) Source: Raty et al. [10]. Reproduced with permission of Springer Nature.

Alkali metals show a "universal" behavior at the equilibrium level. In the SSF $S(k)$ and pair distribution factor $g(r)$ near each melting point of liquid Li, Na, K, Rb, and Cs, there are not broadenings and shoulders, indicating a typical structure property of hard-sphere liquid (Figure 2.1a,b). Moreover, the $S(k)$ and $g(r)$ of all alkali metals coincide with a function of k/k_m and r/r_m (k_m and r_m represent the position of the main peak of $S(k)$ and $g(r)$ in each system), respectively, showing the universality of microscopic structure in alkali metals [8].

In addition, liquid alkali metals show an almost uniform compression under high pressure. As shown in Figure 2.1c, $r_1/r_1(0)$ of liquid alkali metals is proportional to $(V/V_0)^{1/3}$ under high pressure (5.1, 6.1, and 4.3 GPa for Na, Rb, and Cs, respectively), indicating almost uniform compression [8]. This phenomenon is ascribed to the high coordination number of alkali metal at ambient pressure, and the local structure will not change drastically under high pressure. As for the slight deviation from the uniform compression in Rb and Cs, it is probably connected with an s–p electronic transition [9]. However, polyvalent metal presents a nonuniform compression under pressure, such as liquid Ga and Bi. This complicated phenomenon will be discussed next (the structure of liquid Ga).

Apart from the structure change, the electron transition will also occur in liquid metal. Complex structural and electronic transitions of dense liquid Na under pressure are revealed by ab initio calculations [10]. The structure of liquid Na evolves with pressure and temperature, similar to that of solid Na. As shown in Figure 2.1d, the change in the $S(q)$ of liquid Na matches well the bcc, fcc, and *cI*16 phases of solid Na. The parallel between liquid Na and solid Na induced by pressure is further confirmed by the variation of electronic structure (Figure 2.1e). Nonetheless, an abnormal phenomenon, the lower coordination number in liquid Na above 65 GPa, is observed, which is ascribed to the opening of pseudogap at the Fermi level. At the same time, the physical properties of liquid Na change along with the pressure, for instance, the electrical conductivity of liquid Na decreases threefold between 40 and 80 GPa (Figure 2.1f).

Liquid Pb favors a close-packed sphere structure with centrosymmetric interactions, which is consist of fivefold or sixfold symmetries [11]. Numerous studies have been conducted to manifest the local structure of liquid Pb. Chen and Li [12] studied the local structure and orientation of liquid Pb near melting point by molecular dynamics simulations, showing that liquid Pb had cubic symmetry and the long-range correlation of this symmetry was weak. With the help of synchrotron high-energy X-ray technology (Figure 2.2a), Reichert and coworkers [13] observed the fivefold local symmetry in liquid lead adjacent to a silicon wall. The local liquid lead symmetry can be a hard-sphere-like structure with a symmetrical first peak located at 2.18 Å (Figure 2.2b). The local arrangements captured in liquid Pb fitting from the SSF show the fivefold symmetry similar to bulk Pb in-plane structure (Figure 2.2c). The local structure evolution of liquid lead at varied temperatures and pressures have also been studied [14, 15].

The liquid structures of In and Tl and IIB elements show slight deviations from the simple hard-sphere-like structure. In terms of the third type of liquid metal, including liquid Ga, Sn, and Bi, it features with a high-q shoulder of the first peak in $S(q)$, suggesting a more anomalous structure. These two types of complex structures cannot be explained by the simple hard-sphere-like model, and several mechanisms

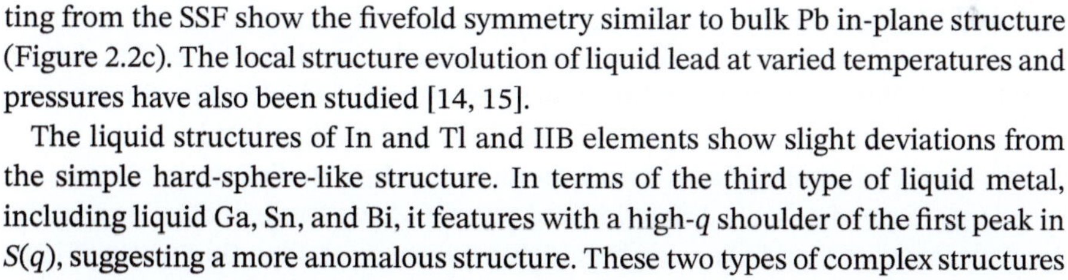

Figure 2.2 The measurement setup and local structure of liquid Pb. (a) Schematic of a synchrotron high-energy X-ray total internal reflectance measurement setup, where the X-rays are transmitted through a silicon crystal. (b) Polytetrahedral arrangement predicted for close-packed liquid Pb. (c) SSF of Pb at the liquid Pb–Si interface after subtracting the Si Signal. Source: Reichert et al. [13]. Reproduced with permission of Springer Nature.

have been proposed up to now. Hafner and coworkers considered that this special structure was determined by the interplay of two characteristic lengths, the effective hard-sphere diameter and the Friedel wavelength, which represented the geometrical requirements of sphere packing and intermediate- and long-range potentials [5, 6]. While Gaspard and coworkers showed that the liquid metal structure could be explained by local anisotropy related to Peierls distortion [16–18]. In general, the anomalous structure featuring with high-q is rather completed, and the origin for the high-q shoulder in the SSF is still controversial. In order to fully understand the cause for the high-q shoulder of these anomalous structures, a picture of atomic structures in three-dimensional space is requested in the future [19].

Among these metals, Ga and Ga-based alloys have wide application value and prospect in the field of energy, flexible device, and biomedicine. Ga presents distinct and unique physical and chemical properties compared with other metals, which is mainly owing to its distinct structure. The structure of solid Ga is multifarious, and at least eight types of different structures of Ga have been discovered, including stable and metastable phases. α-Ga is the ambient pressure-stable phase with a typical orthorhombic oS8 structure with eight atoms in the unit cell [20]. Ga-II and Ga-III are high-pressure-stable phase with body-centered cubic structure with 12 atoms in a unit cell [21], and simple body-centered tetragonal tI2 structure [22, 23], respectively. β-Ga, γ-Ga, δ-Ga, and ε-Ga have also been discovered in ambient pressure, but these phases are metastable [24, 25]. In addition, in the process of solid α-Ga converted to liquid gallium phase, Ga-IV (fcc) and Ga-V (hR6) phases may appear while undergoing different pressure and temperature routes [26, 27]. Figure 2.3a shows the typical phase diagram of Ga and corresponding crystalline structure of Ga phase I, II, and III [28].

As for liquid phase of Ga, the early study suggested Ga and short-lived Ga_2 dimers coexist [30]. However, it is doubtful about the presence of covalent Ga_2 dimers [31]. There are researches suggesting that the local structure of liquid Ga is similar to one or two phases of its solid phases. Yagafarov and coworkers studied the local structure by energy-dispersive X-ray diffraction and reverse Monte Carlo, indicating that the local structure of liquid Ga resembles high-pressure crystalline Ga-I or Ga-II [29]. The high-q shoulder in $S(q)$ and the abnormal high value of the ratio Q_1/Q_2 and r_1/r_2 suggest the complexity of liquid structure of Ga (Figure 2.3b,c). Furthermore, the local structure of liquid Ga changes with pressure. The values of Q_1/Q_2 and r_1/r_2 move to a lower value with the increase of pressure, manifesting Ga turns toward the hard-sphere-like structure (Figure 2.3d). During this process, the coordination number and density of liquid Ga increases (Figure 2.3e). However, the structure transition is not as uniform as the hard-sphere structure owing to the anisotropic structure of liquid Ga. In this study, the local structure of liquid can be considered as the mixture of Ga-I-like and Ga-II-like at low pressure, and Ga-I-like structure decreases with the increase of pressure (Figure 2.3f) [29]. However, another study shows that the local structure of liquid Ga under high pressure and at ambient temperature is made up of Ga-II and Ga-III (Figure 2.3g) [28]. A melting mechanism is proposed to explained this result, in which the topological structure of Ga-I will break up and reconstruct during the melting process owing to the coordination number changes.

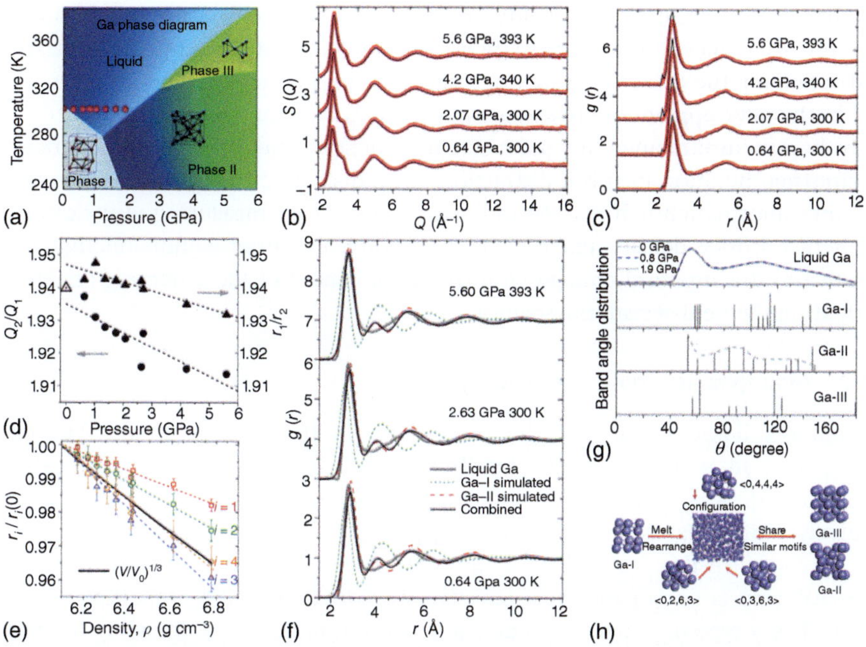

Figure 2.3 Phase diagram of Ga and structure of liquid Ga. (a) Phase diagram of Ga and their corresponding structure. (a) Source: Li et al. [28]. Licensed under CC BY 4.0. $S(q)$ (b) and $g(r)$ (c) of liquid Ga response to pressure 28. (d) The Q_2/Q_1 and r_2/r_1 under different pressures in $S(q)$ and $g(r)$. (e) Density dependence of first four neighbor distances. (f) Structure of liquid Ga evolves with pressure showing the mixture of Ga-I like and Ga-II like. (b–f) Source: Yagafarov et al. [29]. Reproduced with permission of American Physical Society. Source: (g) Structure of liquid Ga showing the mixture of Ga-II like and Ga-III like. (h) Schematic illustration of the melting of liquid Ga [28]. (g, h) Source: Li et al. [28]. Licensed under CC BY 4.0.Source:

As shown in Figure 2.3h, Ga-I breaks the crystalline atomic configuration to form locally denser motifs while meeting topological and chemical requirements, such as the three typical Voronoi polyhedral (like Ga-II and Ga-III). (Ga_2 dimers disappeared because all Ga atoms are equal in the disordered state thermodynamically.) Therefore, the liquid structure of gallium correlates to Ga-II- and Ga-III-like motifs. In addition, Chen and coworkers [1] analyzed the local bond-orientational order parameters of liquid gallium at ambient pressure. Two types of clusters are contained in the liquid structure, one featuring sixfold orientational symmetry similar to hard-sphere fluid and the other having fourfold orientational symmetry similar to β-Ga, and the later cluster is dominant. Furthermore, liquid–liquid phase transitions have been observed in liquid gallium under high pressure [32].

Generally, the metal in liquid state presents a short-range order, and the structure varies with temperature and pressure. The pressure-induced structure changes are dependent on the anisotropy in bonding, and the liquid structure behaves toward the more-close-packed and reduced structure. In addition, liquid–liquid phase transitions are frequently observed in liquid metal, which is usually accompanied by the change of electrical conductivity and density. Up to now, there is still not a comprehensive theory and mode to describe the subtle structures of liquid metals. Understanding the local structure and principle of liquid metal plays an essential part in the development of condensed matter. It is of great significance to deeply study the structure, change law, and mechanism of liquid metal. In spite of advanced synchrotron radiation, neutron diffraction techniques, and theory simulation, the structure information can also be indirectly obtained by measuring electric conductivity, viscosity, density, magnetic susceptibility, and other parameters that are sensitive to the structure change with the development of the emerging analysis methods in the field of material science.

2.1.2 Monophasic Liquid Metal Alloy

Monophasic liquid metal is defined as totally in the state of liquid without solid phase, in which elements distributed homogenously without segregated domains [33]. Apart from the mono-component metal, liquid metal alloy with relatively uniform structures and melting points below 300 °C is classified into this catalogue. In this section, we mainly introduce the monophasic liquid metal alloy. Alloying is one of the most effective ways to change the properties of the metals. Every type of alloy has a eutectic composition, allowing the alloy to melt congruently at a single melting point [34]. The eutectic alloy has the lowest melting point among all different composition alloys. If the composition deviates from the eutectic mixture, it requires higher temperatures to form monophasic liquid metal. Except for copper and nickel, which are completely miscible with each other in liquid state, there is usually some limits to the solubility of one metal in another. The temperature has a great influence on the solubility of added elements in liquid metals, and the solubility follows the general relationship described in Eq. (2.1), where C presents the solubility in atomic percent (at%), A and B are experimentally determined constants, and T is the absolute temperature.

$$\log_{10} C = A - \frac{B}{T} \tag{2.1}$$

Even if a small amount of ingredient is added into the system, it may have great significance in modulating chemical and physical properties of the alloy. Except metals, a small amount of nonmetal elements such as silicon and phosphorous can also be added. The properties of liquid metal alloys are strongly bonded with the microstructures of atoms and electronic intercalations in these melts. In-depth understanding of the local structure of liquid alloys has great significance in uncovering the cause of these performances, preparing alloy, and developing novel materials. Yu and coworkers revealed the structure of liquid EGaIn and EGaInSn [35]. As shown in Figure 2.4a, after Ga alloying with In and Sn, the first peaks of $g(r)$ shift to higher positions, owing to that the atoms in the first coordination shells are further distributed. The reduced height and broad peaks in EGaIn and EGaInSn indicate the disordered structure of the liquid alloys. As for the phenomenon that the first peak in 500 K shifts to lower positions, it can be explained by the decrease in the distance between atoms in the first shell related to the redistribution of the polyhedral clusters [36]. The results are corresponding to the $S(\kappa)$ in Figure 2.4b. The electron

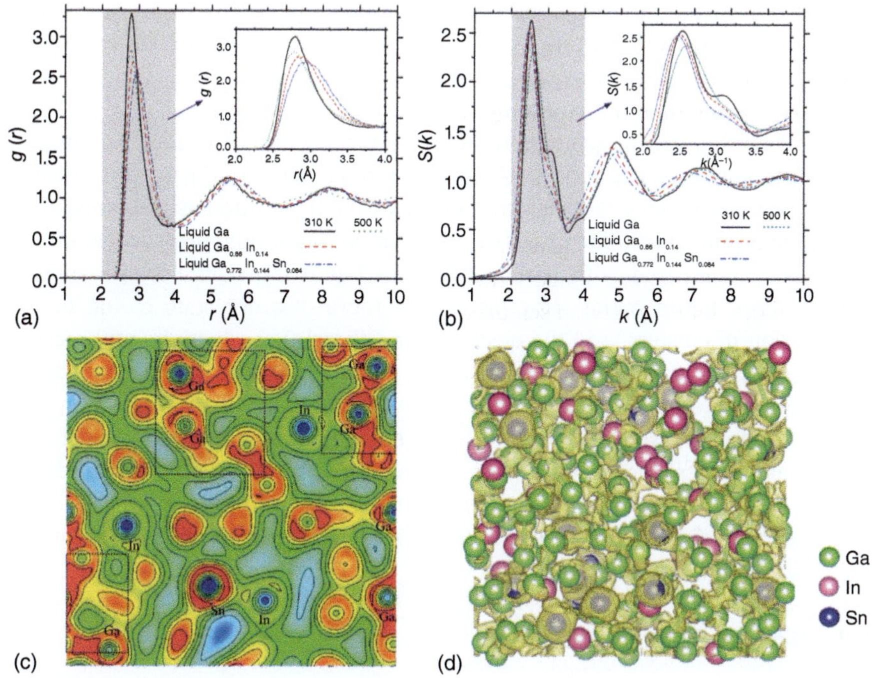

Figure 2.4 Structures of liquid Ga, EGaIn, and EGaInSn. (a) Predicted $g(r)$ of liquid Ga, EGaIn, and EGaInSn at temperature of 310 K, and Ga also at 500 K. (b) Predicted $S(k)$ of liquid Ga, EGaIn, and EGaInSn at temperature of 310 K, and Ga also at 500 K. (c) Contours of charge densities for liquid EGaInSn at 310 K. (d) Structure of the liquid EGaInSn at 310 K with contours of the constant charge density of 0.035 ec$Å^{-3}$. Source: Yu and Kaviany [35]. Reproduced with permission of AIP Publishing.

distribution of EGaInSn alloy in liquid phase is displayed in Figure 2.4c, d, to further understand the properties, including lower melting point and reduced electronic conductivity compared with pure Ga. The charge density of Ga atom is much higher than that of In and Sn, suggesting the strong force in existence among Ga atoms. The rather lower charge density of In atoms leads to reduced electrical transport properties in GaInSn alloy.

Like monocomponent liquid metal, the local structure of liquid metal alloy can also be described by the mixture of local structures [37]. Furthermore, like the structure of monocomponent liquid metal, the structure of alloy will also evolve with temperature and pressure, and plenty of attempts have been conducted to reveal structural evolutions during these processes, including theoretic analyses, experiments, and simulations [37–40]. The structure of liquid GaSb at high pressure up to 20 GPa was revealed by synchrotron X-ray diffraction, which can be described as a mixture of β-Sn-like and bcc-like local structures (similar to high-pressure phase of crystalline GaSb) and these two types of local structures can coexist under 20 GPa [37]. Liquid GaSb shows a nonuniform contraction process different from simple liquids, owing to that the ratio of several local structural changes are under pressure.

Moreover, liquid–liquid phase transition is observed in monophasic liquid metal alloy introduced by temperature or pressure, which is often accompanied by the change of physical properties, such as electronic conductivity, viscosity, and density [40, 41]. Liquid–liquid phase transition in Sn–Bi alloy is observed at 700–800 °C because of cluster rearrangements induced by temperature. Covalent Sn–Sn and Sn–Bi bonds are broken and the number of electrons increases during this process, leading to the improvement of electrical density [41]. In addition, liquid metal alloy may exhibit liquid–crystalline phase transformation under high pressure [42].

Summarily, compared with monocomponent liquid metal, the local structure of monophasic liquid metal alloy presents a more disorder state owing to the redistribution of clusters, and the electrical and thermal conductivity will reduce. Its local structure can be described as the mixture of several local structures, and varies with temperature and pressure. And the liquid–liquid transition and liquid–crystalline phase transition are observed, which are ascribed to the cluster rearrangements introduced by pressure or temperature. The atom arrangement of monophasic liquid metals is still under controversy. Further work is required to better command the microstructure of liquid alloy, and then design new materials with attracting properties.

2.1.3 Biphasic Liquid Metal Mix

Generally, biphasic liquid metal consists of liquid and solid domains, which can be formed by adding extra elements above from the solubility limit into the liquid or deviating from the eutectic composition [33]. In the case, solid phase will appear in the form of nano/microparticles during the cooling process from a monophasic phase, and then liquid–solid matrix comes into being. Moreover, intermetallic compound may form in biphasic liquid metal mix [43, 44]. The composition of biphasic liquid metal is not constant, and changes with temperature and time. On the other hand, biphasic liquid metal can exist stably for a long time if the size of particles is appropriate. The surfactant also influences the stability of biphasic liquid metal [45].

Owing to Ga having a strong ability to alloy with other metals, intermetallic compounds may be formed during the process of preparing biphasic liquid metals. Tang and coworkers [43] obtained a biphasic liquid metal by adding Cu particles into EGaIn at room temperature, presenting the semisolid/semiliquid state. The composition of biphasic liquid metal can be well controlled by the ratio of added Cu particles, resulting in the formation of an intermetallic compound $CuGa_2$ (Figure 2.5a). The content of $CuGa_2$ increases with time (Figure 2.5b), and the corresponding process is revealed by SEM (Figure 2.5c,d). As time goes, the Cu surface near liquid EGaIn first dissolves to form a Cu-rich layer, then $CuGa_2$ nuclei form and gradually grow into large blocks. Profiting from the formation of $CuGa_2$, the biphasic liquid metal obtains the improved thermal, electrical, and mechanical properties. Moreover, Wu and coworkers reported a biphasic structure with reliably stretchable and formable ability by mixing liquid GaIn and Ni particles and following with a heat treatment process, which can be ascribed to the formation of intermetallic compound Ga_4In_3 and $InNi_3$ on the surface of Ni particles [47]. Especially, biphasic magnetofluid may be formed if magnetic nanoparticles are mixed with liquid metal, such as Fe, Cd, and Ni [48].

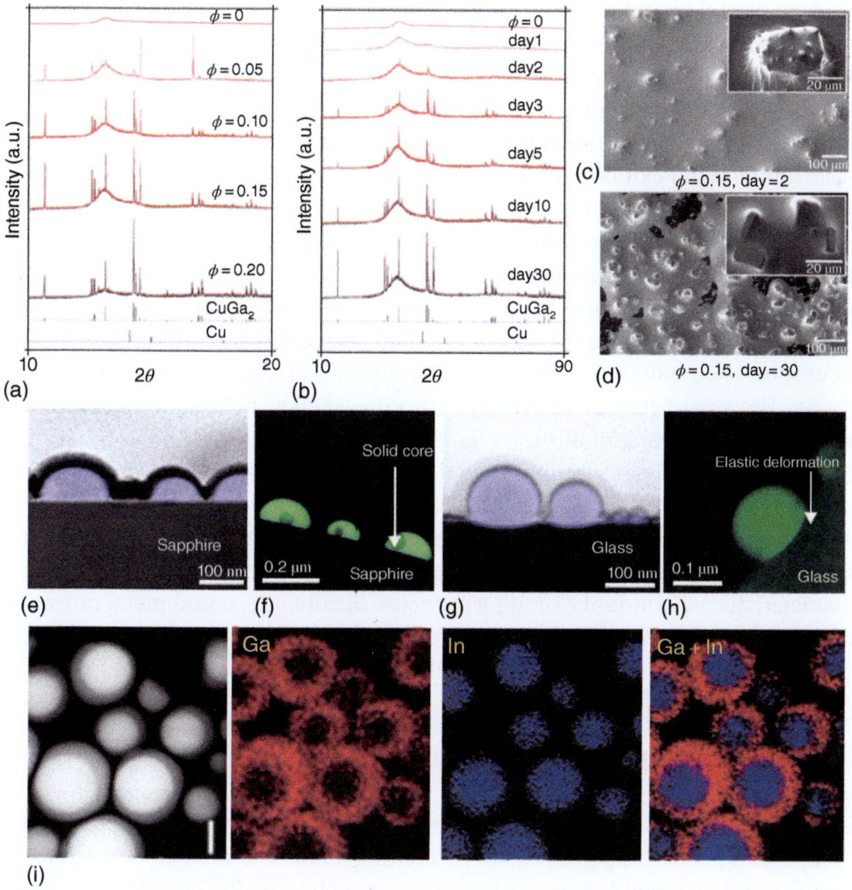

Figure 2.5 The structure of biphasic liquid metal. (a–d) The structure evolutions of biphasic mixture containing liquid EGaIn and Cu particles. XRD measurements of the mixture consisting of EGaIn and (a) different ratio of Cu particles and (b) mixture containing 15 wt.% Cu particles showing the formation process of $CuGa_2$. (a, b) Source: Tang et al. [43]. Reproduced with permission of American Chemical Society. (c) SEM images of the corresponding mixture in (b) on day 2 (c) and day 30 (d). (c, d) Source: Tang et al. [43]. Reproduced with permission of American Chemical Society. (e–h) The structure of solid phase γ-Ga core and liquid Ga shell. Combined high angle annular dark field scanning transmission electron microscopy and energy-dispersive X-ray spectroscopy images of Ga nano particles on sapphire (e) and glass substrate (g). Energy filtered plasmon loss images of Ga nanoparticles deposited on sapphire (f) and glass (h), respectively. (e–h) Source: Losurdo et al. [25]. (i) The HAADF images of GaIn alloy and corresponding EDS elemental maps at temperatures of 241 K. Source: Tang et al. [46].

In addition to the slushes made up of liquid metal and solid nano/macroparticles, core–shell biphasic liquid metal, which contains a liquid metal core and a solid shell can also be synthesized. Zheng and coworkers [49] prepared a core–shell structure, which consisted of liquid EGaIn core and solid Ag shell by in situ chemical plating, endowing with excellent conductivity and self-repairing ability. Moreover, liquid metals such as alkali metals and gallium-based metals are easily oxidized when exposed to air. Taking advantage of this property, a core–shell biphasic structure can be formed, which is composed of a liquid core and solid oxide shell [50]. When the size of liquid Ga is decreased into nanoscale, a new solid phase γ-Ga within liquid Ga on sapphire substrate is observed, and they can stably coexist over a wide range of temperature from 180 to 800 K (Figure 2.5e–h). The formation of this new biphasic liquid metal can be explained by the high Laplace pressure, which drives the nucleation of γ-Ga and the epitaxial relationship between sapphire substrate and γ-Ga [25]. Furthermore, Tang and coworkers reported that a solid core and spherical liquid metal shell could stably coexist without the support of a crystalline substrate at room temperature (Figure 2.5i) [46]. During the cooling process, phase separation will appear in liquid GaIn alloy, and then solid In core and liquid Ga shell come into being since the solubility of In within Ga reduces as the temperature decreases. The size of In core in this biphasic core–shell structure is dependent on the temperature and content of In.

In summary, the phase in biphasic liquid metal mix changes with the temperature variation. Compared with monophasic liquid metal, the rheological, electrical, and thermal properties of biphasic liquid metal have some changes, where viscosity increases, and the electrical/thermal transport property is dependent on the intrinsic characteristic and amount of adding particles. Meanwhile, liquid metal mix may obtain extra properties originating from the suspended particles, such as enhanced electrical conductivity and thermal conductivity.

2.2 Low Melting Point of Liquid Metal

Different from traditional metals, liquid metals can be easily operated at a relatively low temperature benefiting from their low melting point. The melting point of metals is contingent on many factors, such as crystal structure while they are in the state of solid, the number of electrons involved in the bond, the degree of delocalization of valence electrons, and the radius of metals. In this section, we will introduce the origin of low melting point of liquid metal mainly from four parts: zinc group metals, alkali metals, posttransition metals, and liquid metal alloy.

2.2.1 Zinc Group Metals

In terms of zinc group metals possessing low melting, the full electrons in d and s shells lead to a pseudonoble gas electron configuration [51, 52]. The valance electrons of zinc group metals are only two electrons, which are involved in the bonding in their compounds, and the completely filled $(n-1)$ d^{10} subshell is the part of the atomic core with a strong shielding effect. The full ns^2 shell results in a relatively weak bond interaction, thus suggesting low melting and boiling points. Especially, mercury has the lowest melting point among all monomeric metals. Owing to the full d and f shell electrons, as well as significant relativistic effects of "inert $6s^2$ pair," the $6s^2$ electrons do not involve much in the metal–metal bonds [52]. Hence, the van der Waals interactions or probably weak 6p orbital interaction causes the bonding in mercury, leading to such a weak Hg—Hg bond [53].

2.2.2 Alkali Metals

The melting point of alkali metal is relatively low. Alkali metal has a ns^1 valance electron configuration, following the closed shell arrangements of the noble gases, and the outer electron is shielded from nuclear charges. Moreover, the radius of alkali metal is much larger than other elements in the same periodic. As a consequence, the interaction between metal ions and electrons is weak while they are in the solid state, and the weak metallic bond is easily broken leading to the low melting and boiling points.

2.2.3 Post-transition Metals

Posttransition metals present low melting, which can keep liquid state at a wide temperature range. These elements have electronic configurations with completely filled d and partially filled p shells. Unpaired p shell electrons can form either covalent or metallic bonds, and lighter elements feature significant covalent metal–metal bond [54]. The presence of these reactive p shell electrons is responsible for the high boiling points. Gallium has an unusual melting temperature of 29.28 °C, which is the lowest among all posttransition metals. This can be explained by its unique crystal structure of solid phase. Unlike any other simple crystal structure, the crystal structure of gallium is polymorphism. Although many phases of gallium have been discovered in the experiment and studied by theoretical calculation, they are metastable phases. α-Ga (Ga-I) is the stable phase at ambient pressure, with both metallic and covalent character. The structure is commonly considered as face-centered orthorhombic unit cell with eight atoms per cell (Figure 2.6a–c) [33]. In particular, each atom has only one nearest neighbor at a distance of 2.44 Å, and the other six next-nearest-neighbor atoms are arranged with a radius between 2.71 and 2.79 Å [54]. The distance between the nearest-neighbor atom pairs is comparatively short for a normal metallic bond, which is more conformable with the nature of covalent bond. The existence of covalent Ga_2 dimers in solid gallium is further proved by band structure calculations based on pseudopotential approximations [54, 56] and the scanning tunneling microscopy provides direct evidence that Ga_2 dimers are the basic constituents of the crystal [57]. It indicates the highly structural anisotropy of gallium, in which the covalent bond of each Ga atom with the only one nearest neighbor and the metallic bond between each Ga atom and the next nearest neighbor. The anisotropic structure of Ga manifesting the crystal structure can be broken up at a lower temperature in contrast to many other metals. Other elements of posttransition metals also have a similar anisotropic structure, such as In [58], Sn [59], and Bi [60], showing low melting temperatures.

2.2.4 Liquid Metal Alloys

The melting point of metal can be reduced when a metal forms a eutectic composition with the other metal. While both metals are fused in a certain proportion, the melting point is lower than that of either component. The melting point of the alloy is lowest at a particular proportion, and the temperature is the eutectic point. Usually, there are some restrictions to the formation of eutectic alloys that requires a degree of solubility of one metal in another. The solubility of metals within each other can be estimated by the Hum–Rothery rules, which are listed as follows. First, the atomic diameter of the two kinds of atoms must be similar (within about 15%) so that lattice strain will not be excessive. Second, the crystal structures of both metals must be identical. Third, it is necessary for both metals to have the same valance to form a thermodynamic stable system. Fourth, analogous electronegativities are required to minimize the tendency for the metals to form compounds. The solubility of a metal may differ by orders of magnitude in diverse systems. Meanwhile, the solubility between different metals has a strong relationship with the temperature, which has been summarized in Section 2.1.2. Take gallium and gallium-based alloy as an example, when gallium is alloyed with indium, the melting points can be further decreased. The melting points of Ga–In system vary with the proportion of gallium and indium (Figure 2.6d) [55]. As the atomic percentage of indium is 14.2%, the system shows the lowest eutectic temperature of ~16 °C. Moreover, when the

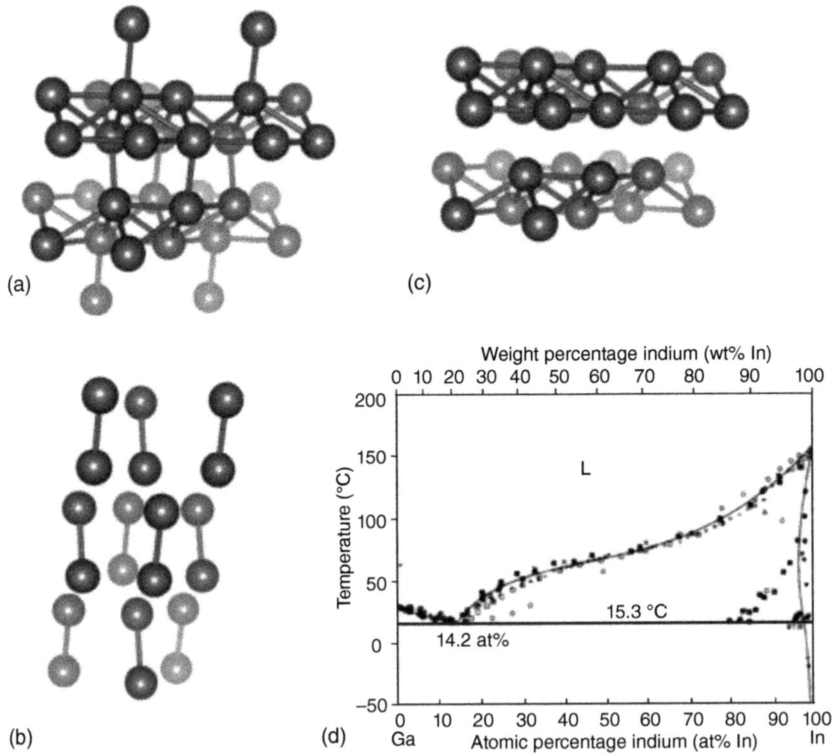

Figure 2.6 (a) Crystal structure of α-Ga. (b) Crystal structure of α-Ga showing only short bonds <2.5 Å. (c) Crystal structure of α-Ga showing only the long bonds 2.5–3 Å (a–c) Source: Daeneke et al. [33]. Reproduced with permission of Royal Society of Chemistry. (d) Phase diagram of the binary Ga–In system. Source: Anderson and Ansara [55]. Reproduced with permission of Springer Nature.

Ga–In system is added other metals and forms a ternary or multicomponent alloy, the melting points can be further reduced. List Ga–In–Sn system as an example, the mechanism of the melting points reducing was explored by ab initio molecular dynamic (AIMD), and could be explained from two aspects [35]. First, as the atomic volume of In and Sn is larger than Ga, the degree of structure disorder increases by expanding the cell volume and broadening their interatomic distance along with alloying process. Second, due to the fact that valance electrons of In atom is strongly bound to the atom core similar to the noble-gas atoms with weak interactions (just like Hg) [52], the charge density around In atoms is found to be significantly lower when compared with Ga and Sn atoms, suggesting that the interactions of these atoms are weak. As a result, compared with Ga (303 K) and EGaIn (288.65 K), EGaInSn (283.65 K) displays lower melting points.

2.3 Rheological Properties of Liquid Metal

Liquid metal has shown great potentials in stretchable and soft electronics [61], electrodes [62, 63], microfluidics [64], and electronics printing [65], which have great relation with their rheological properties. Taking advantages of both liquid and metal, liquid metal presents low viscosity with high mobility and self-healing capacity, as well as favorable electrical and thermal conductivity at low temperatures. Rheological properties are deformation or flow behaviors that respond to the applied forces or stresses, manifesting the evolution of its microstructure [66]. The rheology property of liquid metal is affected by many factors, such as structure, surface oxide, and component. Understanding rheological characteristics of liquid metals is of great significance to study the local structure and solidification behavior of liquid metals, contributing to improving fabricating components and properties of metal and alloy.

2.3.1 Basic Rheology Property of Liquid Metal

Generally, rheological property of a liquid metal reflects the dependence of viscosity on shear stress and shear strain. The viscosity of liquid metals is correlated with surrounding temperatures. In addition, viscosity is sensitive to the microstructure, which can also reflect the structure evolution of liquid metals. Simple liquid is generally taken as a Newtonian fluid, but with in-depth research, non-Newtonian fluid phenomenon is investigated in liquid gallium, bismuth, tin as well as other metals and their alloys (Figure 2.7a,b) [66, 67, 69]. This abnormal characteristic is attributed to the metallic and covalent bonds in these liquid metals, which is consistent with the discussion of liquid structure in Section 2.1.1. The covalent electrons result in the formation of larger clusters or networks, which may be responsible for the non-Newtonian behavior [66].

Usually, owing to the expansion of free volume and decrease of friction among atoms and clusters at higher temperature, the viscosity of liquid metal decreases along with the increase of temperature. However, as the temperature increases,

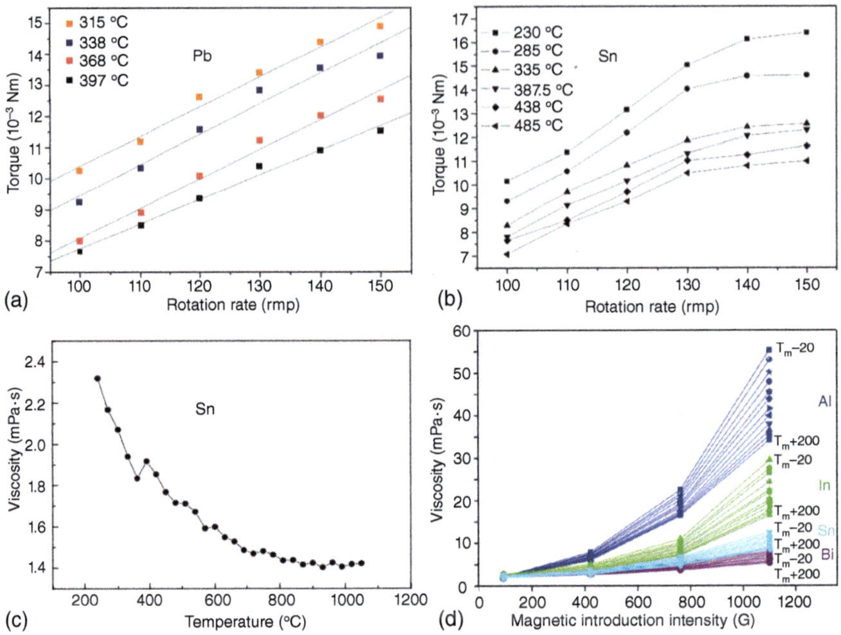

Figure 2.7 (a) Relationship between the torque and the rotational rate of liquid Pb. (b) Relationship between the torque and the rotational rate of liquid Sn. (a, b) Source: Sun et al. [66]. Reproduced with permission of AIP Publishing. (c) Temperature dependence of dynamic viscosity curve of Sn. Source: Sun et al. [67]. (d) The variations of viscosities with the magnetic field intensity in liquid Al, In, Sn, and Bi. Source: Yang et al. [68].

abnormal increase of viscosity in non-Newtonian liquid metal at a certain temperature range may be observed (Figure 2.7c), which probably results from the transition of covalent bond toward metallic bond [67]. In addition, the viscosity of liquid metal is affected by magnetic field, resulting from the Lorentz force generated from the interaction between charged cations and magnetic field [68]. The viscosity value increases as the horizontal magnetic field enhances, and magnetic field has a stronger influence on viscosity of lighter liquid metals than heavier metals (Figure 2.7d).

2.3.2 Effect of Oxide Surface

The rheology of liquid metal is greatly influenced by its oxide surface. Liquid metals, such as gallium, alkali metal and some other active metals are very easily oxidized under environmental conditions, and then a solid oxide skin forms on the surface. The oxide surface is sensitive to the strain, providing an engineering framework to regulate the rheological properties of liquid metals [70]. In terms of Ga-based liquid metal, the surface predominately covered with oxides of gallium was confirmed by XPS [70–72], coincident with the fact that gallium is highly reactive when exposed to oxygen at room temperatures. And the existence of oxide skin prevents the bulk liquid from further oxidation through forming a highly uniform passivation layer-like aluminum. Meanwhile, it facilitates the formation of metastable structures and implements the applications in channel injection, spraying, and 3D printing.

The rigid oxide skin endows the liquid metal with high surface stress, allowing the formation of stable structures in microchannels [70]. As shown in Figure 2.8a, the liquid EGaIn with oxide skin behaves as both elastic and viscous at a large strain range. As a function of surface stress, elastic moduli (G') and viscous moduli (G'') decrease rapidly at about 0.5 N m^{-1}, at which point it flows readily. The EGaIn features an elasticity until the surface stress exceeds a critical point of ~0.5 N m^{-1} [70]. In order to study the cause of this rheology of EGaIn, EGaIn is exposed to HCl solution to remove the oxide skin. As a result, EGaIn in microchannels prefilled with HCl is instable, indicating the rheology of EGaIn origins from the skin rather than inner bulk EGaIn. Furthermore, more detailed study about the rheology response of EGaIn to variable HCl concentrations is displayed in Figure 2.8b. The oxide surface layer can be fully removed at a concentration of 0.1 $molL^{-1}$, and the corresponding mechanical effects result from the disappearance of oxide surface [74]. Apart from pH, electrochemistry can be operated to manipulate the oxide surface under a constant supply of energy [76, 77]. A more simple and convenient strategy, the use of water, provides new opportunities to control the oxide-forming liquid metals [73]. As shown in Figure 2.8c, compared with the EGaIn in air, the surface elastic modulus exposed to water is decreased by an order of magnitude, and the surface yield stress is also reduced by about a factor of five. The changes of rheology performance are ascribed to the composition of oxide, which makes the skin less passivating.

The oxide skin was also observed in alkali metals and their alloys, which are considered as promising storage energy electrode materials. Unlike gallium, the oxide skin Li_2O of liquid lithium is not passivation layer, which can allow the penetration of electrolyte [78]. Similar behavior has also existed in liquid sodium and potassium due to their analogous physicochemical properties [79].

In conclusion, the oxide skin is ubiquitous on most metals, which can be used as the "surfactants" to modulate the rheological properties of liquid metals. Multiple methods including pH, electrochemistry, substrate contact, and so on have been used to manipulate the oxide skin and change the rheological properties of liquid metal and alloy. Liquid metals coupled with switchable surface activity can be harnessed to induce rich opportunities in shape-reconfigurable metallic components in electronic, electromagnetic, and microfluidic devices.

2.3.3 Effect of the Formation of Composite

The oxide surface is dominant in response to rheological properties of gallium-based liquid metal, endowing them with yield stress and elastic modulus. However, the thin oxide surface of liquid metal is easily broken and the shape of liquid metal is unstable because of its high surface tension and low viscosity, which impedes its further application. Changing the composition of liquid metal provides another effective strategy to regulate the deformability and viscoelasticity.

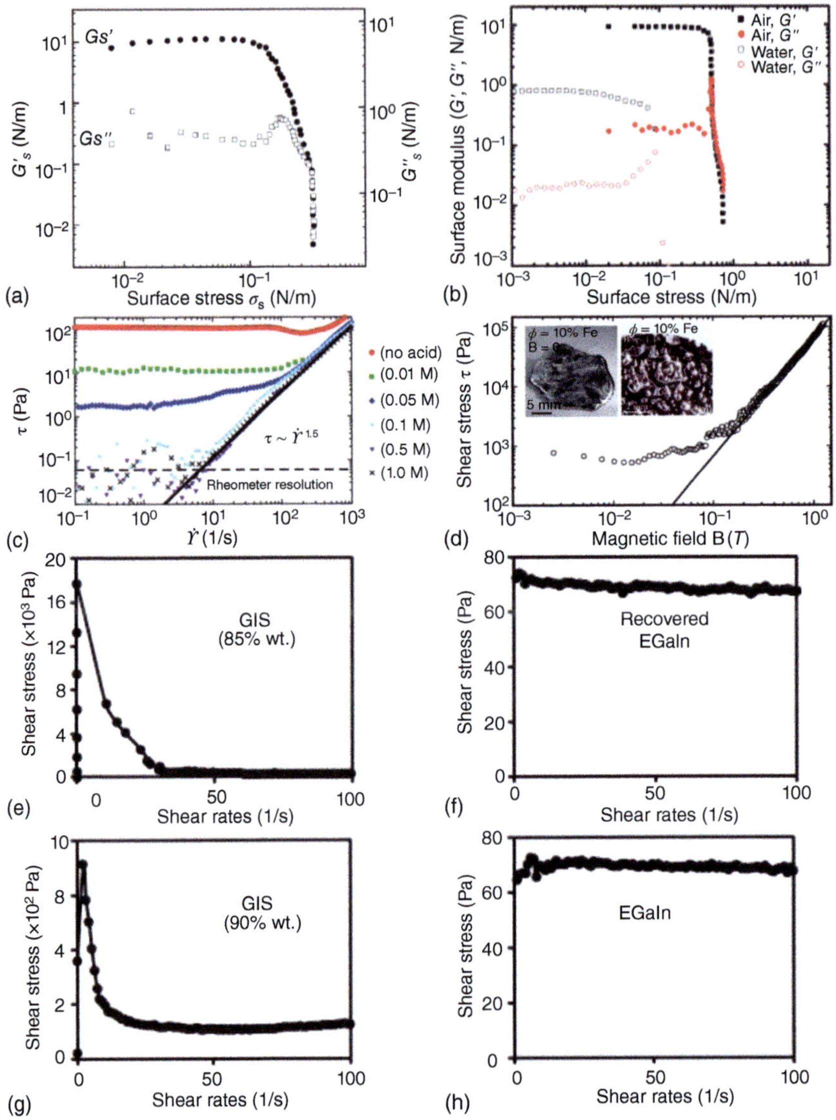

Figure 2.8 Rheological behavior of liquid metal. (a) Rheological property of EGaIn showing the stabilizing effect of the oxide skin below a critical yield stress. Source: Dickey et al. [70]. Reproduced with permission of John Wiley & Sons. (b) Rheological property of EGaIn with oxide skin exposed to water. Source: Khan et al. [73]. Reproduced with permission of American Chemical Society. (c) Rheological property of EGaIn response to different hydrochloric acid concentration. Source: Xu et al. [74]. Reproduced with permission of AIP Publishing. (d) The magnetorheological effect for a suspension of $\varphi = 39\%$ iron particles in EGaIn. Source: Carle et al. [75]. Reproduced with permission of American Physical Society. (e) Flow curve of EGaIn/SiO_2 paste with 85% mass content of EGaIn. (f) Flow curve of recovered EGaIn. (g) Flow curve of EGaIn/SiO_2 paste with 90% mass content of EGaIn. (h) Flow curve of pure EGaIn. Source: Chang et al. [65]. Reproduced with permission of American Chemical Society.

Metal nano/microparticles can be dissolved (or dispersed) in liquid metal to form a suspension liquid, which endows the composition with rheological property of magnetorheological fluid [48, 75, 80, 81]. The particles will arrange directionally and the mixture changes into a reversible, jammed, and paste-like state when a magnetic field is applied. The magnetorheological effect of a liquid metal composition is shown in Figure 2.8d, which contains Fe nanoparticles suspended in EGaIn [75]. The measurement reveals that the shear stress at a fixed shear rate (approximate to the yield stress) increases by nearly two orders of magnitude when a magnetic field between 0.1 and 1 T is applied [75]. Resulting from the high surface tension and poor wettability of liquid metal, it is difficult to mix particles into liquid metal. In order to improve the wettability, SiO_2, which is compatible with liquid metal is coated on the surface of particles before mixing [82]. The deformability and movement of liquid metal composition can be operated under magnetic field, inspiring the flourish in magnetofluid [48] and reconfigurable soft robots [83].

In addition, nickel particles were mixed with EGaInSn via high-energy sonication to form liquid metal paste, showing increased viscosity and elasticity, thus the paste can be used for printing successfully [84]. And the rheological properties of liquid metal paste can be quantitatively altered by controlling the sonication energy as well as proportion and size of nickel particles. G', G'', yield stress and viscosity increase with the increase in percentage of nickel particles. The rheological values of the mixtures become larger when nickel is in the form of nanoparticles than in the form of microparticles, which may be ascribed to the interaction differences between microparticles and nanoparticles in the presence of sonication [84]. Furthermore, nonmetallic SiO_2 power was mixed with EGaIn, obtaining a rheologically reversible liquid metal paste [65]. EGaIn can be easily separated from quartz particles when the paste is immersed in both acidic and alkaline solution, owing to the dissolution of surface oxide. The EGaIn/SiO_2 pastes containing 85% and 90% EGaIn show the non-Newtonian property, and there is a drop in shear stress at high shear rates and in yield stress at high EGaIn content (Figure 2.8e–h). The rheology property of the liquid metal paste is also affected by EGaIn content. The mechanical properties of liquid metal paste behave more like a solid when the EGaIn content is higher [43]. Moreover, the rheological properties of recovered EGaIn are almost the same as those of pure EGaIn [65].

Furthermore, incorporating nano/microparticles of liquid metal into polymer provides an opportunity to combine the favorable flexibility of polymer and good thermal and electric conductivity of liquid metal [85, 86]. The liquid metal (LM) microparticles as the fillers are dispersed with liquid crystal elastomers (LCE) and form a composite named LM-LCE [85]. The EGaIn microparticles exhibit liquid state at room temperature, allowing the composite to possess semisoft elasticity and elastic regime before breaking. The 50 vol.% LM-LCE composite shows similar storage moduli and retains >90% of actuation capabilities relative to unfilled LCEs at the same stress. This composition features good mechanical and shape morphing properties, improved thermal conductivity, and outstanding electrical conductivity, allowing further applications in flexible and stretchable devices.

In this section, we introduce the rheology of liquid metal, which is affected by the local structure, surface oxide, and composition. Liquid metal with a simple hard-sphere-like structure is usually considered as Newtonian liquid, while the liquid metal featuring complicate structure is non-Newtonian liquid. The surface oxide of liquid metal changes the rheology of liquid metal, showing elasticity before reaching yield stress. In addition, liquid metal can form stable composition with inorganic particles and polymer, which can also influence the rheology property of liquid metal.

2.4 Electrical Conductivity of Liquid Metal

The electrical conductivity of metal is a basic topic when we research metals and apply them into industrial production and daily life. Copper is commonly used to convey current because of its high electrical conductivity as well as low cost advantages. A majority of metals have excellent electrical conductivity since they contain free electrons [87]. However, there are some special metals, such as Ge, Pb, and Bi, that show poor conductibility. There are some theories explaining the electrical properties of liquid metals. One is based on the tight binding approximation [4]. The second is the nearly free electron picture [88] and another one is the mean electron-free path theory [89]. Table 2.1 presents these conductive data of liquid metals in both solid and liquid states [10]. There is another interesting phenomenon that the electrical resistance of most liquid metals is larger than their solid counterparts. In this section, we will discuss some distinct electrical conduction properties of liquid metals.

2.4.1 Resistance Change at Melting Point

Table 2.1 shows the resistance change at the melting point. From this table, we can see for most metals that their resistance in solid state is smaller than that in liquid state. Whereas the liquid Ga, Ge, Sb, and Bi are abnormal. According to the simple perturbation theory [88], resistance of metal is connected with the PDF, which is different for solid and liquid. As is well-known, liquid possesses a certain short-range ordered structure (within four atomic diameters) [90]. However, it is different from these ordered structures of solid due to the intense thermal motion. Difference of the order between solid and liquid metals determines that the electron mean free path is different, which is responsible for the resistance of metal [91]. According to the approximate free electron model, resistance of liquid metals is deduced by Ziman, that is

$$\rho = \frac{12\pi V}{e^2 h v_F^2} \int_0^1 |U(Q)|^2 S(Q) \left(\frac{Q}{2k_F}\right)^3 d\left(\frac{Q}{2k_F}\right) \tag{2.2}$$

where e is electric charge, v_F is Fermi velocity, k_F is Fermi wave vector, V is the atom volume, $U(Q)$ is the pseudo-potential, determined by the band gap of metal,

Table 2.1 The electrical resistance of metals both in solid and liquid state.

Metal	T_m	ρ_s	ρ_l	ρ_l/ρ_s	α	β
Li	452					
Na	370.9	6.598	9.573	1.451	0.038	−4.5
K	336.8	8.32	12.97	1.56	0.064	−8.5
Zn	692.6	16.7	37.4	2.24		
Ga	303	56	25.8	0.45	0.0195	19.9
		18		1.46		
		8		3.12		
Ge	1232	900	60	0.067		
		1200	63	0.053		
		1100	85	0.077		
Rb	312.1	13.7	22.0	1.60	0.086	−4.8
Cd	594	17.1	33.7	1.97		
In	427	15.2	33.1	2.18	0.0255	22.2
Sn	505	22.8	48	2.10	0.0249	35.4
Sb	903.6	183	113.5	0.61	0.0270	87.9
Cs	301.6	21.7	36.0	1.66		
Hg	234.2	18.4	90.96	4.94		
		24.3		3.74		
Tl	576	35.5	73.1	2.06	0.0292	56.3
Pb	600.5	49	95	1.94	0.0479	66.6
Bi	544	290.8	130.2	0.045	0.0570	99.2

and the $S(Q)$ is the structure factor. From this equation, electron mean free path of these metals is relative with band gap [92], which is expressed in $U(Q)$. These metals with a relatively large band gap will have shorter electric mean free path and thus increase electrical resistance. On the other hand, according to the $S(Q)$, the electron mean free path will reduce when metals melt, resulting in resistance increase for normal metals. But these abnormal liquid metals are different. Investigating their radial distribution function, for example, liquid Ga, the first peak of the function is asymmetric [93]. The offside of the first peak has a slight shoulder peak. This characteristic peak indicates that there is a directional bond of these atoms in the liquid state, which maintains consistency within the atomic structure discussed in Section 2.1. This special structure leads to the volume shrinking of these metals when they melt. These metal atoms stack compactly in liquid state, resulting in free electrons per unit volume and the decrease of resistance. Figure 2.9a–h shows the PDF and the structural factors of these abnormal metals and the data of volume change when they melt are listed in Table 2.2 [19, 94–96].

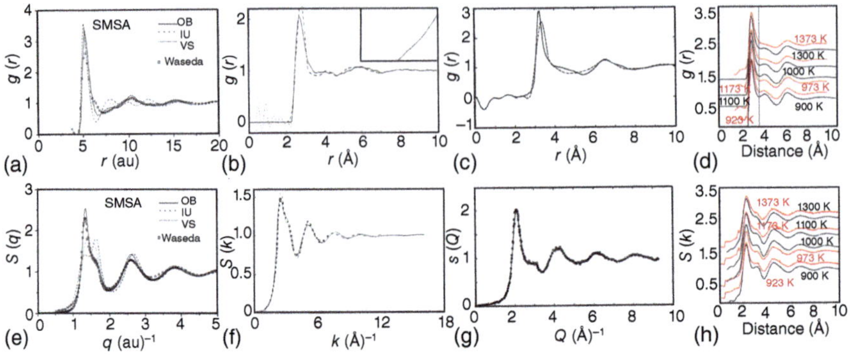

Figure 2.9 (a–d) The pair contribution function of liquid Ga, Ge, Bi, and Sb, respectively. (e–h) The structural factors function of liquid Ga, Ge, Bi, and Sb, respectively. (a) Source: Tsai et al. [19]. Reproduced with permission of AIP Publishing; (b, e, f) Source: Salmon [94]. Reproduced with permission of IOP Publishing; (c, g) Source: Dahlborg and Davidovi [95]. Reproduced with permission of Taylor & Francis; (d, h) Source: Jones et al. [96]. Reproduced with permission of AIP Publishing.

Table 2.2 The data of changing in volume of some metals as they melt.

Metal	$\Delta V_m/V_m$
Li	2.74
Na	2.6
K	2.54
Rb	2.3
Cs	2.6
Tl	2.2
Ge	−5.1
Bi	−3.87
Sb	−0.95
Hg	3.64
Pb	3.81
In	2.6
Sn	4.08
Cd	3.4
Sn	2.4
Ga	−2.9

Source: Based on Yang et al. [31].

2.4.2 Temperature Dependency

Electrical resistance is also relative to temperature since the structure of the liquid depends on temperature. Electrical resistance is connected with the disorder of liquid state. Therefore, it increases with the increase in temperature. The electrical resistance can be calculated as follows,

$$\rho = \alpha T + \beta \tag{2.3}$$

where α is the temperature coefficient, defined as $\alpha(T) = \mathrm{d}R(T)/\mathrm{d}T$.

2.4.3 Resistance of Liquid Alloy

For liquid alloy systems, two rules need to be obeyed. One is the Nordheim rule, which defines that the electrical resistance of alloy is in proportion to the product of percentage of metal in the alloy and concentration [9]. And the other is Linde rule, which defines that the derivative of resistivity with respect to atomic concentration is relative with the variance of variables of metal in the alloy [7]. Generally, electrical resistance of liquid alloy system depends on the free electron behavior. Specific changes are related to the system. For example, the electrical resistance of Cd–Sn liquid alloy will reach peak with the change of Cd concentration, whereas the electrical resistance of In–Hg alloy will decrease monotonically with the increase of In concentration [97].

In this section, we discussed the electrical conduction properties of liquid metals and their alloys. Some electrical data have been listed. Liquid metals are used widely in soft electrical devices such as interconnector, fiber, and antenna due to their fascinating electrical properties. Thus, a full understanding of these electrical properties, such as temperature variability at melting point and temperature dependence, will be instructive for future applications.

2.5 Thermal Property of Liquid Metal

The thermal conductivity of metal relates with the free electrons. Generally, because the mean free path of the electron is shortened, the thermal conductivity of metal decreases with the increase of temperature. When metal is in the liquid state, there is not a general theory to accurately describe the thermal conductivity. As to liquid metals in this book, they possess high thermal conductivities at wide temperature ranges compared with other common fluids (Figure 2.10a), showing great potential as heat transfer fluids [98]. In this section, we will discuss the thermal conductivity of these liquid metals.

2.5.1 Pure Liquid Metal

The thermal conductivity of a molten metal is made up of heat transport resulting from free electrons and phonon exchange determined by structure scattering, and the former is dominant [102]. As for pure metals in liquid state under a temperature near melting point, the lattice conduction can be ignored, and the contribution of lattice conduction to the overall thermal conductivity of liquid metal increases at higher temperatures. According to Wiedemann–Franz law [103], there

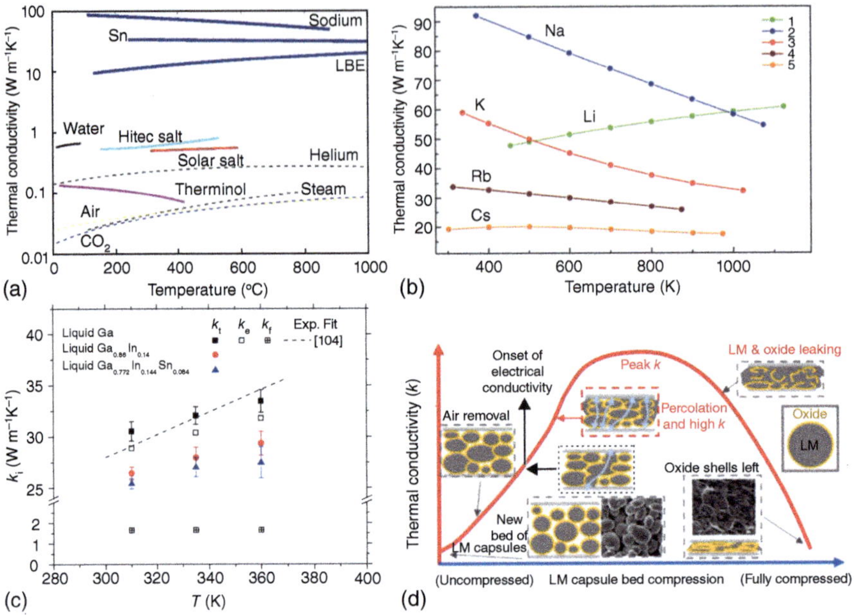

Figure 2.10 Thermal conductivity of liquid metal (a) Thermal conductivity temperature dependence of liquid metal and other common fluids. Source: Heinzel et al. [98]. Reproduced with permission of John Wiley & Sons. (b) Thermal conductivity as a function of liquid alkali metals, where 1, 2, 3, 4, and 5 represent Li, Na, K, Rb, and Cs, respectively [99]. Source: Reproduced with permission of Taylor & Francis. (c) Variations of the total thermal conductivity of the liquid Ga, EGaIn, and EGaInSn with temperature [35, 100]. Source: Yu and Kaviany [35]. Reproduced with permission of AIP Publishing. Reproduced with permission of Cambridge University Press for [100]. (d) The thermal conductivity of liquid Ga capsule under pressure. Source: Uppal et al. [101]. Reproduced with permission of American Chemical Society.

is a linear relationship between the thermal conductivity of a solid metal and its electrical conductivity. Usually, the metal with high electrical conductivity is regarded as possessing high thermal conductivity. The theory is established on the hypothesis that the metallic conductivity can be entirely attributed to the transport of electron gas. When the metal transforms into liquid phase, the theory is under debate and the thermal conductivity at melting point usually shows a deviation within 10% from Wiedemann–Franz law [104]. However, this is the only theoretical treatment of any kind available [105]. Mills and coworkers summarized the thermal conductivity of molten metals near melting points and some studied in this book are listed in Table 2.3 [104]. The thermal conductivity of a metal changes greatly at the melting point. For most metals, their thermal conductivity in liquid state is less than that in solid state, similar to the jump of electrical conductivity at melting points. However, Ga, Sn, and Sb are exceptions, which are related with their abnormal structures (detailed discussion is in Chapter 2.1).

Besides, the thermal conductivity of metal in the liquid state has a complex relationship with temperature. Take alkali metals as examples, the thermal conductivity

Table 2.3 Recommended reference thermal conductivity values of liquid metals at 0.1 Pa.

Metal	λ_s^m (W m^{-1}K^{-1})	λ_l^m (W m^{-1}K^{-1})
Li	71	43
Na	120	88
K	98.5	56
Cs	35.9	20
Ga	24.1	28
In	76	38
Sn	56	27
Pb	30	15
Bi	76	12
Sb	17	26
Zn	90	50
Cd	41	93
Hg	—	8.1

Source: Data from Mills et al. [104].

of liquid Na, K, Ru, and Cs decreases as the temperature increases except for liquid Li (Figure 2.10b). This abnormal phenomenon can be observed in liquid Li, probably because the thermal expansion of Li is small and mean free path of valence electrons of liquid Li changes weaker than other alkali metals [99]. The temperature dependence of other alkali metals can be ascribed to the fact that the mean free path of valence electrons decreases with the increase of temperature.

2.5.2 Liquid Metal Alloy

When metal alloys with another, the thermal conductivity usually decreases owing to the increased disordered structure. The thermal conductivity of the liquid Ga, EGaIn, and EGaInSn as a function of temperature is shown in Figure 2.10c, where the thermal conductivity of liquid GaIn and GaInSn alloys is lower than that of pure liquid Ga. Alloying with gallium leads to more disordered structure by broadening the atomic distance near the In and Sn atoms, which reduces the thermal transport properties [35]. This is also proved by the electron distribution in liquid GaInSn alloy. Electrons in GaInSn alloy are mostly distributed around Ga atoms, and the charge density around In atoms is relatively lower than that around Ga and Sn atoms, which explains the In-induced decrease of thermal conductivity.

2.5.3 Effect of Oxide Surface

In terms of Ga-based liquid metal and alloy, a native oxide skin with a thickness of 0.5–2.5 nm usually excites when they are exposed to environmental condition [101].

Because metal oxide has poor thermal conductivity in contrast to metal, the wrinkled surface hinders the thermal transport between adjacent particles. The thermal conductivity of liquid metal can be controlled by the oxidation extent, which can be achieved by stirring [106], exposing it in acid or alkali solution [74], and electrochemical method [76]. Furthermore, the transport path will be established once the oxide surface of liquid metal fractures. The improvement of the thermal conductivity in whole liquid metal requires the creation of such many conductive pathways.

Applying an external pressure to crack the oxide layer can allow the inner liquid metal to flow out, thus the conductive pathways are built among liquid metal particles. Uppal and coworkers prepared the liquid metal capsule and explored its thermal conductivity variation along with different pressures (Figure 2.10d) [101]. The thermal conductivity of liquid metal capsule improves with the increase of compression pressure at early stages, which is owing to the removal of air and destruction of oxide skin to build conductive pathways between adjacent liquid metal particles. As for the reduced thermal conductivity when the pressure increases further, it is because more bulk liquid metal overflows from the sample test area than oxide fragments.

In addition, when liquid metal is mixed with other particles to form a biphasic liquid metal, the thermal conductivity will also be changed. If the added particles possess good thermal transport properties, such as carbon nanotube (CNT), gold, silver, and copper, the thermal conductivity of liquid metal will be improved [107]. Moreover, the formation of intermetallic composition in biphasic liquid metal also makes contribution to the increase of thermal transport property [43]. Different from the mentioned effects, dispersing liquid metal micro/nanoparticles within polymer matrix has a side effect on the thermal conductivity of liquid metal [108].

2.6 Surface Properties of Liquid Metal

We have introduced these fascinating properties of liquid metals, which originate from their electron-rich metallic core. These advances of metallic and fluidity lay the foundation of liquid metal applications such as electrodes, chip cooling, energy storage, and synthesis of nanoparticles [48, 109, 110]. Beyond that, the surface of liquid metal also provides abundant opportunities for future discoveries and practical applications because of these interesting characteristics, which are easy to be regulated.

The surface of liquid metals is unique because it balances the two completely different phases, constructing a thermodynamic equilibrium between air and inner metal core [111]. The transition from ion–ion interactions to van der Waals atom–atom interaction at the liquid metal surface separates liquid metals from other dielectrical liquids. The later smoothly change their electronic density at surface due to the invariable van der Waals interaction. Furthermore, these electrons at surface of a liquid metal are localized, totally different from their inner metal core, where electrons are free to move. These surface characteristics, such as high surface energy, high surface activity, and ordered surface state, provide many fascinating chances to liquid metal applications.

2.6.1 High Surface Energy

The surface energy of liquid metals is ultrahigh due to the intensive metal atom interaction, which can be defined as a measure of the destruction of intermolecular bonds in the formation of material surfaces. Surface energy can exist at any two-phase interface, which is meaningful for stabilizing system.

2.6.1.1 Surface Tension

The value of surface tension is the same as that of surface energy per unit area of liquid, whereas the physical meaning is different. Surface tension of liquid metals is ultrahigh, dozens of times more than any other liquids. The data of liquid metal surface tension are listed in Table 2.4. The underlying reason for high surface tension is the asymmetrical force at surface due to intensive interaction between metal atoms. Therefore, it is hard to bring metal atoms to surface from inner core phase. The metallic bond differentiates liquid metals from other van der Waals liquids. The surface metal atoms, which lose a half interaction in the air side (remain be equilibrium), are in an unusual energy state, resulting the tendency of shrinking to sphere at surface. Thus, the relative low surface tension of liquid alkali metals is comprehensible because of the weak metal bonds, compared with other group metals. Understanding the origin of liquid metal surface tension provides a way to manipulate surface properties by regulating atomic interaction and apply them into microfluidics [113], electronic printing [114], and drug delivery [115].

The ultrahigh liquid metal surface tension originates from the intense interaction between atoms at surface, which is determined by chemical composition. A common

Table 2.4 Summary of surface tension and temperature coefficient.

Element	γ (mN m^{-1})	κ (Mn m^{-1} K^{-1})
Ga	711	−0.070
Sb	371	−0.045
Bi	382	−0.080
Cd	673	−0.150
Cs	69	−0.047
In	556	−0.090
Pb	457	−0.110
Li	399	−0.150
Hg	489	−0.231
K	110	−0.070
Na	197	−0.090
Tl	459	−0.106
Sn	562	−0.103
Zn	789	−0.021

Source: Based on Kasama et al. [112].

way to regulate the interaction forces inside metals is alloying metals with other elements. The final alloy surface tension will be between the original surface tension of each metal. These added elements will increase disorder of liquid metal and change its electronic state (Figure 2.4c,d). Both atom arrange and electron density can alter the interaction between atoms and change surface tension. Oxygen contamination [116] and other impurity [112] also influence the surface tension, thus the resulting calculated data of surface tension is always larger than experimental measurement.

Besides the influence of extra element addition, the state of metal atoms also depends on temperature. For most liquid metals, the isothermal work required to produce a unit free surface always decreases with the temperature increasing linearly due to the disorder caused by thermal expansion [117]. Thus, the surface tension is decreasing accordingly. When the temperature reaches the critical point, interface between liquid phase and gas phase will disappear and surface tension will reduce to zero. The temperature coefficients are also listed in Table 2.4. It is worth noting that the experimental conditions will influence the accuracy inevitably. For example, systematic error introduced by experimental method will cause different results [112].

Comprehending these effect factors opens a way to control surface tension subsequently, that is generating a surface tension gradient by asymmetric regulation of these factors. This propels liquid metal motion and lays the root for liquid robots and motors.

2.6.1.2 Nonreactive Surface Wetting

Liquid metals always appear spherical due to the ultrahigh surface energy. They still exhibit extremely high curvature when they adhere to a substrate. It means liquid metals will not wet most substrates. However, liquid metals can readily wet metal substrates even though there is no reaction occurring due to the intrinsic high adhering work between liquid metals and substrates. Table 2.5 lists these contact angles in liquid metal/metal substrate.

For a general system, such as liquid metal-adhering polymer substrate, the adhesion is weak. We can regulate wettability by means of applying extra energy fields. Figure 2.11a shows the scheme of electrowetting process. There is a dielectric

Table 2.5 Contact angle of liquid metal on substrate.

Type of solid	Contact angle	Examples
Solid metals	$\theta < 90°$	Pb/Fe, 40°
Semiconductors	$\theta < 90°$	Sn/Ge, 40°
Metallic ceramics	$\theta < 90°$	AgCu/Ti_3SiC_2, 10°
Covalent ceramics	$\theta > 90°$	Au/AlN, ≈134°
Carbon materials	$\theta > 90°$	Au/C, 119°–135°
Iono-covalent oxide	$\theta > 90°$	Cu/Al_2O_3, 120°–130°

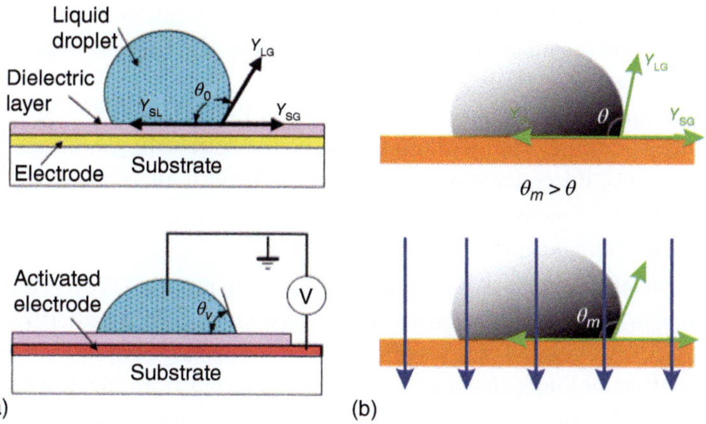

Figure 2.11 (a) Scheme of electrowetting, where $\theta_v < \theta_0$. There is a dielectric layer, which prevents substrate from erosion by liquid metal. Source: Yi et al. [118]. Reproduced with permission of IOP Publishing, Ltd. (b) Scheme of poor wettability of liquid metal under the magnetic field, where $\theta_m < \theta$. Source: Li et al. [119]. Reproduced with permission of American Chemical Society.

layer between the liquid metal and electrode, preventing substrate from erosion by liquid metal. Liquid metals are an excellent conductor, whose surface charges will redistribute under electric field, reducing the surface energy. On this basis, liquid metals will wet any substrates readily and the ultimate contact angle depends on the applied voltage [118]. Similar phenomenon can be found in other type liquids, for example, gelatin drops [120]. Therefore, electrical wetting is a common phenomenon of liquids.

Electrowetting process is of practical importance, and is widely applied in electronic printing [121]. On the other hand, we can exert an external force to roll these adhered liquid metal drops, reducing the contact area between liquid metals and substrates, which can generate charges on the surface of liquid metals. These charges can be harvested, realizing energy conversion from mechanical energy to electric energy [122].

External field can also impose a wetting obstruction for liquid metals, for example, magnetic field. Since the variety of functions are applied by a contactless form, magnetic field is widely used in metallic materials processing, such as material growth [123], crystal array [124], and the solidification processing [125]. In these processes, magnetic field modified the interface condition, for example, interface morphology. The wettability of liquid metal will be poorer under magnetic field, shown in Figure 2.11b [119]. The contact angle in a magnetic field of 5 T is 140.2°, whereas that without a magnetic field is 120.1°. A proposal explaining this phenomenon is that the shape of liquid metal can be adjusted by magnetic force, which can partly balance gravity [126]. The magnetic force can be described as follows:

$$F = \int_0^V \left(\frac{x}{\mu_0} B \frac{dB}{dz} \right) dV \tag{2.4}$$

where V is the volume, x is the volume magnetic susceptibility, μ_0 is the vacuum permeability, and B is the magnetic field intensity. Therefore, the shape change can be attributed to the large susceptibility.

There is another explanation for the hydrophobicity of liquid metal in magnetic field when the magnetic force is too weak to influence the contact angle. Based on the Antonow's rule [127], the contact angle equation can be written as

$$\gamma_{sl} = \frac{1 - \cos\theta}{1 + \cos\theta}\gamma_{sv} \tag{2.5}$$

where γ_{sl} and γ_{sv} are solid surface tension and liquid surface tension [128]. A magnetic field can change the interfacial tension of the system of liquids adhering to substrates [129–131]. The essential reasons of surface tension change are still controversial. Some claim that an increasing surface tension originates from the magnetic dipole interaction near the interface. Some suggest that the two-particle distribution function will change under the magnetic field. Recently, one hypothesis is emerging, that viscosity changes in the magnetic field cause surface tension to increase. Further exploration of the underlying mechanism is needed.

The wettability of liquid metal is a double-edged sword for practical application. On the one hand, it enables liquid metal to roll spontaneously with low friction [122], on the other hand, it makes it difficult to patterning for liquid metal, which is important for liquid metal electronics [121]. Therefore, it is necessary to adjust the wettability of liquid metal according to the work environments.

2.6.2 High Surface Activity

The most prominent surface property of liquid metals is the high surface reactive activity. The metallic nature allows it to react with other chemicals and to be oxidized.

2.6.2.1 Surface Internalization

Liquid metal can readily erode or internalize other metals due to the highly reactive property of surface. The surface internalization process broadens a road for fabrication of functional materials. Figure 2.12a shows the grinding amalgamation process of liquid metal to fabricate functional materials at inertia atmosphere [33]. Liquid metal internalizes Hf powder quickly, which is similar to endocytosis [134]. Right EDX mapping shows uniform distribution of these elements. This method can endow liquid metals with many additional functions. For example, internalizing magnetic gadolinium particles into liquid metals can fabricate magnetocaloric refrigeration [48]. And light-sensitive material titanium dioxide also can be decorated in liquid metals to mimic artificial synapses [135]. Generally, surface internalization turns liquid metals into a versatile material.

We have mentioned earlier (2.6.1.2) that liquid metals are superhydrophobic on the majority of substrates due to the high surface energy. But the highly reactive metal atoms will react with some substrates intensively to improve wettability, termed as reactive wetting. Reactive wetting is an irreversible procedure, which forms intense chemical bonds to increase the adhesion work of liquid metal on

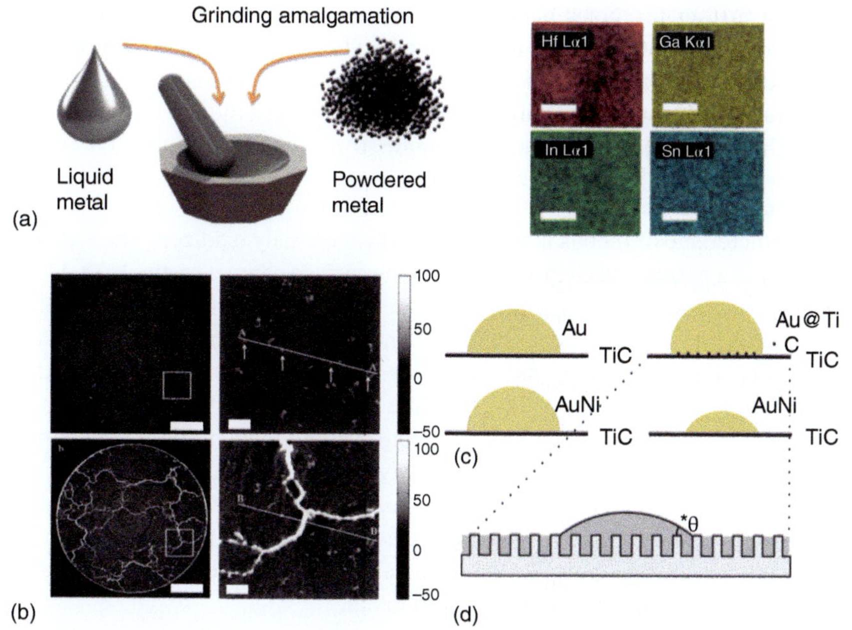

Figure 2.12 (a) Grinding amalgamation process of liquid metal to fabricate functional materials at inertia atmosphere. Right EDX mapping. Source: Daeneke et al. [33]. Reproduced with permission of Royal Society of Chemistry. (b) X-ray microtomography picture of Ga permeating into Al grain boundary. Source: Ludwig and Bellet [132]. (c) Scheme of Au based liquid metal wetting the TiC substrate. (d) Scheme of the real wetting area. (c, d) Source: Based on Bico et al. [133].

substrate. It can also be used to explain why water can wet the aluminum substrate but fail to wet other metal sheets [136]. Liquid metal Ga has high chemical reactivity, which can react with almost all of metals. Figure 2.12b exhibits the Synchrotron Radiation X-ray microtomography technology, which images the penetration process of liquid Ga into Al through grain boundary [132]. Liquid Ga spreads out on Al sheets readily due to reactive wetting, triggering the embrittlement of substrates. Therefore, it is indispensable to pay attention to corrosion when applying liquid metal as electrode, although Schottky contact can be achieved in this system. The reaction wetting between liquid metals and substrates can modify the structure and topography of interface, which can act in turn on the wetting condition [137]. We elaborate this issue through the following example. There is a nonwetting system with 130° contact angle where liquid Au locates onto TiC substrate (Figure 2.12c). Actually, liquid Au can easily react with Ti, but it shows no help for wetting since the interface is not smooth anymore, which was decorated by carbon particles. Liquid Au cannot wet carbon [138]. Moreover, the carbon particles further impede wetting of liquid Au due to the topology and morphology formation. Compared to smooth interface, rough substrates change the actual contact area, which is shown in Figure 2.12d [133]. Therefore, the roughness of contact interface is an influence factor of wetting. Rough substrate can positively feedback the wettability, that means it further promotes or hinders wetting. There is an entirely different situation if adding Ni into liquid Au because Ni will reactively contact carbon and facilitate wetting. Therefore, flexible regulation of the components of liquid metal and substrates can change the wettability bidirectionally since the interface reaction can reduce the wetting barrier as well as increase it.

2.6.2.2 Surface Oxidation

Another characteristic of metals is that their surface is easily oxidized. The surface oxidation skin is a natural surfactant, changing the surface characteristics [139, 140]. Furthermore, the oxidation skin will change the physical properties of liquid metal such as electrical and thermal conductivity. And the rheological property is also changed so that liquid metals transform into a non-Newtonian fluid [70]. At the same time, the fluid will adhere to the substrate relative tightly, which lay the fundamental for constructing components by injecting liquid metal into microchannel or fiber directly [141]. The oxidation skin also gives the liquid metal a certain rigidity, which can fight against gravity and maintain the liquid metal in an irregular shape. It permits liquid metal to be used as inks in 3D printing. Of course, the oxidation skin has its noisome point in the practical application. Thus, it is necessary to comprehend the oxidation process of liquid metal.

Surface Self-Limiting Oxidation For the most liquid metal, they will form an ultrathin oxidation skin rapidly when exposed to air, even though the oxygen concentration is ultralow. Ga is a typical liquid metal with surface self-limiting oxidation. The oxidation skin will further protect the liquid metal from oxidization, and the final thick will reach to tens of Å. *Mott* first put forward a model to explain this process, which was shown in Figure 2.13a [142, 144, 145]. In this model, the topside oxidation layer of metal is termed as Langmuir layer and the oxygen molecule will be absorbed on the Langmuir layer and be ionized by the electrons, which are escaping from the inner core of liquid metal since the tunnel effect. Thus, between the ionized oxygen

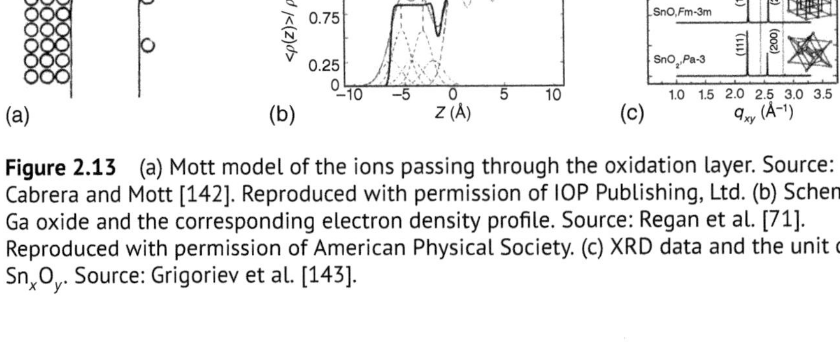

Figure 2.13 (a) Mott model of the ions passing through the oxidation layer. Source: Cabrera and Mott [142]. Reproduced with permission of IOP Publishing, Ltd. (b) Scheme of Ga oxide and the corresponding electron density profile. Source: Regan et al. [71]. Reproduced with permission of American Physical Society. (c) XRD data and the unit cell of Sn_xO_y. Source: Grigoriev et al. [143].

molecular and the liquid metal, a *Mott potential* thus generates. The *Mott potential* helps the metal ions to diffuse toward the free surface, accelerating the oxidation rate so that the oxide layer gradually thickens. At the early stage of oxidation, the growth mode conforms to the Logarithmic law [146]. The oxidation layer is growing thick as time goes on and the impetus also waned [144, 147]. Ultimately, the oxidation skin is stable at room temperature without any external disturbance. The temperature, acidic or basic environment, and partial pressure of oxygen will influence the oxidation skin of liquid metals [77, 148]. Figure 2.13b shows the model of Ga oxidation skin and the corresponding electron density profile [71]. The Mott model is widely used in complaining the oxidation of liquid metal, even though it has some rough approximation, which ignores the concentration gradient and space charges.

Nonlimiting Surface Oxidation The oxidation instance of liquid Sn is totally different from liquid Ga. The surface of liquid Sn will generate SnO, SnO_2, Sn_2O_3, and Sn_3O_4 when sufficient oxygen is supplied, shown in Figure 2.13c. Different oxide formation at same time introduces strain and tears up the oxidation skin. Fresh liquid metal can leak from these seams and be oxidized continually. And these Sn_xO_y compounds grow continuously, leading the oxidation skin to become thick [143, 149].

2.6.3 Ordered Surface State

Different from these molecular liquids, atomic interaction change at surface of liquid metal is intensive, leading to the electronic density oscillation. The immediate result is that there exist several transition layers at surface of liquid metal. This phenomenon is called surface layering. Scientists have researched this phenomenon by using X-ray reflectivity technology. Figure 2.14a,b shows the reflectivity curve of liquid Ga and Sn surface, which clearly indicates the layering phenomenon existing at liquid metal surface and the competition between covalence bond and metallic bond caused by the Ga dimers [150, 151].

It has long been known that there is a transition layer between surface and inner core of liquid metal. In some special conditions, for example, at triple point, the layering thickness is dependent on surface tension γ and isothermal compressibility k_T [155]. The dependence can be roughly described as

$$\gamma k_T \approx 0.07l \tag{2.6}$$

l is the thickness of transition layer.

From the fact that these layering phenomena are discovered in Hg ($\gamma = 500\,\mathrm{mN\,m^{-1}}$), Ga ($\gamma = 750\,\mathrm{mN\,m^{-1}}$), and In ($\gamma = 550\,\mathrm{mN\,m^{-1}}$) earliest, it is tempting to take the large surface tension as basic requirement of surface layering. But in Figure 2.14c, the reflectivity curve of liquid K ($\gamma = 110\,\mathrm{mN\,m^{-1}}$) also presents the layering features [152]. This experiment contradicts the "surface tension domination" assumption, because the similar surface tension of water ($\gamma = 72\,\mathrm{mN\,m^{-1}}$) shows no influence on the surface structure. Therefore, the surface tension fails to explain the surface layering by itself. Another theory persists that surface layering is a general phenomenon of all liquids, as long as the temperature T is smaller than $0.2T_c$, where T_c is the critical temperature of liquids [156]. For liquid metals, this hypothesis perfectly matches due to the ratio between T and T_c is less than 0.2 (0.043 for Ga, 0.066 for Sn, 0.15 for K). But for other liquids, they are already solidifying when the temperature reaches $T = 0.2T_c$, thus cannot verify this concept. But the layering phenomenon indeed can exist in many liquids besides liquid metal if the surface contacts with a hard wall or under some external field. For example, the water molecules will order at electrode–electrolyte interface under voltage [153]. Figure 2.14d,e shows the surface structure of liquid Ga coming in contact with a diamond.

Similarly, there is an ordered surface state in the liquid metal alloy system, whose cause is different from pure liquid metal, terming "surface freezing." The freezing of

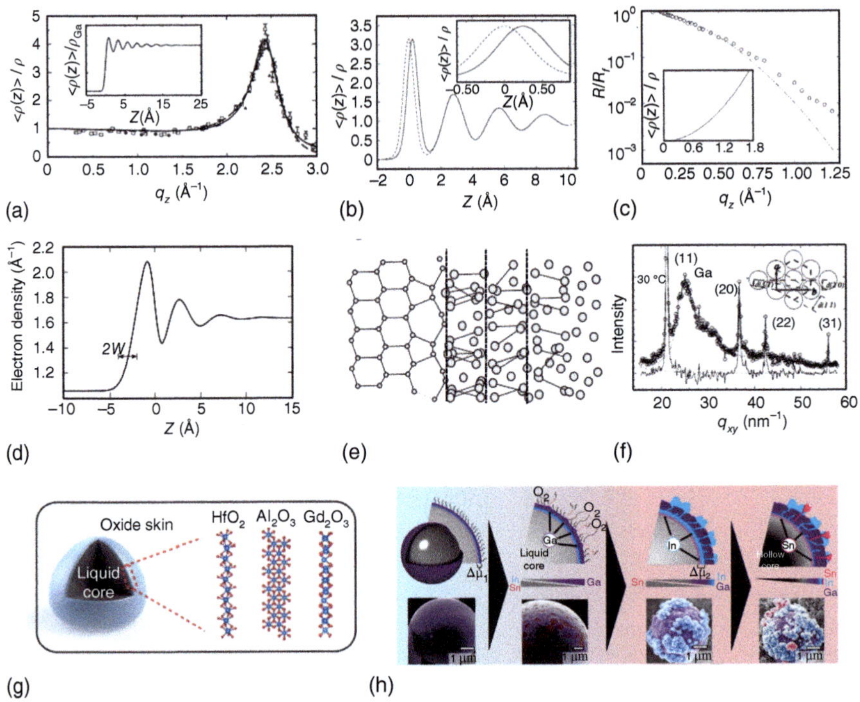

Figure 2.14 (a–c) X-ray reflectivity of the Ga, Sn, and K, which clearly shows the surface layering [150–152]. Source: (a) Regan et al. [150]. Reproduced with permission of American Physical Society; (b) Shpyrko et al. [151]. Reproduced with permission of American Physical Society; (c) Shpyrko et al. [152]. Reproduced with permission of American Physical Society. (d) Electron density of liquid Ga surface contacting with diamond hard wall. (e) Illustration of the Ga dimers converging at the surface, presenting the surface layering. (d, e) Source: Huisman et al. [153]. Reproduced with permission of Springer Nature. (f) Grazing incidence diffraction (GIXD) data for GaPb alloy, indicating the formation of a frozen hexagonal lead in the topmost layer of the alloy. Inset: the atomic arrangement. Source: Yang et al. [154]. Reproduced with permission of American Physical Society. (g) A cross-sectional diagram of a liquid metal droplet, with possible crystal structures of thin layers of HfO_2, Al_2O_3, and Gd_2O_3 as indicated Source: Zavabeti et al. [134]. Reproduced with permission of The American Association for the Advancement of Science. (h) Surface composition inversion of liquid metal oxidation under high temperature. Source: Martin et al. [111]. Reproduced with permission of John Wiley & Sons.

liquid alloy is more similar to a self-sort process, which abides the Gibbs's absorption rule, that is the component with lower surface energy in the liquid alloy will segregate to the surface. And there are some transition layers between the surface and the bulk. For example, at the liquid Ga–Pb alloy, there are at least three atomic layer separations extending from surface to the bulk, in which the topmost surface monolayer is pure Pb, the lower surface tension component. And the closer components get to inner core, the less Pb there is. The inner layer is pure gallium layer (Figure 2.14f). Similar phenomena have also been found in the liquid alloy of Ga–Bi, Ga–In, Bi–In, and so on.

In theory, every component in an alloy can be oxidized at surface, abiding the oxidation rule. And there is an order of the alloy surface oxidation according to the Gibbs free energy variety, similar to the liquid alloy freezing, that means these components with lower surface energy (which are occupying alloy surface) have greater changes in free energy after oxidation and situate at the inner core (Figure 2.14g). Under the external field interaction, this order can be disturbed. For example, at high temperatures gallium will no longer be self-limiting oxidized state since the high temperature promotes ions to migrate in ions migration. Ga will continue to be oxidized until it is depleted, and then other metal components are oxidized and occupy the surface of gallium oxide, achieving surface composition inversion (Figure 2.14h).

In this section, we have discussed the surface properties of liquid metals, including high surface energy, high reactive and the surface order. We have dug out these mechanisms underlying these phenomena. These surface characteristics lay the foundation of emerging applications. Liquid metal soft robots can move by asymmetric regulation of surface tension. Applying the same concept in drug delivery system with surfactant realizes precise treatment. The properties of high reactivity open a way to fabricate functional materials, for example, magnetic materials. The surface layering and surface oxidation skin can be used to fabricate two-dimensional (2D) materials. Furthermore, the skin endows liquid metal a certain rigidity against gravity and surface tension. Thus, liquid metals can exist in an unbalanced shape. In conclusion, the surface of liquid metal holds a wealth of science.

2.7 Liquid Metal in Other Dimensions

In addition to these fascinating properties of liquid metal on the form of bulk, liquid metal can also lay a foundation for study of these ground state properties in other dimensions. These nanomaterials are the forefront of scientific research due to surface/interface properties, quantum confinement effect, and macroquantum tunnel effect. When liquid metals are dimensioned into a nanomaterial, there are unique physical and chemical characters.

2.7.1 One-Dimensional Liquid Metal

One dimensional (1D) materials (nanorods, nanotubes, nanowires, etc.) are widely used in optical, electrical, magnetic, and catalytic materials due to their special morphology and size. Liquid metals will form strange, amorphous, stable atomic structures once the diameter of the ultrathin wire has been reduced below the critical size (the order of the spacing of the atoms) [157, 158]. These structures differ greatly from the bulk LM structures mentioned in the first section and tend to be predominantly icosahedral, as they can expose 100% (111) of the surface and have the lowest surface energy. The final structure depends on the material and wire thickness. These different structures will also bring different properties of liquid metals.

Moreover, 1D nanomaterials have directional electron ion conduction direction, strong stress bearing capacity, and short axial electron ion transport path, especially for 1D metal materials. Quantization of transversal electronic states and resulting quantum ballistic conductance and its stepwise variation in the course of stretching of wires draw many attentions. In this section, we will introduce the superlative physical properties of these liquid metal elemental 1D materials.

2.7.1.1 Ga

Ga is a high-conducting element, which is widely used in soft electrical devices, such as fibers [159], interconnects [160], and antennas [161]. When it exists in 1D form, it can be integrated into nanodevices.

Ga nanowires can be stably encapsulated in CNTs to form a nanothermometer, which is still in liquid state though temperature is as low as 193 K (Figure 2.15a,b) [162]. The special physical space confinement stabilizes the Ga nanowires in the thermodynamically unstable structure, β phase or γ phase, instead of the common α phase. This physical space confinement is relative with the diameter of Ga nanowire and the interaction between the CNT and the Ga nanowire. Figure 2.15c shows three interface profiles of Ga nanowire and CNT, which indicates that the bond length and bond angle have changed [163]. In fact, the atomic structure at the surface of an encapsulated Ga nanowire is distorted significantly. Such distortion rearranges the Ga atoms at the interfacial layer of the Ga nanowire, characterizing the amorphous feature there. Such nanowire has many chain structures. Figure 2.15d shows the three kinds of chain structures of Ga nanowire: the linear chain (LC), the narrow angle zigzag (NZ) chain, and the planar wide angle zigzag (WZ) chain and their binding energy [164]. The binding energy versus lattice parameter for the three kinds of chains of Ga nanowires clearly demonstrates that the latter two zigzag chain structures are more stable in energy than the metallic LC, consistent with previous researches [165, 166]. While LC has the lowest, NZ has the highest cohesive energy per atom. All the three-chain structures are metallic (Figure 2.15e).

2.7.1.2 In

In nanowires are a potential electrical component due to the temperature-dependent electrical properties [167]. It is always used in superconducting quantum interference devices [168]. Besides the prominent electrical properties, In nanowires have attracted much attention due to the fascinating optical properties and it is widely used in sensing devices, storage media, and nanoelectronic probes [169]. Nanorods and nanowires are the most prominent examples of nanoscale entities with surface

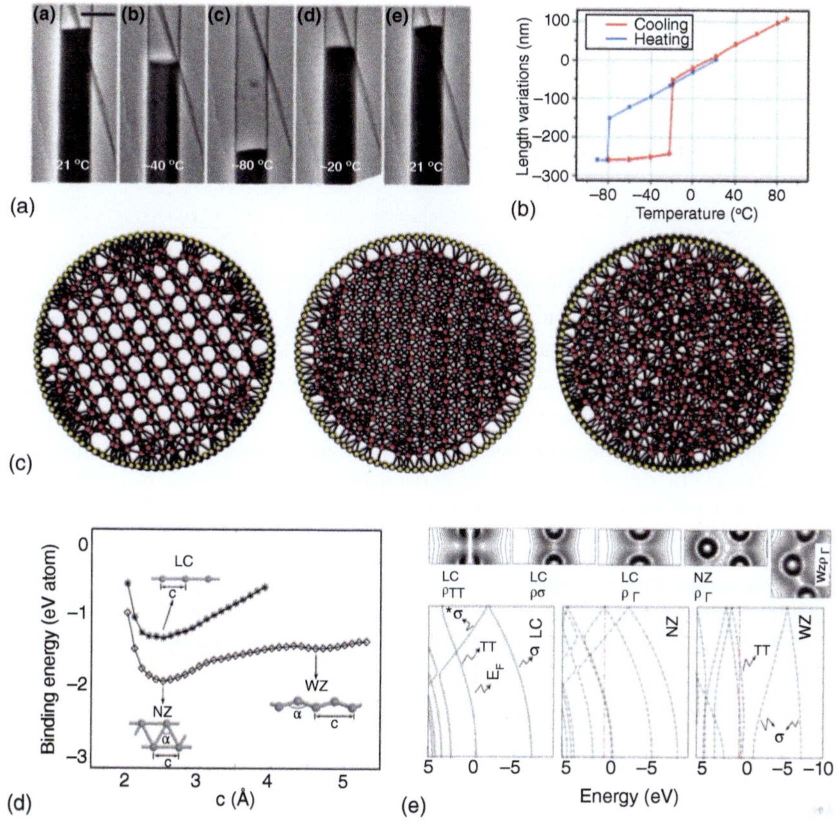

Figure 2.15 (a) Ga nanowire thermometer. Source: Liu et al. [162]. (b) The length variation of thermometer in different temperature. Source: Liu et al. [162]. Reproduced with permission of American Physical Society. (c) The interface of Ga nanowire and CNTs. Source: Li et al. [163]. Reproduced with permission of AIP Publishing. (d) The three-chain structure of Ga nanowires. (e) The band structure of Ga nanowires. (d, e) Source: Durgun et al. [164]. Reproduced with permission of American Physical Society.

plasma resonance (SPR) ranging from the visible to the near-infrared because the variation of the parameters such as shape and geometry enables tuning of the optical resonances. Similar to the electrical properties, the optical properties are also size-dependent (Figure 2.16a). For example, when the diameter of nanowire is increasing, the maximum absorption wavelength will be red shifted. Therefore, In nanowires can be applied to different working environments by controlling the morphology reasonably.

2.7.1.3 Sn

Sn nanowire is also a superconductor. Its temperature-dependent magnetization curve under the magnetic field is shown in Figure 2.16b. The data of Sn nanosquare is shown as a comparison [170]. From this figure, we can see the superconductivity behavior of Sn nanowire (the typical diamagnetic behavior), when the temperature is under 3.7 K, the transition temperature. For bulk β-Sn, the transition temperature

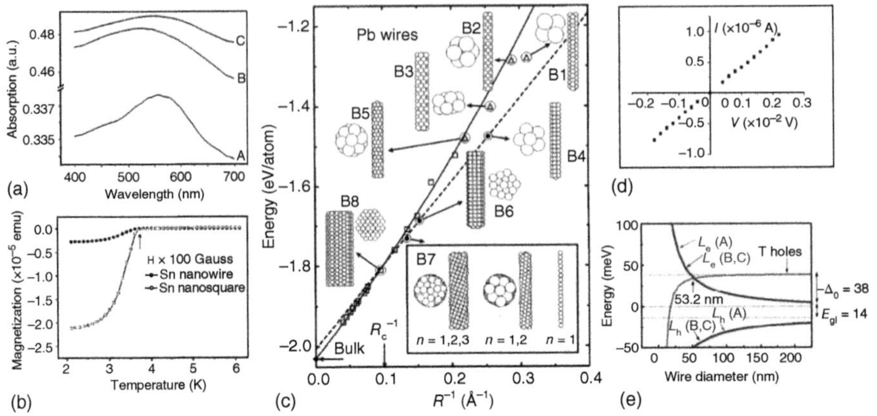

Figure 2.16 (a) Diameter-dependent optical property of In. Source: Li et al. [169]. Licensed under CC BY 2.0. (b) The superconductivity of Bi nanowires. Source: Hsu and Lu [170]. Reproduced with permission of American Chemical Society. (c) The structure of Pb nanowire. Source: Gülseren et al. [171]. Reproduced with permission of American Physical Society. (d) *I*–*V* curve of Pb nanowires. Source: Wang et al. [172]. Reproduced with permission of American Chemical Society. (e) Calculated sub-band energies for Bi nanowires and the corresponding band gap at 77 K as a function of the wire diameter. Source: Lee et al. [173]. Reproduced with permission of IOP Publishing, Ltd.

is 3.73 K [174]. There are no significant changes in the transition temperature for different dimensional metals.

2.7.1.4 Pb

Lead (Pb) nanowire is a classical and important 1D material due to its high reactivity and superconductivity. Researchers have discovered a unique structure of Pb NWs and justified that the tendency of the change from crystalline to amorphous state is general. The molecular dynamics-based "simulated annealing" methodology has been used to study the Pb NWs. Figure 2.16c shows the simulated structures [171]. The main manifestation of Pb NWs is an icosahedral feature but there is still a helical structure. It has the same superconducting critical temperature as its bulk counterpart, whereas its critical field is up to 3000 Oe, determined by the Meissner effect measurement [175]. The superconducting behavior of nanowires has drawn attention due to their potential applications [176].

Figure 2.16d shows the electronic characteristic of Pb nanowire [172]. From this I–V curve, we can infer the resistance of Pb nanowire is 2.5 KΩ at room temperature and the conductivity is 4.0×10^4 S cm^{-1}, which is almost identical to the bulk counterpart (4.8×10^4 S cm^{-1}). It is found that the resistance of a single nanowire has a corresponding response to temperature: the lower the temperature, the lower the resistance. When the temperature drops to 7 K, a sharp change in resistance (five to six times smaller than that at 8 K) is observed, indicating that the nanowires have entered a superconducting state. When the Pb materials are limited in 1D, their electrical performances are strongly influenced by size and morphology. For example, the superconducting transition temperature is dependent on the diameter [177].

2.7.1.5 Sb

Besides Bi nanowires, Sb nanowires are also a kind of excellent thermoelectric material. The higher carrier mobility, lower lattice thermal conductivity, and the higher state density above Fermi level brought about by the low dimension makes the Sb nanowire a potential ZT thermoelectric material [178]. Furthermore, because of the quantum confinement effects, the Seebeck coefficient will elevate on the premise of unchanging of electrical conductivity. And the boundary scattering effect and ballistic effect also lay the foundation of influencing the value of thermal conductivity. As a result, the Sb nanowire is used in thermoelectric applications [179].

2.7.1.6 Bi

There is a long history of the research on Bi due to the semimetal characteristic and the highly anisotropic Fermi surface, indicating the anisotropic carrier transport[180]. Bi could be metals, semimetals, and semiconductors. Its electronic properties are sensitive to the size and geometry of the sample [181] and bipolar electrical resistivity [182]. It is well-known that there is a critical diameter d_c for a single crystal Bi nanowire. Due to the quantum confine effect, there exists a transition from semimetal to semiconductor for Bi nanowire when the diameter is smaller than d_c. The d_c depends on the crystallographic orientation of Bi nanowire [183]. From Figure 2.16e, we can clearly see the transition of the nanowire, which shows band gap of Bi nanowire sub-band energies as a function of diameter. When the diameter is smaller, the conduction band goes up and the band gap goes down [173]. Furthermore, this is a superconductivity evidence for Bi nanowire [183], although the origins of superconductivity are unclear. Initially, Weitzel and Micklitz interpreted the mechanism as quantum size effect. But recently, Zhang and coworkers found that when the size is less than 4 nm, Bi will be transformed into amorphous state from diamond structure [180].

The thermal conductivity of Bi nanowire is also dependent on the wire diameter owing to the limitation of electron and phonon boundary scatter in Bi nanowire [184]. Similar to the carrier transport, the thermal conductivity is also anisotropic. For a Bi nanowire with 100 nm diameter, there is a difference of ~7.0 W m^{-1} 3 K between the [102] and [110] directions. Thus, besides the diameter of nanowire, the direction also plays an important role in the application of Bi nanowire.

It has been demonstrated that the physical properties of liquid metal nanowires depend strongly on not only the diameter but also the crystal orientation, and the controlling of growth of the nanowire orientation is thus very essential for their future applications.

2.7.2 Two-Dimensional Liquid Metal

Since the thermodynamic instability of 2D materials, people generally think it is impossible to fabricate ideal 2D crystals and maintain their stable form under natural condition. But the unique viable birth of graphene breaks this traditional cognition, bringing about the research upsurge of 2D material [17]. These materials present a special system in the atomically thin limits due to the quantum confine effect. There is an ideal 2D material type consisting of monolayer metal sheet on semiconductor substrates, in which electrons are confined in ultrathin metal monolayer [185]. In this section, we will introduce the superlative physical properties of these liquid metal elemental 2D materials.

2.7.2.1 Ga

The gallenene (2D Ga) can exist on substrates in two different crystallographic forms, termed as "gallenene a_{100}" and "gallenen b_{010}," displayed in Figure 2.17a. Both types have four atoms in a unit cell. For gallenene a_{100}, the bonds of Ga–Ga are almost equal, which are covalent in nature, indicating a honeycomb structure of gallenene when the crystal lattice completely relaxes. In the original structure, due to the incomplete sp^2 hybridization, the planar hexagon shows slight distortion with 122.56° and 118.9° interior angles [186]. On the other hand, the Ga–Ga bonds are anisotropy in b_{010} form, of which is covalent in the y direction and shows more metallic nature along other directions. This hybridization bond nature results in more stability of gallenene b_{010} structure. In fact, the formation energy of gallenene of b_{010} (0.36 eV) is much smaller than a_{100} (0.64 eV). From the view of the original crystal structure, the gallenene b_{010} is a buckling structure, similar to these fascinated 2D materials, for example, black phosphorus [189].

The buckled structure in the gallenene endows it abundant physical and chemistry properties, which can reflect in the anisotropic electrical structure. The band structure of gallenene exhibits obvious metallic characteristic, which possesses nonlinear dispersive bands and partially filled Dirac cones. And more interesting is that the magnetoelectric transport characteristic at the ultralow temperature exhibits the transformation of Berezinskii–Kosterlitz–Thouless (BKT) in zero magnetic field and the existence of multiple intersections in resistance curves varying with magnetic field intensity are revealed. These experimental data demonstrate the Griffiths singularity existing in the 2D system [190].

2.7.2.2 In

There is another superconducting mechanism of metal in the 2D limit, which can be simply described as follows: the 2D metal thin films act as a charge reservoir. The metallic bonding provided by these metal atoms, and the interface bonds provided by the metal film and substrate, contributes to the electron–phonon interactions [187, 191]. These combinations will generate a superconductor, for example, the In single atomic layer on the Si (111) surface (Figure 2.17f). Different from the graphene, the freestanding monolayer In film is unstable and needs a substrate to stabilize the structure. There are many dangling bonds at the surface of Si (111), which plays an important role in the superconductor system. Some of the electrons in these dangling bonds bond to indium covalently, and some permeate into their interstices, increasing the electron number of per indium atom to 3.45 from 3. Thus, the 2D metal, whose Fermi surface is similar to the ideal 2D electron gas, is contributed by the In atoms and the dangling bonds originated from Si (111). This special interface effect makes the indium thin films become a strong coupling superconductor.

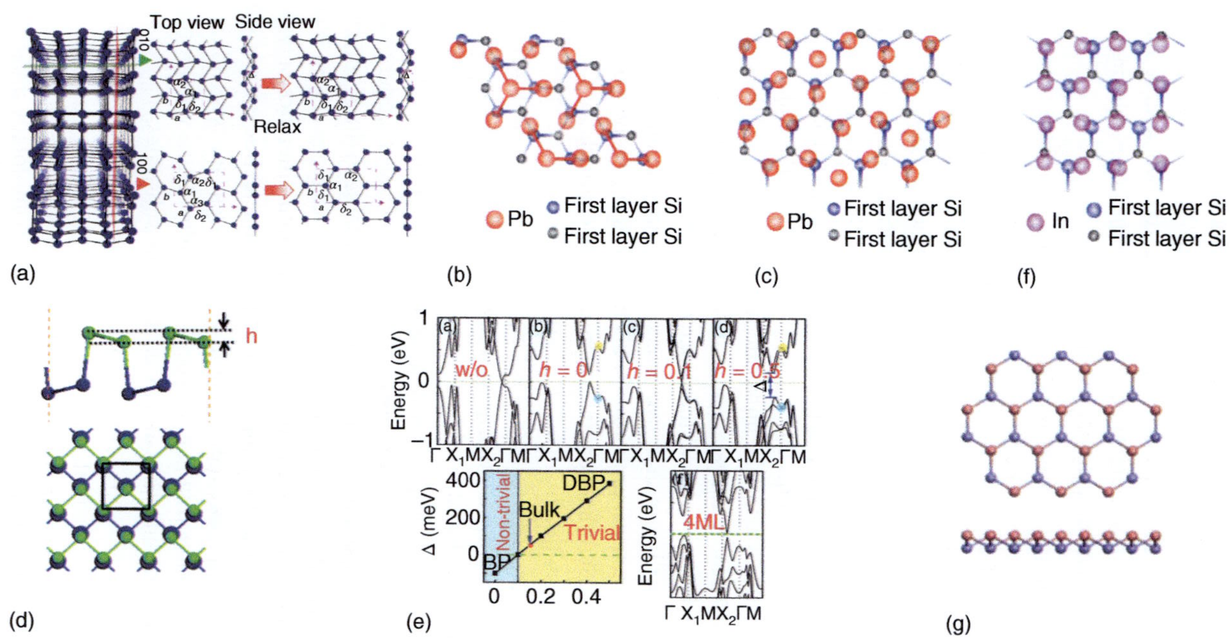

Figure 2.17 (a–c, f) the atomic structure of 2D Ga, Pb, Bi, and In. Source: (a) Kochat et al. [186]. Reproduced with permission of The American Association for the Advancement of Science. ; (b, c, f) Zhang et al. [187]. Reproduced with permission of Springer Nature. (e) The band structure of 2D Bi with different h and the band structure of 4ML Bi. (d, e) Source: Lu et al. [188]. Reproduced with permission of American Chemical Society.

2.7.2.3 Sn

Tin is in the same main group with carbon and its 2D allotrope presents a graphene-like structure [192, 193]. But the atomic mass of Sn is larger than C and the intensive spin-orbit coupling (SOC) makes the band structure of 2D Sn different from graphene, opening a gap at the Dirac point. Similar to the 2D Bi, the 2D Sn forms a buckled honeycomb lattice, which is composed of two layers of α-Sn (111) stacking together and the difference in height between the two layer is about 1 Å, presenting a topologically nontrivial state. The state can be effectively regulated by chemical functionalization and external strain.

2.7.2.4 Pb

The superconductivity of metal in the 2D limit can also exist in the Pb ultrathin film [180]. The superconductivity is relative to the thickness and the structure type of the thin film. According to the quantum growth principle [194], the thickness limit for a film to be stable is two layers, which possesses only one quantum well channel exactly. With the increase of Pb film thickness, the quantum well channel is also increasing and the underlying superconducting mechanism is more complex. There is no three layers of Pb film due to the quantum size effect, only a few four layers and five layers. Comparing the transition temperature (T_c) of these two-layer thickness Pb films (6.7k for four layers and 6.3k for five layers), we can see that the quantum oscillations is continuing in the superconducting gap. And the amplitude is more intensive as four layers than five layers. These phenomena are complying with the fact that the electrical properties of 2D materials is closely bound up with the layer number [195]. But there are two quantum well channels for four and five layering Pb thin film, resulting in higher T_c than that for two layers Pb film. Besides the influence of thickness, the substrate effect is prominent in the 2D limit due to the complex interface environments, such as the roughness and charge transfer. The Pb film deposited on the Si substrate will be strained. In fact, there are two types of Pb film, shown in Figure 2.17b,c. Type I is with underlying 1×1 atomic structure, whose surface lattice parameter is similar to bulk Pb. And the other is with underlying $\sqrt{3} \times \sqrt{3}$ atomic structure, whose surface lattice parameter is same to the Si substrate. In the ordinary environment, the electrical property of the two types of film is similar due to the similar atom and electron density. But the slight difference between the two types of film is reflected in the superconducting characteristic evidently due to the substrate effect. The T_c for type I is about 4.9 K whereas for type II it is about 3.65 K, and the superconducting gap of type II is smaller. Thus, we can tune the electrical properties of metallic thin film by devising the substrates.

2.7.2.5 Sb

Elemental Sb nanosheets are also important 2D materials, whose atomic structure is shown in Figure 2.17g. The electrical properties of Sb are totally different from other 2D metals, even though they have the same buckling structures [196, 197]. Sb consists of three σ bonding orbitals and a π bonding orbital whereas the metal, such as Ge, merely consists of π bonding orbitals. As a result of this special configuration, the Sb acts as an indirect semiconductor (band gap 2.28 eV). In fact, the bulk Sb is a semimetal. This transformation from semimetal of bulk to semiconductor of monolayer can be ascribed to the second-order effects and the quantum confinement effect. From the former we can be simply understood that there is no interlayer interaction in the monolayer system. And in the monolayer system, the latter can be comprehended that the conduct band of Sb is located at the vacuum level. The conduct band bottom is mainly composed of 5p orbitals, which will split into three bonding orbitals and three antibonding orbitals, thus forming a gap around the Fermi level. The indirect band gap restricts Sb in the practical application such as photoelectric device. Exerting a biaxial stress can transform the indirect semiconductor into direct semiconductors, which is one of the commonly used means of electrical property regulation.

2.7.2.6 Bi

Similarly, there are also two different configurations of the 2D Bi thin film on different substrate [184]. A buckled honeycomb structure, termed as Bi (111) will be obtained by using the threefold substrate (Si (111)), sixfold substrate (Bi_2Te_3), and fourfold substrate (Si (001)). And the other is the Bi (110) thin film, which can be gained on the reconstructed Si surface and the highly oriented pyrolytic graphite (HOPG). Due to the charge transfer between the HOPG and Bi (110) thin film, the structure of Bi thin film will distort to be the black-phosphorus-like structure [198, 199]. The difference between the two types is illustrated in Figure 2.17d,e, which mainly reflects in the altitude intercept h between the neatest neighborhood atoms. Bi is a heavy element (atomic number is 83) with strong SOC, which opens the gap between the p_z^- and p_z^+ orbitals, resulting in the topological insulation state existing in the Bi thin film. But the electrical properties of the two types of Bi film are different due to the substrate effect. The freestanding Bi (111) thin film is a topological insulator with a large gap, about 100 meV, whereas the gap of black-phosphorus-like structure Bi (110) is smaller since the inversion symmetry is broken and the corresponding degeneracy of each band goes up. Thus, the altitude intercept h plays an important role in the topological properties of Bi thin film. There will be a Dirac corn in the band structure of Bi when the h equals to 1 Å, thus the Bi (110) film remains a topological nontrivial system with inverted bandgap as long as the h is less than 1 Å, which are shown in Figure 2.17d,e.

The properties of Bi thin film also depend on the thickness. There exist no thin Bi films with odd layer number due to the abound dangling bonds on the surface, for which can be saturated forming an even layer film. The topological gap is as large as 100 meV for two-layer Bi film and reduced to 59 meV when the thickness reaches four layers. For six-layer film, it is a metallic material even though it remains topologically nontrivial. A film of this thickness is getting closer to the bulk Bi thus transferring from semiconductor to semi-metal states. In practical experiments, there are no more than four films of Bi (110) since the thickness of these films is unstable and will change to Bi (111) films.

Bi thin film (less than six layers) is a topological insulator, which is an important electronic material. Its insulate gap can be tuned by the devise of substrate and thickness, thus presenting a more complex edge state.

Generally, these liquid metal elements will present exciting properties when they reduce to the 2D limits. These properties such as superconductivity, large-gap 2D quantum spin Hall state, and enhanced thermoelectricity provide many opportunities to the basic sciences and practical applications, attracting many researchers to jump in.

2.8 Conclusion

It is well-known that the research of liquid metal has a long history. Many novel and fascinating features of liquid metals have been discovered. In this section, we have discussed these physicochemical properties. First, we have introduced the several typical structures of a liquid metal, including hard-sphere-like structure, distorted structure, and stronger anomalies structure. The liquid structure responding to the temperature and pressure are also included. Besides, the structures of monophasic liquid metal alloy and biphasic liquid metal alloy are also discussed. Understanding the local structure and principle of liquid metal plays an essential part in the development of condensed matter. Based on these structures, we can comprehend these unique characteristics and properties. The most prominent characteristic of liquid metals is the ultralow melting point. The melting point of metals is contingent on many factors, such as crystal structure while they are in the solid state, the number of electrons involved in the bond, the degree of delocalization of valence electrons, and the radius of metals. Based on this, we introduce the low melting point origins of four typical liquid metals. When liquid metals form alloys, the melting point decreases due to the disorder structure and the electronic lubrication effect. The structure of liquid metal has been studied for a long time, but there is still no perfect model to simulate liquid metals. In addition to the traditional liquid metal, more complex liquid metal alloys have been developed to meet the requirements of practical applications. For example, $AlSi_8Mg_4$ alloy is used to generate hydrogen, and X-ray tomoscopy has been developed to image the nucleation and growth process in the liquid metal. For these complex alloy systems, the atom interaction and electron density distribution are more intricate. Thus, more sophisticated characterization methods and more interpretation model need to be developed.

The advantages of liquid state make liquid metals totally different from other liquids or solid metals. Taking advantages of both liquid and metal, liquid metal presents low viscosity with high mobility and self-healing capacity, as well as favorable electrical and thermal conductivity at low temperatures. We introduced these rheological properties, electrical conductivity, and thermal property and related influence factors. Besides these whole properties, the surface of liquid metals is also a system worth studying. We introduced three characteristics of liquid metal surface, including high surface energy, high surface activity, and ordered surface state. They provide many fascinating chances to liquid metal applications, such as soft robots, flexible electronic devices, and catalysts. However, there are still some challenges in these domains. The behaviors of liquid metal robots are still very simple and complex operations are difficult to execute, which means we should develop faster and more precise means of surface tension control. The reactivity of liquid metals can help us fabricate functional materials. More functional and stable liquid metal complexes need to be further developed. The reactivity is a double-edged sword. It can erode the devices when liquid metals contact with the working object directly. Thus, anticorrosion is also an important issue in applying liquid metal. The surface order provides many opportunities for fabricating 2D materials. Similar to other materials, Liquid metals will exhibit different properties than those we have mentioned earlier in 2.6 and 2.7 when they are at the nanoscale. These nanomaterials are the forefront of scientific research due to surface–interface properties, quantum confinement effect, and macroquantum tunnel effect. In this section, we simply introduce these properties, such as superconductivity, size-dependent optical properties, thermoelectric properties, and large-gap 2D quantum spin Hall state. Using liquid metals to fabricate some materials, which is difficult to prepare and unstable, is also an important research direction in the future.

We have introduced these properties of liquid metal in this section. But there are still some gaps and concerns in liquid metals, after all, liquid metals are a rich treasure trove still waiting to be explored. These physiochemical properties provide a fertile field for these emerging applications, which will be discussed in later sections.

References

1 Chen, L.Y., Tang, P.H., and Wu, T.M. (2016). Analysis of local bond-orientational order for liquid gallium at ambient pressure: two types of cluster structures. *J. Chem. Phys.* 145: 024506.

2 Wei, S., Oyanagi, H., Liu, W. et al. (2000). Local structure of liquid gallium studied by X-ray absorption fine structure. *J. Non-Crystalline Solids* 275: 160.

3 Waseda, Y. and Suzuki, K. (1972). Structure factor and atomic distribution in liquid metals by X-ray diffraction. *Phys. Status Solidi B* 49: 339.

4 Tsay, S.F. and Wang, S. (1994). Anomalies in the liquid structure of Ga metal. *Phys. Rev. B* 50: 108.

5 Hafner, J. and Kahl, G. (1984). The structure of the elements in the liquid state. *J. Phys. F* 14: 2259.

6 Hafner, J. and Jank, W. (1990). Structural and electronic properties of the liquid polyvalent elements. III. The trivalent elements. *Phys. Rev. B* 42: 11530.

7 Mokshin, A.V., Khusnutdinoff, R.M., Novikov, A.G. et al. (2015). Short-range order and dynamics of atoms in liquid gallium. *J. Exp. Theoret. Phys.* 121: 828.

8 Mokshin, A.V., Khusnutdinov, R.M., Akhmerova, A.R., and Musabirova, A.R. (2017). Universality of microscopic structural and dynamic features in liquid alkali metals near the melting point. *JETP Lett.* 106: 366.

9 Katayama, Y. and Tsuji, K. (2003). X-ray structural studies on elemental liquids under high pressures. *J. Phys. Condens. Matter* 15: 6085.

10 Raty, J.Y., Schwegler, E., and Bonev, S.A. (2007). Electronic and structural transitions in dense liquid sodium. *Nature* 449: 448.

11 Rajesh, C., Majumder, C., Rajan, M.G.R., and Kulshreshtha, S.K. (2005). Isomers of small Pb_n clusters (n = 2–15): geometric and electronic structures based on ab initiomolecular dynamics simulations. *Phys. Rev. B* 72: 235411.
12 Chen, K. and Li, Q. (1993). Local structure and bond orientation order of liquid metal Pb. *Chin. Phys. Lett.* 10: 365.
13 Reichert, H., Klein, O., Dosch, H. et al. (2000). Observation of five-fold local symmetry in liquid lead. *Nature* 408: 839.
14 Urrutia-Bañuelos, E., Posada-Amarillas, A., and Garzón, I.L. (2002). Temperature effect on the local order of liquid Ni, Ag, and Pb: a molecular dynamics study. *Phys. Rev. B* 66: 144205.
15 Cricchio, F., Belonoshko, A.B., Burakovsky, L. et al. (2006). High-pressure melting of lead. *Phys. Rev. B* 73: 140103.
16 Holland-Moritz, D., Inui, M., Kajihara, Y. et al. (2017). Asymmetrical bonding in liquid Bi disentangled by inelastic X-ray scattering. *EPJ Web of Conf.* 151: 5002–5007.
17 Bichara, C., Pellegatti, A., and Gaspard, J. (1993). Properties of liquid group-V elements: a numerical tight-binding simulation. *Phys. Rev. B* 47: 5002.
18 Gaspard, J.P., Marinelli, F., and Pellegatti, A. (1987). Peierls instabilities in covalent structures. *Europhys. Lett. (EPL)* 3: 1095.
19 Tsai, K.H., Wu, T.M., and Tsay, S.F. (2010). Revisiting anomalous structures in liquid Ga. *J. Chem. Phys.* 132: 034502.
20 Sharma, B.D. and Donohue, J. (1962). A refinement of the crystal structure of gallium. *Z. Krist.* 117: 293.
21 Bosio, L. (1978). Crystal structures of Ga (II) and Ga (III). *J. Chem. Phys.* 68: 1221.
22 Bridgman, P.W. (1935). Polymorphism, principally of the elements, up to 50, 000 kg/cm^2. *Phys. Rev.* 48: 893.
23 Jayaraman, A., Klement, W., Newton, R.C., and Kennedy, G.C. (1963). Fusion curves and polymorphic transitions of the group III elements—Aluminum, gallium, indium and thallium—At high pressures. *J. Phys. Chem. Solids* 24: 7.
24 Di Cicco, A. (1998). Phase transitions in confined gallium droplets. *Phys. Rev. Lett.* 81: 2942.
25 Losurdo, M., Suvorova, A., Rubanov, S. et al. (2016). Thermally stable coexistence of liquid and solid phases in gallium nanoparticles. *Nat. Mater.* 15: 995.
26 Kenichi, T., Kazuaki, K., and Masao, A. (1998). High-pressure bct-fcc phase transition in Ga. *Phys. Rev. B* 58: 2482.
27 Li, Z. and Tse, J.S. (2000). High-pressure bct to fcc structural transformation in Ga. *Phys. Rev. B* 62: 9900.
28 Li, R., Wang, L., Li, L. et al. (2017). Local structure of liquid gallium under pressure. *Sci. Rep.* 7: 5666.
29 Yagafarov, O.F., Katayama, Y., Brazhkin, V.V. et al. (2012). Energy dispersive x-ray diffraction and reverse Monte Carlo structural study of liquid gallium under pressure. *Phys. Rev. B* 86: 174103.

30 Gong, X.G., Chiarotti, G.L., Parrinello, M., and Tosatti, E. (1993). Coexistence of monatomic and diatomic molecular fluid character in liquid gallium. *Europhys. Lett. (EPL)* 21: 469.

31 Yang, J., Tse, J.S., and Iitaka, T. (2011). First-principles study of liquid gallium at ambient and high pressure. *J. Chem. Phys.* 135: 044507.

32 Li, R., Li, L., Yu, T. et al. (2014). Study of liquid gallium as a function of pressure and temperature using synchrotron x-ray microtomography and x-ray diffraction. *Appl. Phys. Lett.* 105: 041906.

33 Daeneke, T., Khoshmanesh, K., Mahmood, N. et al. (2018). Liquid metals: fundamentals and applications in chemistry. *Chem. Soc. Rev.* 47: 4073.

34 Agarwal, R., Samui, P., and Kulkarni, S.G. (2013). Thermodynamic investigations of (Bi+Ni) system – Part II. *J. Chem. Thermodyn.* 57: 477.

35 Yu, S. and Kaviany, M. (2014). Electrical, thermal, and species transport properties of liquid eutectic Ga-In and Ga-In-Sn from first principles. *J. Chem. Phys.* 140: 064303.

36 Lou, H., Wang, X., Cao, Q. et al. (2013). Negative expansions of interatomic distances in metallic melts. *Proc. Natl. Acad. Sci. U. S. A.* 110: 10068.

37 Hattori, T., Tsuji, K., Taga, N. et al. (2003). Structure of liquid GaSb at pressures up to 20 GPa. *Phys. Rev. B* 68: 224106.

38 Li, M., Huang, Y., Xue, P. et al. (2019). Temperature-induced atomic structural evolution in a liquid Ga-based alloy. *Vacuum* 170: 108966.

39 Yu, Q., Guo, F.M., Wang, X.D. et al. (2019). Structural evolution of low-temperature liquid GaIn eutectic alloy. *J. Mol. Liq.* 293: 111464.

40 Jia, P., Li, X., Zhang, J. et al. (2018). Liquid-liquid structure transition and its effect on the solidification behaviors and microstructure of $Sn_{75}Bi_{25}$ alloy. *J. Mol. Liq.* 263: 218.

41 Li, M., Geng, H., Long, F. et al. (2015). Discontinuous structural phase transition of Sn–Bi melts. *J. Mol. Liq.* 204: 27.

42 Yu, Q., Ahmad, A.S., Stahl, K. et al. (2017). Pressure-induced structural change in liquid GaIn eutectic alloy. *Sci. Rep.* 7: 1139.

43 Tang, J., Zhao, X., Li, J. et al. (2017). Gallium-based liquid metal amalgams: transitional-state metallic mixtures (trans M (2) ixes) with enhanced and tunable electrical, thermal, and mechanical properties. *ACS Appl. Mater. Interfaces* 9: 35977.

44 Ma, B., Xu, C., Chi, J. et al. (2019). A versatile approach for direct patterning of liquid metal using magnetic field. *Adv. Funct. Mater.* 29: 1901370.

45 David, R. and Miki, N. (2019). Synthesis of sub-micrometer biphasic Au-$AuGa_2$/liquid metal frameworks. *Nanoscale* 11: 21419.

46 Tang, S.-Y., Mitchell, D.R.G., Zhao, Q. et al. (2019). Phase separation in liquid metal nanoparticles. *Matter* 1: 192.

47 Wu, Y.h., Deng, Z.f., Peng, Z.f. et al. (2019). A novel strategy for preparing stretchable and reliable biphasic liquid metal. *Adv. Funct. Mater.* 29: 1903840.

48 de Castro, I.A., Chrimes, A.F., Zavabeti, A. et al. (2017). A gallium-based magnetocaloric liquid metal ferrofluid. *Nano Lett.* 17: 7831.

49 Zheng, R., Peng, Z., Fu, Y. et al. (2020). A novel conductive core–shell particle based on liquid metal for fabricating real-time self-repairing flexible circuits. *Adv. Funct. Mater.* 30: 1910524.

50 Wang, D., Gao, C., Wang, W. et al. (2018). Shape-transformable, fusible rodlike swimming liquid metal nanomachine. *ACS Nano* 12: 10212.

51 Jensen, W.B. (2003). The place of zinc, cadmium, and mercury in the p;eriodic table. *J. Chem. Educ.* 80: 952.

52 Norrby, L.J. (1991). Why is mercury liquid? Or, why do relativistic effects not get into chemistry textbooks? *J. Chem. Educ.* 68: 110.

53 Pyper, N.C., Grant, I.P., and Gerber, R.B. (1977). Relativistic effects on interactions between heavy atoms: the Hg–Hg potential. *Chem. Phys. Lett.* 49: 479.

54 Heine, V. (1968). Crystal structure of gallium metal. *J. Phys. C Solid State Phys.* 1: 222.

55 Anderson, T.J. and Ansara, I. (1991). The Ga-In (Gallium-Indium) system. *J. Phase Equilib.* 12: 64.

56 Reed, W.A. (1969). Band structure and fermi surface of gallium by the pseudopotential method. *Phys. Rev.* 188: 1184.

57 Zuger, O. and Durig, U. (1992). Atomic structure of the alpha-Ga (001) surface investigated by scanning tunneling microscopy: direct evidence for the existence of Ga_2 molecules in solid gallium. *Phys. Rev. B* 46: 7319.

58 Flower, S.C. and Saunders, G.A. (2006). The elastic behaviour of indium under pressure and with temperature up to the melting point. *Philos. Mag. B* 62: 311.

59 Lee, J.A. and Raynor, G.V. (1954). The lattice spacings of binary tin-rich alloys. *Proc. Phys. Soc. Sect. B* 67: 737.

60 Cucka, P. and Barrett, C.S. (1962). The crystal structure of Bi and of solid solutions of Pb, Sn, Sb and Te in Bi. *Acta Crystallogr.* 15: 865.

61 Hirsch, A., Dejace, L., Michaud, H.O., and Lacour, S.P. (2019). Harnessing the rheological properties of liquid metals to shape soft electronic conductors for wearable applications. *Acc. Chem. Res.* 52: 534.

62 Zhang, L., Li, Y., Zhang, S. et al. (2019). Non-Newtonian fluid state K–Na alloy for a stretchable energy storage device. *Small Methods* 3: 1900383.

63 Zhang, L., Xia, X., Zhong, Y. et al. (2018). Exploring self-healing liquid Na-K alloy for dendrite-free electrochemical energy storage. *Adv. Mater.* 30: 1804011.

64 Yu, M., Bian, X., Wang, T., and Wang, J. (2017). Metal-based magnetic fluids with core-shell structure FeB@SiO2 amorphous particles. *Soft Matter* 13: 6340.

65 Chang, H., Zhang, P., Guo, R. et al. (2020). Recoverable liquid metal paste with reversible rheological characteristic for electronics printing. *ACS Appl. Mater. Interfaces* 12: 14125.

66 Sun, C., Geng, H., Ji, L. et al. (2007). Rheological properties of Pb, Sb, Bi, and Sn melts. *J. Appl. Phys.* 102: 034901.

67 Sun, C., Geng, H., Yang, Z. et al. (2005). Viscous and structural behaviors of molten In–Sn alloys. *Mater. Charact.* 55: 383.

68 Yang, J., Bian, X., Yang, C. et al. (2013). Viscous features of single-component melt under horizontal magneticfield. *Phys. B Condens. Matter* 415: 18.

69 Jeyakumar, M., Hamed, M., and Shankar, S. (2011). Rheology of liquid metals and alloys. *J. Non-Newtonian Fluid Mech.* 166: 831.

70 Dickey, M.D., Chiechi, R.C., Larsen, R.J. et al. (2008). Eutectic gallium-indium (EGaIn): a liquid metal alloy for the formation of stable structures in microchannels at room temperature. *Adv. Funct. Mater.* 18: 1097.

71 Regan, M.J., Tostmann, H., Pershan, P.S. et al. (1997). X-ray study of the oxidation of liquid-gallium surfaces. *Phys. Rev. B* 55: 10786.

72 Zhang, W., Ou, J.Z., Tang, S.-Y. et al. (2014). Liquid metal/metal oxide frameworks. *Adv. Funct. Mater.* 24: 3799.

73 Khan, M.R., Trlica, C., So, J.H. et al. (2014). Influence of water on the interfacial behavior of gallium liquid metal alloys. *ACS Appl. Mater. Interfaces* 6: 22467.

74 Xu, Q., Oudalov, N., Guo, Q. et al. (2012). Effect of oxidation on the mechanical properties of liquid gallium and eutectic gallium-indium. *Phys. Fluids* 24: 063101.

75 Carle, F., Bai, K., Casara, J. et al. (2017). Development of magnetic liquid metal suspensions for magnetohydrodynamics. *Phys. Rev. Fluids* 2: 013301.

76 Khan, M.R., Eaker, C.B., Bowden, E.F., and Dickey, M.D. (2014). Giant and switchable surface activity of liquid metal via surface oxidation. *Proc. Natl. Acad. Sci. U. S. A.* 111: 14047.

77 So, J.-H., Koo, H.-J., Dickey, M.D., and Velev, O.D. (2012). Ionic current rectification in soft-matter diodes with liquid-metal electrodes. *Adv. Funct. Mater.* 22: 625.

78 Wang, J., Wang, H., Xie, J. et al. (2018). Fundamental study on the wetting property of liquid lithium. *Energy Storage Mater.* 14: 345.

79 Wang, C., Xie, H., Zhang, L. et al. (2018). Universal soldering of lithium and sodium alloys on various substrates for batteries. *Adv. Energy Mater.* 8: 1701963.

80 Li, F., Kuang, S., Li, X. et al. (2019). Magnetically- and electrically-controllable functional liquid metal droplets. *Adv. Mater. Technol.* 4: 1800694.

81 Ren, L., Sun, S., Casillas-Garcia, G. et al. (2018). A liquid-metal-based magnetoactive slurry for stimuli-responsive mechanically adaptive electrodes. *Adv. Mater.* 30: 1802595.

82 Xiong, M., Gao, Y., and Liu, J. (2014). Fabrication of magnetic nano liquid metal fluid through loading of Ni nanoparticles into gallium or its alloy. *J. Magn. Magn. Mater.* 354: 279.

83 Hu, L., Wang, H., Wang, X. et al. (2019). Magnetic liquid metals manipulated in the three-dimensional free space. *ACS Appl. Mater. Interfaces* 11: 8685.

84 Daalkhaijav, U., Yirmibesoglu, O.D., Walker, S., and Mengüç, Y. (2018). Rheological modification of liquid metal for additive manufacturing of stretchable electronics. *Adv. Mater. Technol.* 3: 1700351.

85 Ford, M.J., Ambulo, C.P., Kent, T.A. et al. (2019). A multifunctional shape-morphing elastomer with liquid metal inclusions. *Proc. Natl. Acad. Sci. U. S. A.* 116: 21438.

86 Markvicka, E.J., Bartlett, M.D., Huang, X., and Majidi, C. (2018). An autonomously electrically self-healing liquid metal-elastomer composite for robust soft-matter robotics and electronics. *Nat. Mater.* 17: 618.

87 Ijaz-ur-Rahman and Ahmed, K. (1978). Electron mean free path in a thin metal film. *J. Appl. Phys.* 49: 3625.

88 Ziman, J.M. (1961). A theory of the electrical properties of liquid metals. I: the monovalent metals. *Philos. Mag.* 6: 1013.

89 Khajil, T. (1999). Electrical resistivity of alkali-lead binary alloys. *Phys. Chem. Liq.* 37: 773.

90 Wei, L.F. and Nori, F. (2004). Coherently manipulating two-qubit quantum information using a pair of simultaneous laser pulses. *Europhys. Lett. (EPL)* 65 (1).

91 Ziman, J.M. (1964). The method of neutral pseudo-atoms in the theory of metals. *Adv. Phys.* 13: 89.

92 Porai-Koshits, E.A. (1976). The structure of noncrystalline materials. *Annu. Rev.* 6: 389.

93 Boulahbak, M., Wax, J.F., Jakse, N., and Bretonnet, J.L. (1997). Ion-ion and electron-ion correlations in liquid gallium. *J. Phys. Condens. Matter* 9: 4017.

94 Salmon, P.S. (1988). A neutron diffraction study on the structure of liquid germanium. *J. Phys. F* 18: 2345.

95 Dahlborg, U. and Davidovi, M. (1986). On the structure of liquid bismuth. *Phys. Chem. Liq.* 15: 243.

96 Jones, R.O., Ahlstedt, O., Akola, J., and Ropo, M. (2017). Density functional study of structure and dynamics in liquid antimony and Sbn clusters. *J. Chem. Phys.* 146: 194502.

97 Cusack, N., Kendall, P., and Fielder, M. (1964). Electron transport properties in a liquid alloy. *Philos. Mag.* 10: 871.

98 Heinzel, A., Hering, W., Konys, J. et al. (2017). Liquid metals as efficient high-temperature heat-transport fluids. *Energy Technol.* 5: 1026.

99 Ewing, C.T., Grand, J.A., and Miller, R.R. (1952). Thermal conductivity of lithium, sodium and potassium in the liquid state. *J. Am. Chem. Soc.* 74: 11.

100 Braunsfurth, M.G., Skeldon, A.C., Juel, A. et al. (1997). Free convection in liquid gallium. *J. Fluid Mech.* 342: 295.

101 Uppal, A., Ralphs, M., Kong, W. et al. (2020). Pressure-activated thermal transport via oxide shell rupture in liquid metal capsule beds. *ACS Appl. Mater. Interfaces* 12: 2625.

102 Prabhuram, S. and P, M. (1982). Thermal conductivity of liquid mercury. *Ind. Eng. Chem. Fundam.* 21: 484.

103 Franz, R. and Wiedemann, G. (1853). Ueber die Wärme-Leitungsfähigkeit der Metalle. *Ann. Phys. Chem.* 165: 497.

104 Mills, K.C., Monaghan, B.J., and Keene, B.J. (1996). Thermal conductivities of molten metals: Part 1 Pure metals. *Int. Mater. Rev.* 41: 209.

105 Wakeham, W.A. and Peralta Martinez, V. (2001). Molten metals: a challenge for measurement. *J. Chem. Thermodyn.* 33: 1623.

106 Gao, Y. and Liu, J. (2012). Gallium-based thermal interface material with high compliance and wettability. *Appl. Phys. A Mater. Sci. Process.* 107: 701.

107 Wang, Q., Yu, Y., and Liu, J. (2018). Preparations, characteristics and applications of the functional liquid metal materials. *Adv. Eng. Mater.* 20: 1700781.

108 Bartlett, M.D., Kazem, N., Powell-Palm, M.J. et al. (2017). High thermal conductivity in soft elastomers with elongated liquid metal inclusions. *Proc. Natl. Acad. Sci. U. S. A.* 114: 2143.

109 Tang, W., Jiang, T., Fan, F.R. et al. (2015). Liquid-metal electrode for high-performance triboelectric nanogenerator at an instantaneous energy conversion efficiency of 70.6%. *Adv. Funct. Mater.* 25: 3718.

110 Zhang, Q. and Liu, J. (2013). Nano liquid metal as an emerging functional material in energy management, conversion and storage. *Nano Energy* 2: 863.

111 Martin, A., Kiarie, W., Chang, B., and Thuo, M. (2020). Chameleon metals: autonomous nano-texturing and composition inversion on liquid metals surfaces. *Angew. Chem. Int. Ed. Engl.* 59: 352.

112 Kasama, A., McLean, A., Miller, W.A. et al. (2013). Surface tension of liquid iron and iron-oxygen alloys. *Can. Metall. Q.* 22: 9.

113 Hu, L., Yuan, B., and Liu, J. (2017). Liquid metal amoeba with spontaneous pseudopodia formation and motion capability. *Sci. Rep.* 7: 7256.

114 Park, Y.G., An, H.S., Kim, J.Y., and Park, J.U. (2019). High-resolution, reconfigurable printing of liquid metals with three-dimensional structures. *Sci. Adv.* 5: eaaw 2844.

115 Lu, Y., Hu, Q., Lin, Y. et al. (2015). Transformable liquid-metal nanomedicine. *Nat. Commun.* 6: 10066.

116 Mills, K.C. and Keene, B.J. (2013). Factors affecting variable weld penetration. *Int. Mater. Rev.* 35: 185.

117 Kasama, A., Iida, T., and Morita, Z.-i. (1976). Temperature dependence of surface tension of liquid pure metals. *J. Jpn. Inst. Metal.* 40: 1030.

118 Yi, U.-C. and Kim, C.-J. (2006). Characterization of electrowetting actuation on addressable single-side coplanar electrodes. *J. Micromech. Microeng.* 16: 2053.

119 Li, C., Cao, Y., Lippmann, S. et al. (2018). Reduced wettability of solids by a liquid Ga–In–Sn alloy in a steady magnetic field. *J. Phys. Chem. C* 122: 27451.

120 Mugele, F., Duits, M., and van den Ende, D. (2010). Electrowetting: a versatile tool for drop manipulation, generation, and characterization. *Adv. Colloid Interf. Sci.* 161: 115.

121 Watson, A.M., Cook, A.B., and Tabor, C.E. (2019). Electrowetting-assisted selective printing of liquid metal. *Adv. Eng. Mater.* 21: 1900397.

122 Krupenkin, T. and Taylor, J.A. (2011). Reverse electrowetting as a new approach to high-power energy harvesting. *Nat. Commun.* 2: 448.

123 Hoshikawa, K. (1982). Czochralski silicon crystal growth in the vertical magnetic field. *Jpn. J. Appl. Phys.* 21: L545.

124 de Rango, P., Lees, M., Lejay, P. et al. (1991). Texturing of magnetic materials at high temperature by solidification in a magnetic field. *Nature* 349: 770.

125 Sun, Z.H.I., Guo, M., Vleugels, J. et al. (2012). Strong static magnetic field processing of metallic materials: a review. *Curr. Opin. Solid State Mater. Sci.* 16: 254.

126 Katsuki, A., Kaji, K., Sueda, M., and Tanimoto, Y. (2007). Magnetic field effects on shape and size of pendent water drop. *Chem. Lett.* 36: 306.

127 Chen, L.J. and Yan, W.J. (1993). Novel interfacial phenomena at liquid–liquid interfaces of the three-component surfactant system water+n-tetradecane+C_6E_2. *J. Chem. Phys.* 98: 4830.

128 Tavana, H. and Neumann, A.W. (2007). Recent progress in the determination of solid surface tensions from contact angles. *Adv. Colloid Interf. Sci.* 132 (1).

129 Iino, M. and Fujimura, Y. (2009). Surface tension of heavy water under high magnetic fields. *Appl. Phys. Lett.* 94: 261902.

130 Sun, Z.H.I., Guo, X., Guo, M. et al. (2012). Strong magnetic field effect on surface tension associated with an interfacial magnetic pressure. *J. Phys. Chem. C* 116: 17676.

131 Li, C., Chen, L., and Ren, Z. (2012). Application of ring method to measure surface tensions of liquids in high magnetic field. *Rev. Sci. Instrum.* 83: 043906.

132 Ludwig, W. and Bellet, D. (2000). Penetration of liquid gallium into the grain boundaries of aluminium: a synchrotron radiation microtomographic investigation. *Mater. Sci. Eng. A* 281: 198.

133 Bico, J., Thiele, U., and Quéré, D. (2002). Wetting of textured surfaces. *Colloi. Surf. A* 206: 41.

134 Zavabeti, A., Ou, J.Z., Carey, B.J. et al. (2017). A liquid metal reaction environment for the room-temperature synthesis of atomically thin metal oxides. *Science* 358: 332.

135 Karbalaei Akbari, M. and Zhuiykov, S. (2019). A bioinspired optoelectronically engineered artificial neurorobotics device with sensorimotor functionalities. *Nat. Commun.* 10: 3873.

136 Azimi, G., Dhiman, R., Kwon, H.M. et al. (2013). Hydrophobicity of rare-earth oxide ceramics. *Nat. Mater.* 12: 315.

137 Kramer, R.K., Boley, J.W., Stone, H.A. et al. (2014). Effect of microtextured surface topography on the wetting behavior of eutectic gallium-indium alloys. *Langmuir* 30: 533.

138 Dezellus, O. and Eustathopoulos, N. (2010). Fundamental issues of reactive wetting by liquid metals. *J. Mater. Sci.* 45: 4256.

139 Ding, Y., Zeng, M., and Fu, L. (2020). Surface chemistry of gallium-based liquid metals. *Matter* 3: 1477.

140 Doudrick, K., Liu, S., Mutunga, E.M. et al. (2014). Different shades of oxide: from nanoscale wetting mechanisms to contact printing of gallium-based liquid metals. *Langmuir* 30: 6867.

141 Dickey, M.D. (2014). Emerging applications of liquid metals featuring surface oxides. *ACS Appl. Mater. Interfaces* 6: 18369.

142 Cabrera, N. and Mott, N.F. (1949). Theory of the oxidation of metals. *Rep. Prog. Phys.* 12: 163.

143 Grigoriev, A., Shpyrko, O., Steimer, C. et al. (2005). Surface oxidation of liquid Sn. *Surf. Sci.* 575: 223.

144 Fromhold, A.T. and Cook, E.L. (1967). Kinetics of oxide film growth on metal crystals: electron tunneling and ionic diffusion. *Phys. Rev.* 158: 600.

145 Litrico, G., Proulx, P., Gouriet, J.B., and Rambaud, P. (2015). Controlled oxidation of aluminum nanoparticles. *Adv. Powder Technol.* 26 (1).

146 Ermoline, A. and Dreizin, E.L. (2011). Equations for the Cabrera–Mott kinetics of oxidation for spherical nanoparticles. *Chem. Phys. Lett.* 505: 47.

147 Zhdanov, V.P. and Norton, P.R. (2012). Kinetics of thin oxide film growth on metal crystals. *Surf. Rev. Lett.* 07: 135.

148 Chabala, J.M. (1992). Oxide-growth kinetics and fractal-like patterning across liquid gallium surfaces. *Phys. Rev. B* 6: 11346.

149 Sangaletti, L., Depero, L.E., Allieri, B. et al. (2011). Oxidation of Sn thin films to SnO_2. micro-Raman mapping and X-ray diffraction studies. *J. Mater. Res.* 13: 2457.

150 Regan, M.J., Kawamoto, E.H., Lee, S. et al. (1995). Surface layering in liquid gallium: an X-ray reflectivity study. *Phys. Rev. Lett.* 75: 2498.

151 Shpyrko, O.G., Grigoriev, A.Y., Steimer, C. et al. (2004). Anomalous layering at the liquid Sn surface. *Phys. Rev. B* 70: 224206.

152 Shpyrko, O., Huber, P., Grigoriev, A. et al. (2003). X-ray study of the liquid potassium surface: structure and capillary wave excitations. *Phys. Rev. B* 67: 115405.

153 Huisman, W., Peters, J., Zwanenburg, M. et al. (1997). Layering of a liquid metal in contact with a hard wall. *Nature* 390: 379.

154 Yang, B., Li, D., Huang, Z., and Rice, S.A. (2000). Structure of the liquid-vapor interface of a dilute alloy of Pb in Ga. *Phys. Rev. B* 62: 13111.

155 Egelstaff, P.A. and Widom, B. (1970). Liquid surface tension near the triple point. *J. Chem. Phys.* 53: 2667.

156 Shpyrko, O., Fukuto, M., Pershan, P. et al. (2004). Surface layering of liquids: the role of surface tension. *Phys. Rev. B* 69: 245423.

157 Ino, S. (1969). Stability of multiply-twinned particles. *J. Phys. Soc. Jpn.* 27: 941.

158 Hall, B.D., Flueli, M., Monot, R., and Borel, J. (1991). Multiply twinned structures in unsupported ultrafine silver particles observed by electron diffraction. *Phys. Rev. B* 43: 3906.

159 Zhu, L., Wang, B., Handschuh-Wang, S., and Zhou, X. (2020). Liquid metal-based soft microfluidics. *Small* 16: e1903841.

160 Zhang, B., Dong, Q., Korman, C.E. et al. (2013). Flexible packaging of solid-state integrated circuit chips with elastomeric microfluidics. *Sci. Rep.* 3: 1098.

161 Adams, J.J., Duoss, E.B., Malkowski, T.F. et al. (2011). Conformal printing of electrically small antennas on three-dimensional surfaces. *Adv. Mater.* 23: 1335.

162 Liu, Z., Bando, Y., Mitome, M., and Zhan, J. (2004). Unusual freezing and melting of gallium encapsulated in carbon nanotubes. *Phys. Rev. Lett.* 93: 095504.

163 Li, K., He, H.Y., Xu, B., and Pan, B.C. (2009). The stabilities of gallium nanowires with different phases encapsulated in a carbon nanotube. *J. Appl. Phys.* 105: 054308.

164 Durgun, E., Dag, S., and Ciraci, S. (2004). Theoretical study of Ga-based nanowires and the interaction of Gawith single-wall carbon nanotubes. *Phys. Rev. B* 70: 155305.

165 Sánchez-Portal, D., Artacho, E., Junquera, J. et al. (1999). Stiff monatomic gold wires with a spinning zigzag geometry. *Phys. Rev. Lett.* 83: 3884.

166 Sen, P., Ciraci, S., Buldum, A., and Batra, I.P. (2001). Structure of aluminum atomic chains. *Phys. Rev. B* 64: 195420.

167 Oh, S.S., Kim, D.H., Moon, M.-W. et al. (2008). Indium nanowires synthesized at an ultrafast rate. *Adv. Mater.* 20: 1093.

168 Giordano, N. (1988). Evidence for macroscopic quantum tunneling in one-dimensional superconductors. *Phys. Rev. Lett.* 61: 2137.

169 Li, H., Liang, C., Liu, M. et al. (2008). Synthesis of indium nanowires by galvanic displacement and their optical properties. *Nanoscale Res. Lett.* 4: 47.

170 Hsu, Y.J. and Lu, S.Y. (2005). Vapor-solid growth of Sn nanowires: growth mechanism and superconductivity. *J. Phys. Chem. B* 109: 4398.

171 Gülseren, O., Ercolessi, F., and Tosatti, E. (1998). Noncrystalline structures of ultrathin unsupported nanowires. *Phys. Rev. Lett.* 80: 3775.

172 Wang, Y., Jiang, X., Herricks, T., and Xia, Y. (2004). Single crystalline nanowires of lead: large-scale synthesis, mechanistic studies, and transport measurements. *J. Phys. Chem. B* 108: 8631.

173 Lee, S., Ham, J., Jeon, K. et al. (2010). Direct observation of the semimetal-to-semiconductor transition of individual single-crystal bismuth nanowires grown by on-film formation of nanowires. *Nanotechnology* 21: 405701.

174 Tian, M., Wang, J., Snyder, J. et al. (2003). Synthesis and characterization of superconducting single-crystal Sn nanowires. *Appl. Phys. Lett.* 83: 1620.

175 Riminucci, A. and Schwarzacher, W. (2014). Magnetic signature of granular superconductivity in electrodeposited Pb nanowires. *J. Appl. Phys.* 115: 223910.

176 Natarajan, C.M., Tanner, M.G., and Hadfield, R.H. (2012). Superconducting nanowire single-photon detectors: physics and applications. *Supercond. Sci. Technol.* 25: 063001.

177 Yi, G. and Schwarzacher, W. (1999). Single crystal superconductor nanowires by electrodeposition. *Appl. Phys. Lett.* 74: 1746.

178 Hicks, L.D., Harman, T.C., and Dresselhaus, M.S. (1993). Use of quantum-well superlattices to obtain a high figure of merit from nonconventional thermoelectric materials. *Appl. Phys. Lett.* 63: 3230.

179 Zhang, X., Ding, Y., Zhang, Y. et al. (2007). Thermal behavior of antimony nanowire arrays embedded in anodic aluminum oxide template. *J. Therap. Anal. Calorim.* 89: 493.

180 Zhang, Z., Sun, X., Dresselhaus, M.S. et al. (2000). Electronic transport properties of single-crystal bismuth nanowire arrays. *Phys. Rev. B* 61: 4850.

181 Gallo, C.F., Chandrasekhar, B.S., and Sutter, P.H. (1963). Transport Properties of Bismuth Single Crystals. *J. Appl. Phys.* 34: 144.

182 Abeles, B. and Meiboom, S. (1956). Galvanomagnetic Effects in Bismuth. *Phys. Rev.* 101: 544.

183 Ye, Z., Zhang, H., Liu, H. et al. (2008). Observation of superconductivity in single crystalline Bi nanowires. *Nanotechnology* 19: 085709.

184 Roh, J.W., Hippalgaonkar, K., Ham, J.H. et al. (2011). Observation of anisotropy in thermal conductivity of individual single-crystalline bismuth nanowires. *ACS Nano* 5: 3954.

185 Mannix, A.J., Kiraly, B., Hersam, M.C., and Guisinger, N.P. (2017). Synthesis and chemistry of elemental 2D materials. *Nat. Rev. Chem.* 1: 14.

186 Kochat, V., Samanta, A., Zhang, Y. et al. (2018). Atomically thin gallium layers from solid-melt exfoliation. *Sci. Adv.* 4: e1701373.

187 Zhang, T., Cheng, P., Li, W.-J. et al. (2010). Superconductivity in one-atomic-layer metal films grown on Si (111). *Nat. Phys.* 6: 104.

188 Lu, Y., Xu, W., Zeng, M. et al. (2015). Topological properties determined by atomic buckling in self-assembled ultrathin Bi (110). *Nano Lett.* 15: 80.

189 Li, L., Yu, Y., Ye, G.J. et al. (2014). Black phosphorus field-effect transistors. *Nat. Nanotechnol.* 9: 372.

190 Xing, Y., Zhang, H.M., Fu, H.L. et al. (2015). Quantum Griffiths singularity of superconductor-metal transition in Ga thin films. *Science* 350: 542.

191 Rotenberg, E., Koh, H., Rossnagel, K. et al. (2003). Indium square root 7 x square root 3 on Si (111): a nearly free electron metal in two dimensions. *Phys. Rev. Lett.* 91: 246404.

192 Liu, C.-C., Jiang, H., and Yao, Y. (2011). Low-energy effective Hamiltonian involving spin-orbit coupling in silicene and two-dimensional germanium and tin. *Phys. Rev. B* 84: 195430.

193 Broek, B.v.d., Houssa, M., Scalise, E. et al. (2014). Two-dimensional hexagonal tin: ab initio geometry, stability, electronic structure and functionalization. *2D Mater.* 1: 2053.

194 Zhang, Z., Niu, Q., and Shih, C.-K. (1998). "Electronic Growth" of metallic overlayers on semiconductor substrates. *Phys. Rev. Lett.* 80: 5381.

195 Mak, K.F., Lee, C., Hone, J. et al. (2010). Atomically thin MoS (2): a new direct-gap semiconductor. *Phys. Rev. Lett.* 105: 136805.

196 Gibaja, C., Rodriguez-San-Miguel, D., Ares, P. et al. (2016). Few-layer antimonene by liquid-phase exfoliation. *Angew. Chem. Int. Ed. Engl.* 55: 14345.

197 Zhang, S., Yan, Z., Li, Y. et al. (2015). Atomically thin arsenene and antimonene: semimetal-semiconductor and indirect-direct band-gap transitions. *Angew. Chem. Int. Ed. Engl.* 54: 3112.

198 Hirahara, T., Nagao, T., Matsuda, I. et al. (2006). Role of spin-orbit coupling and hybridization effects in the electronic structure of ultrathin Bi films. *Phys. Rev. Lett.* 97: 146803.

199 Walker, E.S., Na, S.R., Jung, D. et al. (2016). Large-area dry transfer of single-crystalline epitaxial bismuth thin films. *Nano Lett.* 16: 6931.

3

Preparation of Liquid Metal

Wisdom lies in diligence, and genius in accumulation.
Source: Luogeng Hua (1910–1985).

3.1 Introduction

Liquid metal (LM) elements include posttransition (Ga, In, Tl, Sn, Al, and Bi), zinc group metals (Zn, Cd, and Hg), and alkali metals (Li and Na), which are attractive superconductor, semiconductor, and thermoelectric materials [1–3]. Ga is mainly obtained as a by-product when extracting Al from bauxite or Zn from zinc ore, and a small amount of Ga comes from the recovery of the associated element Ga in coal. More than 90% of the world's primary Ga is currently extracted from the seed stock

Liquid Metals: Properties, Mechanisms, and Applications, First Edition. Lei Fu and Mengqi Zeng.

solution used to produce alumina [4]. There are currently two types of In refining, namely primary In and regenerated In. Primary In is mainly extracted from the original ore, and it is also the main source of smelting In. Recycled In is the smelting of recycled metal, and the main by-products recovered from the smelting process of Pb, Zn, Cu, Sn, and other ores, but the total amount of recycled In is not large [5]. In addition to producing primary Sn directly from concentrates, recycled Sn can also be obtained from tin-containing waste metallurgy [6]. Bi is also obtained by two main methods, including hydrometallurgy of Bi-containing materials and refining Bi–Pb alloy [7]. Zn hydrometallurgy is the most important Zn smelting method in the world today, which accounts for more than 85% of the world's total Zn production [8]. Nowadays, it is necessary to synthesize and explore their properties because of their various nanoscale morphologies. Tailoring the morphology of LM can greatly enrich its applications in various aspects. In this chapter, we review the preparation approaches to obtain the LM materials: LM droplets, LM nanowires (NWs), and LM nanosheets (NSs). In general, unstable LMs like Ga, In, and Sn are always obtained in a confined condition. Active LMs such as Al and Zn, prefer to grow by a redox reaction. Because of the considerable plastic deformation of Bi, it is easier to obtain Bi-based materials by introducing external force. First, the preparations of LM droplets were discussed and were mainly divide into liquid-phase preparation and vapor-phase preparation, which are available for nano- and microdroplets, respectively [9–11]. The key challenge of preparing LM droplets is overcoming the high surface tension of LMs, leading to the merger of adjacent droplets. Then, we introduced various methods, which were used to grow NWs, including catalyst-assisted methods [12], catalyst-free methods [1, 13], and others [14–16]. During the process to obtain NWs, successfully controlling the diameter and orientation of NWs plays a momentous role in directing their properties. As for LM NSs, MBE [17], exfoliation [18], hot pressing [19], and confinement heteroepitaxy (CHet) [20] are common measurements to obtain atomic thin 2D LM materials for applications. Nowadays, there is an upsurge of producing high-quality and large-scale stable LM NSs, which will vigorously promote the applications for devices like FETs. Finally, these main research advances are summarized and a few research directions that might be worthwhile for exploration are also listed.

3.2 Preparation of Liquid Metal Droplets

The underexplored bulk LM acts as an ideal material for preparing LM droplets due to its strong surface tension and easy phase control property. The nano- and microsized droplets play an important part in LM-based electrical systems and flexible mechanical [21]. With increasing the surface area to volume ratio and reducing the physical size, LM droplets are crucial for certain applications including electrochemical sensors [22], fluidic actuators [23], switches [24], and reconfigurable/stretchable electronics and biomedicine [25, 26]. It is significant to prepare the monodisperse, uniform, and microsized LM droplets in large scale. In this chapter, we mainly introduce the liquid-phase and vapor-phase methods for preparing the LM droplets.

3.2.1 Liquid-Phase Preparation

The liquid-phase preparation method for fabricating the LM droplets has been an important field from either academics or industries perspective. To date, several methods have been developed to produce micro-sized LM droplets, including solution-phase chemistry, sonication, microfluidic flow focusing, and channelless fluidic jetting. These methods significantly accelerate the advances of the LM science and art.

Solution-phase chemistry provides an access to LM droplets, which can be stabilized by the long-chain organic capping ligands [9]. Solution-phase chemistry utilizes the redox reaction among the reaction precursor to prepare LM droplets in liquid state. Melendrez and coworkers reported the chemical liquid deposition method for synthesizing the colloidal LM NPs [27]. In this system, Ga and organic solvent (2-propanol) mixture were heated and co-deposited for Ga colloidal NPs in a crucible under a liquid nitrogen (N_2) atmosphere. The mean diameter of colloidal Ga particles could reach to 5 nm. However, the Ga particles exhibited a wide distribution in the apparent size. In order to increase the size uniformity, Yarema and coworkers reported a conventional solution-phase chemistry method for preparing the colloidal Ga NPs [9]. In this work, they put Ga, tris-(dimethylamide) dime, di-*n*-octylamine, and 1-octadecene into a reaction flask, which was filled with Argon (Ar_2) and heated to 280 °C. The Ga NPs were successfully prepared with size ranging from 12 to 46 nm (Figure 3.1a). Owing to the chemical passivation of the thin native oxide shell and colloidal stabilization with long-chain organic capping ligands, this Ga NPs exhibited narrow size distributions with the size difference among 7–8%. After freezing the Ga NPs, the phase structure turned into delta (δ)-Ga (Figure 3.1b). It is worthy to mention that the complex additives utilized in solution-phase chemistry are hard to clean completely and always make the experimental steps tedious. Yan and coworkers used the ultrasmall water-soluble melanin nanoparticles as the capping agent for the synthesis of LM nanodroplets, which could avoid the use of complex additives in the synthesis process and make the process simple [30].

Sonication makes a confluence of mechanical and chemical characteristics, which is an effective approach for translating bulk LM into 0D NPs in organic solvents [31]. Sonication uses the acoustic wave to impact the LM immersed in a solution. Adding ligands in the system can make LM functionalized during the process. The droplet size can be reduced by prolonging process time [22]. Filtration and centrifugation can act as postprocessing methods for narrowing the size distribution [32]. Hohman and coworkers prepared the LM NPs via ultrasonication. In typical synthesis process, an ultrasound probe was used for introducing the cavitation in the solution, resulting in extreme values of local pressure and temperature within an ultrashort time. Under this effect, the oxidation layer on bulk LM surface was destructed, and bulk LM would change into some LM NPs under the oscillating shear force. At the same time, thiolated ligands in the solution assembled onto the surface of the newly formed NPs immediately, which was prior to the reoxidation process. Owing to the protection from the thiolate ligands [26, 32], this process can yield spherical LM ($\approx$110 nm) stably during the continuous sonication. Tang and coworkers used an

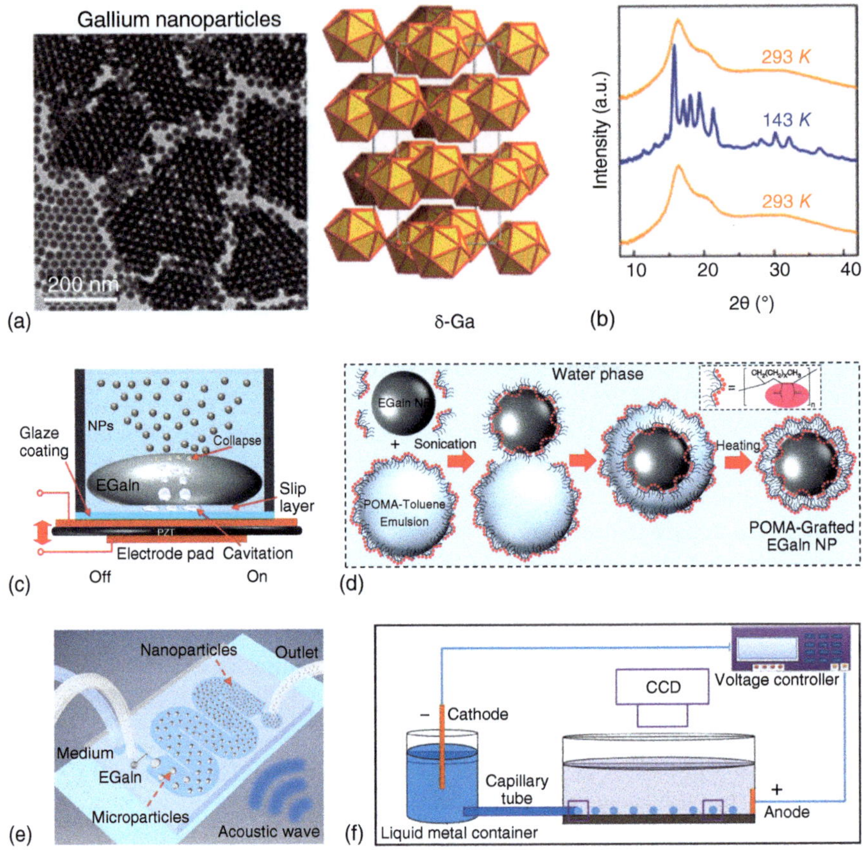

Figure 3.1 Liquid-phase preparation of LM NPs. (a) The TEM image and the crystal structure of Ga NPs. Source: Yarema et al. [9]. (b) Wide-angle powder X-ray diffraction pattern of Ga NPs. (a right; b) Source: Yarema et al. [9]. Reproduced with permission from American Chemical Society. (c) The mini ultrasonic platform for production of GaIn NPs. Source: Tang et al. [28]. Reproduced with permission from John Wiley & Sons. (d) The mechanism of preparing the stable LM NPs. Source: Lin et al. [10]. Reproduced with permission of Royal Society of Chemistry. (e) Schematic representation of the working mechanism. Source: Tang et al. [11]. Reproduced with permission from John Wiley & Sons. (f) Schematic diagram of the electro-hydrodynamic shooting. Source: Fang et al. [29]. Reproduced with permission of AIP Publishing.

acoustic wave-induced force to prepare the LM droplets with a series of controllable sizes [33]. In this work, they constructed an acoustic-based, on-chip LM micro-droplet production system in which the acoustic fields promoted LM to vibrate. Because the oscillating shear forces caused the non-Newtonian liquid to flow, the thin oxide skin of the LM was fractured and separated, which made the LM unstable and started to break into microsized droplets with transducer activation. Applying different voltages to change the surface tension of the LM could control the size of LM droplets. With the development of technique, magnetic and microwave-assisted heating also could provide the improved control over the separated nucleation and growth stages, which favored a narrower size distribution of particles [34]. Lin and coworkers discovered that the prepared LM NPs possess a strong stability for etchant [10]. The LM was put into a mixture of poly(1-octadecene-alt-maleic anhydride) (POMA) and toluene. Sonication enabled the POMA coat on the GaIn NPs. When the NPs with POMA contacted with toluene, a C18 chain could be formed on the surface of GaIn NPs, which encapsulated the GaIn NPs by the hydrophobic insulation layer (Figure 3.1d). The energy dispersive spectrometer (EDS) mapping of GaIn NPs indicates that the elements of Ga and In distribute uniformly. Because the hydrophobic insulating layer could prevent NPs from dissolution by etchant, such as a strong acid, the NPs could be kept stable for at least 60 days without aggregation. Generally, the dispersive LM NPs tend to oxidize, precipitate, and coalesce in aqueous solutions. The balance between the coalescence and dispersion can be tuned by the natural surface oxide layer on the surface of LM NPs and the surfactant [35]. Tang and coworkers discovered that the thickness of the oxide shell on the surface of NPs could be controlled by ultrasonic system [28]. A vapor cavity generating ultrasonic platform for fabricating LM NPs could be realized by triggering the vapor cavities and collapsing them at the LM–medium interface (Figure 3.1c). Taking advantage of this platform, the thickness of the oxide shell on the surface of 0D NPs increased with the increasing oxidative potential [36]. Chen and coworkers found that a controllable switching between dispersion and reunion of LM could be tuned by the surface oxide layer of LM NPs, where the surface oxide layer was controlled by the vibration frequency and amplitude in sonication system [37]. This work opened a feasible and universal method for fabricating dispersed LM NPs.

Microfluidic flow focusing is the most common fabrication means for LM microdroplets. Microfluidic approach can obtain LM droplets with improved control over droplet size when compared with sonication and other conventional methods [38–40]. The structure of microfluidic system is a cross-like microchannel configuration by using polydimethylsiloxane material. The channel is filled with aqueous polyethylene glycol (PEG) solution or deoxygenated silicone oil. The aqueous PEG solution or deoxygenated silicone oil is used to create high viscosity fluid. The sodium dodecylsulfate in the reservoir can avoid the coalescence and retain the shape of prepared droplets [41]. After the LM stream is extruded from the channel into a reservoir by flow focusing, the LM stream will cause a disturbance to the solution in the reservoir and tend to be droplets soon later due to the surface tension of LM [42]. Hutter and coworkers prepared the spherical LM droplets in microfluidic channels at room temperature [41]. They found that the size of droplets was mainly controlled by the channel dimension and flow rate, that is, wider channels always produced spherical microdroplets, while narrower channels favored the rodlike drops. However, LM exhibits a large surface tension relative to water, which always causes a significant barrier to microdroplet formation. Based on it, Tang and coworkers applied an acoustic wave in microfluidic system to settle the difficulty of LM microdroplet formation [11]. When flowing through the microchannel toward the outlet, the bulk LM broke into nanoscale LM NPs due to the additional oscillating shear forces provided by acoustic wave (Figure 3.1e). As for the microfluidic approach, the main problem is that high-pressure drop and the corresponding high shear rate within the microchannel can cause hydrodynamic instability in the droplet formation process. It is difficult to obtain highly monodisperse LM droplets. To address this problem, one possible strategy is to develop a glass capillary microfluidic approach, which owns the advantages in accurately controlling the droplet distribution of LM droplets. Hu and coworkers added a stainless steel microneedle into the inner glass capillary to obtain large-scale highly monodispersed LM droplets [43]. In this device, the inner and outer syringes, which were installed onto syringe pumps, contained two different solutions with diverse phase structures. After the solutions passed the injection tube and interstice caused by the injection tube and square capillary, the two phases were injected into a public glass device with a constant flow rate. By carefully manipulating the flow rate of different phase, uniformly monodispersed droplets are produced from the tip because the high viscosity solution squeezed the low viscosity solution. This strategy overcame the hydrodynamic instability of the LM stream when the inner phase was pinched off, which showed great advantage in the stable fabrication of droplet with precisely controlled size. Benefiting from the delicate design of device, this method exhibited highly controllable flexibility for the high-yield fabrication of uniformly droplet emulsion.

Chanelless fluidic jetting was developed on the basis of microfluidic flow focusing system. By using external force to jet the LM flow with a certain velocity into another fluid from a small hole, LM can form multiple droplets with diameter to be hundreds of micrometers. This approach can make the liquid stream break into highly monodispersed LM droplets under the surface tension, which is potential for large-scale preparation. Yu and coworkers revealed this jetting process [42]. First, LM was turned into a continuous thin stream in the needle. Then, after the stream was jetted into the reservoir, the speed of this stream was retarded. A LM neck appeared at the front end, which helps us to separate it from the main jetting stream. Finally, the stream at the front end keeps growing until it turns from irregular shape into spherical droplets because of the LM surface tension. The LM would break into droplets because the surface tension was stronger than the shear stress, which was made for the highly monodispersed LM droplets. However, the droplet size was influenced by many factors, including the needle diameter, the viscosity and density of fluid metal, and surfactants. Based on it, Fang and coworkers used the electrohydrodynamic jetting method to fabricate LM droplets stably [29]. Figure 3.1f shows the experimental setup, where the capillary is used as a channel to connect the LM and the sodium hydroxide (NaOH) solution reservoir. The surface tension and pressure on the interface between NaOH solution and LM were broken immediately when electric field was introduced in the system. First, the electric field served as traction force that drove the LM out of the capillary nozzle and entered the NaOH solution. Then, the stream splits to form droplets continuously, owing to the surface tension of LM. Because the traction force of this system was derived from electric field and was not influenced by external factors, the fabrication of LM droplets was stable and can be applied to the large-scale preparation.

In a word, liquid-phase preparation for LM droplets is a conventional method. Solution-phase chemistry can achieve excellent size distribution by adjusting the reaction parameters. Nevertheless, some organic solutions used in reaction process still remain on the surface of LM droplets, which makes the reaction steps complex and influence the application of LM droplets. Sonication dispersion is a simple and fast strategy for preparing the droplets including Ga, In, Bi, Sn, and Zn. However, the droplet sizes change from dozens of nanometers to several micrometers, which are always inhomogeneous. Additionally, the acoustic wave generation technique for producing droplets is very costly. The droplet production equipment is bulky, which makes it impossible to apply in the integrated systems for miniaturized lab-on-a-chip applications. In microfluidic system, the glass capillary microfluidic technique can optimize the uniformity of size, but this approach is not suitable for large-scale preparation because the fabrication of microfluidic channels is costly and technically complex. Chanelless fluidic jetting under electric field is a novel means to achieve highly monodispersed LM droplets with controllable size. This strategy for fabricating the LM droplets is stable, which is expected to be applied to large-scale preparation.

3.2.2 Vapor-Phase Preparation

Vapor-phase preparation is a strategy that can synthesize surfactant-free LM NPs with controlled particle size on various substrates. The ordered arrangements of LM NPs can be realized by selecting appropriate substrates, which fulfill their wide applications in flexible, stretchable, and printable electronics. In this section, we introduce the preparation method of thermal deposition and MBE for preparing LM NPs.

Thermal deposition is a fast, simple, and effective method for fabricating LM NPs. This approach can utilize many conductive and insulating substrates to regulate the morphology of LM NPs. The principle of thermal deposition is to change the liquid or solid metal into gas state by various means, making it undergo physical changes or chemical reactions in the gas state, and finally condense to form nanoparticles on substrate. Generally, the operating temperature is higher than the volatile point of the metal. The alternative substrates were aluminum (Al), silicon (Si), and glass. Yu and coworkers discovered a bottom-up vapor deposition to prepare the GaIn LM NPs [44]. In this work, GaIn alloy was heated in a tungsten boat to supply LM source under a vertical vacuum chamber, where the vaporized Ga and In atoms diffused and condensed on the above cold target silicon substrate. It was found that GaIn NPs were homogeneously deposited on the silicon substrate. Figure 3.2a shows the nucleation, growth, and merging process. The main process is at early stage, the evaporated Ga and In atoms were condensed on the Si substrate to form adatoms, which will turn into stable nuclei later. After the initial nucleation, the nuclei grew into isolated islands by aggregating the surrounding adatoms. With further deposition, these islands were transformed into the uniformly distributed NPs due to the Volmer–Weber mechanism. The size of NPs is regulated by a coalescence mechanism that can be controlled by the mass of LM and the deposition time [46, 47]. It is worth noting that small NPs tend to consolidate with big one due to the Ostwald ripening effect, which lead to the inhomogeneous particle size distribution of NPs when prolong the deposition time [48]. Catalan-Gomez and coworkers found that the nanopatterned Al substrate could be applied to synthesize the highly

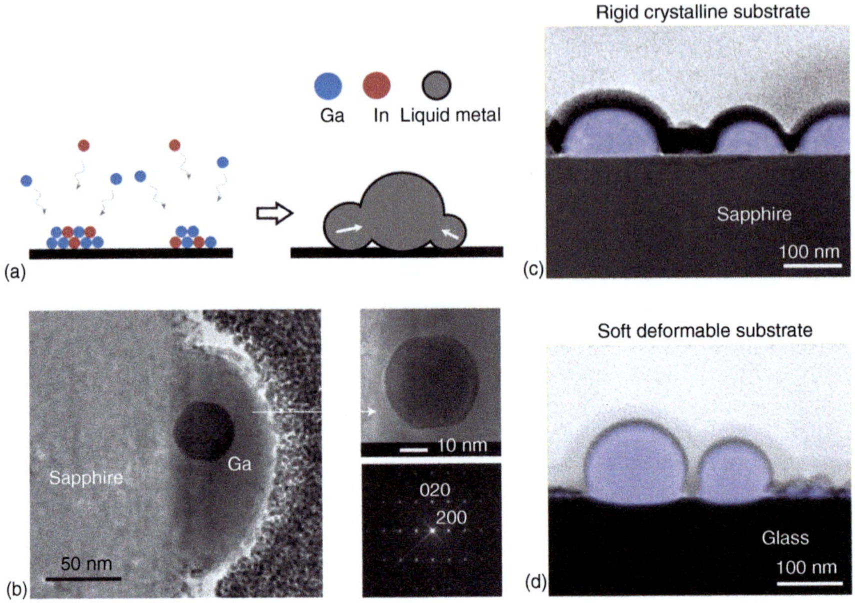

Figure 3.2 Vapor-phase preparation of LM NPs. (a) A schematic of the synthesis process of LM NPs. Source: Yu et al. [44]. (b) TEM image exhibiting coexistence of two Ga forms in a single nanoparticle. (c) HRTEM images of Ga NPs deposited on the sapphire substrates. (d) HRTEM images of Ga NPs deposited on the glass substrates. Source: Losurdo et al. [45].

ordered spherical Ga NPs by thermal deposition [49]. They used Al and Si substrates to deposit the Ga NPs. On the Si substrate, because the different resonance mode (transversal mode and longitudinal mode) along different axis caused two dimensions of the axes of the NPs, the shape of Ga NPs was hemispherical. On the Al substrate, the shape of Ga NPs was spherical because the convexity of the nanopatterned template tends to form one single mode. In this work, they adjusted the shape of LM NPs by tuning the mode to match the dimensions of the template, resulting in a unimodal distribution of ordered spherical NPs. Moreover, selecting some 2D materials to modify the substrate, including graphene and molybdenum disulfide (MoS_2), are also good choices to prepare Ga NPs [46, 50, 51]. Pau and coworkers utilized graphene substrate to grow nanoscale Ga NPs. Owing to the weak bonding charge between graphene and Ga, Ga NPs with graphene substrate behaved at a larger interparticle distance and exhibited an improved size uniformity when compared with Ga NPs on Si substrate [51].

MBE can be applied to yield the Ga NPs. The molecular beam epitaxy system is deployed with an in situ spectroscopic ellipsometer and is operated in the range of 1.5–6.5 eV in an ultrahigh-vacuum chamber [52]. LM can be deposited onto selected substrates at a constant Ga beam equivalent pressure. LM NPs can be obtained by self-assembling LM via a process of adsorption, surface diffusion, and coalescence through Ostwald ripening [53]. The size of the NPs is adjusted by the LM flux and deposition time. Losurdo and coworkers prepared LM Ga NPs via MBE method. In this work, they selected sapphire and glass substrates to deposit Ga NPs [45]. Owing to the high density, low surface energy, and near lattice-match to the rigid sapphire, Ga NPs on the rigid sapphire generated an unexpected two-phase solid γ-phase core and a liquid shell in the liquid nanoparticle (Figure 3.2b,c). However, on the soft glass substrate, the Ga NPs exhibited a single-phase liquid phase because there was no chemical reaction between substrate and Ga NPs (Figure 3.2d). Knight and coworkers prepared a densely packed Ga NP arrays with 72 ± 15 nm in diameter via MBE, where the plasmonic modes of interacting Ga NPs were hybridized for producing strong hot spots in the ultraviolet [54]. In addition, MBE with light regulation is an effective way to prepare LM NPs. Light can adjust the process of LM NPs assembly and the particle growth. MacDonald and coworkers discovered that light could dramatically adjust the process of nanoparticle self-assembly [55]. When exposed to light, Ga atoms could be deposited on the substrate, which finally grew into NPs with a relatively narrow size distribution. In the absence of light, the Ga atom turned into layer structure. Optical excitation of covalent bond structures in elemental Ga NPs could explain this phenomenon. The transition of the covalent structures' bonding–antibonding was stimulated by light, reducing the stability of the surrounding crystalline structure and the activation energy for desorption. In addition, the covalent structure can make Ga uniquely susceptible to another laser-induced adsorption suppression. Therefore, the Ga alloy can transform into NPs under light.

In this section, three methods are introduced to prepare LM particles. As for thermal deposition, the substrate plays an important role to adjust the morphology of LM NPs. LM NPs transferred from the substrate can be functionalized by ion implantation or chemical and thermal treatments in the chosen solvent. The drawbacks of chemical process, such as additive contamination and low process rate, can also be avoided. However, the LM particles prepared by thermal deposition method need to be transferred for its further application. As for MBE, many closely spaced LM NPs obtained by other approaches can form clusters with strong interaction, and some new resonant modes were hybridized and formed. MBE with light regulation can stimulate the covalent bond structures in LM structure, which makes for the nanoparticle self-assembly.

3.2.3 Others

Other than the above-mentioned means for fabricating the LM NPs, there are many alternative ways, such as shearing liquids into complex particles (SLICE), coulomb explosion, and molding. SLICE is an almighty method, which can be precisely tuned to make particles of various sizes, compositions, shapes, and surface morphologies. Coulomb explosion is a method to synthesize NPs by charging explosion, and the particle size can be controlled by the irradiation time. Molding forces the LM into reservoirs that modulate the size and location of the final LM droplets. The process involved would be detailed in this section.

SLICE is used to create LM droplets by shearing the low melting point metal and alloys into small-sized particles in a carrier fluid [56, 57]. The shearing apparatus are made up of a variable speed rotary tool and shearing implement. The speed of rotary tool is modulated by a variable transformer, which can tune the rotational speed of the devices. The shearing device uses a cross-shaped polytetrafluoroethylene (PTFE) with a steel rod core. Generally, the shearing apparatus is 110 V and the rotation speed is 17700 rpm. The operating environment was in an acetic acid solution. Tevis and coworkers fabricated the soft core–shell LM NPs via the SLICE method [56]. With tuning the rotational speeds and shearing time of the variable transformer, GaIn alloy could be transformed into a series of tunable particle sizes (Figure 3.3a). With the assistance of the ion beam, GaIn particles were milled to remove the top layer and expose the newly formed smooth surfaces. The bare top portion of the GaIn NPs coalesced, which could form different shapes of GaIn particles. However, when the melted BiInSn alloy suffered gravitational and centrifugal forces in a fluidic flow, Sn tended to segregate to the surface and Bi would precipitate due to the density differences, which finally caused a phase separation (Figure 3.3b). Applying the SLICE can slice the encapsulating layer for manipulating the LM particles. This technique is expected to make intricate three-dimensional structures and surfaces through selectively gathering the adjacent LM particles [57, 59]. Beyond that, Martin and coworkers fabricated GaInSn particles followed by the SLICE method [58]. Then, the GaInSn particles were treated by heat treatment to introduce fractal-like structures on the metal surface by exploiting underlying stochastic fracturing process (Figure 3.3c).

Furthermore, coulomb explosion is an unconventional technique for fabricating LM NPs by employing electron beam to irradiate the microscale metal balls. Chen and coworkers obtained Ga balls via a coulomb explosion [60]. In this work, they used an electron beam to irradiate the silica-coated Ga spheres in a TEM instrument. With prolonged irradiation, the charge began to accumulate on the Ga ball surface. Coulomb explosion would occur as the electric field strength is higher than the critical field strength (15 V/nm for Ga) of field evaporation, then many small LM droplets were ejected from the surface of the original silica-shelled ball. The particle size was controlled by the irradiation time, that is, the ball size decreased as the irradiation time increased. On the whole, the size of the exploded particles decreased almost linearly with increasing distance from the center ball. The particle size exhibited a wide distribution.

In addition, Van der Waals exfoliation enables the production of LM droplets with mean radius below 100 nm [61]. This method uses the top substrate to press onto the LM droplet on the bottom substrate, squeezing the droplet into a thin LM film. Then, the upper substrate was lifted and separated vertically, and an ultrathin Ga_2O_3 layer was exfoliated onto the surface of both substrates. Meanwhile, nano-scaled LM droplets were found on these Ga_2O_3 layer. The formation of LM droplets is because of the dynamic wetting failure in liquid coating process [62]. When lifting the top substrate with a speed higher than critical speed, the wetting line between the LM and Ga_2O_3 layer will spontaneously break into a sawtooth shape. With increasing speed, the trailing tips of the contact line will break into small sessile droplets due to Rayleigh-like instability. In this work, 75% LM droplets have a mean radius below 50 nm.

In a word, SLICE is a green and simple method to rapidly synthesize nanoscale or microscale patchy particles. This method will induce phase separation owing to the surface tension under fluidic flow and density differences, which can form particles with various compositions. Coulomb explosion is a novel approach to obtain metal

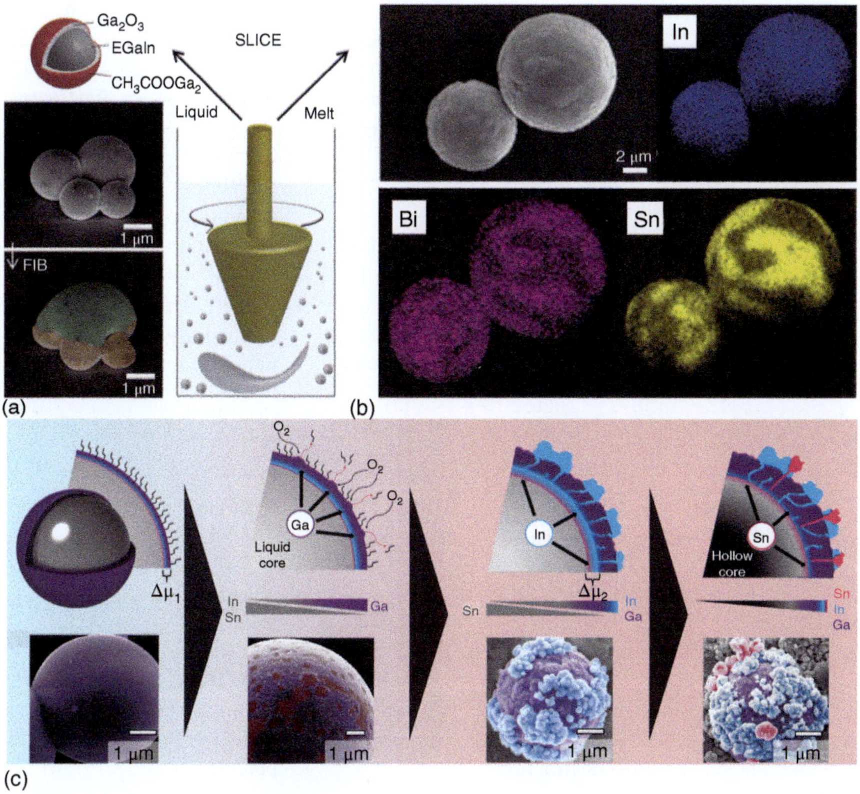

Figure 3.3 Other preparation methods of LM NPs. (a) Schematic illustration of the SLICE process. (a inset) Source: Tevis et al. [56]. Reproduced with permission of American Chemical Society.(b) EDS mapping of InBiSn NPs. (a, b) Source: Tevis et al. [56]. (c) The GaInSn particles were oxidated into chameleon metals. Source: Martin et al. [58]. Reproduced with permission from John Wiley & Sons.

particles with low melting point. However, the LM NPs always exhibit different diameters, which is not desired for the application. Van der Waals exfoliation can fabricate the LM droplets without any stabilizing surfactant, while the LM droplets exhibit a substantial inhomogeneity. The characteristic of method for preparing LM droplets is summarized in Table 3.1.

3.3 Preparation of Liquid Metal Nanowires

NWs have drawn immerse attention for decades because of their potential use as building blocks to fabricate nanoscale devices and architectures. Compared to their bulk counterparts, the unique properties they possessed, such as quantum confinement, allows their applications in optoelectronic and photovoltaic devices [63, 64]. It is urgent to study the NWs because of the continued miniaturization of electronic components based on Moore's law [65] and the upcoming device size constraints.

Table 3.1 Methods for preparation of LM droplets.

Method		LM characteristics	Advantage	Disadvantage	References
Liquid-phase preparation	Solution-phase chemistry	Wide size distributions	Facile	Complex additives	[9]
	Sonication	Nanometers to several micrometers	Large-scale production	(i) Wide size distribution; (ii) high-cost consumption	[22]
	Microfluidic flow focusing	Small size distributions	Controllable distribution	(i) Hydrodynamic instability; (ii) complex preparation of channels	[41]
	Chanelless fluidic jetting	Highly monodispersed LM droplets	High dispersity	Complex droplet fabrication	[42]
Vapor-phase preparation	Thermal deposition	Different morphologies	Adjustable morphology	Complex further process	[44]
	MBE	Densely packed Ga NPs arrays	LM NPs assembly	Complex further process	[52]
Others	SLICE	Different components	Various compositions	Phase separation	[56]
	Coulomb explosion	Small LM droplets	Simple operation	Wide distribution	[60]
	Van der Waals exfoliation	Below 100 nm	No surfactant	Inhomogeneity	[61]

LMs consist of Zn, Ga, In, Sn, and Bi, which are attractive superconductor, semiconductor, and thermoelectric materials [1–3]. Several methods are used to synthesize LM NWs. These approaches include catalyst-assisted methods, catalyst-free methods, and others, which are introduced in detail as subsequently.

3.3.1 Catalyst-Assisted Methods

Catalyst-assisted methods for growing NWs generally adhere to the vapor–liquid–solid (VLS) mechanism. During this process, the LM material (usually provided in the form of vapor) is absorbed by nanodroplets of an appropriate catalytic material. The droplets at the tip of the NW are considered as seed to synthesize NWs, and also determine the NW diameter [66, 67]. Cai and coworkers synthesized high quality and pure Sn NWs via a physical vapor transport (PVT) method. Si was used as the substrates and Au nanoparticles (with diameter ranging from 10 to 15 nm) were used as the catalyst [12]. Here, Selenium (Se) atoms were also introduced to assist the rapid growth of Sn NWs. In the temperature range from 650 to 800 °C, Sn NWs exhibit uniform morphology (Figure 3.4a). The important effect of Se atoms was also explored. First, low eutectic Se–Au alloy droplets were formed because of the mixture of Se atoms and Au nanoparticles, which then continuously absorb gaseous Sn atoms due to their high eutectic coefficient at VLS interface. Then, Sn crystallized because of the highly supersaturated concentration of Sn atoms in alloy and formed Sn NWs. However, the growth temperature in externally heated tube furnaces is strongly coupled to the sublimation temperature of the solid precursor materials, which limits the parameter settings for NW synthesis. To address this problem, Kast and coworkers used a cold-wall physical vapor deposition (CWPVD) chamber at high vacuum conditions and growth temperatures of 150 °C to synthesize Zn NWs [69]. Different substrates were used during the process. Si substrates covered with Au seeds exhibit NWs of two different morphologies (woollike and rodlike NWs), while on the Pt covered substrate, only rodlike NWs, which are homogeneous and uniform. The reason is that Pt/Si substrates exhibit a much lower surface roughness compared to that of Au/Si substrates. During the growth of Zn NWs, the density of the Zn nuclei is influenced by the Zn adatom surface mobility, which is effected by the substrate surface roughness. The smoother the surfaces are, the more homogeneous Zn nuclei are favored and more unidirectional NWs can be obtained. However, higher and lower solubility of LM materials in metal catalysts will have an adverse effect on the electrical properties of NWs [70]. Thus, other methods without using catalyst should be explored.

3.3.2 Catalyst-Free Methods

It is of paramount importance to control the NW length and orientation because of the dimensional-dependent properties [71]. Ga NWs can be obtained by encapsulating liquid Ga in carbon nanotubes (CNTs). Liu and coworkers found that the encapsulated Ga can crystallize in either β phase or γ phase rather than the common α phase upon freezing [72]. By employing calculations, Li and coworkers demonstrated that Ga NW with α phase has the highest insertion energy which indicates that the encapsulated Ga NW does not favor α phase at all. Otherwise, β phase or γ phase are favorable for the encapsulated Ga NW in the CNT. The calculated energies of the encapsulated Ga NWs and the insertion energies do strongly demonstrate that the favored encapsulated Ga NW in a CNT characterizes either β phase or γ phase, rather than α phase. They also proposed that the length of the NWs can be plotted as the function of temperature [13] (Figure 3.4b).

Though metal seeded methods can control the diameters of NWs, it is still impossible to realize the diameters of the NWs to precisely match the diameter of the nanoparticles. Comparatively, template method is an effective method to control the diameters and preferential orientation without changing the NW morphology. Single-crystal Sn NWs with diameters of 40–160 nm were produced by electrochemical deposition in porous polycarbonate membranes [2]. Wang and

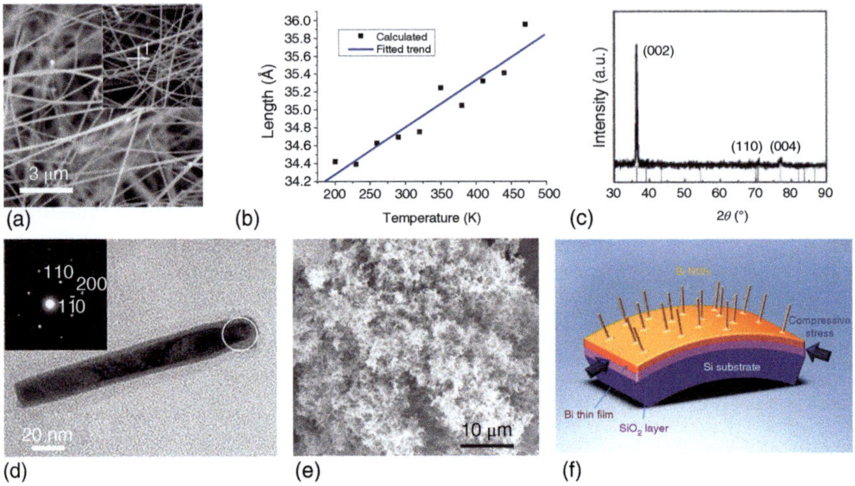

Figure 3.4 Synthesis of LM NWs. (a) Se seeded growth of Sn NWs. Source: Cai et al. [12]. (b) (Color online) The length of the encapsulated Ga NW as a function of the heated temperature. The solid line guides the trend. Source: Li et al. [13]. Reproduced with permission of AIP Publishing. (c) The XRD pattern of Zn NW arrays by template method. Source: Wang et al. [1]. Reproduced with permission of AIP Publishing. (d) TEM image of In NW grown via galvanic displacement method. Source: Li et al. [14]. (e) SEM micrographs of Zn NWs by reducing ZnO via boron. Source: Chen et al. [68]. (f) Illustration of the growth mechanism of Bi NWs. Source: Roh et al. [16]. Reproduced with permission of American Chemical Society.

coworkers realized the growth of ordered Bi NW arrays by electrodeposition using nanoporous aluminum oxide membranes (AAMs) rather than polycarbonate membranes [73]. Compared to NWs in polycarbonate membranes, those in AAMs have an advantage that they are completely parallel and exactly perpendicular to the membrane surface, besides, the NWs exhibit a uniform diameter along their entire length. Figure 3.4c shows the XRD pattern (JCPDF 04-0831) of Zn NW arrays with the diameter of 45 nm by using AAMs, which refers to the *hcp* phase. The high intensity of (002) planes indicates that the Zn nanowires preferred to grow along the (001) direction [1]. Although using physical templates can obtain predefined morphology and controllable wire diameters, there are also many limitations: First, the total number of produced wires in the process is restricted by the density of pores in a membrane. Second, challenge still remains to produce NWs longer than 20 μm. Third, the quality of obtained NWs need to be improved. Fourth, perfect template should be applied otherwise NWs with kinks or nonuniform diameters will be produced. Last but not least, to measure properties or fabricate devices, additional process are needed to separate the NWs from the template [74].

3.3.3 Others

Throughout the methods mentioned above, either catalyst or template are used to assist the growth of LM NWs. However, problems also appear like poor purity of

NWs and complex process. Here, some other methods such as galvanic displacement method, thermal evaporation and stress-induced method have also been employed for the growth of high-purity NWs.

Galvanic displacement is a useful manufacturing method suitable to many systems because positive redox potential leads to spontaneous reaction. Different from electrodeposition and chemical bath deposition, here the substrates act as the reducing agents instead of the electric power or other reductant. Thus, high purity products can be obtained. On the other hand, it doesn't need equipment as compared with the standard evaporation techniques. As shown in Figure 3.4d, In NWs along (100) direction are synthesized on a Zn sheet after being treated in concentrated sulfuric acid. After treatment with concentrated sulfuric acid, many holes will form on the surface of Zn substrate, in which surfactant SDS was absorbed. The anionic surfactant SDS could further attract In^{3+} ions, In^{3+} ions contacted Zn atoms at the bottom of the hole (Zn substrate) and were reduced to In atoms, which were confined by the hard template. Based on the SAED, In NWs were growing along (100) facets and SDS was speculated to absorb on the (100) facets of In seeds that occur. In^{3+} ions obtain electrons, which were transferred by In atoms from Zn substrate and piled along (100) facets. In this case, SDS acted as a soft template for the growth of In NWs. Therefore, this process was considered as the synergic effect of treated Zn substrate and SDS [14]. This offers the opportunity for the large-scale production of other LM NWs with high purity at a mild condition.

Metallic Zn shows promising electric, optical, and catalytic properties, and Zn NWs possess special thermoelectric and magnetoresistance [75, 76] as well as superconducting properties [77]. Thermal evaporation of Zn metal beads [78] or Zn sheets [79] and reduction ZnO via carbon source [15] or NH_3 [80] were used to synthesize Zn NWs. In Figure 3.4e, by introducing boron as the reductant, zinc oxide (ZnO) powders can be reduced and zinc NWs can be synthesized at 1050 °C under a nitrogen atmosphere [68]. Single-crystalline Sn NWs with diameters of 10–20 nm has also been prepared without any catalyst and template by the thermal evaporation of bulk Sn balls on Si substrate [81].

Bi NWs grown by a stress-induced method exhibited uniform diameter, and the aspect ratios exceed 1000 [16]. After heat treatment of the deposited Bi thin film onto a thermally oxidized Si(100) substrate at 270 °C for 10 hours, Bi NWs were extruded from the surface to relax compressive stress on the Bi film (Figure 3.4f). The mechanism can be according to the different thermal expansion coefficient for Bi films and substrate during the process [82]. Based on this mechanism, Sn NWs were also synthesized by initiating phase separation of Sn embedded in a matrix of amorphous Si [83]. In addition, Si is immiscible in Sn, which is important to generate large stresses and form highly pure Sn NWs. By tailoring the nanostructure of Si–Sn composite coatings, the diameter and length of Sn NWs can be readily regulated. Higher Sn concentrations resulted in Sn NWs with larger diameters (~10 μm) but lower density and shorter length. On the contrary, smaller diameter Sn NWs (~50 nm) with higher density and longer length were synthesized with a lower Sn composition. This approach provides additional flexibility for making desirable metallic NWs with controlled dimensions.

In this part, we discussed the methods to produce LM NWs. Catalyst-assisted method is a facile method to obtain NWs. The high reaction temperature allows the use of various metal catalyst leading to the alteration of growth conditions such as temperature and pressure. However, realizing the growth of uniform diameter of NWs is challenging. As for the catalyst-free methods, encapsulating LM in CNTs is a method, which can control the length by tuning the temperature. Template method is an important method in controlling the diameter and orientation of NWs. The diameter of NWs can be regulated by controlling the pore of the template. To avoid the contamination of catalyst and the complex processes, other methods are also introduced. Galvanic displacement can obtain many LM NWs without any equipment and can realize the high field and purity at the same time. Thermal evaporation can be applied to synthesize any LMs with low melting point. Stress-induced method is a novel approach which can realize the high purity and provide more selectivity by control the composition of LM and foreign elements. We also summarized the method for preparing LM NWs in Table 3.2.

3.4 Preparation of Liquid Metal Nanosheets

As mentioned in Chapter 2, ultrathin LM NSs will present special properties due to the quantum confine effect. For example, different from its bulk counterpart, the 2D Bi offering an opportunity to explore quantum confinement [84], semimetal to semiconductor transition [85], and large magnetoresistance [86], which is a promising material for applications such as thermoelectrics [87], photocatalysis [88], and electrochemical sensors [89]. Confinement of the metal atoms in a 2D structure results in the opening of a bandgap, band inversion, and the formation of topologically protected edge states [90]. 2D Ga [91] and Sn [17] are obtained and predicted to exhibit superconducting properties.

By employing MBE, 2D Sn samples were grown on several metallic substrates (Bi_2Te_3 (111), Cu (111), and Ag (111)) since 2015. However, the interaction with the substrates limits the transport researches and electronic applications of 2D Sn [92]. It is investigated that Sn grown on insulating substrates provides the only possible to explore their properties. PbTe is an insulator with a bandgap of 0.3 eV and shows similar lattice constant in (111) plane of Sn. Zang and coworkers prepared PbTe (111) with MBE by co-evaporating Pb and Te to the surface of Bi_2Te_3(111) films on Si (111). Then 2D Sn was produced by evaporating Sn onto the PbTe (111) substrate at 150 K following the annealing of the sample at 400 K [93]. Employing the method, Falson and coworkers realized the growth of atomic structure of trilayer Sn on PbTe substrates with low-temperature MBE as the scheme illustrated in Figure 3.5a. And they found that the position of the Fermi level will decrease with the decrease of the buffer layer, which can be attributed to the reduced donation of carriers from PbTe. This phenomenon leads to a lower zero-field superconducting transition temperature $T_{c,0}$ for trilayer Sn on six-layer PbTe [17]. Though MBE is a promising method, its poor accessibility, harsh conditions, special requirements for the substrate, and

Table 3.2 Methods for preparation of LM NWs.

Methods		LM characteristics	Advantage	Disadvantage	References
Catalyst-assisted method		Sn (with diameter ranging from 10 to 15 nm)	Facile; alterative growth conditions	Nonuniform diameter; contamination of catalyst	[12]
Catalyst-free methods	Encapsulating LM in CNTs	Ga (with stable β or γ phase)	Length controllable	CNTs need to be removed	[13]
	Template method	Zn, Bi, Sn (with diameters of 40–160 nm)	Obtain NW arrays with controlled diameter and orientation	Complex process, template contamination	[1]
Others	Galvanic displacement	In	Large-scale production; high purity products; no equipment	Solvent residual	[14]
	Thermal evaporation	Zn, Sn (10–20 nm)	High quality and purity; for LMs with low melting point	High cost	[68]
	Stress-induced method	Bi, Sn	Uniform diameter; high purity and tunable addition	Time-consuming	[16]

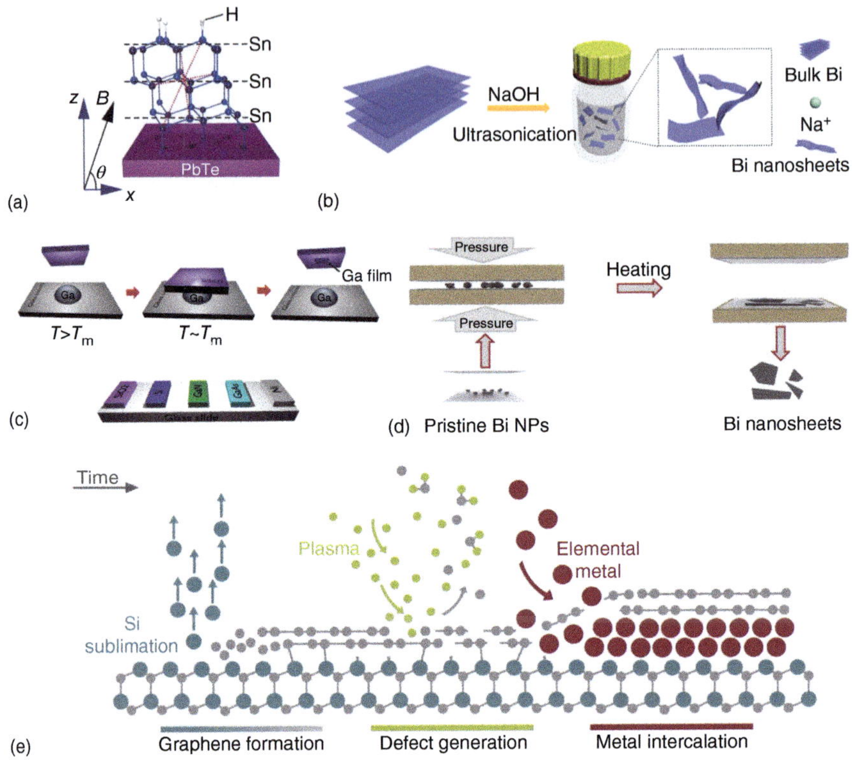

Figure 3.5 Synthesis of LM NSs. (a) Atomic structure of hydrogen decorated trilayer stanine on a PbTe substrate. Source: Falson et al. [17]. Reproduced with permission from The American Association for the Advancement of Science. (b) The process to obtain ultrathin Bi nanosheets by liquid-phase exfoliation. Source: Zhang et al. [18]. (c) Schematic to exfoliate gallenene onto Si/SiO_2 wafers. Source: Kochat et al. [91]. Reproduced with permission of the American Association for the Advancement of Science. (d) A Schematic diagram of hot-pressing method. Source: Hussain et al. [19] Reproduced with permission from John Wiley & Sons. (e) Schematic of CHet showing EG growth, O_2/He plasma treatment, and intercalation steps. Source: Briggs et al. [20]. Reproduced with permission from Springer Nature.

the interface between the material and the substrate have changed the inherent performance of the material and severely restricted its production efficiency.

Gu and coworkers proposed a useful approach to synthesize Bi NSs by liquid-phase exfoliation in an isopropyl alcohol (IPA) solution with a constant concentration of NaOH (Figure 3.5b). It is beneficial for the exfoliation process when NaOH was introduced in the dispersion system because as long as sodium ions reach the edges, the charge transfer will act as a driving force to promote the continuous insertion of Na^+ into the interlayer of bulk Bi [94]. It is worth noting that the synthesized metal Bi NSs are ultrathin (1.2–1.5 nm) and exhibit much superior activity, selectivity, and stability than bulk Bi [18].

Because of the weak interface between solid and molten phases of Ga, Kochat and coworkers demonstrated a facile exfoliation method and obtained atomically thin 2D Ga [91]. This novel strategy uses an in situ nanochemical testing system combined with an SEM. The liquid Ga droplet on the Si/SiO_2 substrate was cooled down to 30 °C under ~10^{-8} mbar. To realize the exfoliation process, the tensile and compress tests of Ga were employed using the indenter. Figure 3.5c showed the Ga exfoliation process on Si/SiO_2 wafers. At first, to uniformly melt the whole droplet, a Ga droplet was heated to 50 °C on a hot plate. Then, after the temperature is lowered to 30 °C, a little higher than the melting point (T_m) of Ga, a clean Si/SiO_2 substrate is placed on top of the Ga droplet. The lower temperature at the SiO_2–Ga interface leads to the solidification of the surface Ga layers and realized the exfoliation of 2D Ga. Ga was also exfoliated on other substrates like Si(111), GaN(0001), GaAs(111), and polycrystalline Ni by using this method. Averagely, the exfoliation on Si and GaN yielded flakes that are thinner than that on GaAs and Ni, which means that the exfoliation process is highly influenced by the interaction between the substrate and Ga. Besides, due to the strong interaction, the electronic properties are strongly influenced by the substrate. The unique interaction between Ga and substrate is also validated by a scratch test with a diamond tip. The forces measured on different substrates (Si, SiO_2, GaAs, and Ag) show different loads. Ga has the highest adhesion to Si, followed by GaN, SiO_2, and Ag (lowest). The change in adhesion (energy) is consistent with the change in the interaction energy of Ga on different substrates.

Though LM NSs can be obtained by MBE and other wet chemical methods, complex procedures and the used surfactants in solution-based method hinder further applications, including electronic devices, spintronic, and catalysis. Hussain and coworkers reported a novel, simple, and low-cost method named "hot-pressing" (HP) for the production of ultrathin Bi NSs [19]. First, Bi NPs were placed between two Si substrates. Then as the temperature of substrates increases up to 150 °C, the pressure was changed from atmospheric pressure to 0.54 GPa at the same time. Pressing the Si substrates for 30 minutes at 150 °C squeezes the NPs into a 2D NSs. After the temperature was cooled down to room temperature, crystallized 2D sheets with thicknesses as low as 2.5 nm were obtained (Figure 3.5d). It is reported that metals/semimetals like Bi can withstand considerable plastic deformation. During the deformation process, because of the variation in the shape of the grains, the microstructure of the metal changes, leading to an increase in the grain boundary area. However, deformations, as the increasing temperatures indicate, lead to even sizable expansion at the grain boundaries according to thermal expansion, making them soft enough to be shaped into the desired shape. Thus, during the process, the agglomeration of Bi on the Si substrate starts to soften from 25 to 150 °C, and validly transforms into NSs. The mechanism was demonstrated that as the temperature up to 150 °C, the Bi agglomerates on the Si substrate begins to soften and effectively transform into NSs. The whole process may include the plastic deformation of metal bismuth at high temperature and a certain degree of recrystallization.

Although various methods mentioned earlier are already used to fabricate 2D materials, the complex operation, high cost, and incompatibility with other technology limit the applications for 2D materials [95]. And the fabrication of 2D materials-based equipment to wafer-scale attracts the attention of scientists. However, it is still challenging for large-scale integration because of the sensitivity of 2D metals to external environmental influences. CVD are verified to be an effective method to produce large-scale and high-quality 2D materials, because of the precise control on structure, shape, and size of the obtained materials [20, 96, 97]. Metals can barely resist the degradation of their top few atomic layers when exposed to environment, and for an atomic thin 2D metal, the situation is even worse. Thus, developing a general approach for producing environmentally stable and wafer-scale 2D metals would be a significant issue. Briggs and coworkers demonstrated a CHet approach and crystalline 2D Ga, In, and Sn can be stabilized by the high-energy interface between epitaxial graphene (EG) and silicon carbide (SiC) [20]. As shown in Figure 3.5e, defects are generated on monolayer EG synthesized by silicon sublimation from 6H–SiC (0001) when exposed to oxygen plasma. Ga, Sn, and In precursors were situated directly beneath the EG and then the temperature was up to 700–800 °C. The LM vapor diffuses from EG defects to reach the EG/SiC interface and creates continuous films of 2D metals. Because the EG on SiC is inert, it only passivates part of the SiC surface underneath, thus maintaining the high-energy intercalation-driven interface. Besides, it facilitates 2D morphologies by lowering the energy of the top surface of the metal. Moreover, it can avoid the oxidation of ultrathin non-noble metals. STEM images display that 2D Ga, In, and Sn are one to three atomic layers thick, corresponding to the DFT calculations. The stability of these 2D metals in air for months greatly promotes ex situ characterization and makes it possible to device fabrication.

In a word, several methods were employed to synthesize LM NSs. MBE is a promising method, however, the harsh operation conditions and the interaction between the material and the substrate will severely restrict the application of materials. Exfoliation is a method, which can avoid the harsh conditions and realize the mass production of LM NSs. Yet, the introduction of solvent or surfactant will influence the properties of materials. Hot pressing is a novel approach, which obtains the LM NSs by LM NPs. Developing a universal approach for synthesizing stable and wafer-scale 2D metals is essential for the fabrication of devices. The proposed CHet approach can obtain crystalline 2D Ga, In, and Sn, which were stabilized by the high-energy interface between EG and SiC. Recently, CVD is demonstrated to be an effective strategy to produce high-quality 2D materials, because of the precise control on structure, shape, and size of the grown materials. As shown in Table 3.3, we summarized the methods for producing LM NSs.

3.5 Conclusion and Prospect

In a word, this chapter provides an overview of the recent developments involved in preparing LM materials, involving LM droplets, LM NWs, and LM NSs. Dimensional engineering has achieved great success in LMs, which has given people great inspiration to design more new LMs in different dimensions and different applications.

Liquid-phase and vapor-phase preparation are two mainly methods for getting LM droplets. As for liquid-phase preparation, there are three ways to prepare LM droplets, including solution-phase chemistry, sonication, and microfluidic

Table 3.3 Methods for preparation of LM NSs.

Methods	LM characteristics	Advantage	Disadvantage	References
MBE	Trilayer Sn on PbTe substrates	Common method	Poor accessibility, harsh conditions, special requirements for the substrate, strong interaction between substrate and LMs	[17]
Liquid-phase exfoliation	Ultrathin Bi NSs (1.2–1.5 nm)	Avoid the harsh conditions and realize the mass production, high yield	Solvent or surfactant residual	[18]
Van der Waals exfoliation	2D Ga	Facile, avoid solvent contamination	Strong interaction between substrate and LMs	[91]
Hot-pressing method	Bi NSs (as thin as 2.5 nm)	Facile and low-cost	High pressure	[19]
Confinement heteroepitaxy	2D Ga, In, and Sn	Large-scale, good stability, layer controllable	High cost	[20]

flow focusing. Solution-phase chemistry can achieve small-sized LMs NPs (5 nm) by tuning the reaction parameters, and the particle size distribution is narrow. However, the products are always enshrouded with complex additives that make the process complex to clean the additives. Even though sonication is efficient, simple, and possess potential for large-scale applications, the size distribution of LMs always changes from dozens of nanometers to several micrometers. For microfluidic flow focusing technique, the uniformity of LM NPs can be improved. However, the main problem is the hydrodynamic instability in the droplet formation process, which causes the poor monodispersion of NPs. Chanelless fluidic jetting with the aid of electric fields can prepare the LM NPs with nanoscale-size and a relatively narrow size distribution. Besides, this system can fabricate the highly monodisperse LM droplets stably, which is potential for large-scale application. Vapor-phase tuning strategy includes thermal evaporation and MBE for LM droplets. Thermal deposition can control the morphology of LMs and can avoid the chemical processing drawbacks. MBE can realize the LM array. However, this means it always generates the LM NPs with strongly interaction, where the LMs particles cannot be hybridized to form some new resonant modes. MBE with light regulation stimulates the covalent bond structures in LM structure, which is in favor of the nanoparticle self-assembly. Some other tuning strategies, such as SLICE and coulomb explosion, are useful for LM droplets. SLICE fabricates the LM droplets by shearing the low melting point metal alloys into small-sized particles in a carrier fluid. The LM particles with different shapes and compositions are achievable. But it is easier to induce phase separation owing to the surface tension under fluidic flow and the density difference between alloys. Coulomb explosion is novel but is unable to control the particle size, whereas the particle size changed linearly with the distance from the center ball. Van der Waals exfoliation can produce the isolated and randomly distributed LM droplets with mean radius below 100 nm. However, the inhomogeneity of droplets remains to be improved. Thus, more effective and controllable methods should be developed to obtain LM droplets with demanded size and high dispersity.

Until now, there have been many methods to obtain LM NWs, including catalyst-assisted method, template method, and others. It is investigated that alternative metal catalysts can be employed in this method, which provides a preferred way than seedless methods. Nevertheless, the introduction of metal catalysts will more or less lead to the contamination of the materials and influence their properties. Thus, seedless or self-seeded growth is a novel approach and still needs to be explored in the future to improve the quality of materials. Yet, realizing the growth of uniform diameter of NWs is still challenging. As for this, template method shows promising potential in controlling the diameter and orientation of NWs. The diameter of NWs can be regulated by controlling the core of template and the orientations are controlled by different substrates. To avoid the contamination of catalyst and improve the quality of NWs, other methods are also introduced. Galvanic displacement can obtain many LM NWs without any equipment and can realize high field and purity at the same time. Thermal evaporation can be applied to the synthesis of any LMs with low melting point. Stress-induced method is a novel approach, which can realize high purity and provide more selectivity by controlling the composition of LM and foreign elements.

To obtain LM NSs, MBE was developed at first; however, complex operation conditions and the interaction between the material and the substrate will seriously hinder the application of materials. Exfoliation was further employed to synthesize LM NSs, which can avoid the harsh conditions and realize the mass production of LM NSs. However, the properties of materials will be affected by the introduction of solvent or surfactant. A novel developed approach named hot pressing can obtain the LM NSs by LM NPs. Nowadays, it is essential to develop a universal approach to grow highly stable and wafer-scale 2D metals to fabricate devices. Recently, a CHet approach was proposed which can obtain crystalline 2D Ga, In, and Sn stabilized by the high-energy interface between EG and SiC. This approach extremely provides an opportunity to investigate the properties and further applications. It is well-known that CVD is an effective way to produce high-quality 2D materials because of the precise control on structure, shape, and size of the obtained materials. Although some progress has been achieved, barriers for synthesizing LM NSs still remain. The acquisition of LM NSs is still in its early days. First, the current production yield, quality, and quantity of LM NSs are still far from the criteria limiting the applications for industry or commercialization. Second, current obtained LM NSs with poor stability and durability strongly restrict their potential applications. Third, the properties of LM NSs are highly dependent on their structures, which determine their performances in their applications. Therefore, the preparation of LM NSs with large areas, desired structures, high quality, and satisfactory stability is a vital issue in the future.

Except for the LM nanomaterials we introduced here, LM compositions are drawing the attentions of researchers, and some landmark advances are also being made, which will be introduced in later chapters. Because of the intrinsic low melting point, low vapor pressure, and excellent combination of conductivity and deformability, controllable integration of LM with functional materials (e.g. metal nanoparticles, polymers, and drug molecules) would significantly regulate their properties and provide potential for applications in biomedicine, chemical catalysis, flexible electronics, and soft robots.

As mentioned earlier, the preparation of LMs in different dimensions has made great progress but some challenges still remain. Since these LMs are predicted to have excellent performance for various applications, this chapter will draw a promising blueprint for obtaining other new LM materials and exploring their properties and applications.

References

1 Wang, X.W., Fei, G.T., Zheng, K. et al. (2006). Size-dependent melting behavior of Zn nanowire arrays. *Appl. Phys. Lett.* 88: 173114.

2 Tian, M., Wang, J., Snyder, J. et al. (2003). Synthesis and characterization of superconducting single-crystal Sn nanowires. *Appl. Phys. Lett.* 83: 1620.

3 Liu, K., Chien, C.L., Searson, P.C., and Yu-Zhang, K. (1998). Structural and magneto-transport properties of electrodeposited bismuth nanowires. *Appl. Phys. Lett.* 73: 1436.

4 Bokstein, B.S., Banin, V.E., Kletskina, E. et al. (2000). Unusual solid gallium penetration in polycrystalline aluminum. *Solid State Phenom.* 72: 245.

5 Li, X., Deng, Z., Li, C. et al. (2015). Direct solvent extraction of indium from a zinc residue reductive leach solution by D_2EHPA. *Hydrometallurgy* 156: 1.

6 Fuerstena, M.C. and Wang, G. (1997). Selective separation of tin from a chloride leach solution. *Hydrometallurgy* 46: 229.

7 Wang, Z.-j., Ding, F.-h., Zhan, J., and Zhang, C.-f. (2017). Solvent extraction mechanism and precipitation stripping of bismuth (III) in hydrochloric acid medium by tributyl phosphate. *J. Cent. South Univ.* 23: 3085.

8 Lupi, C. and Pilone, D. (2020). Effectiveness of saponified D_2EHPA in Zn(II) selective extraction from concentrated sulphuric solutions. *Miner. Eng.* 150: 106278.

9 Yarema, M., Worle, M., Rossell, M.D. et al. (2014). Monodisperse colloidal gallium nanoparticles: synthesis, low temperature crystallization, surface plasmon resonance and Li-ion storage. *J. Am. Chem. Soc.* 136: 12422.

10 Lin, Y., Genzer, J., Li, W. et al. (2018). Sonication-enabled rapid production of stable liquid metal nanoparticles grafted with poly(1-octadecene-alt-maleic anhydride) in aqueous solutions. *Nanoscale* 10: 19871.

11 Tang, S.Y., Qiao, R., Yan, S. et al. (2018). Microfluidic mass production of stabilized and stealthy liquid metal nanoparticles. *Small* 14: 1800118.

12 Cai, H., Wang, W., Liu, P. et al. (2015). Enhanced synthesis of Sn nanowires with aid of Se atom via physical vapor transport. *J. Cryst. Growth* 420: 42.

13 Li, K., He, H.Y., Xu, B., and Pan, B.C. (2009). The stabilities of gallium nanowires with different phases encapsulated in a carbon nanotube. *J. Appl. Phys.* 105: 054308.

14 Li, H., Liang, C., Liu, M. et al. (2008). Synthesis of indium nanowires by galvanic displacement and their optical properties. *Nanoscale Res. Lett.* 4: 47.

15 Araújo, J.V.D.S., Ferreira, R.V., Yoshida, M.I., and Pasa, V.M.D. (2009). Zinc nanowires synthesized on a large scale by a simple carbothermal process. *Solid State Sci.* 11: 1673.

16 Roh, J.W., Hippalgaonkar, K., Ham, J.H. et al. (2011). Observation of anisotropy in thermal conductivity of individual single-crystalline Bismuth nanowires. *ACS Nano* 5: 3954.

17 Falson, J., Xu, Y., Liao, M. et al. (2020). Type-II Ising pairing in few-layer stanene. *Science* 367: 1454.

18 Zhang, W., Hu, Y., Ma, L. et al. (2018). Liquid-phase exfoliated ultrathin Bi nanosheets: uncovering the origins of enhanced electrocatalytic CO_2 reduction on two-dimensional metal nanostructure. *Nano Energy* 53: 808.

19 Hussain, N., Liang, T., Zhang, Q. et al. (2017). Ultrathin Bi nanosheets with superior photoluminescence. *Small* 13: 1701349.

20 Briggs, N., Bersch, B., Wang, Y. et al. (2020). Atomically thin half-van der Waals metals enabled by confinement heteroepitaxy. *Nat. Mater.* 19: 637.

21 Fassler, A. and Majidi, C. (2015). Liquid-phase metal inclusions for a conductive polymer composite. *Adv. Mater.* 27: 1928.

22 Zhang, W., Ou, J.Z., Tang, S.-Y. et al. (2014). Liquid metal/metal oxide frameworks. *Adv. Funct. Mater.* 24: 3799.

23 Tang, S.-Y., Sivan, V., Petersen, P. et al. (2014). Liquid metal actuator for inducing chaotic advection. *Adv. Funct. Mater.* 24: 5851.

24 Wissman, J., Dickey, M.D., and Majidi, C. (2017). Field-controlled electrical switch with liquid metal. *Adv. Sci.* 4: 1700169.

25 Liang, S., Li, Y., Chen, Y. et al. (2017). Liquid metal sponges for mechanically durable, all-soft, electrical conductors. *J. Mater. Chem. C* 5: 1586.

26 Lu, Y., Hu, Q., Lin, Y. et al. (2015). Transformable liquid-metal nanomedicine. *Nat. Commun.* 6: 10066.

27 Melendrez, M.F., Cardenas, G., and Arbiol, J. (2010). Synthesis and characterization of gallium colloidal nanoparticles. *J. Colloid Interface Sci.* 346: 279.

28 Tang, S.-Y., Qiao, R., Lin, Y. et al. (2019). Functional liquid metal nanoparticles produced by liquid-based nebulization. *Adv. Mater. Technol.* 4: 1800420.

29 Fang, W.-Q., He, Z.-Z., and Liu, J. (2014). Electro-hydrodynamic shooting phenomenon of liquid metal stream. *Appl. Phys. Lett.* 105: 134104.

30 Yan, J., Zhang, X., Liu, Y. et al. (2019). Shape-controlled synthesis of liquid metal nanodroplets for photothermal therapy. *Nano Res.* 12: 1313.

31 Zhang, W., Srichan, N., Chrimes, A.F. et al. (2016). Sonication synthesis of micro-sized silver nanoparticle/oleic acid liquid marbles: a novel SERS sensing platform. *Sensors Actuators B Chem.* 223: 52.

32 Hohman, J.N., Kim, M., Wadsworth, G.A. et al. (2011). Directing substrate morphology via self-assembly: ligand-mediated scission of gallium-indium microspheres to the nanoscale. *Nano Lett.* 11: 5104.

33 Tang, S.-Y., Ayan, B., Nama, N. et al. (2016). On-chip production of size-controllable liquid metal microdroplets using acoustic waves. *Small* 12: 3861.

34 Chikan, V. and McLaurin, E.J. (2016). Rapid nanoparticle synthesis by magnetic and microwave heating. *Nanomaterials (Basel)* 6: 85.

35 Yamaguchi, A., Mashima, Y., and Iyoda, T. (2015). Reversible size control of liquid-metal nanoparticles under ultrasonication. *Angew. Chem. Int. Ed. Engl.* 54: 12809.

36 Khan, M.R., Eaker, C.B., Bowden, E.F., and Dickey, M.D. (2014). Giant and switchable surface activity of liquid metal via surface oxidation. *Proc. Natl. Acad. Sci.* 111: 14047.

37 Chen, S., Ding, Y., Zhang, Q. et al. (2018). Controllable dispersion and reunion of liquid metal droplets. *Sci. China Mater.* 62: 407.

38 Hong, Y., Ding, S., Wu, W. et al. (2010). Enhancing heat capacity of colloidal suspension using nanoscale encapsulated phase-change materials for heat transfer. *ACS Appl. Mater. Interfaces* 2: 1685.

39 Blaiszik, B.J., Kramer, S.L.B., Grady, M.E. et al. (2012). Autonomic restoration of electrical conductivity. *Adv. Mater.* 24: 398.

40 Wang, Y. and Xia, Y. (2004). Bottom-up and top-down approaches to the synthesis of monodispersed spherical colloids of low melting-point metals. *Nano Lett.* 4: 2047.

41 Hutter, T., Bauer, W.-A.C., Elliott, S.R., and Huck, W.T.S. (2012). Formation of spherical and non-spherical eutectic gallium-indium liquid-metal microdroplets in microfluidic channels at room temperature. *Adv. Funct. Mater.* 22: 2624.

42 Yu, Y., Wang, Q., Yi, L., and Liu, J. (2014). Channelless fabrication for large-scale preparation of room temperature liquid metal droplets. *Adv. Eng. Mater.* 16: 255.

43 Hu, Q., Ren, Y., Zheng, X. et al. (2019). A micro-needle induced strategy for preparation of monodisperse liquid metal droplets in glass capillary microfluidics. *Microfluid. Nanofluid.* 23: 13.

44 Yu, F., Xu, J., Li, H. et al. (2018). Ga-In liquid metal nanoparticles prepared by physical vapor deposition. *Prog. Nat. Sci: Mater.* 28: 28.

45 Losurdo, M., Suvorova, A., Rubanov, S. et al. (2016). Thermally stable coexistence of liquid and solid phases in gallium nanoparticles. *Nat. Mater.* 15: 995.

46 Catalán-Gómez, S., Garg, S., Redondo-Cubero, A. et al. (2019). Photoluminescence enhancement of monolayer MoS_2 using plasmonic gallium nanoparticles. *Nanoscale Adv.* 1: 884.

47 Garcia Marin, A., Hernandez, M.J., Ruiz, E. et al. (2015). Immunosensing platform based on gallium nanoparticle arrays on silicon substrates. *Biosens. Bioelectron.* 74: 1069.

48 Campbell, C.T. (1997). Ultrathin metal films and particles on oxide surfaces: structural, electronic and chemisorptive properties. *Surf. Sci. Rep.* 27: 1.

49 Catalán-Gómez, S., Bran, C., Gordillo, N. et al. (2018). Self-assembly of highly ordered plasmonic gallium nanoparticles driven by nanopatterning. *Nano Futures* 2: 041001.

50 Yi, C., Kim, T.H., Jiao, W. et al. (2012). Evidence of plasmonic coupling in gallium nanoparticles/graphene/SiC. *Small* 8: 2721.

51 Pau, J.L., García-Marín, A., Hernández, M.J. et al. (2016). Optical biosensing platforms based on Ga-graphene plasmonic structures on Cu, quartz and SiO_2/Si substrates. *Phys. Status Solidi B* 253: 664.

52 Wu, P.C., Kim, T.-H., Brown, A.S. et al. (2007). Real-time plasmon resonance tuning of liquid Ga nanoparticles by in situ spectroscopic ellipsometry. *Appl. Phys. Lett.* 90: 103119.

53 Pae C Wu, M.L., Kim, T.-H., Giangregorio, M. et al. (2009). Plasmonic gallium nanoparticles on polar semiconductors: interplay between nanoparticle wetting, localized surface plasmon dynamics, and interface charge. *Langmuir* 25: 924.

54 Knight, M.W., Coenen, T., Yang, Y. et al. (2015). Gallium plasmonics: deep subwavelength spectroscopic imaging of single and interacting gallium nanoparticles. *ACS Nano* 9: 2049.

55 MacDonald, K.F., Fedotov, V.A., Pochon, S. et al. (2002). Optical control of gallium nanoparticle growth. *Appl. Phys. Lett.* 80: 1643.

56 Tevis, I.D., Newcomb, L.B., and Thuo, M. (2014). Synthesis of liquid core-shell particles and solid patchy multicomponent particles by shearing liquids into complex particles (SLICE). *Langmuir* 30: 14308.

57 Cinar, S., Tevis, I.D., Chen, J., and Thuo, M. (2016). Mechanical fracturing of core-shell undercooled metal particles for heat-free soldering. *Sci. Rep.* 6: 21864.

58 Martin, A., Kiarie, W., Chang, B., and Thuo, M. (2020). Chameleon metals: autonomous nano-texturing and composition inversion on liquid metals surfaces. *Angew. Chem. Int. Ed.* 59: 352.

59 Martin, A., Chang, B.S., Martin, Z. et al. (2019). Heat-free fabrication of metallic interconnects for flexible/wearable devices. *Adv. Funct. Mater.* 29: 1903687.

60 Chen, Y., Huang, Y., Liu, N. et al. (2015). Fabrication of nanoscale Ga balls via a Coulomb explosion of microscale silica-covered Ga balls by TEM electron-beam irradiation. *Sci. Rep.* 5: 11313.

61 Shi, Y., Liu, L., Su, X., and Wang, S. (2019). Printing EGaIn nanodroplets out of liquid metal. *Mater. Lett.* 253: 9.

62 Podgorski, T., Flesselles, J.M., and Limat, L. (2001). Corners, cusps, and pearls in running drops. *Phys. Rev. Lett.* 87: 036102.

63 Peng, K., Xu, Y., Wu, Y. et al. (2005). Aligned single-crystalline Si nanowire arrays for photovoltaic applications. *Small* 1: 1062.

64 Duan, X., Huang, Y., Cui, Y. et al. (2001). Indium phosphide nanowires as building blocks for nanoscale electronic and optoelectronic devices. *Nature* 409: 66.

65 Coffman, K.G. and Odlyzko, A.M. (2002). Internet growth: is there a “Moore’s Law” for data traffic? In: *Handbook of Massive Data Sets*, Massive Computing, vol. 4 (eds. J. Abello, P.M. Pardalos and M.G.C. Resende), 47–93. Boston, MA: Springer https://doi.org/10.1007/978-1-4615-0005-6_3.

66 Schmidt, V. and Gösele, U. (2007). How nanowires grow. *Science* 316: 698.

67 Aghazadeh Meshgi, M., Biswas, S., McNulty, D. et al. (2017). Rapid, low-temperature synthesis of germanium nanowires from oligosilylgermane precursors. *Chem. Mater.* 29: 4351.

68 Chen, Y.J., Chi, B., Zhang, H.Z. et al. (2007). Controlled growth of zinc nanowires. *Mater. Lett.* 61: 144.

69 Kast, M., Schroeder, P., Hyun, Y.J., and Pongratz, P. (2007). Synthesis of single-crystalline Zn metal nanowires utilizing cold-wall physical vapor deposition. *Nano Lett.* 7: 2540.

70 Gu, G., Burghard, M., Kim, G.T. et al. (2001). Growth and electrical transport of germanium nanowires. *J. Appl. Phys.* 90: 5747.

71 O’Regan, C., Biswas, S., Petkov, N., and Holmes, J.D. (2014). Recent advances in the growth of germanium nanowires: synthesis, growth dynamics and morphology control. *J. Mater. Chem. C* 2: 14.

72 Liu, Z., Bando, Y., Mitome, M., and Zhan, J. (2004). Unusual freezing and melting of gallium encapsulated in carbon nanotubes. *Phys. Rev. Lett.* 93: 095504.

73 Wang, X.F., Zhang, J., Shi, H.Z. et al. (2001). Fabrication and temperature dependence of the resistance of single-crystalline Bi nanowires. *J. Appl. Phys.* 89: 3847.

74 Wang, Y., Jiang, X., Herricks, T., and Xia, Y. (2004). Single crystalline nanowires of lead: large-scale synthesis, mechanistic studies, and transport measurements. *J. Phys. Chem. B* 108: 8631.

75 Li, Q., Kwong, K.W., Ozkaya, D., and Cockayne, D.J.H. (2004). Self-assembled periodical polycrystalline-ZnO/a-C nanolayers on Zn nanowire. *Phys. Rev. Lett.* 92: 186102.

76 Li, Q. and Wang, C. (2003). Cu nanostructures formed via redox reaction of Zn nanowire and Cu^{2+} containing solutions. *Chem. Phys. Lett.* 375: 525.

77 Wang, J.-G., Tian, M.-L., Kumar, N., and Mallouk, T.E. (2005). Controllable template synthesis of superconducting Zn nanowires with different microstructures by electrochemical deposition. *Nano Lett.* 5: 1247.

78 Yuvaraj, D., Narasimha Rao, K., and Barai, K. (2009). Synthesis of platestacks and microtowers of zinc by thermal evaporation. *Solid State Commun.* 149: 349.

79 Tong, Y., Shao, M., Qian, G., and Ni, Y. (2005). Large-scale fabrication of high-purity and uniform Zn nanowires by thermal evaporation. *Nanotechnology* 16: 2512.

80 Li, J. and Chen, X. (2004). Large-scale and catalyst-free synthesis of zinc nanotubes and nanowires. *Solid State Commun.* 131: 769.

81 Hsu, Y.-J. and Lu, S.-Y. (2005). Vapor-solid growth of Sn nanowires: growth mechanism and superconductivity. *J. Phys. Chem. B* 109: 4398.

82 Shim, W., Ham, J., Lee, K.-i. et al. (2009). On-film formation of Bi nanowires with extraordinary electron mobility. *Nano Lett.* 9: 18.

83 Xiao, X., Sachdev, A.K., Haddad, D. et al. (2010). Stress-induced Sn nanowires from Si–Sn nanocomposite coatings. *Appl. Phys. Lett.* 97: 141904.

84 Tian, M., Wang, J., Zhang, Q. et al. (2009). Superconductivity and quantum oscillations in crystalline Bi nanowire. *Nano Lett.* 9: 3196.

85 Hoffman, C.A., Meyer, J.R., Bartoli, F.J. et al. (1993). Semimetal-to-semiconductor transition in bismuth thin films. *Phys. Rev. B* 48: 11431.

86 Yang, F.Y., Liu, K., Hong, K. et al. (1999). Large magnetoresistance of electrodeposited single-crystal Bismuth thin films. *Science* 284: 1335.

87 Son, J.S., Park, K., Han, M.-K. et al. (2011). Large-scale synthesis and characterization of the size-dependent thermoelectric properties of uniformly sized Bismuth nanocrystals. *Angew. Chem. Int. Ed.* 50: 1363.

88 Zhao, J., Han, Q., Zhu, J. et al. (2014). Synthesis of Bi nanowire networks and their superior photocatalytic activity for Cr(vi) reduction. *Nanoscale* 6: 10062.

89 Safardoust-Hojaghan, H., Salavati-Niasari, M., Motaghedifard, M.H., and Hosseinpour-Mashkani, S.M. (2015). Synthesis of micro sphere-like bismuth nanoparticles by microwave assisted polyol method; designing a novel electrochemical nanosensor for ultra-trace measurement of Pb^{2+} ions. *New J. Chem.* 39: 4676.

90 Yu, X.-L., Huang, L., and Wu, J. (2017). From a normal insulator to a topological insulator in plumbene. *Phys. Rev. B* 95: 125113.

91 Kochat, V., Samanta, A., Zhang, Y. et al. (2018). Atomically thin gallium layers from solid-melt exfoliation. *Sci. Adv.* 4: e1701373.

92 Zhu, F.-f., Chen, W.-j., Xu, Y. et al. (2015). Epitaxial growth of two-dimensional stanene. *Nat. Mater.* 14: 1020.

93 Zang, Y., Jiang, T., Gong, Y. et al. (2018). Realizing an epitaxial decorated stanene with an insulating bandgap. *Adv. Funct. Mater.* 28: 1802723.

94 Gu, J., Du, Z., Zhang, C. et al. (2017). Liquid-phase exfoliated metallic antimony nanosheets toward high volumetric sodium storage. *Adv. Energy Mater.* 7: 1700447.

95 Li, L., Lu, S.-z., Pan, J. et al. (2014). Buckled Germanene formation on Pt(111). *Adv. Mater.* 26: 4820.

96 Chen, Y., Liu, K., Liu, J. et al. (2018). Growth of 2D GaN single crystals on liquid metals. *J. Am. Chem. Soc.* 140: 16392.

97 Xu, D., Chen, W., Zeng, M. et al. (2018). Crystal-field tuning of photoluminescence in two-dimensional materials with embedded lanthanide ions. *Angew. Chem. Int. Ed.* 57: 755.

4

External Field Tuning of Liquid Metal

All change is a miracle to ponder, every moment of the things can be a miracle.
Source: Henry David Thoreau (1817–1862). Public Domain.

4.1 Introduction

Liquid metals that own many unique and excellent properties simultaneously, such as low melting point, excellent liquidity, high electrical conductivity, and good thermal conductivity, have recently attracted scientific community's immense

Liquid Metals: Properties, Mechanisms, and Applications, First Edition. Lei Fu and Mengqi Zeng.

attention [1, 2]. Among the various excellent properties, rheological property is the most essential and concerned, which endows liquid metals with the ability of motion including deformability and mobility. Similar to the movie series of Terminator, liquid metals can deform to any shape and are capable of autonomous healing after suffering damage, thus becoming a competitive candidate in robotics, machines, intelligent flexible devices, and biomedical engineering [3]. Inspired by this, numerous efforts have been devoted to exploring the motion of liquid metals, which have gained lots of achievements.

To achieve the motion of liquid metals, an external energy source supply is required. Liquid metals can be powered by external fields including electric field, magnetic field, chemicals, force, temperature, light, and humidity. Liquid metals can perform deformability and electrical conductivity due to the integration of characteristics of the liquid and metals [1]. Based on this, Sheng and coworkers have investigated the diversely novel behaviors of liquid metal droplets under electrical field and explained the basic driving mechanism of motion [4]. Owing to the high electrical conductivity, liquid metals can also be manipulated by electromagnetic field [5]. For example, Yu and coworkers discovered that liquid metals can deform, migrate, and even suspend under alternating magnetic field (AMF) [6]. Besides, liquid metals can easily combine with ferromagnetic materials, which endows liquid metals with macroscopic magnetic properties. The liquid metals endowed with magnetic guidance ability are capable of actuating in plane space or anti-gravity climbing motion. In addition, chemical control is one of the most versatile ways to tune liquid metals. For instance, constructing galvanic cell between liquid metals and other chemicals including Al [7], Pt [8], graphite [9], and the like realizes all sorts of novel behaviors. And for force, owing to the liquid properties, liquid metals can deform for granted under the stimulation of force, which has been widely applied in flexible and stretch electronics [10]. With regard to other stimuli, such as temperature, light, and humidity, most of them need to combine with other responsive materials [11, 12]. Up to now, there are increasing research interests and opportunities in the motion of liquid metals under the manipulation of external fields. It is expected that the Terminator in movies will become reality and liquid metals-based motion will yield unusually brilliant results in all fields in the future. In addition, these tuning strategies of external fields will have a great effect on the properties of liquid metals. For instance, liquid metals have inductive heating property upon the electromagnetic field due to the high electrical conductivity, which has been widely applied in photothermal therapy [13]. These property changes under external fields will bring a new perspective to the application of liquid metals, which should also be considered.

This chapter's aim is to highlight the motion of liquid metals including deformation and locomotion, as well as the variation of properties under the stimulation of various external fields such as electric field, magnetic field, chemicals, force, temperature, light, and humidity. By summarizing the progresses that have been made in liquid metals, we intend to review important aspects with regard to the diverse motions under the stimulation of external fields, the mechanisms underlying these phenomena, the changes of properties, and potential applications. Finally, the opportunities and challenges concerning these stimuli to tune liquid metals will be illustrated and interpreted.

4.2 Electrical Field Tuning

Liquid metals can be regarded as the combination of liquid and metals, and it can perform the integrated characteristics of deformability and electrical conductivity [3]. Therefore, numerous novel behaviors of liquid metals can be observed after undergoing the stimulation of electrical field, which makes liquid metals a favorable candidate for future intelligent flexible devices [1]. Moreover, the superiorities of electrical field with easily controlled magnitudes and directions also make the electrical field tuning method a feasible approach to manipulate liquid metals and construct desired devices. In this section, the tuning outcomes of liquid metals are mainly classified into two categories, including locomotion and deformation, which are defined in consideration of the movement of their barycenter.

4.2.1 Electric-Induced Locomotion

If a liquid metal droplet is immersed in the water and then two electrodes are placed in the water at the opposite sides of the droplet with applying a direct-current (DC) voltage, the droplet would directionally move to the anode, as illustrated in Figure 4.1a [4]. The driving force of this electrical-field-induced locomotion of the droplet originates from the surface tension difference between its left and right sides, which can overcome the friction effects between the droplet and its surrounding water as well as the substrate surface (Figure 4.1b). After applying a 12 V DC on the two sides of a liquid meal droplet of gallium$_{67}$indium$_{20.5}$stannum$_{12.5}$ ($Ga_{67}In_{20.5}Sn_{12.5}$), an average moving rate of about six body lengths per second could be generated in water, as suggested by the sequential snapshots in Figure 4.1c.

Sheng and coworkers [4] explained that the electrical-field-induced surface tension difference was based on the formation of an electrical double layer (EDL) on the GaInSn surface. Regarding the huge difference between the surface tension of GaInSn and water, opposite charges would accumulate on their interface and an EDL would form [1]. The applied DC voltage would lead to a nonuniform charge distribution at the EDL and further cause a surface tension difference at the opposite sides based on the continuous electrowetting behavior [14]. Here, the relationship between the potential difference (V) across the EDL and the surface tension (γ) of the liquid metal can be described by the Lippmann equation:

$$\gamma = \gamma_0 - \frac{1}{2}cV^2 \tag{4.1}$$

where γ_0 is the intrinsic surface tension of the liquid metal (i.e., $V = 0$) and c is the capacitance of EDL per unit area [15]. Thus, it can be inferred that the surface tensions of the liquid metal droplet surface facing the cathode (γ_c) and the surface facing

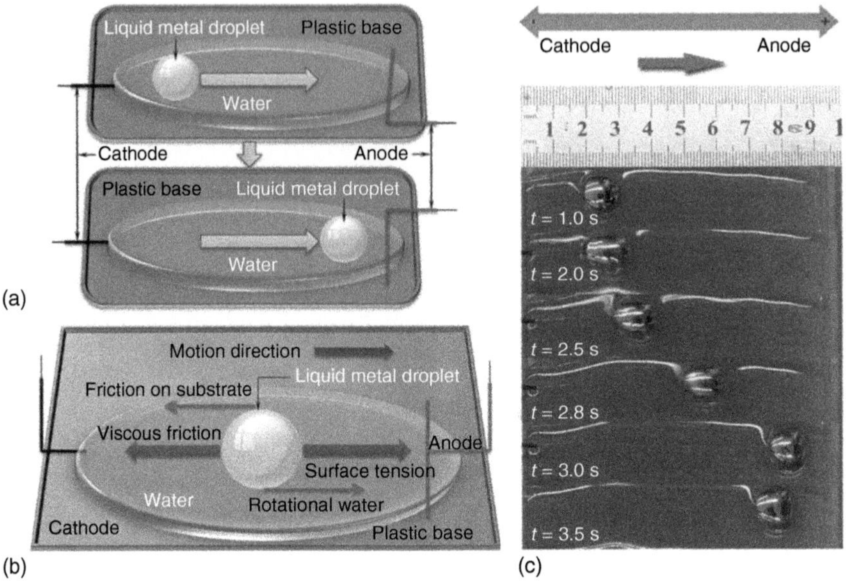

Figure 4.1 Continuous-electrowetting-induced locomotion of liquid metals. (a) The schematic of the directional movement of the droplet triggered by electrical field. (b) The schematic of force analysis of the moving liquid metal droplet. Source: (a, b) Sheng et al. [4]. Reproduced with permission from John Wiley & Sons. (c) Sequential snapshots of the droplet moving toward the anode. Source: Sheng et al. [4].

the anode (γ_a) would be different and the surface tension difference ($\Delta\gamma$) can be expressed as

$$\Delta\gamma = \gamma_c - \gamma_a = \frac{c}{2}({V_a}^2 - {V_c}^2) \tag{4.2}$$

where V_c and V_a are the potential differences across the EDL facing the cathode and anode, respectively. As V_a is larger than V_c owing to the increased charges on this side, γ_c is then larger than γ_a. A pressure difference (Δp) between the two sides would be generated according to the Young–Laplace equation

$$\Delta p = p_c - p_a = \frac{2}{r}(\gamma_c - \gamma_a) = \frac{c}{r}({V_a}^2 - {V_c}^2) \tag{4.3}$$

where p_c and p_a are the pressures at the surface of the droplet facing the cathode and anode, respectively, and r is its radius. The directional locomotion of the liquid metal droplet toward the anode can, therefore, be observed due to that p_c is larger than p_a.

The locomotion of the liquid metal droplet upon the electrical field can be further tuned by surface coating of nanoparticles. Tang and coworkers have investigated the influence of surface-coated tungsten oxide (WO_3) nanoparticles in the locomotion behavior of liquid GaInSn [16]. Different from the locomotion of bare liquid GaInSn, the droplet coated by WO_3 can move toward the cathode side after being stimulated by an appropriate DC voltage. This abnormal behavior is explained by the effect of WO_3 nanoparticles on the charge distribution of the liquid metal.

Besides the directional motion, the liquid metal can also exhibit electrical-field-induced rotational movements after being placed in a cylindrical chamber with closed diameter value as the droplet. Based on this, Tang and coworkers reported a liquid-metal-enabled pump [17]. Here, the channel was filled with sodium hydroxide (NaOH) solution and two graphite electrodes were used for applying the voltage. After placing the liquid GaInSn, the metal elements reacted with the electrolyte and gallates (e.g., $[Ga(OH)_4]^-$) and would form on the surface in this alkali solution, which led to a negatively charged surface followed by the formation of EDL. A potential difference across the EDL would be generated after a voltage was applied, which could induce a nonuniform charge distribution on the droplet surface. It further led to a driving force to promote the motion of the droplet, as discussed earlier. However, in this case, the motion was impeded by the smaller neck of the chamber and the applied electric power turned in the driving force of the mechanical movement of the surrounding electrolyte, thus making the liquid metal droplet an effective pump. Notably, a thin Ga oxide skin would be electrochemically generated on the droplet surface facing the cathode after applying the electric field, which would halt the pumping effect. Thus, an alternating voltage was employed in this work to avoid the formation of this oxide layer, further realizing a continuous pumping effect.

Based on the electric-induced locomotion of the liquid metal droplet, Yao and coworkers reported the fabrication of miniaturized vehicles wheeled by liquid metals [18]. This vehicle was constructed as a hybrid structure, which consisted of soft wheels (liquid metal droplets) and a rigid body. Similarly, the droplet was immersed in the NaOH solution and an electrical field was applied to drive the soft wheels. The continuous motion process of a typical vehicle with four wheels could be seen in Figure 4.2a, in which the movement speed could be adjusted by the size of the droplet and the intensity of the applied electrical field. However, the motion of the vehicles is still limited in the electrolyte environment. To develop a feasible liquid metal robot that can move outside the liquid environment, Wu and coworkers reported a wheeled robot with a liquid metal droplet as the driving core [19]. Liquid eutectic gallium indium (EGaIn) droplet was placed in the NaOH electrolyte to construct an enclosed wheel system with two electrodes positioned at both sides of the droplet (Figure 4.2b). This enclosed system was employed as the core of the driving module, which enabled the wheel to move in a general environment without liquids. Here, the droplet would move directionally after an electrical field was applied. The center of gravity of the whole system would be then altered, which further, in turn, generated a rolling torque (Figure 4.2c). After that, continuous locomotion of the wheeled robot with a steady speed could be realized (Figure 4.2d), in which the maximum angular velocity could be reached to 2.5 rad s^{-1}. To enable steady locomotion, the contact between the droplet with the cathode electrode should be avoided, thus the two electrodes were placed at a suitable distance between them. Moreover, the presence of the resistive forces, including the viscous resistance between the liquid metal and surrounding solution, and the friction between the liquid metal and the wheel, also contribute to the equilibrium of the wheeled system by balancing the driving force.

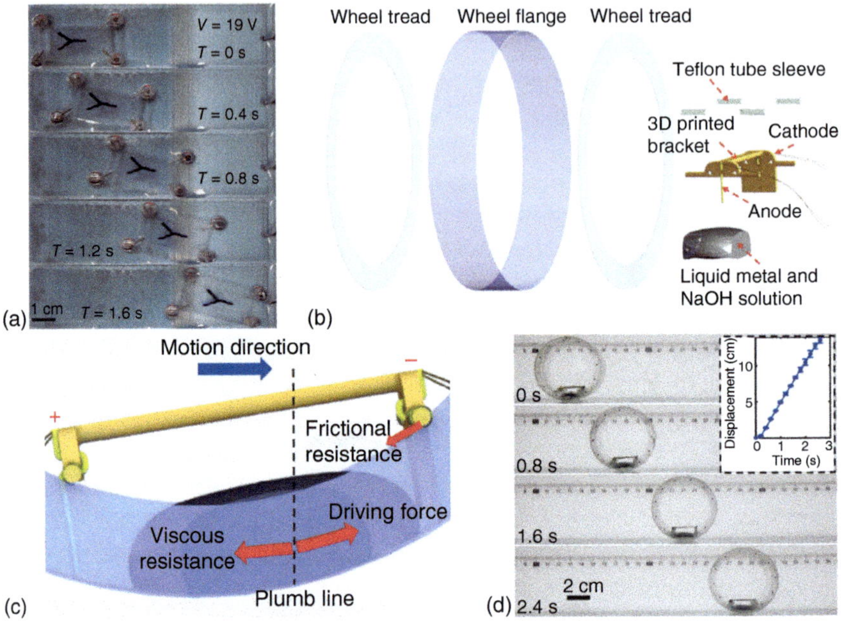

Figure 4.2 Liquid metal-enabled moving robots. (a) Sequential snapshots of the locomotion of the miniaturized vehicle wheeled by liquid metal. Source: Yao and Liu [18]. (b) The exploded view of the wheeled robot enabled by an EGaIn droplet. (c) The illustration of the moving mechanism of the wheeled robot. (d) Sequential snapshots of the motion of the wheeled robot. Source: (b–d) Wu et al. [19]. Reproduced with permission from John Wiley & Sons.

Besides the EDL-driven locomotion behaviors, the liquid metal placed on a conductive substrate can also be induced to flow upon an electrical field owing to the electromigration behavior [20]. Here, the migration of the liquid metal flow is caused by the transportation of the metal ions owing to the momentum transfer between the directionally migrated electrons and the diffused ions. Regan and coworkers reported the observation of the atomic scale mass transport of liquid metal along carbon nanotubes (CNTs) based on the electromigration mechanism [21]. The experiments were carried out inside the chamber of the transmission electron microscope (TEM) to characterize the atomic placement and manipulation capability of this CNT delivery systems, in which the In particles were thermally evaporated onto the CNTs and two electrodes were made to physically contact with the two ends of the nanotubes. After applying a voltage, the local temperature was elevated and the particles decorating the CNTs would be molten via Joule heating. Further altering the voltage, mass transport between these liquid metal particles could be observed. Here, a typical TEM image of a CNT sample with an obvious In nanoparticle decorated on the left of the nanotube could be observed (i of Figure 4.3a). A voltage (1.8 V) was applied between the two ends of the CNT and, after that, the original nanoparticles vanished and with two new smaller nanoparticles appeared in the middle of the CNT (ii of Figure 4.3a). Moreover, after the direction of the electrical

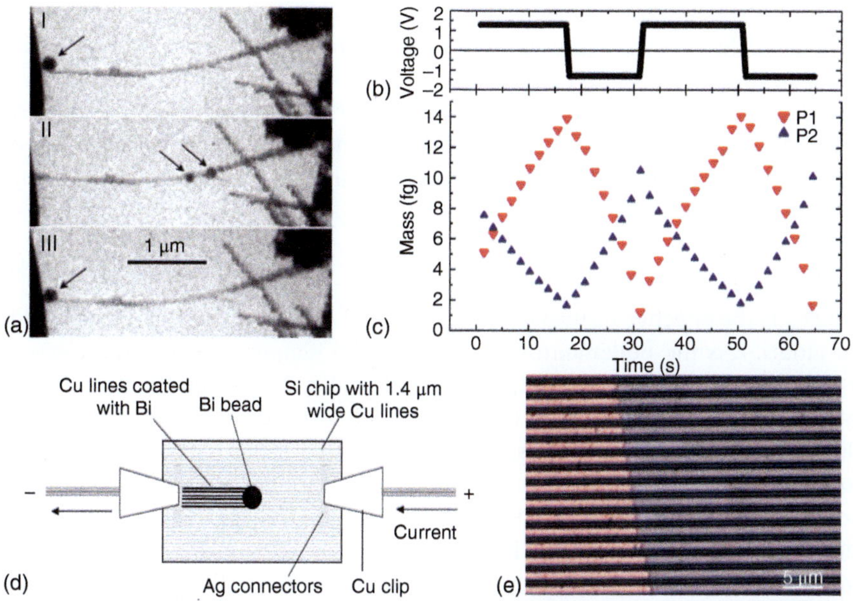

Figure 4.3 Electromigration-induced locomotion of the liquid metals. (a) Sequential TEM images showing the reversible movement (from i to iii) of liquid In nanoparticles on a CNT. Source: Regan et al. [21]. (b) The applied voltage as a function of time. (c) The corresponding mass of two In nanoparticles versus time. Source: (b, c) Regan et al. [21]. Reproduced with permission from Springer Nature. (d) The schematic of the experimental setup. Source: Dutta and Kumar [22]. Reproduced with permission from AIP Publishing. (e) A typical OM image showing the partially coated Cu lines. Source: Dutta and Kumar [22].

field was reversed without changing the numerical value to maintain the temperature, the original nanoparticle appeared at a similar position with original size, while the two smaller nanoparticles disappeared (iii of Figure 4.3a). Thus, the mass transport of liquid In followed the direction of the voltage gradient, which was again toward the cathode. Precise control of the rate and direction of the mass transport could be realized by altering the external electrical field. The two In nanoparticles could alternately grow or shrink (Figure 4.3c) owing to the mass exchange of them under the stimulation of a square-wave forcing function (Figure 4.3b). Here, the migrated In atoms between nanoparticles are believed to constitute as a two-dimensional (2D) gas on the CNT surface, in which the mass transport is derived by a thermally activated diffusion process along the direction of the applied electrical field.

Dutta and coworkers reported the electric-current-induced flow of liquid metals in macroscopic systems [22]. Owing to the electromigration phenomenon, a continuous liquid metal stream can be induced by an electrical field, which is applied to the underlying conductive path. Here, the conductive copper (Cu) lines were deposited on a Si substrate to serve as the current path and a Bi droplet was placed on them, in which the experiments were performed in the chamber of a scanning electron microscope (SEM) at a high vacuum environment (Figure 4.3d).

After applying the electrical field, a current with a value of 6 A was passed through 1500 parallel Cu lines resulting in the Bi droplet reaching the molten stage due to Joule heating. Then long-range continuous flows of liquid Bi streams were observed on the conductive Cu lines with the direction toward the cathode. A typical optical micrograph (OM) image showed that the Cu lines partially coated by liquid Bi was presented in Figure 4.3e. Here, the liquid metal flow (J) induced by the electromigration is given by

$$\gamma = \frac{D_{\text{eff}}}{\Omega kT} Z^* eE \tag{4.4}$$

where D_{eff} is the effective diffusivity of liquid metal ions, Ω is the volume of liquid metal atoms, k is the Boltzmann constant, T is the temperature, e is the electric charge, E is the strength of the applied electrical field, and Z^* is the effective charge number of the liquid metal ion. In liquids, Z^* analysis yields to be a positive value, while that of solids is negative. Thus, the flow direction of liquid metals is opposite to the solid, which migrates toward the anode [23, 24]. Kumar and coworkers further explored the deep mechanism of the electromigration behavior based on the studies of various pure liquid metals [25]. They found that the electric-current-induced flow of liquid metals was diffusion-controlled and the rates of the flows could be positively adjusted by the passing current density, thus confirming the electromigration mechanism. The Z^* of several pure liquid metals were deduced, where the values were close to the valency of the corresponding elements and all the Z^* values were positive except that of liquid Pb. It indicates that the electromigration of liquid metal flow generally migrates along the same direction of the passing electric current. Moreover, they also found that the electric-current-induced flow of liquid metals could be affected by the surface condition of the underlying conductive substrates. In the case of a smooth surface, the better wettability of liquid metals enables a flow with thin and continuous stream, while for a rough surface, the liquid metals form as discrete beads.Kumar and coworkers then reported a theoretical model to predict the direction of liquid metal flows induced by electromigration [26]. Here, the proposed formulation for calculating the Z^* value of liquid metals was based on the cell model [27], in which the Lennard–Jones potential was introduced to represent the interaction of liquid metal atoms. The outcomes of practical experiments based on several liquid metals agree well with the results of theoretical predictions.

Besides the locomotion behaviors based on the continuous electrowetting and electromigration, the liquid metals can also locomote by redox-reaction-induced electrical field [28] and levitation of electrolyte [29] under the stimulation of an electrical field. Gough and coworkers reported the self-actuation of Ga-based liquid alloy based on the redox reaction, in which the droplet was immersed into the NaOH electrolyte [28]. Here, the liquid metal was connected with the anode and a Cu electrode was served as the cathode. The two electrodes were connected by an external conductive channel to enable the spontaneous redox reaction of a liquid metal droplet. As a result, an oxidizing reaction occurred on the anode and the surface of liquid metals facing the conductive channel transformed to Ga_2O_3. The generation of an oxide layer on the surface lowered the localized surface tension of the liquid metal, which allowed it to flow into the conductive channel (Figure 4.4a,

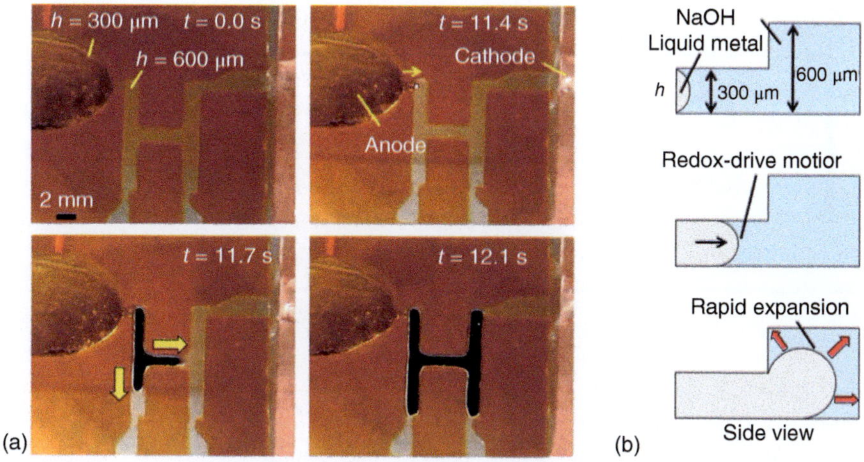

Figure 4.4 The locomotion of the liquid metals based on other driving forces. (a) Sequential snapshots of the flow of the liquid metal based on the redox reaction. Source: Gough et al. [28]. (b) The schematic of the side view of this locomotion [28]. Source: Gough et al. [28]. Reproduced with permission from American Chemical Society.

b). Since the Ga_2O_3 layer can be dissolved in the NaOH electrolyte, the formation of a thick oxide film can be prevented, further leading to a reduced surface tension of liquid metal at the channel entrance to enable it to flow into the channel.

In addition, Zhao and coworkers reported the surfing behavior of EGaIn based on the levitation of electrolyte [29]. Here, the cathode was inserted into the bulk of the EGaIn alloy, while the anode was placed in the NaOH electrolyte without contacting the liquid metal (Figure 4.5a). After applying a voltage between the two electrodes, an interfacial flow could be observed. Once a liquid metal droplet deposited on this rapid-flowing interface, it could levitate and surf on the interfacial flow (Figure 4.5b). An impacting droplet could also be held on the interfacial flow, which indicated a strong levitation effect provided by this surfing phenomenon (Figure 4.5c). Since the surfing of the droplet was maintained by the electrical field across the interface between liquid metal bulk and electrolyte, the levitating droplet would immediately merge into the bulk after the voltage was removed. Here, the realization of this electrical-field-induced surfing was based on the formation of an electrolyte film between the liquid metal bulk and the droplet. The measured values of electrical resistance between the bulk and the droplet before and after surfing were totally different, thus demonstrating the presence of the film.

4.2.2 Electric-Induced Deformation

Besides the locomotion behaviors, the deformation behaviors of liquid metals can also be tuned by the electrical field. Generally, the root cause of the deformation lies in the alteration of the surface tension of liquid metals. Tsai and coworkers reported the transformations of liquid metals based on the electrocapillarity phenomenon [30], which was used to describe the surface tension change of liquids after applying

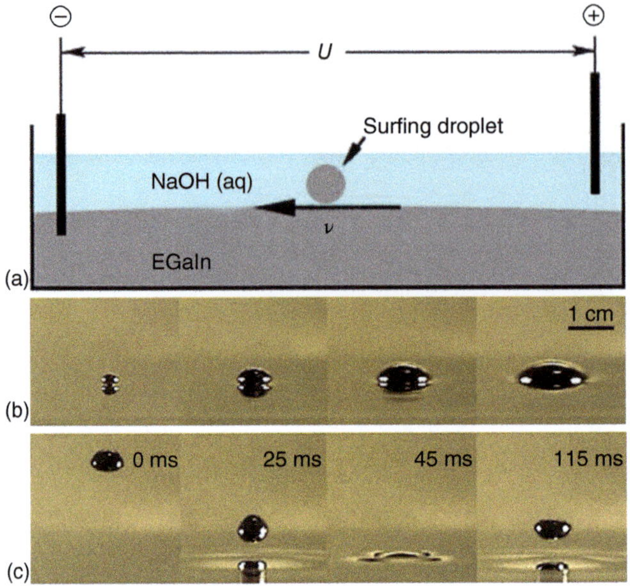

Figure 4.5 (a) The schematic of the side view of the surfing liquid metal droplet. Source: Zhao et al. [29]. Reproduced with permission from AIP Publishing. (b) The images of the surfing EGaIn droplets of different sizes. (c) Sequential snapshots of the impacting droplet. Source: (b, c) Zhao et al. [29].

a voltage [31]. They found that a liquid Ga droplet could transform into a thin film on a conductive In Sn oxide (ITO) as the anode. The overall experiments were conducted in the hydrogen chloride (HCl) electrolyte and a Cu wire merging in the electrolyte acted as the cathode. Based on the electrocapillarity-induced liquid Ga film, they have fabricated a prototype light valve, which could exhibit a backlight utility of 96% with a contrast ratio of $10^6 : 1$. Sheng and coworkers reported various transformations of liquid metals induced by the electrical field [4]. Here, the liquid alloy $Ga_{67}In_{20.5}Sn_{12.5}$ was spread on a plastic substrate (i of Figure 4.6a). Water was employed as the electrolyte and covered the liquid metal and a Cu-wire anode was connected with water, whereas the cathode was connected with the spread liquid metal (ii of Figure 4.6a). After applying a 12 V voltage, the liquid metal quickly deformed (iii of Figure 4.6a) and finally transformed into a sphere (iv of Figure 4.6a). These transformation processes indicate an obvious variation of the surface tension of the liquid metal induced by the electrocapillarity phenomenon.

Based on the electrocapillarity-induced deformation of liquid metals, some novel phenomena can be generated. Tang and coworkers reported the generation of chaotic advection based on liquid metals [32]. Here, a liquid Ga–In–Sn alloy droplet was employed as an actuator to induce harmonic Marangoni flow, where the Marangoni effect was a term used to describe the mass transfer on the interface of two fluids induced by a gradient of the surface tension [33]. Thus, similar to the locomotion of liquid metals discussed earlier, the external electrical field was applied to generate the surface tension gradient of the droplet surrounded by NaOH

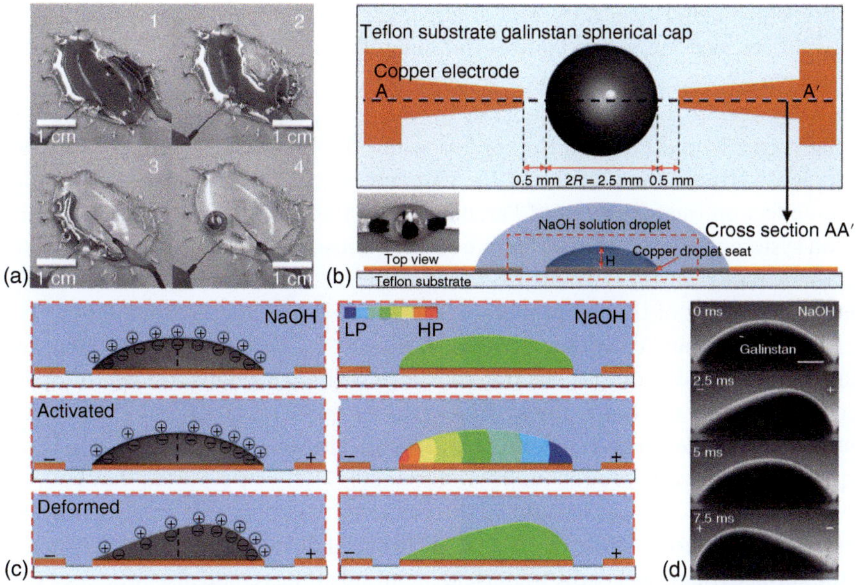

Figure 4.6 Electrocapillarity-induced deformation of liquid metal. (a) Sequential snapshots of the electrical-field-induced deformation of $Ga_{67}In_{20.5}Sn_{12.5}$ form a spreading state to a sphere state (from i to iv). Source: Sheng et al. [4] (b) The schematic of the GalnSn-based actuator for generating the chaotic advection. (c) The schematic of the electrocapillarity-induced deformation mechanism. Source: (b, c) Tang et al. [32]. Reproduced with permission from John Wiley & Sons. (d) Sequential snapshots of the periodically varied polarity of the GalnSn droplet, the scale bar is 500 μm. Source: Tang et al. [32].

electrolyte. But an alternating current (AC) voltage was employed here instead of a DC voltage to induce chaos. The liquid alloy droplet was placed on a Cu seat and the amalgamation effect enabled the droplet to wet the Cu surface and spread as a cap shape. The overall experiment setup could be seen in Figure 4.6b, in which two Cu electrodes were placed on opposite sides of the droplet. After applying the 4 V AC voltage of 100 Hz, a surface tension gradient was induced at the two sides of the droplet, which further induced a pressure difference to drive the deformation of the liquid droplet, that is polarization toward the anode (Figure 4.6c). Because of the alternating signal input, the two electrodes switched periodically, which further induced a periodically varied polarity of the droplet (Figure 4.6d). In addition, they also investigated the influences of AC voltage frequencies and the NaOH solution concentrations, and the deformation of the droplet could be adjusted by the two factors. Moreover, the liquid metal actuator was also introduced in a microfluidic system, in which the as-induced chaotic advection could cause a high mixing efficiency even at a high flow rate. Tang and coworkers also reported the controlled direction of liquid metal in a complex microfluidic network [34]. As the surface tension of the liquid metal could be tuned by the electrical field via electrocapillarity, the flow of it toward an outlet with electrodes could be facilitated. In addition, the surface oxidation of it could be induced by the electrochemical reaction, which could also be utilized to impede the movement. Thus, the electrical field could act as a “valve” to control the direction and the pathway of the liquid flow. Zhao and coworkers also reported the tuning of liquid-metal-based Faraday waves via altering the surface tension by electrocapillarity [35]. Generally, the Faraday waves would form on the liquids when the upper surface was subjected to sufficient vertical vibration. Thus, the characteristics of the liquid system would exert an effect on the formation of Faraday wave patterns. The liquid metal–NaOH solution system with electrically tunable surface tension was therefore employed to tailor the Faraday waves.

The environment of the liquid metal can also influence the deformation behaviors. Bormashenko and coworkers reported the electrical-field-induced deformation of liquid metal placed on a dielectric substrate without using electrolyte [36]. A modest variation of the liquid metal droplet under the stimulation of a DC voltage could be observed. Yuan and coworkers investigated the electrocapillarity behaviors of liquid metals in various aqueous environments [37]. It was reported that both of the components and concentrations of the electrolytes could exert an effect on the electrical-field-induced deformation of a liquid metal droplet. They also discovered that the surface roughness of the underlaid metal substrates could also influence the deformation process.

Considering the surface tension of the surface oxide layer is different from that of the intrinsic liquid metal, the wet behavior of the liquid metal droplet can also be tuned via electrochemically altering the surface oxidation. Khan and coworkers reported the alteration of the surface tension of a Ga-based liquid alloy in a large range via electrochemically formatting or removing the oxide layer on the surface [38]. The experiments were performed on the NaOH electrolyte and the voltages were given by relative potential values with a saturated Ag-AgCl electrode as the reference electrode, which exhibited an open-circuit potential of about −1.5 V. After applying an electrical field of −1.5 V on the liquid alloy droplet, no oxide layer formed on the surface and the deformation behavior of the droplet was in line with the earlier mentioned electrocapillarity phenomenon. Here, a maximum surface tension value was measured as approximately 509 mJ m^{-2}. As the electrical potential was modified to a more oxidative value of −1.0 V, the formation of the oxide layer on the surface started and a sudden change of the surface tension was induced, which facilitated the spread of the liquid metal droplet. Further, increased relative potentials would lead to an enhanced wettability of the liquid alloy. It was believed that the formed oxide layer on the surface could act as a surfactant to decrease the surface tension of the liquid metals by replacing the original high-energy metal surface. In addition, the modification of the surface tension could be completely reversible and this fast switch could be realized in less than 1 s. Compared with the traditional molecular surfactants, the tuning range of this electrochemical behavior was improved by one magnitude and more facile. Zhang and coworkers also reported the controlled deformation of liquid metal based on the electrochemical mechanism [39]. They found that a large-scale deformation of a liquid Ga droplet could be realized in a reversible way via combining the dissolution and formation effect of the oxide layer. Five-time-enlargement of the surface area of liquid Ga compared to the original size could be realized by consuming low electrical energy. Hu and coworkers further reported the electrochemical-induced deformation behavior of liquid metals on the surface of graphite [40]. It was found that an oxide layer would form on the surface of liquid metals at the area contacting with graphite. The oxidation occurred owing to the electrochemical reaction, which was induced by the potential elevation. Based on the interaction between the liquid metal and graphite, the deformation of the liquid metal could also be observed during the electrical-field-induced locomotion process.

Based on the electrical-field-induced deformation in an electrochemical way, Khan and coworkers reported a withdrawal behavior of liquid alloy in microfluidic channels [41]. Since the oxide layer on the surface of liquid metals could be removed by the electrochemical reduction under the stimulation of a reductive potential, a capillary behavior could be induced. If this process occurred in a microfluidic channel filled with liquid metal, the reductive potential would cause a withdrawal behavior of the liquid metal and leave the channel filled with electrolyte. Thus, in a complex conductive path fabricated by liquid metal, a specific path could be removed on demand by the electrochemically induced capillary (Figure 4.7a). Moreover, it was observed that the withdrawal behavior of liquid metal occurred along with the shortest electrical path. Wissman and coworkers reported a field-controlled electrical switch based on the coalescence and separation of two liquid metal droplets [43]. Here, the two droplets were placed very close together and immersed in the NaOH electrolytic solution. Two Cu pads were attached to the droplets to serve as the conductive electrodes and another two pads were placed at the opposite sides without contacting the droplets to serve as the switching electrodes. After applying a voltage on the two switching electrodes, the surface oxidation and reduction of the liquid droplets could be altered. The as-induced surface tension variation could further control the coalescence and separation states of the two droplets, thus enabling the access and open of the conductive electrodes. Tan and coworkers reported the realization of an electrochemically induced Marangoni flow of liquid Ga based on a Ga–Cu galvanic corrosion couple [44]. The formation of the galvanic corrosion couple originated from the amalgamation effect of liquid Ga when a liquid Ga droplet was connected with the Cu pads. A surface tension gradient was then generated to promote the Marangoni flow.

Tang and coworkers reported the dynamical fabrication of liquid metal microdroplets with controlled size via electrically tuning the surface tension of a liquid metal stream [42]. The electrochemistry and electrocapillarity were both utilized to alter the surface tension and a typical OM image of the fabrication system could be observed in Figure 4.7b. The size of the as-synthesized microdroplets was inversely proportional to the capillary number (Ca) [45], which is related to the surface tension of the liquid EGaIn alloy and can be described as

$$Ca = \frac{\mu Q W_{EGaIn}}{2h\gamma W_{\text{glycerol–NaOH}}}\left(\frac{1}{W_{\text{orifice}}} - \frac{1}{2W_{EGaIn}}\right) \tag{4.5}$$

where the μ is the viscosity of the glycerol–NaOH solution acting as electrolyte, Q is its flow rate, h is the height value of the microchannel, γ is the interface tension between liquid alloy and the solution, W_{EGaIn}, $W_{\text{glycerol–NaOH}}$, W_{orifice}, are the widths of the liquid EGaIn inlet microchannel, solution inlet microchannel and the orifice.

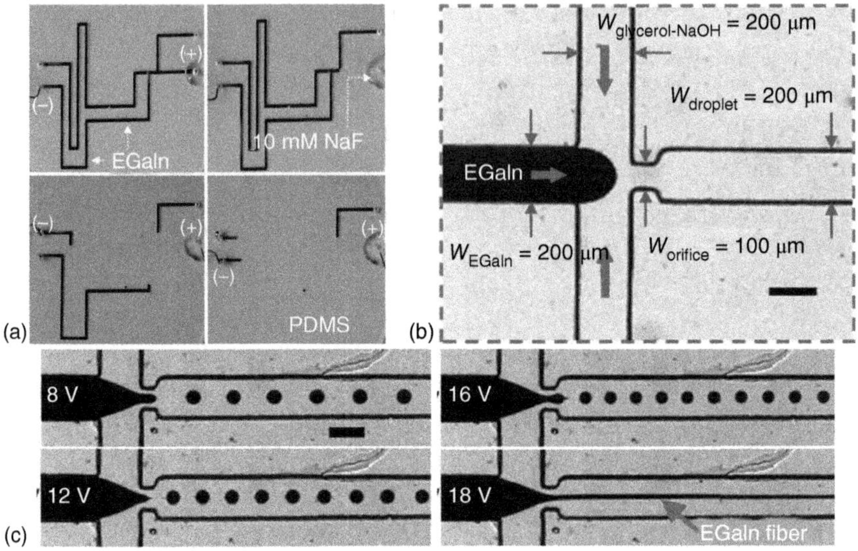

Figure 4.7 (a) The electrochemically induced withdrawal of EGaIn in microfluidic channels. Source: Khan et al. [41]. (b) The OM image of the dynamical fabrication system of liquid-metal microdroplets with electrically controlled size, the scale bar is 200 μm. (c) Fabrication of liquid metal microdroplets after applying various voltages, the scale bar is 200 μm. Source: (b, c) Tang et al. [42].

For this fabrication system, the three width values were fixed (Figure 4.7b). After setting the flow rate of the solution, the size of the fabricated microdroplets could be well controlled by the applied electrical field (Figure 4.7c). Here, a positive DC voltage was applied to the liquid EGaIn alloy for the formation of an oxide layer and altered the interface tension in an electrochemical way. A negative DC voltage could also be applied to lower the interface tension via electrocapillarity, further leading the formation of EGaIn microdroplets with reduced sizes. Furthermore, it was found that the sizes of the microdroplets responded much faster than the changes achieved by adjusting the flow rates. Thus, it is capable to dynamically alter the sizes of the microdroplets and obtain a liquid droplet stream with uniform distribution.

Based on the electrical-filed-induced defamation of liquid metals, various applications could be fabricated. Chrimes and coworkers reported the fabrication of reconfigurable optical reflectors based on the electrochemical deformation of liquid Ga [46]. Here, an AC voltage with a DC offset was applied to the liquid metal to avoid the surface oxidation and oscillating effect [32]. After applying a voltage with high DC offset (more than +1.5 V), obviously surface oxidation would generate and the surface roughness was increased owing to the formation of surface textures. Therefore, an ideal +1.1 V DC offset voltage was employed to spread the liquid metal. Then the optical properties of the as-fabricated reconfigurable reflector based on liquid Ga was tested with a 532-nm laser (Figure 4.8a). Obvious bright spots could be observed on the charge-coupled device (CCD) camera (Figure 4.8b), confirming the reflect ability of this Ga-based reflector. Jin and coworkers reported the fabrication of a

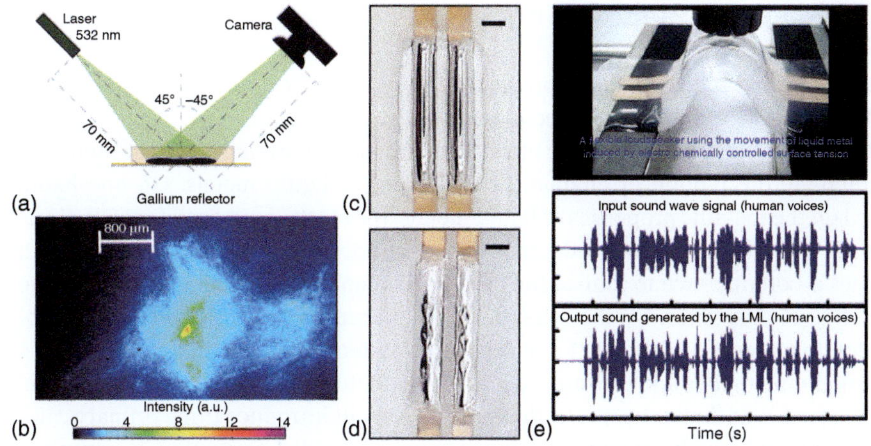

Figure 4.8 Applications based on the electrical-field-induced deformation of the liquid metals. (a) The schematic of the testing system of the reconfigurable optical reflector. Source: Chrimes et al. [46]. Reproduced with permission from American Chemical Society. (b) The intensity image of the reflected laser acquired from the CCD camera. Source: Chrimes et al. [46]. The optical images of the liquid metal-based loudspeaker (c) without and (d) with applied AC voltage, the scale bar is 0.3 cm. Source: (c, d) Jin et al. [47]. (e) The optical image of the bent loudspeaker (upper) and the corresponding input and output signal of human voices (lower). Source: Jin et al. [47]. Reproduced with permission from John Wiley & Sons.

loudspeaker based on the liquid metal with electrochemically tuned surface tension [47]. As the shape of the liquid Ga-based alloy was stabilized after equilibrium was established between the surface tension and its gravity (Figure 4.8c). The formation and removal of the surface oxide layer would lead to a surface tension variation of the liquid alloy, and further enable us to tailor its shape via applying a voltage. If an AC voltage was applied, the shape variation would be continuously proceeded and cause the liquid alloy to vibrate (Figure 4.8d). Based on this, if a sound wave signal was imputed to alter the electrical-field-induced deformation process, the liquid metal would vibrate according to the signal and then output sound (Figure 4.8e). It was found that the human sound with 40–50 dB could be produced while the applied voltage was lower than 1 V. Moreover, they also found that the frequency response of the loudspeaker agreed with the input information when the device was bent, thus demonstrating its flexibility feature.

In conclusion, the electrical-field-induced locomotion of liquid metals can be derived by continuous electrowetting in the electrolyte or by the electromigration with a conductive underlay. In addition, the electrical-field-induced deformation can be realized by the electrocapillarity and electrochemistry mechanisms. Both approaches aim at altering the surface tension of the liquid metals. Although the realization of the locomotion or deformation of liquid metals requires specific environments, such as electrolyte and conductive underlay, practical applications can still be realized by designing appropriate systems integrating the environments.

4.3 Magnetic Field Tuning

Magnetic-based methods are favorable tuning strategy for liquid metals due to the fact that the magnitude and direction can be precisely controlled. Magnetic field derives from the electromagnetic effect and magnetic materials. Both of them can manipulate the motion or change the properties of liquid metals. On one hand, based on the metallic properties of liquid metals, the electrons in liquid metals directionally migrate to generate inductive currents under electromagnetic field, which creates a Lorentz force to induce the deformation and locomotion behaviors or the variation of properties. On the other hand, liquid metals can easily mix with ferromagnetic materials. Magnetic force derived from ferromagnetic materials can also deform liquid metals and the introduction of magnetic materials is capable of tuning the properties of liquid metals. In this section, we will introduce three primary categories of liquid metals under the tuning of magnetic field: deformation, locomotion, and properties.

4.3.1 Magnetic-Induced Deformation

As is well known, AMF can generate eddy currents in a conductor with closed loops and then the created currents will generate another magnetic field opposed to the external AMF, further giving rise to a Lorentz force on the conductor [5]. The created repulsion force can be served as a drive force to induce the deformation or locomotion behaviors. Based on it, Yu and coworkers found a liquid metal (EGaIn) droplet transformed into "stand-up" shape under the AMF, where the height had nine orders of magnitude increase [6]. As demonstrated in Figure 4.9a, an EGaIn droplet (100 μl) was placed above the magnetic coil center. When subjected to an AMF with the intensity of 250 A m^{-1} and the frequency of 245 kHz, the droplet immediately elongated into a thin stick (Figure 4.9b). This phenomenon was attributed to the imbalanced force and the oxide layer of the liquid metal. The magnetic field at the coil center was most intense. Thus, the generated eddy currents in the liquid metal droplet offered an upward repulsive force with sufficient magnitude to overcome the gravity of the droplet. Besides, the oxide layer increased the adhesion of the liquid metal on the surrounding oxygen-containing substrates in virtue of weak but surface dispersed weak van der Waals forces [49]. This affinity was crucial to the mobility of the liquid metal, which induced the deformation rather than locomotion behavior. These magnetic deformation behaviors absolutely depend on the original properties of liquid metals.

As for deformation, a further intriguing strategy is combing liquid metals with ferromagnetic materials. Hu and coworkers [48] combined liquid metal (GaInSn) with iron (Fe) particles in virtue of acid-assisted method to endow the liquid metal with macroscopic magnetic properties (Figure 4.9c). Due to the polarizable surface and weak liquid–liquid attraction of the liquid metal, Fe particles could easily suspend in liquid metal [50]. The magnetic liquid metal afforded concomitantly the remarkable properties of Fe and liquid metal, which could integrate many emerging functionalities, especially magnetic guidance ability and deformability. As displayed

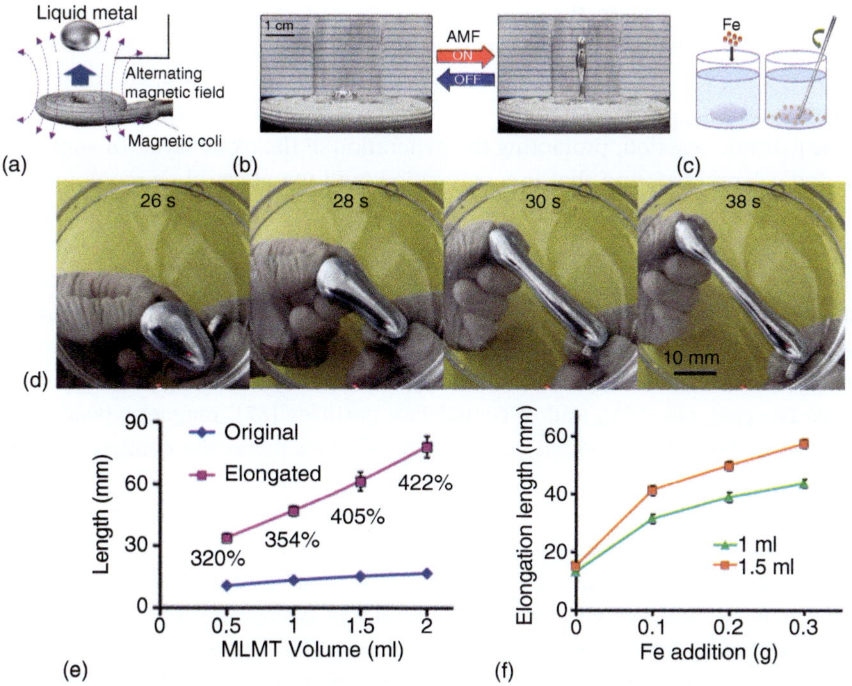

Figure 4.9 (a) Schematic of an EGaIn droplet under an AMF. (b) AMF-induced deformation behavior of the liquid metal droplet. Source: (a, b) Yu and Miyako, [6]. (c) Schematic of the preparation process of GaInSn/Fe composites. (d) The photographs of the stretching behaviors of the GaInSn/Fe composites under magnetic field. Source: Hu et al. [48]. (e) The stretching ability of the GaInSn/Fe droplet with various size. The pink and blue curves represent the diameter of the GaInSn/Fe composites at an initial state and elongated state. (f) The largest elongation length of the GaInSn/Fe composites under the different amounts of Fe particles. Source: (c, e, f) Hu et al. [48]. Reproduced with permission from American Chemical Society.

in Figure 4.9d, when directly positioning two magnets of neodymium–iron–boron (NdFeB) with two opposite directions under the magnetic liquid metal droplet, the droplet could be elongated into a strip shape. This deformation behavior was because the dispersed Fe particles were attracted and aggregated to the bottom of the droplet by the magnet. By moving the magnet, the aggregated Fe particles would drive and deform the droplet to a continuous shape [51]. More deeply, the underlying mechanism of the deformation phenomenon could be explained by the mechanic equilibrium among the magnetic force induced by the magnet (f_m), the friction force (f) derived from the substrate, the resistance from the surrounding medium (f_r), and the capillary force (f_c) from the deformation of the droplet [52]. Thereinto, f_m was the driving force while f and f_r cooperated to cause the obstacle. When f_m exceeded the maximum value of f_c, Fe particles would be extracted out, inducing the deformation behavior. The stretching ability was able to be tuned by the size of the droplet. The result illustrated in Figure 4.9e suggested that the droplet had a better stretching ability at a larger size. The magnitude of stretching could reach to 420% when the volume was over 2 ml. Such a large deformation ability was due to the oxide layer, which significantly reduced the surface tension. Moreover, a higher content of Fe particles could further promote the deformation ability (Figure 4.9f). This was attributed to the fact that more Fe particles resulted in intense galvanic reaction, promoting the generation of the oxide layer. In addition, Jeon and coworkers found that this magnetic liquid metal could be manipulated, merged, and separated in microfluidic channels using magnetic field [53], showing deformation behaviors of sorts. This deformation behavior allows direct patterning of liquid metal on planar and nonplanar surfaces, exhibiting a promising application in stretch electronic [51, 54]. Compared with AMF-based approach, this method is capable of realizing deformation behavior in a large scale. And the modulation of magnetic field is more convenient. Similarly, mixing with ferromagnetic materials such as Ni- [55], Gd- [56], and Si-coated FeB particles [57], magnetic bead [52], and liquid metal can also exhibit deformation behavior under the manipulation of magnetic field.

In conclusion, the magnetic field-induced deformation of liquid metals is derived from the electromagnetic effect and ferromagnetic materials. Based on the electromagnetic effect, liquid metals are able to generate a Lorentz force upon alternating electromagnetic field, which can deform the liquid metals. Another approach is that liquid metals combine with ferromagnetic materials. Under the stimulation of magnetic field, magnetic particles aggregate and then deform liquid metals. Both approaches can realize the deformation of liquid metals. But compared with electromagnetic effect, combing with ferromagnetic materials is a more efficient way and can deform liquid metals on-demand.

4.3.2 Magnetic-Induced Locomotion

Besides deformation, magnetic field can also induce the locomotion of liquid metals. As for the bare liquid metals, the oxide layer together with magnetic control is able to induce the deformation behavior. However, if the interfacial oxide layer is removed or a slip layer is added into the interface between liquid metals and substrate, the locomotion of liquid metals will occur under magnetic field. Yu and coworkers added water between the liquid metal droplet and glass vessel, which has implemented the jumping motion under an AMF (250 A m^{-1}, 245 kHz) [6]. As displayed in Figure 4.10a, by modulating the on–off state of the AMF, the droplet jumped from the substrate. The magnetic field at the magnetic coil center was intensest. Thus, the induced eddy currents in the liquid metal produced a sufficient Lorentz force to overcome the gravity of its own (Figure 4.10b). In contrast, when the droplet shifted 1 cm away from the center of the magnetic coil and when simultaneously the same power of the AMF was added to the droplet, the jumping behavior could not happen. This result was attributed to the inclined direction of the AMF, resulting in the decreased strength exerted on the droplet. The vertical component of the Lorentz force was neutralized by the gravity of the droplet and the planar component of the Lorentz force drove the droplet to sideways. Remarkably, the droplet could also float stably

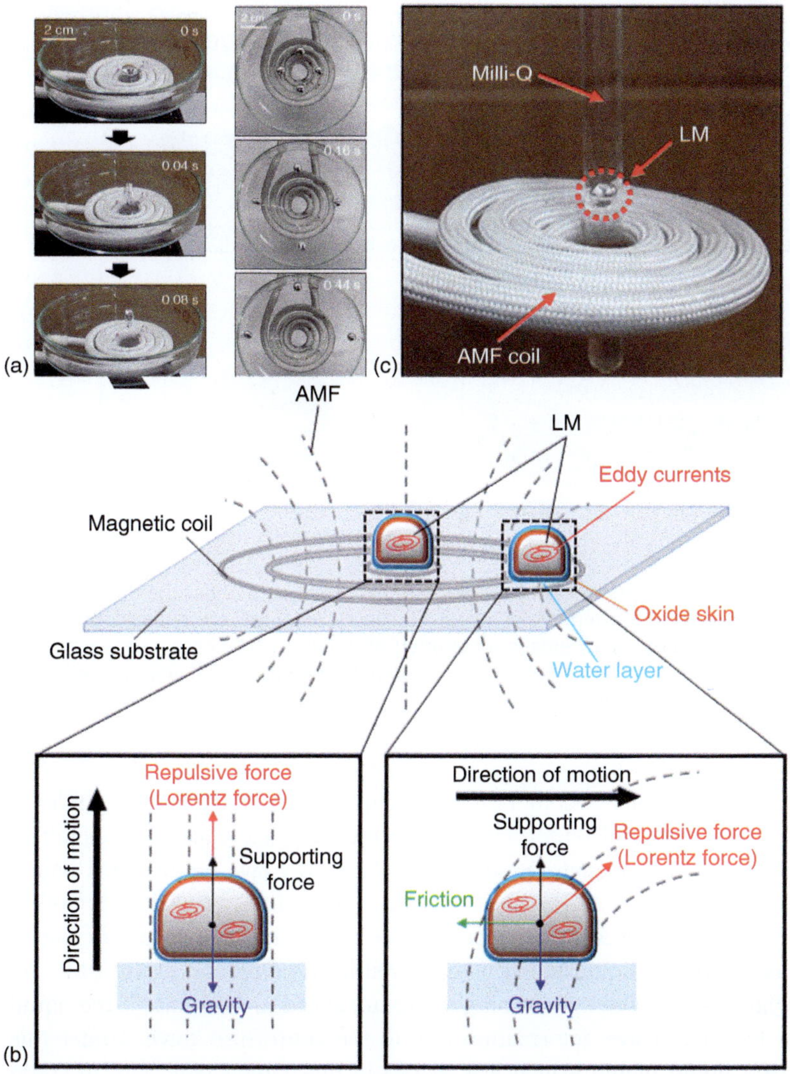

Figure 4.10 (a) Sequential snapshots of the locomotion behaviors of the EGaIn in horizontal and vertical directions under the AMF. (b) The force analysis of the EGaIn droplet under the manipulation of the AMF with the different direction. (c) The magnetic levitation behaviors of the EGaIn droplet in water after placing the droplet in the center of the coil. Source: Yu and Miyako [6].

in the water upon the continuous exposure to the AMF (Figure 4.10c). Such a levitation behavior was ascribed to the Brownian motion of the surrounding water of the liquid metal, which suppressed the horizontal oscillation of the droplet. Thus, the Lorentz force in the vertical direction could overcome the gravity of the liquid metal.

What's more, as for the bare liquid metal, Zhao and coworkers discovered a surprising phenomenon that a liquid metal (EGaIn) droplet could maintain

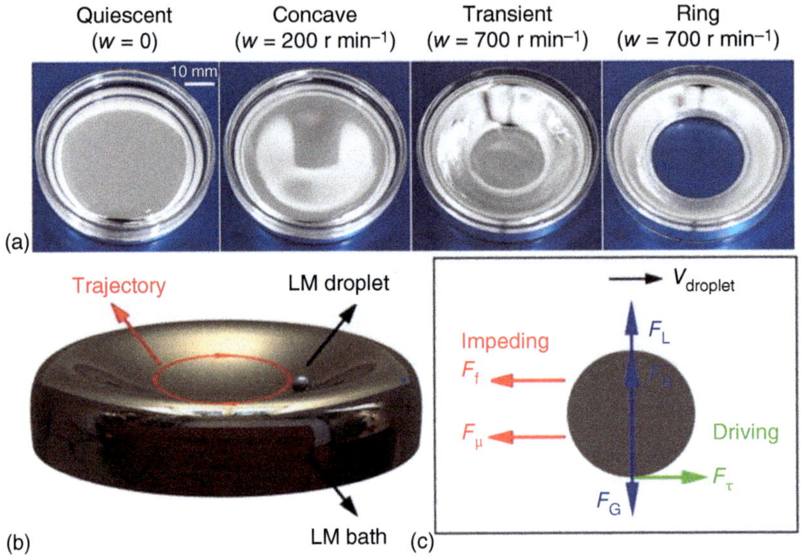

Figure 4.11 (a) The different morphology of the liquid metal bath under the different rotary speed of magnet. (b) Schematic of a delayed-coalescent liquid metal droplet on the concave liquid metal bath. (c) Schematic illustration of the noncoalescence mechanism and force analysis of a levitating liquid metal droplet. Source: Zhao et al. [58]. Reproduced with permission from AIP Publishing.

noncoalescent and rotary locomotion on a liquid metal/NaOH bath under a rotating magnetic field (RMF) [58]. The RMF formed on the x–y plane was obtained by the rotation of two magnets. Based on the Faraday's law, liquid metal would generate currents in z-axis under the RMF. The created currents produced another magnetic field that opposed RMF, appearing as a Lorentz force perpendicular to the currents. Thus, the liquid metal bath flowed in the direction of the RMF. As shown in Figure 4.11a, when increasing the rotary speed to 200 r min^{-1}, the liquid metal bath had a concave deformation due to the centrifugal force. Under this condition, if a liquid metal droplet was placed in the liquid metal bath, it could rotate along the concave surface for several seconds (Figure 4.11b). The reason for the noncoalescence was that the droplet and bath were separated by a film of the NaOH solution. This film provided a viscous shear force (F_τ) on the droplet, which served as a driving force to induce the rotational motion (Figure 4.11c). Moreover, the droplet also suffered from the resistance of the surrounding NaOH solution (F_μ) and the viscous friction between the solution and container wall (F_f). First, the speed of F_τ was the largest because it increased with the velocity of the bath. When the speed of the droplet continually increased, the decreased velocity difference between the droplet and bath would result in an increased resistance and decreased driving force. Finally, the droplet reached a stable state when the force on the droplet was counterbalanced each other. This noncoalescent effect is of great interests in fluid dynamic, providing a possibility in self-actuated liquid metal motors.

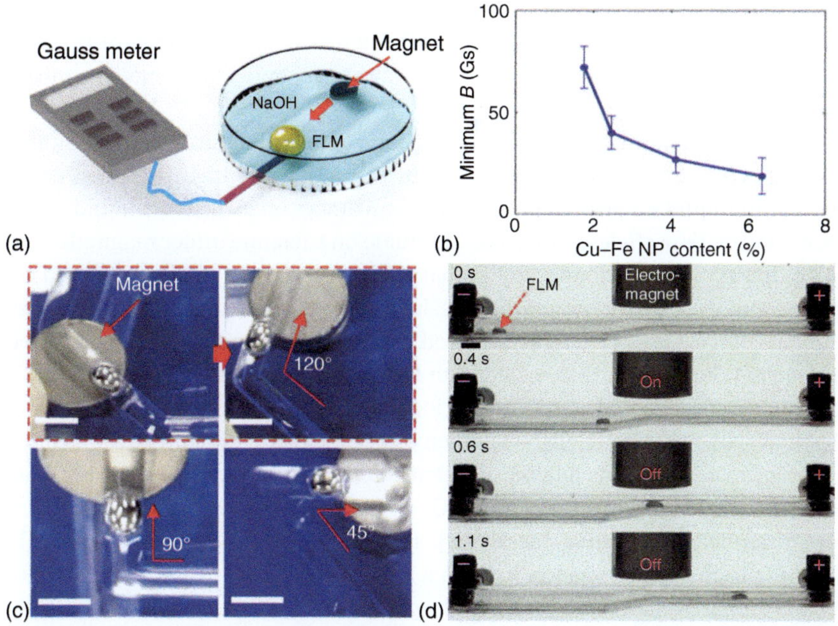

Figure 4.12 (a) Schematic of the experimental method used for magnetic manipulation process. (b) The minimum magnetic field intensity required to drive the magnetic liquid metal droplets (bottom). Source: (a, b) Li et al. [60]. Reproduced with permission from John Wiley & Sons. (c) Magnetic locomotion process of the droplet in crooked channels. (c) The climbing locomotion of the magnetic liquid metal droplet under magnetic field [60]. (d) The climbing locomotion of the magnetic liquid metal droplet under magnetic field. Source: (c, d) Li et al. [60].

Combining liquid metals with ferromagnetic materials is a more efficient and promising way to realize the locomotion of liquid metals. Locomotion is an essential ability for magnetic materials-based motors to navigate in specified trajectories. As aforementioned, when the f_m overcomes the f_r but less than the maximum value of f_c, the locomotion will occur rather than deformation behavior. Moreover, in the absence of the oxide layer, liquid metals flow into a sphere that minimizes the interfacial free energy [59], which is beneficial to the locomotion of liquid metals under a magnetic field. Based on this, Li and coworkers [60] reported that magnetic liquid metal comprising GaInSn and Cu–Fe particles enabled precise actuation within a cross-linked channel on a NaOH solution. Figure 4.12a showed the minimum magnetic field required to drive the GaInSn/Cu–Fe droplet (100 μl) as a function of the content of Cu–Fe particles. The minimum magnetic field decreased with the increased amounts of Cu–Fe particle since the droplet was subjected to a larger magnetic force (Figure 4.12b). The locomotion of the droplet within crooked channels was demonstrated in Figure 4.12c, in which it was clearly seen that the droplet could easily pass through the turns toward the different angles. These results revealed that the magnetic liquid metal possessed similar mobility, deformation capacity to those of bare liquid metal. Due to the precise controllability of magnetic field, magnetic force was also able to exert on the droplet to overcome the own gravity and further implement the climbing motion (Figure 4.12d). Such an approach offers the promise to develop liquid metal-based soft robots for a variety of applications such as control tools for transport carrier and drug delivery.

The locomotion of liquid metals upon magnetic field can be further tuned by surface coating, which endows liquid metals a superior mobility with infinite degrees of freedom in nonliquid environment. Jeon and coworkers investigated a liquid metal marble coated by Fe particles, which could actuate on substrate under magnetic field [61]. The driving force generated by Fe particles under magnetic field exerted on the marble and then drove the marble. Besides, composite coating can also realize on-demand control of the liquid metal-based marbles. Chen and coworkers reported a liquid metal (GaInSn) marble coated with ferronickel (FN) and polyethylene (PE) microparticles, which could actuate on various substrate such as fiber, glass, Cu, and PMMA under magnetic field [62]. The schematic of the fabrication process was illustrated in Figure 4.13a. It was noteworthy that NaOH solution was used to eliminate the oxide layer and later the created $[Ga(OH)_4]^-$ antioxidation layer could prevent further oxidation of the liquid metal [63]. Hence, the liquid metal droplet could maintain the high surface tension to form a spherical shape. The contact angles of the marble prepared in this method were larger than 150° on the different substrate (Figure 4.13b), exhibiting nonwetting properties. Similar to the case of aqueous environment, the marble could move with the magnet. In addition, Figure 4.13c showed that the marble was able to be actuated by magnet along the inner side of the wheel, changing the center of gravity and then causing the wheel to roll. Nevertheless, in the matter of this approach, Fe particles easily fell off from the surface of marbles after repeatedly actuating under magnetic field, causing the instability or even the loss of drive capability of the marble. Another approach was that ferromagnetic particles were suspended in liquid metal matrix and then hydrophobic particles were coated on the surface of the liquid metal droplet. Zhang and coworkers reported a graphite-coated EGaIn/Fe marble [64]. When the marble was confined to a narrow conductive gradient channel, it could reversibly move along the channel over 500 cycles by changing the magnetic field. After integrating this reversible locomotion behavior into batteries, the on-demand switching of batteries could be realized.

In addition, ferromagnetic liquid metal composites are not merely limited to the droplet-based locomotion. Cao and coworkers coated the EGaIn/ferromagnetic NdFeB on the paper, which carried out an inchworm-like locomotion under the manipulation of RMF [65]. Because of the oxidation of liquid metal, EGaIn/NdFeB had a good wetting with the paper. The magnetic polarity of the magnetic paper was designed to ternary form, where the polarity in middle part was reversed and the direction was consistent with the magnetized microparticles under a strong magnetic field. As demonstrated in Figure 4.14a, the synergistic and alternate anchor-pull and anchor-push actions triggered the paper to complete the motion. As for the anchor-pull actions, the magnetic repulsion force induced by the applied the magnetic field raised the body, thus deforming and contracting the body. However, the magnetic attraction force weakened the front and back feet anchor and slide. When the magnet was positioned near the bottom back feet, the magnetic

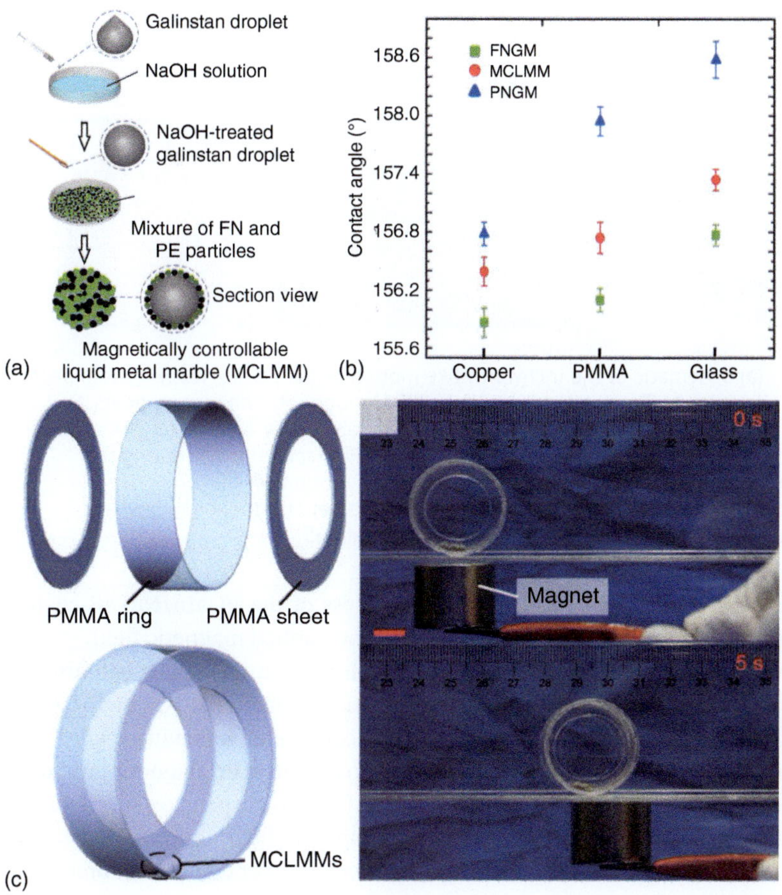

Figure 4.13 (a) Schematic of the fabrication process of the magnetic liquid metal marble. (b) The contact angles of three types marble including ferronickel-coated NaOH-treated GaInSn marble, magnetically controllable liquid metal marble, and PE-coated NaOH-treated GaInSn marble in the different substrates. (c) The motion of the wheel under the manipulation of magnet. Source: Chen et al. [62]. Reproduced with permission from John Wiley & Sons.

repulsion exerted on the body disappeared and meanwhile its deformation was released to cause the movement of the feet on both sides. The magnetic attraction between the magnet and the back feet was enhanced to prevent its movement toward the back. This moment, the front feet anchor was released to prompt its movement toward the front. To further achieve the continuous locomotion, an RMF was designed to actuate the magnetic paper (Figure 4.14b). The magnetic flux density distribution on the surface confirmed that the magnetic polarity in the middle part was opposite with that of the front and back under the RMF (Figure 4.14c), thus inducing the inchworm behaviors of the magnetic paper (Figure 4.14d). Remarkably, the crawling speed of the paper could reach 29.7 mm s^{-1} (Figure 4.14e).

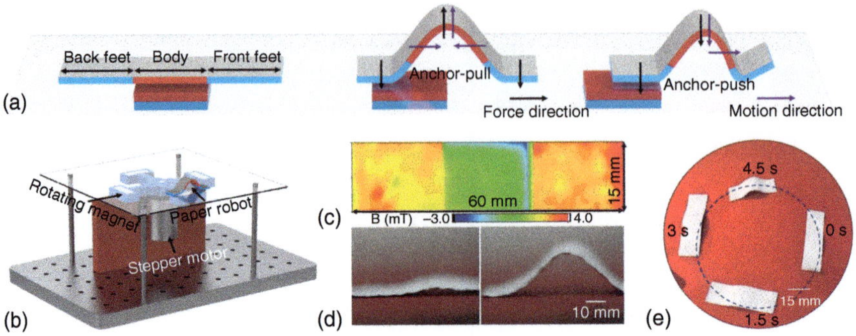

Figure 4.14 (a) Schematic of the inchworm-like motion of the EGaIn/NdFeB-coated paper under magnetic field. (b) The production of the RMF through rotating four magnets. Source: (a, b) Cao et al. [65]. Reproduced with permission from John Wiley & Sons. (c) The magnetic flux density distribution on the surface of the EGaIn/NdFeB-coated paper during the inchworm process. (d) The anchor-pull and anchor-push motions of the EGaIn/NdFeB-coated paper during the inchworm process. (e) The circling movement induced by an RMF [65]. Source: (c, d, e) Source: Cao et al. [65]

And the speed is mainly determined by the stiffness of the EGaIn/NdFeB-coated paper and the interaction between the paper and the applied magnetic field.

In summary, like the deformation, the locomotion of liquid metals can also be achieved by electromagnetic effect and ferromagnetic materials. As for the electromagnetic effect, the generated Lorentz force is capable of achieving planar and rotary locomotion. And for another approach, due to the intrinsic magnetic guidance ability of ferromagnetic materials, liquid metals after the suspension of magnetic particles can be actuated as on-demand under magnetic field. Remarkably, based on this method, liquid metals after coating a nonwetting layer can also locomote on various substrates.

4.3.3 Magnetic-Tuned Properties

Liquid metals are an electron-rich metallic conductor [66], consisting of the charged clusters, charged atoms, and free electrons. Under an external magnetic field, liquid metals will produce inductive current, which is able to change the properties of liquid metals. For example, AMF can induce eddy currents in a conductor with closed loops. The current flowing through the resistance of liquid metals can dissipate energy, generating the inductive heat. And the inductive heating property can be mediated by the AMF. Wang and coworkers characterized the electromagnetic heating process of EGaIn [13]. As displayed in Figure 4.15a, the thermal infrared (IR) images showed that the EGaIn was clearly heated and the temperature changed from 21.1 to 41.6 °C under the exposure of the AMF (100 kHz). Nevertheless, the temperature of the oxide layer without the liquid metal metallic core remained unchanged. This result demonstrated that the electromagnetic heating effect was because the electron-rich metallic cores induced the eddy currents under an AMF. Besides, Wang and coworkers reported that the temperature of the liquid metal (EGaIn) could be

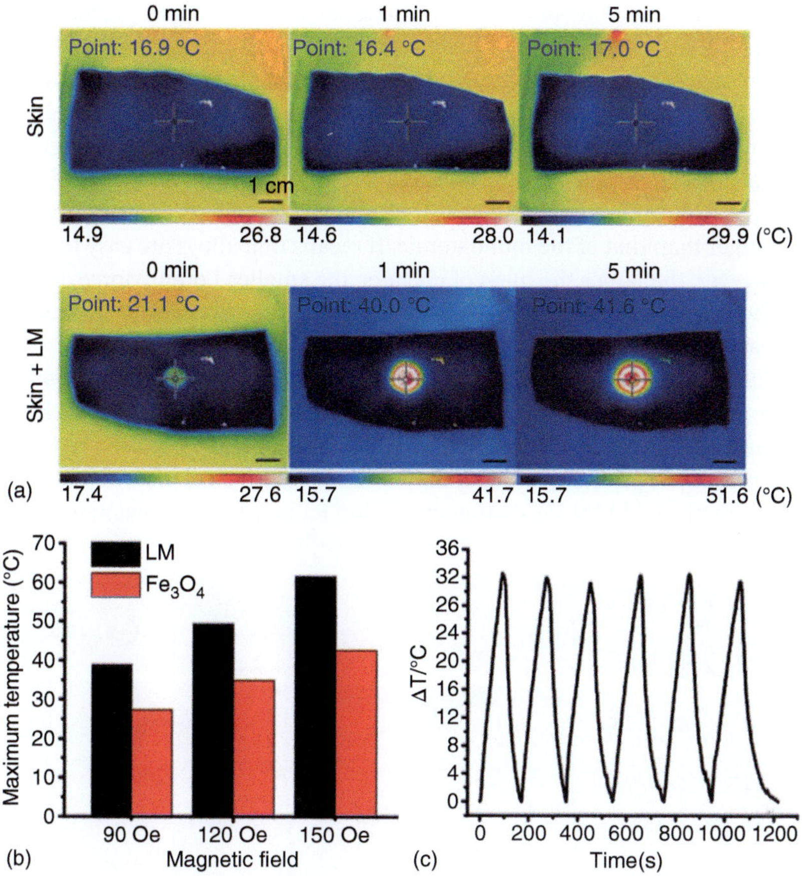

Figure 4.15 (a) Inductive heat effect with and without EGaIn on the pigskin surface under AMF [13]. Source: Wang et al. [67]. (b) The maximum temperature of EGaIn and Fe_3O_4 dependence on magnetic field. (c) The temperature variation under on-and-off AMF. Source: (b, c) Wang et al. [67]. Reproduced with permission from John Wiley & Sons.

tuned by the intensity of the AMF (Figure 4.15b) [67]. The temperature increased with the increased intensity of the AMF. Compared to the traditional superparamagnetic thermal agents of Fe_3O_4 particles with the same agent content, the higher temperature of the liquid metal could be obtained. Moreover, the continuous on/off the power of the AMF had no influence on inductive heating ability (Figure 4.15c), exhibiting a stable property. In addition, Yu and coworkers found the temperature of the liquid metal (40 μl) increased from 25.6 to 198 °C for 3 seconds under the AMF (250 A m^{-1}, 245 kHz), showing a fast speed [6]. All these extraordinary properties of liquid metals have sparked considerable interests in cancer magnetic hyperthermia, computed tomography (CT) imaging, and so on.

The viscosity can also be tuned by external magnetic field. At the superheated temperatures of liquid metals, in the flowing process of monoatomic liquid metals or alloys, the generated Lorentz force upon external magnetic field is opposite to the direction of motion. The stronger Lorentz force as a drag force prevents the shear flow, resulting in the weakly flowing ability of liquid metals. Thereby, the viscosity increases at the macrolevel. Bian et al. verified that the viscosity of pure Cu and Cu-10% Sn increased with the increasing strength of magnetic field [68]. This phenomenon also existed in other liquid metals such as Ga, In, Bi, Ag, Sb, Al, and so on [69–71]. Moreover, the viscosity increment of the monoatomic liquid metals is larger than that of the alloys. The binding force between the different atoms is stronger than that of the monoatomic. It means that alloys are easy to form clusters. However, the larger the mass of clusters, the smaller Lorentz force can be obtained. Thus, magnetic field is of a greater influence on the viscosity increment of the monoatomic liquid metals.

When liquid metals mix with the ferromagnetic materials, the stiffness, viscosity, and viscoelastic is able to be reversibly tuned and subtly controlled over a wider range. Ren and coworkers developed a liquid metal-based magnetoactive fluid prepared by dispersing Fe particles in a GaInSn matrix to describe the viscosity properties under magnetic field [72]. As demonstrated in Figure 4.16a, the magnetoactive

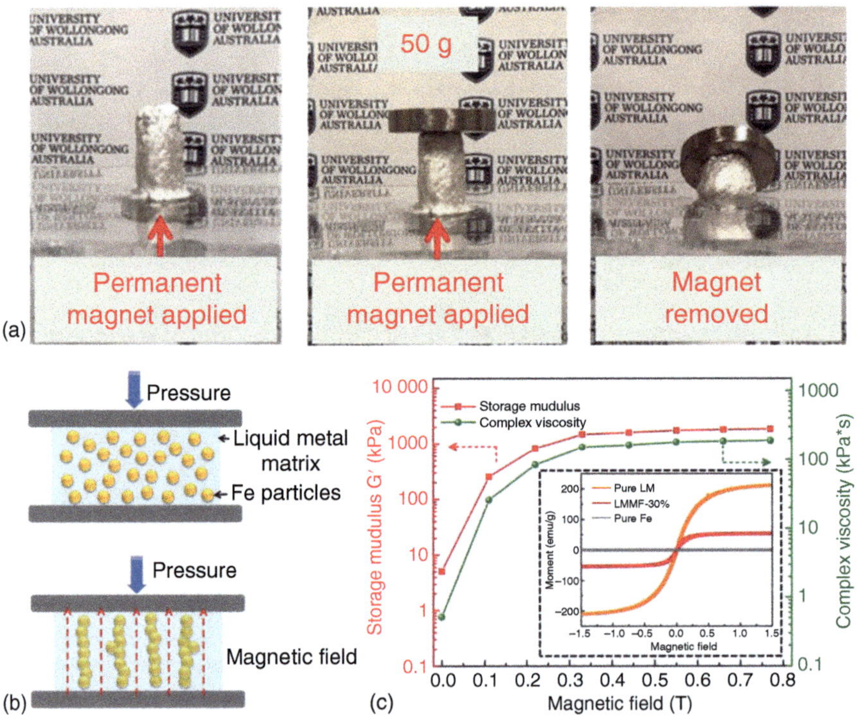

Figure 4.16 (a) The viscosity variation of the GaInSn/Fe mixture under magnetic field. Source: Ren et al. [72]. (b) Schematic of the migration of Fe particles in GaInSn/Fe mixture with and without magnetic field applied. (c) The storage modulus G' and complex viscosity at 0.2% strain amplitude versus magnetic field for the sample containing GaInSn/Fe mixture with Fe amounts of 30 wt.%. The inset diagram is magnetization curves of pure GaInSn, pure Fe, and GaInSn/Fe mixture with the Fe amounts of 30 wt.%. Source: Insert (b, c) Ren et al. [72]. Reproduced with permission of Wiley-VCH.

fluids molded into a cylinder with a surface area of 1.13×10^{-4} m^2 was able to support a weight above 50 g under the magnetic flux density of 0.2 Tesla (T). However, when the magnetic field was removed, the stiffness of this cylinder disappeared and the fluids recovered to its initial soft characteristics, revealing that the viscosity obviously increased under magnetic field. This viscosity change was ascribed to the dipolar interactions induced by the magnetic force, generating the chain-like alignments in the liquid metal matrix (Figure 4.16b) [73]. To further explore the viscoelastic properties of the fluids, oscillatory tests containing amplitude sweep and magnetosweep were presented on the sample (Figure 4.16c). The storage modulus and complex viscosity showed an obvious increase, suggesting that the liquid state of the fluids converted into a solid state. It was worth noting that the elastic modulus of the magnetoactive fluids with Fe content of 30 wt.% could be adjusted from 5 kPa to more than 1 MPa, revealing a mechanical adaptive response behavior. The similar properties could also be observed in liquid metal/Ni magnetoactive fluids. But liquid metal dispersing Fe particles showed a larger increment than that of Ni particles due to a more dispersing stability of Ni particles [9, 74]. This adjustable stiffness and viscosity of the liquid metal-based magnetoactive fluids are capable of applying in multifunctional-implanted electrodes in biosystems.

To sum up, magnetic-tuned properties mainly focus on the bare liquid metals and liquid metal composite. With regard to the bare liquid metals, the inductive heating property and viscosity can be adjusted by electromagnetic field. And for the liquid metal composite, the extra added magnetic particles will generate dipolar interactions under magnetic field, which can change the properties of liquid metals such as stiffness, viscosity, and viscoelastic. These property changes tuned by magnetic field exhibit a great potential in biomedical field.

4.4 Chemical Tuning

Chemical tuning introduces the chemical reagent, which constructs a dynamic interaction with liquid metals to realize the motion including deformation and locomotion or the properties change. Owing to the diversity and designability, chemical tuning is one of the most versatile approaches. On the one hand, chemical tuning is capable of spontaneously converting chemical energy into mechanical activity to induce the autonomous motion including deformation and locomotion. The detailed mechanisms usually involve the tuning of interfacial tension, phase transition, electric potential, and bubble propulsion. On the other hand, liquid metals are easy to tune the own properties by means of self-limiting reaction, doping, or surface modification. In this section, we will discuss typical strategies of chemical tuning from the perspective of deformation, locomotion, and properties.

4.4.1 Chemical-Induced Deformation

Chemical-induced deformation of liquid metals utilizes interfacial tension difference or phase change as the mechanism. Constructing galvanic cell is a main and effective way to tune the deformation behavior. Wang and coworkers designed a graphite/GaInSn galvanic cell, where the GaInSn droplet was capable of spreading on graphite in the NaOH solution [9]. Before their immersion in NaOH solution, GaInSn directly contacted with graphite surface. The expression of the contact potential difference (V_{lg}) between the GaInSn and graphite showed as follows:

$$V_{lg} = V_l - V_g = -\frac{1}{e}(\Phi_l - \Phi_g) \tag{4.6}$$

where V_l and V_g is the potential of the GaInSn and graphite, respectively, Φ_l and Φ_g is the work function of the GaInSn and graphite, respectively, and e is the electron charge. The work function of the GaInSn (each component smaller than 4.4 eV) [75] is lower than the graphite (4.7 eV) [9]. Thus, the voltage between the GaInSn and the graphite was a positive value, which accords with the voltage curve in Figure 4.17a. When the GaInSn was submerged in NaOH solution, the oxide layer was removed and the $[Ga(OH)_4]^-$ was produced. In contrast, the graphite possessed positive charges. Thus, the opposite charges would respectively accumulate on the GaInSn and graphite to form an EDL (Figure 4.17b). In this condition, the measured voltage between the GaInSn and graphite was negative (Figure 4.17a). That means the surface potential of the GaInSn was lower than that of graphite. The electrons

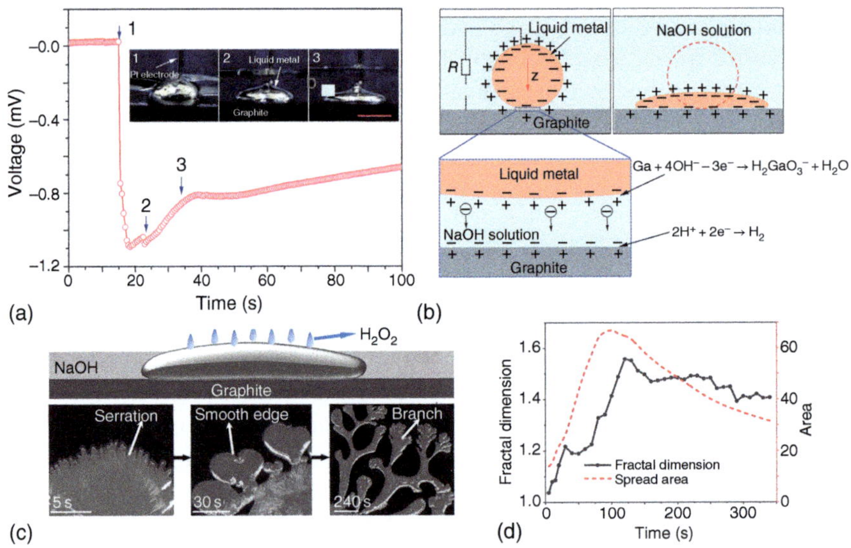

Figure 4.17 (a) The dynamic change of the voltage between the GaInSn and graphite during the immersion process from air to NaOH solution. The illustrations are GaInSn images at the time points of 1, 2, and 3, respectively. Scale bar is 3 mm. (b) Schematic of the charge distribution on the GaInSn surface in the original state. The enlarged image is the charge distributions, as well as the chemical reactions on the graphite/NaOH solution and the GaInSn/NaOH solution EDLs. Source: (a, b) Wang and Liu [9]. Reproduced with permission of Royal Society of Chemistry. (c) The deformation process consists of serrated protrusions, smooth edge, and dendritic structure. (d) The fractal dimension of the liquid metal fractals during the dynamic process in virtue of box-counting calculation. Source: (c, d) Chen et al. [76].

were prone to transfer from the GaInSn to graphite through the electrolyte of the NaOH solution. The GaInSn, graphite, and NaOH electrolyte served as cathode, anode, and electrolyte, respectively, to form a galvanic cell. A downward direction of the potential gradient existed on the GaInSn droplet. On the basis of the Lippmann's equation, the surface tension can be expressed as:

$$\gamma = \gamma_0 - \frac{1}{2}cV^2 \tag{4.7}$$

where γ and γ_0 is the surface tension and the intrinsic surface tension of the GaInSn (i.e., $V = 0$), c is the capacitance of EDL per unit area [77], and V is the potential difference across the EDL. Taking the derivation on both sides of the above equation can be obtained:

$$\frac{d\gamma}{dz} = -cV\frac{dV}{dz} \tag{4.8}$$

Supposing the direction of V is from the NaOH electrolyte to the GaInSn, and then V is positive. Therefore, dV/dz is negative while $d\gamma/dz$ is positive, which indicates that the surface tension of the GaInSn on the upper hemisphere is lower than another part, inducing the spreading phenomenon. In addition, this spreading behavior could be also realized in other substrates, such as Cu and other electric conductive substrates. But the deformation behavior could not happen in insulating substrate such as Teflon because no galvanic effect was generated.

Based on the galvanic cell effect of graphite/liquid metal, by introducing the hydrogen peroxide into systems, a more complex fractal phenomenon can be implemented. Chen and coworkers first reported an interesting liquid metal (EGaIn) fractal phenomenon driven by synergistic oxidation [76]. Figure 4.17c showed that the evolution process of the fractal phenomenon consisted of three stages. When the hydrogen peroxide (H_2O_2) was added on the EGaIn droplet, the serrated protrusions appeared in the edge of the droplet, then changed into a dendritic structure, and finally became a fractal shape. The main reason for the fractal was that H_2O_2 and graphite cooperated to oxidize the upper and bottom of the droplet, respectively, which effectively reduced the surface tension. And the oxidative stress caused by H_2O_2 played a key role in fractal phenomenon. Generally, liquid metal was prone to keep an ellipsoid because of the high surface tension. Whereas the oxidative stress tended to spread the liquid metal. Thus, the oxidative stress and the surface tension competed with each other, which was beneficial to the generation of serrated pattern. After the addition of H_2O_2, some serrated protrusions developed in virtue of synergistic oxidation, but other parts of serrated protrusions disappeared owing to the high surface tension of the liquid metal. Later, these developed serrated protrusions further spread into a smooth edge or even branch. Figure 4.17d expressed that the largest fractal dimension (reflected the validity of the dendritic structure in space) could reach to 1.56 when the spreading area of the liquid metal attained the maximum value. Compared to another way to implement fractal phenomenon (electric field) [78], this approach yielded a larger spread dimension.

The abovementioned graphite/liquid metal galvanic cell is based on local surface or point contact. Constructing a galvanic cell with full surface contact can carry

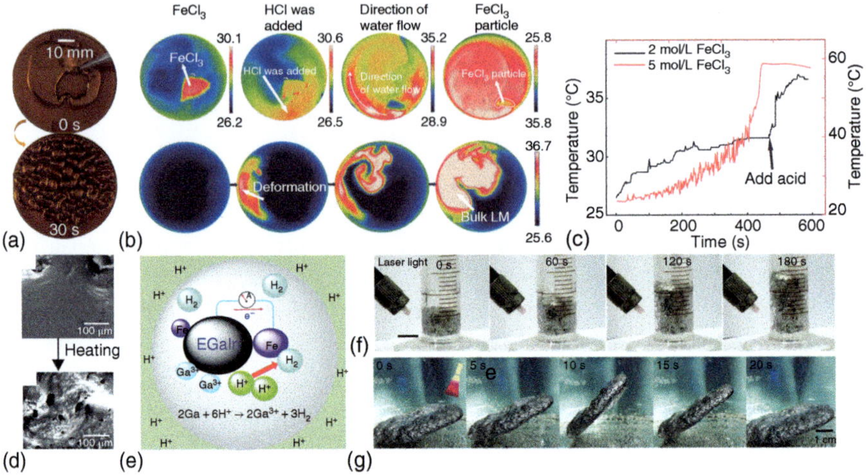

Figure 4.18 (a) Spontaneous dispersion behavior of the liquid metal triggered by 4 mol l^{-1} $FeCl_3$ solution. (b) Infrared images of convection and deformation of the liquid metal in the 5 mol l^{-1} $FeCl_3$ solution. Source: (a, b) Chen and Liu [79]. (c) Temperature variation of the system with the different $FeCl_3$ concentrations during the deformation process. Source: Chen and Liu [79]. Reproduced with permission of American Chemical Society. (d) The SEM image of the EGaIn/Fe before and after reaction. (e) Schematic of galvanic cells between Ga and Fe in a 2 mol l^{-1} HCl solution. Source: Wang et al. [80]. Reproduced with permission from Royal Society of Chemistry. (f) The expansion process of liquid metal/Fe under the laser light irradiation. (g) Floating behaviors of liquid metal/Fe under the local laser light irradiation. Source: (f, g) Wang et al. [80].

out a deformation behavior in a large-scale. Chen and coworkers discovered a spontaneous dispersion and deformation behavior when the bulk EGaIn was absolutely immersed in ferric chloride ($FeCl_3$) solution (Figure 4.18a) [79]. Notably, experiments found that the amount of added acid had a great effect on this deformation behavior. The temperature distribution and variation of the liquid metal was employed to observe the deformation process, which was demonstrated in Figure 4.18b,c. It was clearly seen that the temperature significantly increased and the intense oscillation of the liquid metal appeared after the addition of HCl. The infrared image with the red and white parts owned the higher temperature, which corresponded to the oscillation and the deformation of the liquid metal. This represented that the deformation was connected with the chemical reactions. Ga had a stronger reducibility than In owing to the lower standard electrode potential (ε). In addition, compared to Fe^0/Fe^{3+}($\varepsilon = -0.037$ V), Ga^0/Ga^{3+} ($\varepsilon = -0.549$ V) had a more negative standard potential [81]. Thus, the chemical reaction could happen:

$$Fe^{3+} + Ga \rightarrow Ga^{3+} + Fe \tag{4.9}$$

However, owing to the acid environment, the produced Fe could be transformed into Fe^{2+}.

$$Fe + 2H^+ \rightarrow Fe^{2+} + H_2 \uparrow \tag{4.10}$$

$$Fe + 2Fe^{3+} \rightarrow 3Fe^{2+} \tag{4.11}$$

As the Fe^{3+} was consumed completely, another chemical reaction happened:

$$3Fe^{2+} + 2Ga \rightarrow 2Ga^{3+} + 3Fe \tag{4.12}$$

According to the earlier analysis , the precipitated Fe could not swallow or cover on the surface of the liquid metal, ensuring the continuity of the deformation. On account of the space–time distribution of $FeCl_3$ solution and galvanic effect, a nonuniform interfacial tension occurred, which led to the spontaneous dispersion behaviors. If the liquid metal was removed from an open space, the snakelike locomotion could be realized. Chen and coworkers had published similar works previously [82].

Regulating the interfacial tension is an effective way to realize the deformation. Another phase-change-based method is capable of obtaining an extraordinary deformation behavior. A porous liquid metal foam through loading EGaIn alloys with chemically reactive Fe particles was obtained by Wang and co-workers [80]. Figure 4.18d showed that liquid metal foam had a porous and paste-like appearance. This structure was attributed to the gas filling and phase changes. The gas composed of two parts. On the one hand, Fe particles on the surface of EGaIn in contact with the HCl solution could produce a mass of bubbles (H_2). On the other hand, when EGaIn/Fe was immersed in HCl solution, Ga tended to form a galvanic cell with Fe. Ga and Fe were served as cathode and anode, respectively. The schematic of a Ga–Fe galvanic cell was demonstrated in Figure 4.18e. The chemical reaction on the liquid metal could occur:

$$2Ga + 6H^{+} \rightarrow 2Ga^{3+} + 3H_2 \uparrow \tag{4.13}$$

The above created gases dispersed in the bulk phase of the liquid metal, which could induce the porous structure of the liquid metal. Meanwhile, this structure of the liquid metal would increase the contact area with air, increasing the degree of the Ga oxidation [83]. The liquid metal transformed from the liquid into solid state. The generated mixture including Ga_2O_3, Ga_2O, and metallic Ga could prevent the bubbles from escaping owing to the high viscosity, which prompted the formation of the porous liquid metal foam. More importantly, liquid metal would deform in a larger scale at high temperature because of a faster reaction process. When exposed to laser light, liquid metal significantly expanded and the expansion coefficients of the liquid metal could reach the upper limit (685%) at 100 °C (Figure 4.18f). This expansion effect was able to induce the liquid metal float upward (Figure 4.18g). Moreover, the porous foam could be transformed into the initial liquid state again by decreasing the temperature and stirring.

In summary, chemical-induced deformation usually involves the tuning of interfacial tension and phase change. The deformation of liquid metals derived from the tuning of interfacial tension is normally in the electrolyte environment. Most of them are realized by constructing a galvanic cell between liquid metals and chemical reagent. Chemical-based approach can realize the diverse deformation behaviors of liquid metals such as spreading phenomenon, fractal phenomenon, droplet dispersion, and porous liquid metal foam.

4.4.2 Chemical-Induced Locomotion

In addition to the deformation, chemical tuning can also induce the locomotion behaviors. The most studied mechanisms of locomotion was triggered by chemical including interfacial tension, bubble propulsion, self-electrophoresis, and self-diffusiophoresis [84]. First, for the design of interfacial tension that causes locomotion, a typical approach is Marangoni flow. For example, two different electrolytes with varying pH values were set on two sides of a GaInSn droplet, causing symmetry breaking that generates motion, which was proposed by Zavabeti and coworkers [85]. Here, these two electrolytes placed on both sides of the droplet would induce the ionic imbalance surrounding the GaInSn droplet. And later it could approach equilibrium state through the formation of an EDL (Figure 4.19a) The EDL could be simulated as a parallel-plate capacitor. According to the Lippmann's equation, we have:

$$\gamma = -\frac{C}{2}(\psi - \psi_0)^2 + \gamma_0 \tag{4.14}$$

where C is the capacitance of EDL per unit area, ψ and ψ_0 are the electrode potential and the potential (zero charge), respectively, and γ_0 is surface tension (i.e., $\psi = \psi_0$) [87]. If γ is known, the pressure difference (ΔP) across a GaInSn hemisphere can be expressed as:

$$\Delta P = \gamma \left(\frac{1}{R_1} + \frac{1}{R_2} \right) \tag{4.15}$$

where R_1 and R_2 are the curvature radii at the interface between the electrolyte and liquid metal [87]. From the earlier equations we observe that the surface tension offered a way to change the ΔP that in turn could induce a locomotion of the liquid metal. In addition, changes in surface tension could be triggered by modifying the electrolyte surrounding the liquid metal droplet. Based on it, Figure 4.19b showed the self-propulsion locomotion of the droplet in a HCl/NaOH solution with the concentration of 1.2 and 0.6 mol l^{-1}, respectively. This motion was caused by Marangoni flow that the interfacial region of NaOH with lower surface tension moved to the region of HCl with higher surface tension. In addition, the velocity of locomotion could be tuned by the HCl and NaOH solution concentration. As displayed in Figure 4.19c, a minimum pH difference to trigger the locomotion of a liquid metal droplet was 13. The average velocity of the liquid metal droplet increased with the enhancing HCl concentration until the concentration reached 1.2 mol l^{-1} and then the velocity decreased. The maximum velocity of locomotion could reach up to 25 mm s^{-1}, which was related to the NaOH concentration of 0.6 mol l^{-1}.

Besides the unidirectional locomotion, another back-and-forth motion can be also realized by chemical control. Yi and coworkers found an oscillation phenomenon when both sides of the liquid metal droplet were exposed to ambient air and immersed in the NaOH solution, respectively [86]. Such a unique phenomenon

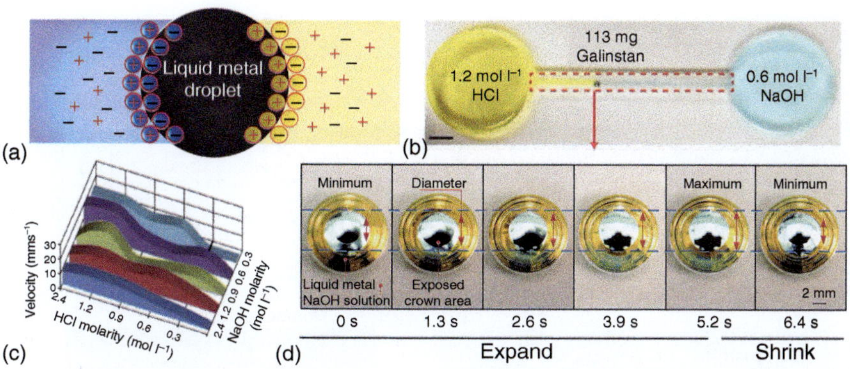

Figure 4.19 (a) Schematic of the formed EDL on the liquid metal droplet after its immersion in NaOH and HCl solution. (b) The propulsion phenomenon of the droplet with the direction from HCl ($1.2\,mol^{-1}$) to NaOH ($0.6\,mol^{-1}$) reservoir. Source: Zavabeti et al. [85]. (c) Average velocity of the liquid metal droplets under the different concentrations of HCl and NaOH solutions. Source: (a, c) Zavabeti et al. [85]. Licensed under CC BY 4.0. (d) The periodic oscillation of the liquid metal droplet. The red arrows represent the exposed area diameter of the liquid metal. The blue dotted lines are maximally exposed area diameter at different times. Source: Yi et al. [86].

resulted from Marangoni gradients comprising of the Ga oxide on the interface of the liquid metal droplet and Ga hydroxide at the interface of NaOH solution/air, offering the impetus source for the oscillation. The variations of the droplet dimension during the oscillation process were exhibited in Figure 4.19d. The exposed area of the droplet in air was capable of enlarging and shrinking spontaneously in the solution, showing a beating heart behavior. Besides, this behavior was able to be repeated several times. The mechanism for the oscillation process was due to the fact that the surface tension gradient perpendicular to the liquid metal/air/NaOH solution contact line, causing the liquid metal droplet and the surrounding NaOH solution to oscillate.

For locomotion, bubble propulsion and self-electrophoresis are also widespread mechanisms for liquid metals. A typical type is liquid-metal-based micron and nanomotors, which have chemically responsive segments with an inherent asymmetric nature. The motors can show motion, propulsion, and collective behaviors [84]. Because of the particle internalization effect of Ga-based liquid metals, they tend to combine with metals such as Al, Mg, and Zn [88]. These active metals are able to react with water or other fuels, resulting in microbubbles (e.g. O_2, H_2, CO_2) or a concentration gradient of molecules. These generated bubbles or concentration gradient of molecules can be regarded as energy sources to autonomously propel micro/nanomotors in a directional manner. Zhang and coworkers developed a self-fuel EGaIn motor in milli/centimeter scale, which was realized by feeding a small Al flake on one side of the motor as "energy sources" [7]. As shown in Figure 4.20a, liquid metal motor was capable of continuous swimming in different geometrical open-up channels that are filled with NaOH solution, such as circular channel, U-shaped channel, and zigzag channel. The diameter of channels was smaller than the liquid metal droplet to disconnect the solution. Thus, the liquid

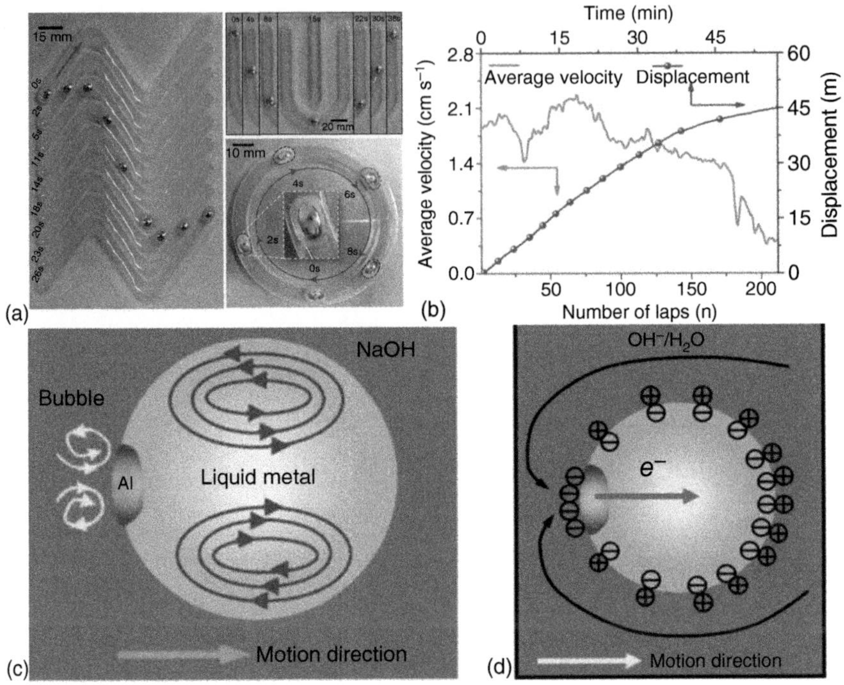

Figure 4.20 (a) Sequential snapshots of the locomotion of a liquid metal droplet in U-shaped, zigzag, and circular open-top channel. Source: Zhang et al. [7]. (b) The average velocity of the EGaIn as a function of laps and displacement of the EGaIn as a function of time. (c) Schematic of the locomotion of the EGaIn droplet and bubble departure. (d) Schematic of the electromigration and EDL change. Source: (b–d) Zhang et al. [7] Reproduced with permission from John Wiley & Sons.

metal motor needs to squeeze itself into an ellipse depending on its own deformability. Additionally, in circular channel, liquid metal motor could swim over 212 cycles with the average velocity of 1.35 cm s^{-1} and the lifetime sustained more than one hour (Figure 4.20b). In the back movement, the average velocity significantly decreased because the integrated Al flake cracked into a lot of small flakes and then covered the liquid metal surface, and finally breaking the asymmetry. The locomotion of the motor was ascribed to two reasons. On one hand, when liquid metal contacted with Al in electrolyte solution, liquid metal wetted aluminum (Al), destroyed the oxide layer on Al surface, and shifted the charges on the Al surface to more negative values. Thus, it would dramatically accelerate the redox reaction between the solution and Al to generate more bubbles [89, 90]. The chemical reaction on the Al side could be expressed as:

$$2Al + 2OH^- - 6e^- + 2H_2O \rightarrow 3AlO_2^- + 3H_2 \uparrow \quad (4.16)$$

Bubbles spiraled and departed from the motor to generate a recoil force, propelling the locomotion of the motor (Figure 4.20c). On the other hand, liquid metal combined with Al in electrolyte to form a galvanic cell, where Al and EGaIn were the cathode and anode, respectively. The lost electrons flowed from Al to

EGaIn. Additionally, Ga would react with NaOH, producing the $[Ga(OH)_4]^-$ and then leading to the surface of liquid metal with more negative charge. Thus, the positive charge accumulated on the surface of the liquid metal and then formed an EDL (Figure 4.20d). Importantly, the electrons derived from Al to EGaIn partially deoxidate the Ga oxide near the Al, altering the charge distribution (Figure 4.20c). Thus, a potential gradient appeared along the liquid metal motor surface, in which the potential of Al side was lower than that of the uncoated liquid metal motor side. Accordingly, the surface tension in Al side was smaller than the uncoated liquid metal side, inducing the locomotion from Al toward liquid metal side [17]. Later, based on this design, Liu's group found that the motor could be served as single or discrete micromotors to propulsion [91]. During the collision and coalescence process, micromotors could combine with each other and recover to its initial state, showing a potential in motors or micro-motors.

As the size of liquid metal-based motors drop down to micro- or nanoscale, it must overcome Brownian motion and the high viscosity environment. Utilizing the chemical or electrochemical energy provided by the Al reaction is difficult to obtain for a long lifetime. Another method is catalytic decomposition of fuels, for example, the widely reported H_2O_2. Wang and coworkers [8] presented a Janus GaInSn-based motors with the dimensions ranging from several hundred nanometers to several micrometers, in which the motors were produced by ultrasonic dispersion of bulk GaInSn and then coated with platinum (Pt) layer on one hemisphere (Figure 4.21a). Taking advantage of the high catalytic activity of Pt to H_2O_2, the chemical or electrochemical energy afforded by the catalytic decomposition reaction could drive the motor in an H_2O_2 solution without any supply of external energy. Besides, the oxide layer on the surface prevented the coalescence and meanwhile allowed electron transfer [92]. As demonstrated in Figure 4.21b, the velocities of the motors increased linearly from 1.0 ± 0.5 to $32.9 \pm 11\,\mu m\,s^{-1}$ when the H_2O_2 concentrations ranged from 0.25 to 30 wt.%. And, the insets in diagram were the trajectories of moving motor, which showed an obvious self-propulsion rather than Brownian-like motion. For the mechanism, bubble propulsion and the Marangoni effect could be ruled out because no bubbles were observed during the locomotion process and Marangoni effect could not occur in micro/nanoscale particles [93]. This propulsion mechanism could be explained as self-electrophoresis. Pt catalyzed the oxidation of H_2O_2 and produced excess protons. And meanwhile the electrons transferred from the Pt side to uncoated side and later reduced H_2O_2 (or O_2) into water. This asymmetric distribution of protons on both sides of liquid metal motors formed an electric field, which could serve as a driving force to propel the motors by self-electrophoresis. To further confirm the mechanism of self-electrophoresis, a current measurement between the Pt and liquid metal electrode was provided (Figure 4.21c). The current density was linearly proportional to H_2O_2 concentration, which was in accord with the relation between the velocities and H_2O_2 concentration. Thus, this results further confirmed that the electric field on the motors had a great influence on self-propulsion behaviors. Additionally, these motors could be used for the precise circuit repair by means of adhering at junction (Figure 4.21d),

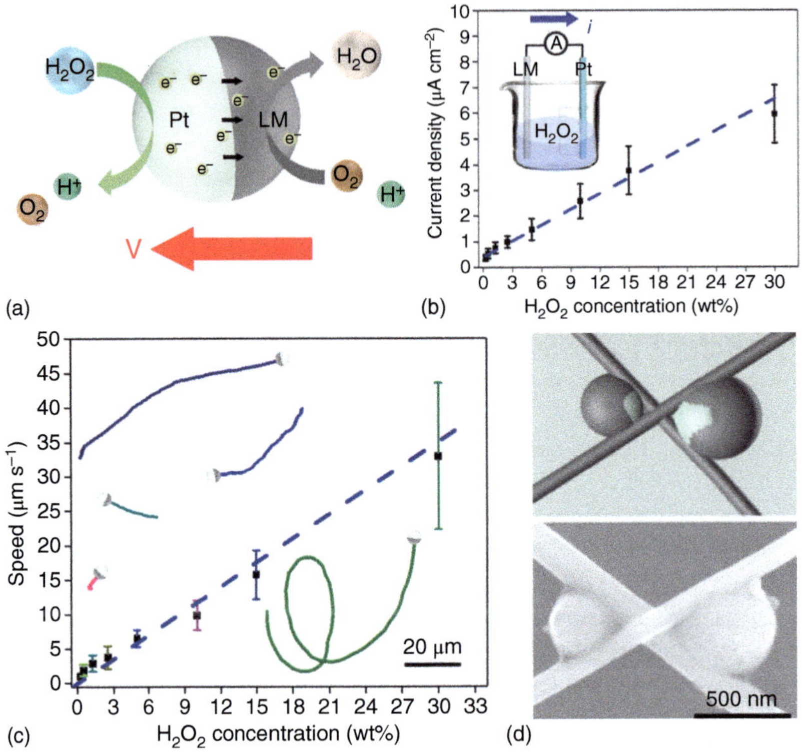

Figure 4.21 (a) Schematic of the Pt-coated GaInSn motor and self-propulsion mechanism. (b) Average velocity of Pt-coated GaInSn motor swimming in different H_2O_2 concentrations of and the inset is the corresponding trajectories during 6 s. (c) Electrochemical test of the current density of the GaInSn–Pt system for the different concentrations of H_2O_2; the inset is the experimental setup. (d) SEM images of Pt-coated GaInSn motors stuck at a silver nanowires junction before and after micro-welding. Source: Wang et al. [8]. Reproduced with permission from John Wiley & Sons.

which provided the inspiration to create innovative applications in fields such as material sciences, robotics, and drug delivery.

All in all, the locomotion of liquid metals upon the stimulation of chemical field was generally realized by designing a segment with an inherent asymmetric nature, for instance, different surface tension of chemical environment surrounding a liquid metal droplet, chemically active metals, or catalytically active coatings deposited on one side of liquid metal droplet. All these approaches should be achieved within liquid environment. However, chemical-based locomotion is of great potential in self-powered robots.

4.4.3 Chemical-Tuned Properties

For properties tuning, the most universal and common chemical approach is surface modification. Thereinto, oxidation is the main way reported to date. Researchers found that the oxide layer will decrease the surface tension [4, 39, 94] and damage

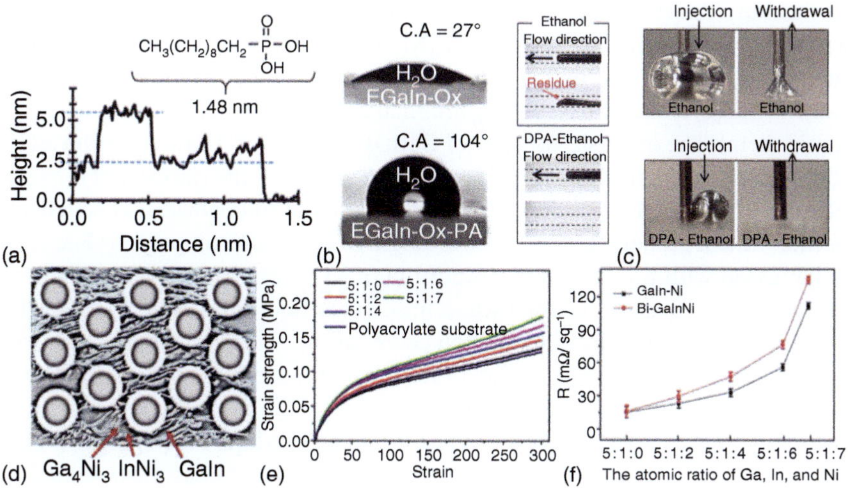

Figure 4.22 (a) Height profile in AFM image of the DPA on Ga oxide. Source: Ilyas et al. [95]. Reproduced with permission from John Wiley & Sons. (b) Contact angle of water on untreated and DPA treated EGaIn surfaces. And EGaIn flow through an ethanol-filled (top) and ethanol/DPA-filled (bottom) quartz tubes. (c) The extrusion and withdrawal measurement of the EGaIn closely contacted with quartz substrates in ethanol (top) and DPA/ethanol (bottom). Source: Ilyas et al. [95]. (d) Schematic of GaInNi structure. (e) The strain strength of biphasic GaInNi film coated on the substrate of polyacrylate versus strain under the different atomic ratios. (f) Initial resistance as function of atomic ratios of Ga, In, Ni before and after heat treatment. Source: (d–f) Wu et al. [96]. Reproduced with permission from John Wiley & Sons.

the electric conductivity of liquid metals. In addition, Ilyas and coworkers [95] proposed a surface modification approach, where the surface of EGaIn was modified by 1-decylphosphonic acid (DPA) to control interfacial properties. The phosphate groups adsorbed to Ga_2O_3 through the connection of hydrogen and covalent bonding, producing a highly stable organic layer on the surface of the liquid metal. The height profile in AFM showed that the thickness of a molecular bilayer equivalent film on the liquid metal was 1.48 nm and accompanied with a little multilayer structure (Figure 4.22a). After modification, liquid metal exhibited an obvious nonwetting state on the various substrates, where the contact angle changed from 27° to 104° (Figure 4.22b). Flowing tests presented that the modified liquid metal has no any physical adhesion within a quartz microchannel. But without the modification, a large amount of liquid metal remained in the channel walls. Besides, the extrusion and withdrawal experiments also confirmed that the presence of DPAs could effectively eliminate physical adhesion (Figure 4.22c), revealing the feasibility of this chemical modification. It is noteworthy that the different properties of liquid metals can be obtained by changing the surface coating.

Besides the surface modification, chemical additive can also adjust the properties of liquid metals. Wu and coworkers proposed a Ni doping strategy, where the amorphous liquid EGaIn was able to turn into a biphasic (liquid–solid) structure after the heat treatment process, which changed the viscosity, electric conductivity, and stretching characteristics [96]. Ni particles reacted with EGaIn and produced the solid phase of intermetallic compound (Ga_4Ni_3 and $InNi_3$). Thus, the biphasic structure consisted of solid phase of Ga_4Ni_3 and $InNi_3$, and liquid phase of EGaIn and Ga (bi-GaInNi) (Figure 4.22d). These structures had a great influence on the viscosity of the liquid metal. The intrinsic viscosity of GaIn was 1.99×10^{-3} Pa s. After the addition of Ni with the atomic ratios of nGa : nIn : nNi = 5 : 1 : 7, the viscosity of mixture increased to 159 ± 6.6 Pa s. Additionally, the formation of solid phase was beneficial to the mechanical properties. The strain strength of bi-GaInNi on polyacrylate substrate increased with the enhanced proportion of Ni atoms (Figure 4.22e). But for the electric conductivity, bi-GaInNi was a little lower than GaIn/Ni (Figure 4.22f). And the electric conductivity of bi-GaInNi decreased with the increased proportion of Ni. These results were due to the fact that electron transmission of bi-GaInNi was weaker than GaIn. Notably, this approach is not merely limited to Ni. Other materials including Cu, Al, Fe, and the like. can also be doped into liquid metal matrix. This doping strategy greatly tunes the properties of liquid metal and has the potential to apply in stretchable electronic.

To sum up, chemical-tuned properties of liquid metals are mainly divided into two ways including surface modification and chemical additive. For surface modification, self-limiting reaction is a common approach because the oxide skin of liquid metals can effectively change the properties of liquid metals. Moreover, liquid metals can also be modified by other materials to change the interfacial properties. In addition, another chemical additive way is adding particles into the bulk of liquid metals. And it mainly tunes the viscosity and mechanical properties of liquid metals. These properties tuning ways have been widely applied in the stretch electronic, flexible robot, and so on.

4.5 Force Field Tuning

External force is the most direct tuning to deform liquid metals. Due to the rheological property of liquid metals, the shape and position of liquid metals can be easily controlled. As long as external force is large enough to rupture their oxide layers, liquid metals can flow out and form a new shape coated with new oxide films. In this section, based on the relative direction of the applied forces and liquid metals, the forces are mainly classified into tensile force and pressure. The phenomena of force-induced deformation, as well as their mechanisms are discussed.

Tensile force is one of the most common external force that can deform liquid metals. As shown in Figure 4.23a, the EGaIn was injected into the core of stretchable hollow fibers [97]. It was noteworthy that the injection of EGaIn had a negligible effect on the mechanical properties of the fibers and the shape of EGaIn was absolutely determined by the fiber. When stretched, the EGaIn that was injected into a stretchable hollow fibers could flow and deform. From the cross section of fibers, it was clearly seen that the cross section of EGaIn was transformed from a triangle to circle, indicating that the rheologic EGaIn could adapt to the changes in external forces. After the tensile force was removed, the EGaIn could shrink with the fiber

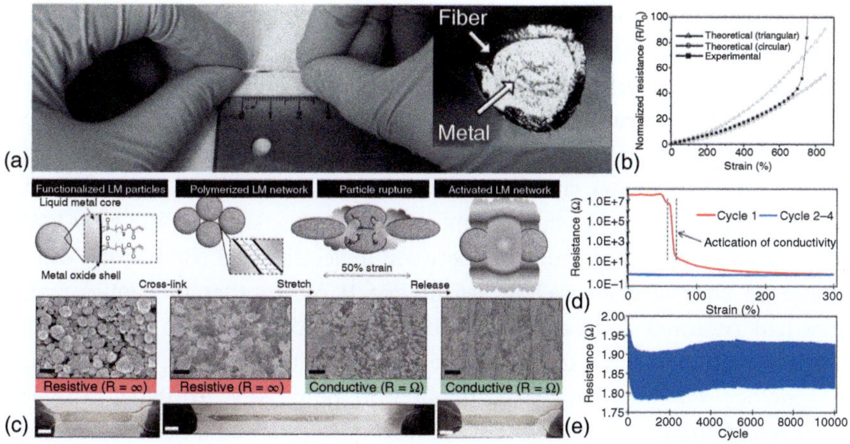

Figure 4.23 (a) An EGaIn-injected stretchable fiber and the insets is the crosssection. Source: Zhu et al. [97]. (b) Theoretical and experimental resistance of fibers versus strain measurements. Source: Zhu et al. [97]. Reproduced with permission from John Wiley & Sons. (c) Schematic and SEM images of the stretch process for poly-LMN. Source: Thrasher et al. [98]. (d) Resistance of poly-LMN as a function of strain. The cycle 1 and cycles 2–4 are the activation process and constant behavior, respectively. (e) Normalized resistance of the activated poly-LMN as a function of strain. Source: (c–e) Thrasher et al. [98]. Reproduced with permission from John Wiley & Sons.

and recover to its initial state. The oxidized skin of the EGaIn played a key role in this process, which could adhere to the inside walls of the fibers and prevent the EGaIn from flowing out of the fiber. Remarkably, the fiber could be reversibly stretched for more than ten cycles. These results indicated that liquid metal could reversibly deform in the fibers. Additionally, these deformation behaviors would cause the resistance changes of fibers. As shown in Figure 4.23b, when the applied force was increased, the resistance gradually increased until the strain reached to the upper limit of 700–800%. The increased resistance was due to the increased length and decreased cross-sectional area of fibers. This conductive fiber can be applied in various applications such as flexible electronics, smart fabrics, stretchable wires, and sensors. However, there is a drawback to this kind of conductive fibers. When the fibers are punched, the resistance will increase remarkably because liquid metals are easy to flow in fibers.

Based on the ability of surface chemical modification of the oxide skin on the surface of liquid metal, Thrasher and coworkers demonstrated another way to tune the deformation of the liquid metal (EGaIn) by tensile force [98]. As shown in Figure 4.23c, the oxide skin of liquid metal and polymer were interconnected by covalent bonding, forming the highly cross-linked liquid metal particle–polymer networks (poly-LMN). Before stretching, the liquid metal particles were kept isolated from each other due to their surface oxide skin and polymer. When stretched, the force would transfer directly by means of the thin oxide skin of liquid metal within the polymer network, causing the rupture of liquid metal particles in the network with uniaxial elongation. These ruptured oxide skin released liquid metal cores to flow and coalescence into a percolated network. Moreover, when the external force was removed, poly-LMNs would retain their interconnected network. But the percolated networks of liquid metal and their newly generated oxides were compressed when the strain was released. This phenomenon had been confirmed by SEM images during the stretching process (Figure 4.23c). In addition, the electric conductivity measurements could also reveal these deformation behaviors. As displayed in Figure 4.23d, when the applied strain exceeded 61%, the poly-LMN would undergo an electrical transition between dielectrics and conductors, where the resistance sharply decreased by ~6 orders. It means that the liquid metal particles had ruptured and percolated within the networks. With continuously increasing the strain to 300%, the resistance had no obvious changes, revealing that the morphology remained largely set. In addition, poly-LMNs exhibited excellent stability even at strains of 300%, where the resistance tests could be repeated over 10 000 cycles (Figure 4.23e). This result revealed that high strain could maintain a stable percolating network. These deformation behaviors could be applied in various fields, for example, electronic strain memory, circuit, and stretch electronic, which was capable of mitigating the deleterious impacts of strain on electronic performance and producing emergent capability.

In addition to the tensile force, pressure can also deform liquid metals. As shown in Figure 4.24a, when the applied pressure exceeded the pressure required to rupture the oxide layer, the bulk liquid metal could be pushed into microfluidic channels [99]. Figure 4.24b,c exhibited liquid EGaIn and Hg in microchannel and their dependence on pressure. When applied a pressure of 85 and 54 kPa, respectively, liquid EGaIn and Hg rapidly filled into the inlet channel but stopped at the entrance to the narrow channel. After increasing the pressure, EGaIn and Hg flowed through the narrow channel toward the outlet, in which the critical pressures of EGaIn and Hg were 89 and 57 kPa, respectively. The pressure differences were derived from their different surface energy. In addition, when the pressure at the inlet was relieved, the EGaIn remained in shape while Hg is reflowed to the inlet (Figure 4.24b,c). The stability in EGaIn was due to the adhesion of Ga oxide on channels. When EGaIn was injected into the channel, the newly formed oxide film will rapidly hold it. Based on this injection technique, the patterning of liquid metal can be achieved by controlling the shape of the different channels.

Based on the aforementioned deformation behaviors, Matsuzaki and coworkers further explored the deformation of liquid metal (GaInSn) electrodes under pressure, where the electrode was obtained by injecting the GaInSn into the polydimethylsiloxane (PDMS) (Figure 4.24d) [100]. When applied the pressure, the electrode would bend, inducing the bending deformation and flow of liquid metal inside. This deformation had a great influence on the connection of the liquid metal and played an important role in resistance of electrode. Furthermore, as the bending deflection increased, the changes in resistance increased as well due to the changes of the cross section and effective length of liquid metal. Therefore, the bending deflection of GaInSn could be reflected by measuring the resistance

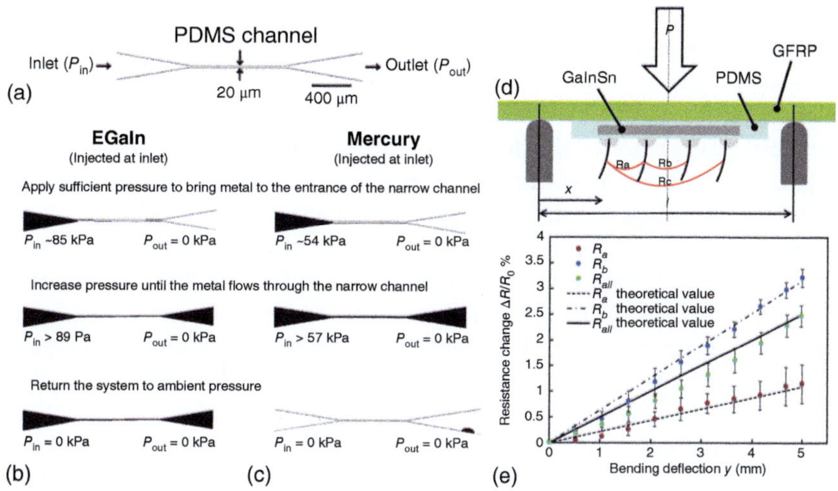

Figure 4.24 Deformation induced by pressure. (a) The schematic of structural composition of the PDMS channel. (b) The backlit, top-down optical photographs of EGaIn in PDMS microfluidic channels. (c) The backlit, top-down images of Hg in PDMS microfluidic channels. Source: (a–c) Dickey et al. [99]. Reproduced with permission from John Wiley & Sons. (d) Schematic of the bending measurements of the liquid metal electrode. (e) The resistances of the liquid metal electrode versus the bending deflection. Source: (d, e) Matsuzaki and Tabayashi [100]. Reproduced with permission from John Wiley & Sons.

(Figure 4.24e). Taking advantage of the relationship of pressure-induced deformation and resistance, liquid metals can be applied in smart electronics such as stress sensors and logic electronics [101].

In summary, force is the most direct tuning way for the deformation of liquid metals, which can be divided into tensile force and pressure. As for the researches on tensile force, it is mainly in the form of injecting liquid metals into hollow channels such as stretched fiber. The applied tensile force was transduced directly through the stretched fiber, inducing the deformation of liquid metals. Another way is surface modification with the aid of the oxide skin on the surface. Under the applied force, the oxide skin will rupture and the core permeates, causing the deformation behaviors. With regard to the pressure, based on the rheological property, liquid metals are able to be injected into microchannel under pressure, realizing the patternings with various shapes.

4.6 Others

Besides the abovementioned stimuli, other stimuli such as temperature, light, and humidity can also tune liquid metals. In this section, we will discuss these stimuli-induced motion and properties variation.

Thermal energy generally refers to the internal energy of a system, which can be manifested by the temperature of this system. As the motion of internal molecules, atoms, and particles of a system can be regulated by tuning its thermal energy, thermal-field tuning can be employed to alter the properties of liquid metals. Chhabra and coworkers reported the dependence of the diffusion coefficient of liquid metals with temperature [102]. They found that the outcomes were similar to that of simple liquids, which could be described by Hildebrand's fluidity model. Itami and coworkers reported the thermally tuned magnetic susceptibility of liquid metals [103]. Here, the temperature-dependence molar magnetic susceptibility of various metals, including Na, K, Cs, Zn, Cd, Hg, Ga, In, Tl, Sn, Pb, and Bi, in their liquid state was measured. They also found that the electron-ion interaction would exert a large effect on the magnetic susceptibility of liquid alkali metals, but little for liquid polyvalent metals. Blairs reported the dependence of sound velocity in liquid metals with the temperature [104]. Generally, the sound velocity decreased slowly when the temperature of liquid metals increased, except a few metals, including Si, Ge, Sb, Te, Ce, and Pu. Although a few theoretical models were developed, they found that the approaches using vaporization enthalpies could estimate the sound velocities better than the procedures based on the hard-sphere model. In addition, based on the thermally tuned properties of the liquid metal, various applications could be fabricated. Li and coworkers reported the fabrication of thermocouple by printing liquid Ga and matching metals [105]. By combining the thermoelectric effect and printable characteristic of liquid Ga, a tiny device could be fabricated. The measured thermoelectric voltage exhibited great linear dependence with temperature with the highest measuring temperature of 200 °C.

Moreover, Wang and coworkers reported that the fabrication of transformable composites was based on liquid metal by thermal-field tuning [11]. Here, the liquid metal composites were fabricated by shearmixing liquid metal, uncured silicone elastomer, and liquid ethanol to construct a liquid metal elastomer embedding with liquid ethanol droplets. After a thermal stimulation, part of the embedded ethanol droplets would change into the gas phase, further inducing a large volume variation. A soft robot with the ability of thermal-field-induced movement could be fabricated based on the composites. Here, the two ends of the composites were inserted in a special socket (Figure 4.25a), which enabled it to move directionally via adjusting the thermal field (Figure 4.25b). Moreover, a robot with the capability to grab objects can also be realized based on the thermal-field-induced deformation of the composites.

As a common external stimulus, light has the unique characteristics of remote control, clean, and contactless, which makes it an attractive candidate for the tuning of liquid metals. Due to the superiorities of light with easily controlled magnitudes and directions, it can precisely focus on micro- or nanoarea, leading to a great potential for applications of liquid metals on the micro/nanodimensions. Light tuning of liquid metals is usually realized by coating on the surface of liquid metals with photoresponsive materials. The abilities to convert light energy or the photocatalytic reactions of these coatings can induce the deformation or locomotion behaviors of liquid metals. Next, we will discuss typical light-induced motions of liquid metals as well as their mechanisms.

The main type of light-induced deformation of liquid metals is converting light energy to heat energy, which further changes the internal molecules to induce

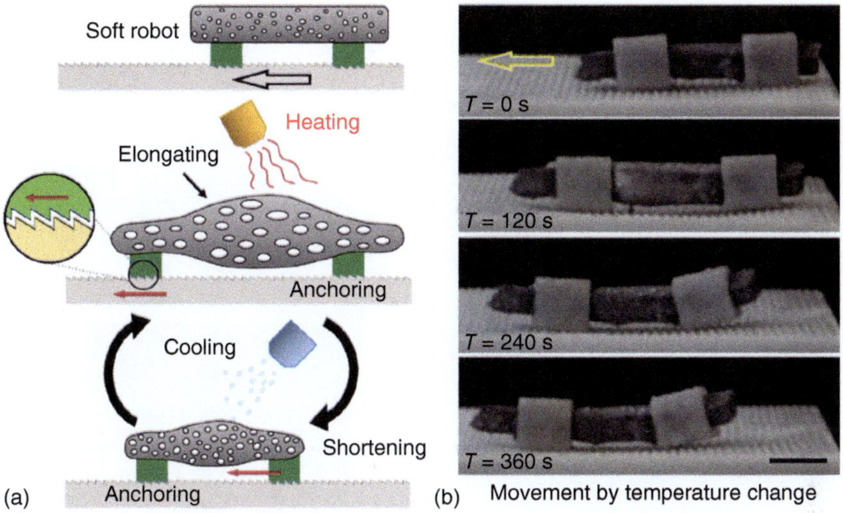

Figure 4.25 Thermal-field tuning of the liquid metals. (a) The schematic of locomotion mechanism of the liquid metal composites. Source: Wang et al. [11]. Reproduced with permission from American Chemical Society. (b) Sequential snapshots of the thermal-field-induced movement of the composites, the scale bar is 2 cm. Source: Wang et al. [11].

the deformation behaviors. Thereinto, coating on the surface of liquid metals with photothermal conversion material is an effective way to realize it. To date, some nanotransformers consisting of liquid metals and polymer photothermal materials have been reported [12, 106, 107]. Gan and coworkers designed the liquid metal nanodroplets coated with polydopamine (PDA) [12]. As shown in Figure 4.26a, liquid metal nanodroplets were fabricated by sonication in dopamine hydrochloride aqueous solution. Afterwards, 2-amino-2-(hydroxymethyl)-1,3-propanediol (Tris) was introduced into above suspension to trigger the self-polymerization of dopamine. Then the PDA-coated liquid metal nanodroplets were obtained, where the PDA deposited onto the surface of liquid metal nanodroplets forming a core–shell nanostructure and the diameters of PDA-coated nanodroplets ranged from 30 to 350 nm (Figure 4.26b). Under near-infrared (NIR) laser irradiation, the PDA-coated liquid metal nanodroplets could be transformed from sphere to ellipsoid with larger size (Figure 4.26c). In contrast, there was no obvious change in morphology when the nanodroplets without PDA coating. These results were attributed to the fact that light-generated heating triggered the dealloying of liquid nanodroplets and induced oxidation reaction to generate Ga oxide monohydroxide (GaOOH) nanorods, causing the shape to change. Moreover, the numbers of ellipsoids increased with the increased irradiation time, indicating that the longer light irradiation could promote the deformation of liquid metal. In addition, the thickness of the coating also had an influence on the deformation of liquid metal. The shape of liquid metal nanodroplets with thicker PDA remained unchanged because thicker and stronger PDA shell could stabilize the shape of liquid metal

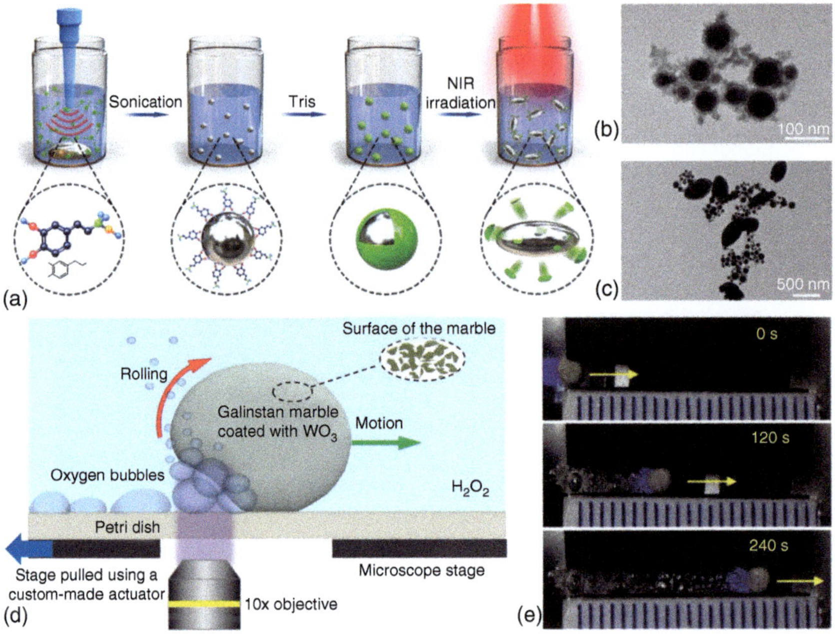

Figure 4.26 Light tuning for liquid metals. (a) Schematic of the production process of PDA-coated liquid metal nanodroplet and their deformation upon the light radiation. Source: Gan et al. [12]. Reproduced with permission from John Wiley & Sons. TEM images of nanodroplet upon the NIR exposure for (b) 0 min and (c) 20 min, respectively. Source: (b, c) Gan et al. [12]. (d) Schematic diagram of the locomotion mechanism of the WO_3-coated marble. Source: Tang et al. [108]. Reproduced with permission from AIP Publishing.(e) Sequential snapshots of the locomotion process of the WO_3-coated marble at the different time. Source: Tang et al. [108].

nanodroplets or prevent oxidation of Ga because the oxygen and water was impeded. In addition to polymer, black carbon materials such as CNTs, graphene, and graphene quantum dots (GQDs) are also known as photothermal materials, which can combine with liquid metal to realize the deformation behaviors as well [109]. For instance, a light-fueled nanoscale deformable body consisting of liquid metal nanospheres coated with GQDs was reported by Lu and coworkers [110]. Under light irradiation, different morphologies of deformation could be observed.

During the deformation process, the nanostructure of these liquid metal systems was damaged due to the shape change and volume increase. Therefore, this kind of shape-morphing nanosystem can be applied in drug delivery. For example, Chechetka and coworkers reported a unique polymeric nanocapsule, which consisted of EGaIn photopolymerized with functional phospholipids [107]. The polymeric nanocapsule was fabricated by loading drugs into EGaIn alloys and then coating with phospholipids. Under laser irradiation, the nanocapsules partially and completely collapsed in 1 and 3 minutes, respectively, which caused the destruction of nanocapsules and released the drugs controllably. The released drugs could act on the tumor site, effectively eliminating cancer cell and controlling intercellular calcium ion flux.

In addition to the deformation, light can also induce the locomotion of liquid metals. As stated earlier in chemical tuning, liquid metals droplet coated with catalytic coating can locomote because the generated bubbles from the process of catalytic decomposition process provide the drive force for it. Similarly, when liquid metals are coated with the photocatalyst, the locomotion behaviors can also be realized under light irradiation. Tang and coworkers demonstrated photochemically induced actuation of liquid metal marbles by virtue of the photocatalytic reaction of WO_3 coating [108]. As displayed in Figure 4.26d, a liquid metal marble was formed by coating on the surface of liquid metal with WO_3 particles. In H_2O_2 solution, when UV light was applied on one side of the marble, the locomotion of the marbles would happen owing to the propulsion of the asymmetrical gas bubbles generated on the marble. On account of the advantages of precisely tuned energy input and reversible on/off state of light radiation, the speed and direction of motion could be adjusted in a controllable way. As demonstrated in Figure 4.26e, by moving the light source, the marbles could move in the direction of light. Furthermore, faster speed would be obtained via increasing the light intensity and the H_2O_2 concentration because more oxygen bubbles were produced on the marble. However, with the increased light intensity, the predicted maximum speed would gradually become faster than the measured speed. This result ascribes that the more bubbles or larger bubbles induced the light scatting effect or reduced the H_2O_2 concentration.

Besides, liquid metals can also be tuned by humidity. In oxidative environments, a solid oxide skin will form at the surface of liquid metals. The oxide films play an important role in their surface morphology. In addition, humidity can also tune the mechanical properties and rheological properties of liquid metals. In the following sections, humidity-induced deformation and property changes of liquid metals are discussed.

Ding and coworkers discovered an evident volume expansion phenomenon of Ga-based liquid metal after exposure to humidity [111]. One of the constituent of the oxide layer was less stable Ga_2O [112], which could be further oxidized by water to generate hydrogen. The generated hydrogen dispersed in the liquid metal, causing the composition and structure changes of the liquid metal. As exhibited in Figure 4.27a,b, when liquid Ga was stirred under humid conditions, an evident volume expansion phenomenon accompanied by loose and porous structures was observed. And the expansion phenomenon was more obvious with the increase of stirring time because of the increased contact area between humidity and liquid metal. As shown in Figure 4.27c, with higher humidity or longer time in air, liquid Ga exhibited more obvious expansion because more hydrogen could be produced. Moreover, thermal conductivity of this Ga-based materials decreased obviously after the expansion process. Therefore, ambient humidity should be precisely controlled during the application of liquid metals.

Besides the adhesion, the mechanical properties of liquid metals can also be tuned by water. As exhibited in Figure 4.27d,e, the surface elastic modulus and the surface

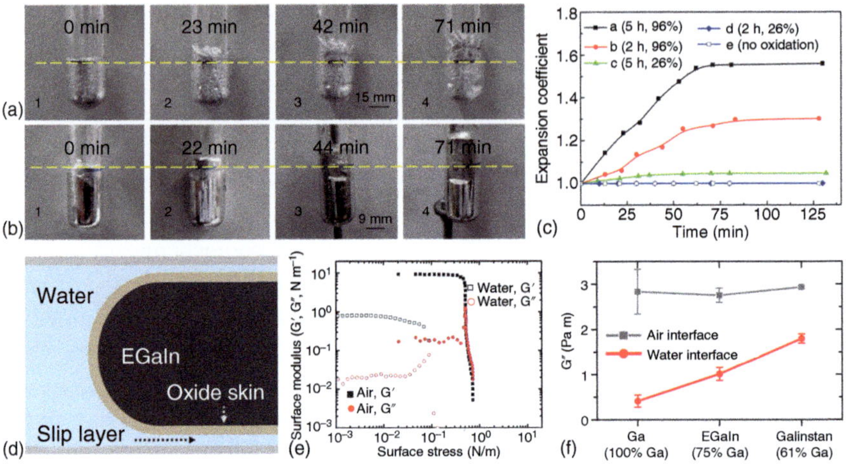

Figure 4.27 Humidity tuning for liquid metals. Volume expansion phenomenon of Ga-based liquid metal under 96% relative humidity for (a) 5 h and (b) 2 h, respectively. Source: (a, b) Ding et al. [111]. (c) Expansion coefficient of Ga-based liquid metal versus time in air. Source: Ding et al. [111]. Reproduced with permission from Springer Nature. (d) Schematic of the behavior of the liquid metal in a water-filled capillary. (e) The mechanical properties of the liquid metal under the exposure of water and air. Source: (d, e) Khan et al. [113]. Reproduced with permission from American Chemical Society. (f) The interfacial storage modulus at air and water for Ga, EGaIn, and GaInSn. Source: Jacob et al. [114]. Reproduced with permission from American Chemical Society.

yield stress of EGaIn significantly decreased after exposure to water, which was possibly attributed to the transformation from Ga oxide to GaOOH [113]. The elasticity of the generated GaOOH was weaker than the Ga oxide. Jacob and coworkers further proved this hypothesis in virtue of a comparison of viscoelastic experiments of the Ga-based liquid metal with and without water. As shown in Figure 4.27f, when submerged in water, Ga and its alloys demonstrated lower values of the storage modulus. In addition, the greater the proportion of Ga, the value of storage moduli will increase, which attributed that more GaOOH generated [114].

In summary, liquid metals can be tuned by various external fields including temperature, light, and humidity. Temperature is mainly employed to alter the properties of liquid metals by changing the internal molecules, atoms, and particles of a system. Besides, the locomotion of liquid metals upon thermal field can be realized by mixing liquid metals with elastomer and liquid ethanol. For light tuning, the deformation or locomotion behaviors of liquid metals are usually realized by coating on the surface of liquid metals with photothermal or photocatalytic materials, respectively. And humidity can also tune the deformation behaviors of liquid metals via chemical reaction between the oxide skin and water. In addition, water can be served as a slip layer to change the interfacial properties of liquid metals. All in all, these external fields tuning liquid metals are less reported, which urgently require further research.

4.7 Conclusion and Prospect

Liquid metals have attracted a great deal of research interests owing to their excellent properties such as high electrical conductivity, good thermal conductivity, and rheological property. These remarkable advantages of liquid metals tremendously broaden their scope of application. Recently, the deformability and mobility of liquid metals have become the hotspots, providing new insights into soft robots, biomedical engineering, and stretching electronics. This section summarizes the recent advances in tuning strategies of the motion of liquid metals integrated with the properties and structure adjustment.

Deformation and locomotion can be triggered by external stimuli including electric field, magnetic field, chemical, force, light, temperature, and humidity. Thereinto, except for chemical, a majority of these stimuli allow for remotely intervening in the direction and speed of the motion, as well as on-demand controlling "on/off" states. Especially, electric field, magnetic field, and chemical tuning are the most widely used tuning methods, which can achieve the diverse deformation and locomotion behaviors. Electric field drives liquid metal droplets in a high speed with almost no effect on the properties of liquid metals. However, the locomotion or deformation of liquid metals often relies on a gradient of interfacial tension on the liquid metal within electrolytes, which limits their application outside a liquid environment. Magnetic field represents one of the most efficient methods to realize the remote control of the motion of liquid metals. The great advantages of magnetic field are strong penetrability, fast response speed, and directionality. Magnetic field can achieve horizontal and even anti-gravity climbing motion in aqueous and nonaqueous solutions, which shows a great promise in fuel-free nanomotors and biomedical applications. Whereas the addition of ferromagnetic material will compromise the intrinsic properties such as deformability and self-healing capability. Chemical tuning spontaneously converts chemical energy into mechanical activity to induce autonomous locomotion, which is an excellent candidate to develop self-powered machines and novel robots. But this approach is restricted to the short lifetime and cannot control the on/off state of motion. In addition, the use of external chemical sources in chemical-powered motors normally results in incomplete fuel degradation, hindering their applications in biological environments. For the stimulus including light and temperature, liquid metals generally need to combine with other materials to obtain the response ability and achieve the deformation or locomotion behaviors. And the development of tuning in these stimuli such as light, force, temperature, and humidity is relatively slow and the form of motion is monotonous.

For properties and structure tuning, the widely used strategies are magnetic field and chemical and force. For magnetic field, because of the high electrical conductivity, liquid metals exposed to magnetic field are capable of generating the magnetothermal effect, which has major advantages in thermochemotherapy. With regard to the chemical, utilizing doping and surface modification strategies are efficient to regulate the various properties of liquid metals. The doped materials can be selected to tune properties as on-demand, for example, viscosity, surface tension, wettability, and the like. Moreover, pressure will induce the phase structure transition of liquid metals. The broad tunable properties and structures offered by chemical tuning will lead to intriguing possibilities.

However, in spite of great achievements on the abovementioned motion and properties tuning, many challenges still remain for future research.

(i) The universality and further exploration of these stimuli. Most driving methods of liquid metals (e.g., electric field, chemical, and light) need to be in aqueous solution or electrolyte environment, which may restrict its application situations. Besides, due to the sticky surface, liquid metals are hard to apply in high viscosity or biological environment. More importantly, for the currently reported researches on motion of liquid metals, most of stimuli concentrate on electric and magnetic fields. The scope of operations offered by liquid metals-based motors is limited. Thus, developing other stimuli is very urgent, especially temperature and light.

(ii) More intelligent liquid metal-based motors. Up to now, the development of liquid metal-based motion is still in the primary stage. The propulsion of liquid metal-based micro/nanomotors is simply controlled and navigated using magnetic fields. It needs to rely on the external manipulation. However, a more complex liquid metal-based motion is hard to achieve such as collective intelligence effect to perform cooperative tasks. Besides, liquid metals tend to aggregate and coalesce with each other, which is ought to be solved.

(iii) The dimension of liquid metals in motion. When the size of bulk liquid metal decreases from millimeter to micrometer or even nanometer, maintaining the continuous movement is difficult. Utilizing chemical fuels is a great solution. However, the direction, speed, and switching of liquid metal-based motors cannot be controlled. This approach greatly limits the scope of application. Thus, the motion of liquid metals at micronanometer scale should further be considered.

(iv) The reversible and controllable properties tuning. Most of stimuli such as chemical and magnetic cannot reversibly tune the properties of liquid metals. In addition, various tuning methods cannot merely tune the single property of liquid metals. This may damage other inherently excellent properties of liquid metals.

Although the various challenges need to be addressed, it is believed that liquid metals will play more roles in various fields in the future. The motion based on liquid metals will become more and more intelligent and diverse.

References

1 Wang, X., Guo, R., and Liu, J. (2018). Liquid metal based soft robotics: materials, designs, and applications. *Adv. Mater. Technol.* 4: 1800549.

2 Zeng, M.Q. and Fu, L. (2018). Controllable fabrication of graphene and related two-dimensional materials on liquid metals via chemical vapor deposition. *Acc. Chem. Res.* 51: 2839.

3 Ding, Y.R., Zeng, M.Q., and Fu, L. (2019). A bright future for liquid functional materials? *Matter* 1: 1099.
4 Sheng, L., Zhang, J., and Liu, J. (2014). Diverse transformations of liquid metals between different morphologies. *Adv. Mater.* 26: 6036.
5 Thess, A., Votyakov, E.V., and Kolesnikov, Y. (2006). Lorentz force velocimetry. *Phys. Rev. Lett.* 96: 164501.
6 Yu, Y. and Miyako, E. (2018). Alternating-magnetic-field-mediated wireless manipulations of a liquid metal for therapeutic bioengineering. *iScience* 3: 134.
7 Zhang, J., Yao, Y., Sheng, L., and Liu, J. (2015). Self-fueled biomimetic liquid metal mollusk. *Adv. Mater.* 27: 2648.
8 Wang, Y., Duan, W., Zhou, C. et al. (2019). Phoretic liquid metal micro/nanomotors as intelligent Filler for targeted microwelding. *Adv. Mater.* 31: 1905067.
9 Wang, L. and Liu, J. (2016). Graphite induced periodical self-actuation of liquid metal. *RSC Adv.* 6: 60729.
10 Dickey, M.D. (2017). Stretchable and soft electronics using liquid metals. *Adv. Mater.* 29: 1606425.
11 Wang, H., Yao, Y., Wang, X. et al. (2019). Large-magnitude transformable liquid-metal composites. *ACS Omega* 4: 2311.
12 Gan, T., Shang, W., Handschuh-Wang, S., and Zhou, X. (2019). Light-induced shape morphing of liquid metal nanodroplets enabled by polydopamine coating. *Small* 15: 1804838.
13 Wang, X., Fan, L., Zhang, J. et al. (2019). Printed conformable liquid metal e-skin-enabled spatiotemporally controlled bioelectromagnetics for wireless multisite tumor therapy. *Adv. Funct. Mater.* 29: 1907063.
14 Eaker, C.B. and Dickey, M.D. (2016). Liquid metal actuation by electrical control of interfacial tension. *Appl. Phys. Rev.* 3: 031103.
15 Lee, J. and Kim, C.J. (2000). Surface-tension-driven microactuation based on continuous electrowetting. *J. Microelectromech. Syst.* 9: 171.
16 Tang, S.Y., Sivan, V., Khoshmanesh, K. et al. (2013). Electrochemically induced actuation of liquid metal marbles. *Nanoscale* 5: 5949.
17 Tang, S.Y., Khoshmanesh, K., Sivan, V. et al. (2014). Liquid metal enabled pump. *Proc. Natl. Acad. Sci. U. S. A.* 111: 3304.
18 Yao, Y.Y. and Liu, J. (2016). Liquid metal wheeled small vehicle for cargo delivery. *RSC Adv.* 6: 56482.
19 Wu, J., Tang, S.Y., Fang, T. et al. (2018). A wheeled robot driven by a liquid-metal droplet. *Adv. Mater.* 30: 1805039.
20 Yasunaga, H. and Natori, A. (1992). Electromigration on semiconductor surfaces. *Surf. Sci. Rep.* 15: 205.
21 Regan, B.C., Alonil, S., Ritchie, R.O. et al. (2004). Carbon nanotubes as nanoscale mass conveyors. *Nature* 428: 924.
22 Dutta, I. and Kumar, P. (2009). Electric current induced liquid metal flow: application to coating of micropatterned structures. *Appl. Phys. Lett.* 94: 184104.
23 Tu, K.N. (1992). Electromigration in stressed thin films. *Phys. Rev. B Condens. Matter.* 45: 1409.

24 Blech, I.A. (1976). Electromigration in thin aluminum films on titanium nitride. *J. Appl. Phys.* 47: 1203.

25 Kumar, P., Howarth, J., and Dutta, I. (2014). Electric current induced flow of liquid metals: mechanism and substrate-surface effects. *J. Appl. Phys.* 115: 044915.

26 Kumar, S., Kumar, P., and Pratap, R. (2017). A model for electromigration induced flow in liquid metals. *J. Phys. D. Appl. Phys.* 50: 39LT02.

27 Magee, J.E. and Wilding, N.B. (2002). The Lennard-Jones-Devonshire cell model revisited. *Mol. Phys.* 100: 1641.

28 Gough, R.C., Dang, J.H., Moorefield, M.R. et al. (2016). Self-actuation of liquid metal via redox reaction. *ACS Appl. Mater. Interfaces* 8: 6.

29 Zhao, X., Tang, J., and Liu, J. (2017). Surfing liquid metal droplet on the same metal bath via electrolyte interface. *Appl. Phys. Lett.* 111: 101603.

30 Tsai, J.T.H., Ho, C., Wang, F., and Liang, C.T. (2009). Ultrahigh contrast light valve driven by electrocapillarity of liquid gallium. *Appl. Phys. Lett.* 95: 251110.

31 Grahame, D.C. (1947). The electrical double layer and the theory of electrocapillarity. *Chem. Rev.* 41: 441.

32 Tang, S.Y., Sivan, V., Petersen, P. et al. (2014). Liquid metal actuator for inducing chaotic advection. *Adv. Funct. Mater.* 24: 5851.

33 Sternling, C.V. and Scriven, L.E. (1959). Interfacial turbulence: hydrodynamic instability and the marangoni effect. *AIChE J.* 5: 514.

34 Tang, S.Y., Lin, Y., Joshipura, I.D. et al. (2015). Steering liquid metal flow in microchannels using low voltages. *Lab Chip* 15: 3905.

35 Zhao, X., Tang, J., and Liu, J. (2018). Electrically switchable surface waves and bouncing droplets excited on a liquid metal bath. *Phys. Rev. Fluids* 3: 124804.

36 Bormashenko, E., Pogreb, R., Stein, T. et al. (2012). Electrically deformable liquid marbles. *J. Adhes. Sci. Technol.* 25: 1371.

37 Yuan, B., He, Z.Z., and Liu, J. (2018). Effect of electric field on the wetting behavior of eutectic gallium-indium alloys in aqueous environment. *J. Electron. Mater.* 47: 2782.

38 Khan, M.R., Eaker, C.B., Bowden, E.F., and Dickey, M.D. (2014). Giant and switchable surface activity of liquid metal via surface oxidation. *Proc. Natl. Acad. Sci. U. S. A.* 111: 14047.

39 Zhang, J., Sheng, L., and Liu, J. (2014). Synthetically chemical-electrical mechanism for controlling large scale reversible deformation of liquid metal objects. *Sci. Rep.* 4: 7116.

40 Hu, L., Wang, L., Ding, Y. et al. (2016). Manipulation of liquid metals on a graphite surface. *Adv. Mater.* 28: 9210.

41 Khan, M.R., Trlica, C., and Dickey, M.D. (2015). Recapillarity: electrochemically controlled capillary withdrawal of a liquid metal alloy from microchannels. *Adv. Funct. Mater.* 25: 671.

42 Tang, S.Y., Joshipura, I.D., Lin, Y. et al. (2016). Liquid-metal microdroplets formed dynamically with electrical control of size and rate. *Adv. Mater.* 28: 604.

43 Wissman, J., Dickey, M.D., and Majidi, C. (2017). Field-controlled electrical switch with liquid metal. *Adv. Sci.* 4: 1700169.

44 Tan, S.C., Yang, X.H., Gui, H. et al. (2017). Galvanic corrosion couple-induced Marangoni flow of liquid metal. *Soft Matter.* 13: 2309.

45 Lee, W., Walker, L.M., and Anna, S.L. (2009). Role of geometry and fluid properties in droplet and thread formation processes in planar flow focusing. *Phys. Fluids* 21: 032103.

46 Chrimes, A.F., Berean, K.J., Mitchell, A. et al. (2016). Controlled electrochemical deformation of liquid-phase gallium. *ACS Appl. Mater. Interfaces* 8: 3833.

47 Jin, S.W., Jeong, Y.R., Park, H. et al. (2019). A flexible loudspeaker using the movement of liquid metal induced by electrochemically controlled interfacial tension. *Small* 15: 1905263.

48 Hu, L., Wang, H., Wang, X. et al. (2019). Magnetic liquid metals manipulated in the three-dimensional free space. *ACS Appl. Mater. Interfaces* 11: 8685.

49 Carey, B.J., Ou, J.Z., Clark, R.M. et al. (2017). Wafer-scale two-dimensional semiconductors from printed oxide skin of liquid metals. *Nat. Commun.* 8: 14482.

50 Carle, F., Bai, K., Casara, J. et al. (2017). Development of magnetic liquid metal suspensions for magnetohydrodynamics. *Phys. Rev. Fluids* 2: 013301.

51 Ma, B., Xu, C., Chi, J. et al. (2019). A versatile approach for direct patterning of liquid metal using magnetic field. *Adv. Funct. Mater.* 29: 1901370.

52 Liu, H., Li, M., Li, Y. et al. (2018). Magnetic steering of liquid metal mobiles. *Soft Matter.* 14: 3236.

53 Jeon, J., Lee, J.B., Chung, S.K., and Kim, D. (2016). On-demand magnetic manipulation of liquid metal in microfluidic channels for electrical switching applications. *Lab Chip* 17: 128.

54 Guo, R., Sun, X., Yuan, B. et al. (2019). Magnetic liquid metal (Fe-EGaIn) based multifunctional electronics for remote self-healing materials, degradable electronics, and thermal transfer printing. *Adv. Sci.* 6: 1901478.

55 Zhang, J., Guo, R., and Liu, J. (2016). Self-propelled liquid metal motors steered by a magnetic or electrical field for drug delivery. *J. Mater. Chem. B* 4: 5349.

56 de Castro, I.A., Chrimes, A.F., Zavabeti, A. et al. (2017). A gallium-based magnetocaloric liquid metal ferrofluid. *Nano Lett.* 17: 7831.

57 Yu, M., Bian, X., Wang, T., and Wang, J. (2017). Metal-based magnetic fluids with core-shell structure FeB@SiO_2 amorphous particles. *Soft Matter.* 13: 6340.

58 Zhao, X., Yang, L., Ding, Y. et al. (2019). Noncoalescent liquid metal droplets sustained on a magnetic field-circulated liquid metal bath surface. *Appl. Phys. Lett.* 115: 083702.

59 Hirsch, A., Dejace, L., Michaud, H.O., and Lacour, S.P. (2019). Harnessing the rheological properties of liquid metals to shape soft electronic conductors for wearable applications. *Acc. Chem. Res.* 52: 534.

60 Li, F., Kuang, S., Li, X. et al. (2019). Magnetically- and electrically-controllable functional liquid metal droplets. *Adv. Mater. Technol.* 4: 1800694.

61 Jeon, J., Lee, J.B., Chung, S.K., and Kim, D. (2016). Magnetic liquid metal marble: characterization of lyophobicity and magnetic manipulation for switching applications. *J. Microelectromech. Syst.* 25: 1050.

62 Chen, R., Xiong, Q., Song, R.Z. et al. (2019). Magnetically controllable liquid metal marbles. *Adv. Mater. Interfaces* 6: 1901057.

63 Yan, J., Lu, Y., Chen, G. et al. (2018). Advances in liquid metals for biomedical applications. *Chem. Soc. Rev.* 47: 2518.

64 Zhang, J., Zhu, X., Zeng, M., and Fu, L. (2020). Magnetically controlled on-demand switching of batteries. *Adv. Sci.* 7: 2000184.

65 Cao, L., Yu, D., Xia, Z. et al. (2020). Ferromagnetic liquid metal putty-like material with transformed shape and reconfigurable polarity. *Adv. Mater.* 32: 2000827.

66 Daeneke, T., Khoshmanesh, K., Mahmood, N. et al. (2018). Liquid metals: fundamentals and applications in chemistry. *Chem. Soc. Rev.* 47: 4073.

67 Wang, D., Xie, W., Gao, Q. et al. (2019). Non-magnetic injectable implant for magnetic field-driven thermochemotherapy and dual stimuli-responsive drug delivery: transformable liquid metal hybrid platform for cancer theranostics. *Small* 15: 1900511.

68 Shein, I.R. and Ivanovskii, A.L. (2007). First principle prediction of vacancy-induced magnetism in non-magnetic perovskite $SrTiO_3$. *Phys. Lett. A* 371: 155.

69 Yang, J., Bian, X., Yang, C. et al. (2013). Viscous features of single-component melt under horizontal magneticfield. *Physica B* 415: 18.

70 Sun, C., Geng, H., Zhang, N. et al. (2008). Viscous feature of Sb-Bi alloy under magnetic field. *Mater. Lett.* 62: 73.

71 Liu, T., Wang, Q., Gao, A. et al. (2011). Effects of a high magnetic field on the phase equilibria of Mn-Sb system during solidification process. *J. Alloys Compd.* 509: 5822.

72 Ren, L., Sun, S., Casillas-Garcia, G. et al. (2018). A liquid-metal-based magnetoactive slurry for stimuli-responsive mechanically adaptive electrodes. *Adv. Mater.* 30: 1802595.

73 Park, B.J., Fang, F.F., and Choi, H.J. (2010). Magnetorheology: materials and application. *Soft Matter* 6: 5246.

74 Ito, R., Dodbiba, G., and Fujita, T. (2005). Mr fluid of liquid gallium dispersing magnetic particles. *Int. J. Mod. Phys. B* 19: 1430.

75 Ibragimov, K.I. and Korolkov, V.A. (2001). Temperature dependence of the work function of metals and binary alloys. *Inorg. Mater.* 37: 567.

76 Chen, S., Wang, L., Zhang, Q., and Liu, J. (2018). Liquid metal fractals induced by synergistic oxidation. *Sci. Bull.* 63: 1513.

77 Beni, G., Hackwood, S., and Jackel, J.L. (1982). Continuous electrowetting effect. *Appl. Phys. Lett.* 40: 912.

78 Eaker, C.B., Hight, D.C., O'Regan, J.D. et al. (2017). Oxidation-mediated fingering in liquid metals. *Phys. Rev. Lett.* 119: 174502.

79 Chen, S. and Liu, J. (2019). Spontaneous dispersion and large-scale deformation of gallium-based liquid metal induced by ferric ions. *J. Phys. Chem. B* 123: 2439.

80 Wang, H., Yuan, B., Liang, S. et al. (2018). PLUS-M: a porous liquid-metal enabled ubiquitous soft material. *Mater. Horiz.* 5: 222.

81 Lide, D.R. (2005). *CRC Handbook of Chemistry and Physics*, Internet Version., 85. *CRC Press.*

82 Chen, S., Yang, X., Cui, Y., and Liu, J. (2018). Self-growing and serpentine locomotion of liquid metal induced by copper ions. *ACS Appl. Mater. Interfaces* 10: 22889.

83 Kim, D., Thissen, P., Viner, G. et al. (2013). Recovery of nonwetting characteristics by surface modification of gallium-based liquid metal droplets using hydrochloric acid vapor. *ACS Appl. Mater. Interfaces* 5: 179.

84 Ortiz-Rivera, I., Mathesh, M., and Wilson, D.A. (2018). A supramolecular approach to nanoscale motion: polymersome-based self-propelled nanomotors. *Acc. Chem. Res.* 51: 1891.

85 Zavabeti, A., Daeneke, T., Chrimes, A.F. et al. (2016). Ionic imbalance induced self-propulsion of liquid metals. *Nat. Commun.* 7: 12402.

86 Yi, L., Ding, Y., Yuan, B. et al. (2016). Breathing to harvest energy as a mechanism towards making a liquid metal beating heart. *RSC Adv.* 6: 94692.

87 Liu, T., Sen, P., and Kim, C.J. (2012). Characterization of nontoxic liquid-metal alloy galinstan for applications in microdevices. *J. Microelectromech. Syst.* 21: 443.

88 Tang, J., Zhao, X., Li, J. et al. (2017). Liquid metal phagocytosis: intermetallic wetting induced particle internalization. *Adv. Sci.* 4: 1700024.

89 Ilyukhina, A.V., Ilyukhin, A.S., and Shkolnikov, E.I. (2012). Hydrogen generation from water by means of activated aluminum. *Int. J. Hydrog. Energy* 37: 16382.

90 Flamini, D.O., Saidman, S.B., and Bessone, J.B. (2006). Aluminium activation produced by gallium. *Corros. Sci.* 48: 1413.

91 Sheng, L., He, Z., Yao, Y., and Liu, J. (2015). Transient state machine enabled from the colliding and coalescence of a swarm of autonomously running liquid metal motors. *Small* 11: 5253.

92 Ma, J., Zheng, M., Chen, C. et al. (2018). Efficient and stable nonfullerene-graded heterojunction inverted perovskite solar cells with inorganic Ga_2O_3 tunneling protective nanolayer. *Adv. Funct. Mater.* 28: 1804128.

93 Chen, Y.J., Nagamine, Y., and Yoshikawa, K. (2009). Self-propelled motion of a droplet induced by Marangoni-driven spreading. *Phys. Rev. E* 80: 016303.

94 Hu, L., Li, J., Tang, J., and Liu, J. (2017). Surface effects of liquid metal amoeba. *Sci. Bull.* 62: 700.

95 Ilyas, N., Cook, A., and Tabor, C.E. (2017). Designing liquid metal interfaces to enable next generation flexible and reconfigurable electronics. *Adv. Mater. Interfaces* 4: 1700141.

96 Wu, Y.H., Deng, Z.F., Peng, Z.F. et al. (2019). A Novel strategy for preparing stretchable and reliable biphasic liquid metal. *Adv. Funct. Mater.* 29: 1903840.

97 Zhu, S., So, J.H., Mays, R. et al. (2013). Ultrastretchable fibers with metallic conductivity using a liquid metal alloy core. *Adv. Funct. Mater.* 23: 2308.

98 Thrasher, C.J., Farrell, Z.J., Morris, N.J. et al. (2019). Mechanoresponsive polymerized liquid metal networks. *Adv. Mater.* 31: 1903864.

99 Dickey, M.D., Chiechi, R.C., Larsen, R.J. et al. (2008). Eutectic gallium-indium (EGaIn): a liquid metal alloy for the formation of stable structures in microchannels at room temperature. *Adv. Funct. Mater.* 18: 1097.

100 Matsuzaki, R. and Tabayashi, K. (2015). Highly stretchable, global, and distributed local strain sensing line using GaInSn electrodes for wearable electronics. *Adv. Funct. Mater.* 25: 3806.

101 Jin, Y., Lin, Y., Kiani, A. et al. (2019). Materials tactile logic via innervated soft thermochromic elastomers. *Nat. Commun.* 10: 4187.

102 Chhabra, R.P. and Sridhar, T. (1983). Temperature dependence of self diffusion in liquid metals. *Phys. Chem. Liq.* 13: 37.

103 Itami, T., Shimoji, M., and Shimokawa, K. (1988). The temperature dependence of the magnetic susceptibility of liquid metals. *Z. Phys. Chem. (N. F.)* 157: 587.

104 Blairs, S. (2006). Temperature dependence of sound velocity in liquid metals. *Phys. Chem. Liq.* 44: 597.

105 Li, H., Yang, Y., and Liu, J. (2012). Printable tiny thermocouple by liquid metal gallium and its matching metal. *Appl. Phys. Lett.* 101: 073511.

106 Kim, D., Hwang, J., Choi, Y. et al. (2019). Effective delivery of anti-cancer drug molecules with shape transforming liquid metal particles. *Cancers* 11: 1666.

107 Chechetka, S.A., Yu, Y., Zhen, X. et al. (2017). Light-driven liquid metal nanotransformers for biomedical theranostics. *Nat. Commun.* 8: 15432.

108 Tang, X., Tang, S.Y., Sivan, V. et al. (2013). Photochemically induced motion of liquid metal marbles. *Appl. Phys. Lett.* 103: 174104.

109 Fujii, S., Yusa, S.I., and Nakamura, Y. (2016). Stimuli-responsive liquid marbles: controlling structure, shape, stability, and motion. *Adv. Funct. Mater.* 26: 7206.

110 Lu, Y., Lin, Y., Chen, Z. et al. (2017). Enhanced endosomal escape by light-fueled liquid-metal transformer. *Nano Lett.* 17: 2138.

111 Ding, Y., Deng, Z., Cai, C. et al. (2017). Bulk expansion effect of gallium-based thermal interface material. *Int. J. Thermophys.* 38.

112 Scharmann, F., Cherkashinin, G., Breternitz, V. et al. (2004). Viscosity effect on GaInSn studied by XPS. *Surf. Interface Anal.* 36: 981.

113 Khan, M.R., Trlica, C., So, J.H. et al. (2014). Influence of water on the interfacial behavior of gallium liquid metal alloys. *ACS Appl. Mater. Interfaces* 6: 22467.

114 Jacob, A.R., Parekh, D.P., Dickey, M.D., and Hsiao, L.C. (2019). Interfacial rheology of gallium-based liquid metals. *Langmuir* 35: 11774.

Part II

Emergent Applications of Liquid Metal

5

Liquid Metal as a New Reaction Medium

In all the wealth that can benefit mankind, I have found that nothing can be greater than new technologies, new contributions, and new inventions that improve human life.
Source: Bacon (1561–1626). Public Domain.

Liquid Metals: Properties, Mechanisms, and Applications, First Edition. Lei Fu and Mengqi Zeng.

5.1 Introduction

Liquid metal as a reaction medium has developed intensively in material synthesis, which not only realizes the dimension control, but also assists in the composition tuning and morphology control of alloys or metals. Except for defined liquid metal, the liquid metal mentioned in this part also includes the metal that is present in liquid state during reaction process.

When it assists in the synthesis of materials in different dimensions, liquid metals usually attribute to the catalyst for the decomposition of the precursor. And the advantages of liquid metals, such as high surface uniformity, fast precursor transferring, and materials growth rate, also confers benefits to the synthesis of nanomaterials [1, 2]. The low melting point and surface energy of liquid metal are helpful to create confined space and realize the morphology control of zero-dimensional (0D) materials [3–5]. And liquid metals as an effective catalyst to control the growth behaviors (e.g. the composition, diameter, growth direction, crystal structure, and morphology) of one-dimensional (1D) nanomaterials are highly expected for years [6–14]. As for two-dimensional (2D) materials, liquid metals can realize the 2D growth of materials and induce the changes of their growth behaviors, such as self-limited growth [15], morphology control [16], orientated-stitching [17], homogeneous nucleation density, and grain size distribution [18]. And the low surface energy of liquid substrates is helpful in the lateral diffusion of clusters across the surface to find binding sites, which contributes to the reduction of grain boundary and lead to the synthesis of large-area, high-quality 2D crystals. Besides, the liquid metals with lower melt point have potential to allow the low-temperature growth of the high-quality nanomaterials, which satisfies the low-cost request of industrial applications [19]. Moreover, the atomically thin oxides of the liquid metals can serve as 2D materials by the van der Waals (vdW) exfoliation technique, which is highly scalable for creating the target oxide nanosheets [20].

As for the assisted synthesis of alloys, liquid metals can participate in the formation of different structures, including traditional ordered phase (e.g. intermetallic compounds) [21], amorphous phase (e.g. amorphous alloys) [22], and emerging solid-solution phase (e.g. high-entropy alloys) [23]. Moreover, the addition of liquid metal elements would improve the physical properties of alloys, including magnetics, mechanical, and corrosion resistance properties.

While used for morphology control of alloys or metals, liquid metal would act as a special solvent to dissolve other metal elements. After the dealloying process, the remaining components could form a self-organized nanopore structure [24]. And the liquid metal dealloying strategy is appropriate for almost entire elements of the periodic table [25]. The room temperature fluidity and surface tension controllability of liquid metals and alloys could promote their deformation ability. When applying an external DC current, the surface tension of the liquid alloy is changed, which leads to suppression of heterogeneous nucleation and promotion of fluidity [26].

Here, recent progress in the liquid metals as reaction medium is discussed in detail. In particular, the dimension control of materials is emphasized in three parts (i.e. 0D, 1D, and 2D nanomaterials). Among them, the liquid metal-assisted synthesis of 2D materials is expounded from the following two aspects: substrate-assisted synthesis and the controlling growth behavior, which have shown great potential for the synthesis of 2D materials in a controllable manner. As for alloys or metals, the composition and morphology tuning by the liquid metals are also designed to broaden application prospect. The current challenges and future directions of liquid metals as new reaction medium are also discussed.

5.2 Nanomaterial Synthesis

5.2.1 Current Status of Nanomaterial Synthesis

Nanomaterials with surface effects and quantum confinement effects [27] have various highly important utilizations in electronics [28], optoelectronics [29], photonics [30], catalysis, and energy conversion [31, 32]. The sizes of nanomaterials would cause tremendous influences on their quantum confinement effects, so it is crucial to control the size along certain directions. Additionally, crystal orientation, chemical constituents, and crystalline quality of nanomaterials will also have influences on their characteristics and applications. Therefore, effective strategies are required to control the growth behaviors of the nanomaterials. Current synthesis strategies of the nanomaterials are faced with many limitations, which would impede their further applications.

The 0D quantum dots obtained by the solution-based synthesis exhibited undesired diameters [33, 34]. The liquid metal with a low melting temperature and modest interaction between the atoms can be easily tailored to form nanoclusters, thus creating a confined space to control the size and distribution of the 0D nanomaterial [4]. And the liquid metal droplets, having a minimization of surface energy, can serve as an ideal reactant to form liquid-based with desired morphologies [5].

As for the 1D materials, the liquid metals and their alloys have always been utilized as the effective reaction media to control the growth behaviors of 1D nanomaterials. Representatively, the vapor–liquid–solid (VLS) strategy based on the metal droplets is versatile, allowing a precise axis-controlled catalytic growth along certain directions. Based on the fluidity and tunable surface tension of the liquid metals, the external field can be easily introduced to the liquid metals to further realize the adjustable growth orientation and diameter [35–37]. Apart from the growth direction and diameter, the branching, kinking, and periodic twinning of the 1D nanowires can also be controlled by employing the catalytic properties of the liquid metal-based VLS growth strategy [6, 8, 9, 11, 12]. Therefore, the liquid metal-assisted approaches to control the growth behaviors of the 1D nanomaterials are highly reliable.

The size, morphology, layer number, growth rate, growth direction, and crystalline quality play key roles in dominating the characters of 2D nanomaterials. Currently, the strategies of 2D materials mainly include mechanical exfoliation [38, 39], liquid-phase exfoliation [40, 41], and chemical vapor deposition (CVD) [42–44]. However, the mechanical exfoliation cannot realize a scaled-up production of 2D materials, and the liquid exfoliation usually causes lattice defects. CVD is an efficient approach for high-yield production of highly crystalline 2D materials [45]. However, the growth strategies utilized by the traditional solid substrates during the CVD process are still far away from the request. Various problems such as the random nucleation and grain distribution, uncontrollable layer number and shape transformation, disorder stacking and stitching, and inevitable grain boundaries on the solid metal substrates need to be conquered. Recently, the liquid metal-assisted synthesis has been regarded as a promising approach to solve these problems. For example, compared with solid substrates, the surfaces of the solid metals are rigid and have stages, while the liquid metals have excellent properties such as fluidity, high surface uniformity, and low surface energy. On the one hand, these properties can induce the changes of the growth behaviors, such as controllable shape transformation [16], orientated-stitching [17], and homogeneous nucleation [18]. On the other hand, they are in favor of reducing the grain boundary, further improving the size and quality of the 2D materials [19]. In addition, adequate vacancies in the liquid metal bulk allow the dissolution ability of the heteroatoms, which could precisely control the segregation of heteroatoms. Therefore, the self-limited growth of the 2D crystals [1, 46] and various 2D heterostructures can be realized [47, 48]. And the surficial self-passivating oxide layers of the liquid metal can provide a confined space in synthesizing wafer-scale 2D films [20, 49–52].

Therefore, after taking these advantages of the liquid metal, we believe that liquid metal-assisted synthesis is a highly scalable strategy and is therefore suitable for creating high-quality, low-dimensional nanomaterials as well as precisely controlling of their growth behaviors.

5.2.2 Liquid Metal-Assisted Synthesis of Zero-Dimensional Nanomaterials

Based on the modest interaction between the atoms of the liquid metal, it is handy to fabricate liquid metal droplet with 0D sizes, which can be employed to control the size, morphology, density, and element composition of the 0D nanomaterials.

Gherasimova and coworkers reported the fabrication of III-nitride quantum dots via pre-depositing III-group liquid metallic nanoclusters based on metal-organic chemical vapor deposition (MOCVD) [3]. Here, the liquid metal nanoclusters were deposited at the melting temperature on a semiconductor surface to act as intermediate stones. The size and distribution of the nanoclusters could be regulated by growth parameters, then the final 0D III-nitride nanostructures can be synthesized by exposing the liquid nanoclusters to ammonia atmosphere. The liquid metal can also induce the morphological changes in the 0D nanomaterials. Kumar and coworkers proposed the fabrication of flower-like III-nitride nanomaterials through plasma-assisted molecular beam epitaxy (MBE) method with liquid metal droplets as reactants [5]. The formation of liquid-in-based nano-flower structures was originated from the minimization of surface energy. Yu and coworkers realized the controllable synthesis of gallium nitride (GaN) nanodots on Si substrate via liquid-droplet-based MBE method [53]. They found that the density of the nanodots

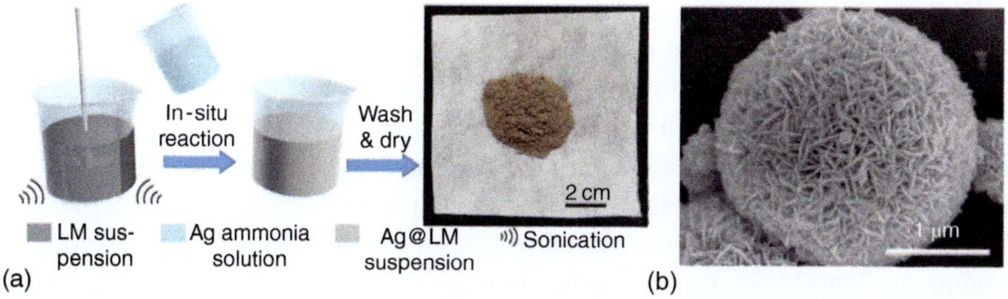

Figure 5.1 Liquid metal-assisted synthesis of 0D nanomaterials. (a) The schematic of the synthesized process of 0D Ag–liquid metal core–shell particles. (b) Typical SEM image of the as-synthesized core–shell particles. Source: Zheng et al. [55].

could be controlled by adjusting the formation temperature of liquid Ga droplets, the surface of the substrates, and the nitridation durations. Choi and coworkers reported the fabrication of spherical 0D III-nitride quantum dots with a high symmetric structure [54]. The sphere-shaped nanostructures were grown on a SiN/Si substrate with curved surface, and liquid III-group droplets were pre-deposited to act as nucleation sites to control the size and morphology of the quantum dots. The emission property of the 0D quantum dots could be improved through a thermal annealing in the liquid-metallic-nanocluster-based MOCVD growth [4]. It was found that an enhanced nitridation of the as-synthesized GaN quantum dots could be realized after the thermal annealing treatment, which could induce an obvious PL emission of the quantum dots.

Besides the fabrication of 0D III–V compounds, liquid metal droplets can also be employed to synthesize 0D metal composites. Zheng and coworkers reported the synthesis of Ag–liquid metal particles with core–shell structure via coating Ag nanostructures on the liquid metal Ga particles [55]. Here, liquid metal particles were synthesized via sonication to form a suspension with glucose. When silver ammonia solution was then introduced, in situ chemical reduction occurred on the surface of particles to form the core–shell particles after drying (Figure 5.1a). The synthesized 0D composites were constructed by flaky Ag shells and spherical Ga cores (Figure 5.1b) and exhibited great electrical conductivity. The rigid and stable Ag shells could maintain the core–shell structure and prevent the chemical etching of Ga cores. In addition, a real-time self-repairing of flexible circuit could be realized under the stimulation of external stress due to the release of liquid metals after the rupture of Ag shells. The fabrication of flexible self-repairing circuit based on the Ag–liquid metal core–shell particles confirms the application potential of liquid metal. Shortly, the liquid metal plays an important role in confining the size, regulating the ingredients, and improving the properties of the 0D nanomaterials.

5.2.3 Liquid Metal-Assisted Synthesis of One-Dimensional Nanomaterials

The liquid metal-assisted synthesis of 1D nanomaterials is always based on the VLS growth mechanism, which is the most common method for synthesizing both compositionally uniform and heterogeneous nanowires [56–59]. During the VLS process, the nanosized liquid droplet is formed by an eutectic reaction between a metal catalyst solvent (usually Au) and the semiconductor solute, and the solute usually is provided by a vapor-phase precursor in CVD system. While the liquid is in supersaturation, the semiconductor solute would precipitate at the interface of the liquid and substrate and finally forms the nanowire. Besides, the VLS method has the ability of controlled growth, which also meets the adjustable requirements of nanowire lengths for device applications [6–9, 11–13, 60, 61]. Notably, except for the VLS method, the solution–liquid–solid (SLS) method is also proved as an important approach to obtain high-quality 1D nanomaterials. Similar to the VLS method, the SLS method also employs the liquid metal nanoparticles as catalyst to realize the nucleation and growth of 1D nanomaterials [62–68]. However, the introduced approach of chemical precursors to metal catalysts is disparate between the two methods. Precursors are provided in a one-time fashion via a flask in SLS method. Differently, the precursors in VLS are supplied by vapor through the reaction chamber and the by-products also could be eliminated by the time. Hence, compared with the SLS method, the VLS synthesis is good at the control in introduction of reactants and removal of by-products, which enables a certain degree of controllable preparation [14].

Laocharoensuk and coworkers designed a SLS growth strategy in a microfluidic reactor to promote the controllability of SLS approach [14]. To be specific, they designed a resealable microfluidics "chip" to hold substrate-containing nanoparticle catalysts in place, which also enabled reactants and coordinating ligands to flow in by a carrier solvent, and this flow–SLS approach also could transport the by-products away during the growing process. Whereafter, the CdSe and ZnSe were synthesized with this flow–SLS approach, the Bi films with different thickness were used to form the liquid catalyst–droplet with manageable sizes, and further controlled the diameter of nanowires (Figure 5.2a). Thicker films lead to larger droplet sizes and thicker nanowires, for instance, the CdSe nanowires with diameters of 16.0 ± 3.2, 21.4 ± 4.6, and 37.6 ± 6.8 nm were corresponding to the 2-, 5-, and 10-nm-thick Bi films. Figure 5.2b revealed an exponential growth trend of CdSe semiconductor nanowire length from 1 to 30 min, which confirmed that the flow–SLS growth was unblocked by solution–solid contributions during all measured growth times. Besides, as shown in Figure 5.2c, the eight-segment semiconductor nanowire, which consisted of single-composition CdSe and ZnSe nanowires, was prepared to develop the controllability of the flow–SLS approach. This work provides a novel flow–SLS approach with controllability and universality and has potential to serve as a convenient platform for further researches.

While fabricated via the VLS method, liquid metal nanoparticles absorb vapor-phase precursors and result in the crystalline nanowires sediments at the interface between liquid and solid. Misra and coworkers reported Sn-assisted Si nanowires growth by the VLS mode, which was realized in a capacitively coupled plasma enhanced chemical vapor deposition (PECVD) [69]. The diameter of the obtained Si nanowires was smaller than 50 nm, and the gradually tapered Si nanowires could continue to grow without visible droplets on the top. This unique

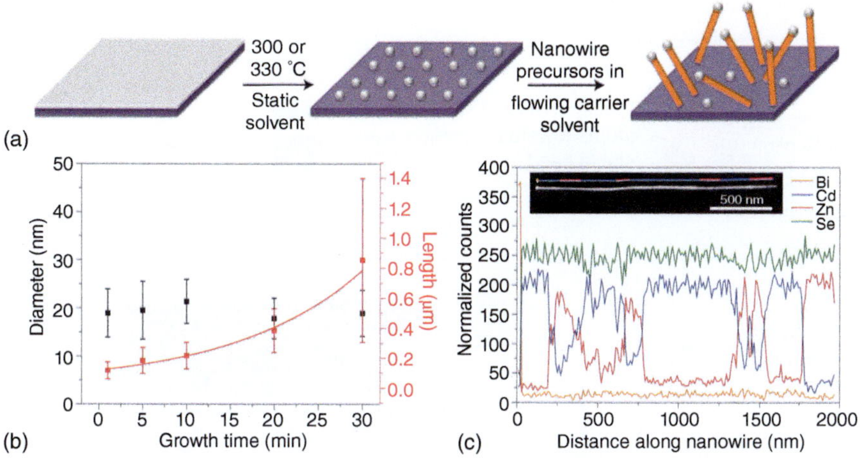

Figure 5.2 Liquid metal-assisted synthesis of 1D nanomaterials. (a) Schematic of flow–SLS synthesis of semiconductor nanowires. (b) Line graph of CdSe semiconductor nanowire dimensions on growth time and reaction temperature. (c) High-angle annular dark-field scanning transmission electron microscopy (HAADF-STEM) image and energy dispersive X-ray (EDX) analysis line-scan profile of the total nanowire length (inset) of eight-segment CdSe–ZnSe semiconductor nanowire. Source: Laocharoensuk et al. [14].

growth behavior was attributed to the ultrathin Sn liquid layer on the sidewall of Si nanowires. This liquid Sn layer assisted to promote the axial prolongation and suppressed the radial growth of the Si nanowires (Figure 5.3a). And the liquid Sn layer eventually would be exhausted during the axial and radial growth of Si nanowires. As for the heterogeneous nanowires, the VLS growth via a liquid metal-assisted catalyst is also an ideal approach. Perea and coworkers realized the growth of axial Ge–Si heterojunction nanowires with sharp interface via Au–Ga alloy catalyst (Figure 5.3b) [70]. The sharp heterointerface of Ge and Si was shown in Figure 5.3b (right). The Ge–Si heterojunction width could be systematically controlled through in situ tuning of the catalyst composition by changing the content of Ga in Au–Ga alloy catalyst. Except for the Si, Ge nanowires [13, 71–73], the metal oxides [61, 74–77] and many III–V nanowires [78–83] are realized by VLS method with various liquid metal catalysts, which indicates the importance of liquid metal in the preparation of 1D materials.

5.2.4 Liquid Metal-Assisted Synthesis of Two-Dimensional Nanomaterials

For the past few years, 2D materials represented by graphene and transition metal dichalcogenides (TMDs) have developed into a big family. More generally, materials (including layered and nonlayered materials) with thickness less than 5 nm (lateral size larger than 100 nm or even up to tens of micromeres) can be called "2D materials" [84, 85]. Compared with the corresponding bulk materials, dimensional thinning brings many novel physical and chemical properties [86]. To date, the

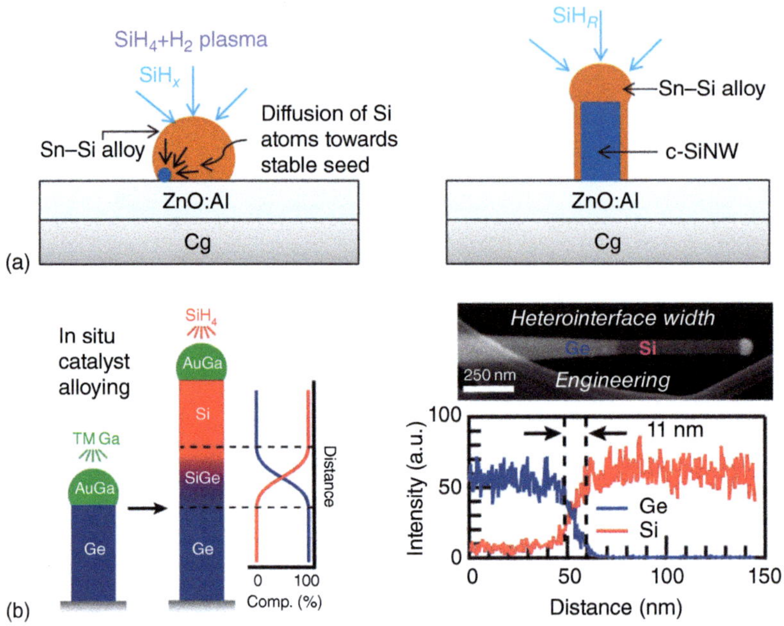

Figure 5.3 Liquid metal-assisted synthesis of 1D nanomaterials. (a) Schematic of the Sn-assisted Si nanowire growth via plasma-assisted VLS mode. Source: Misra et al. [69]. (b) Schematic illustrating the in situ Au–Ga alloy-assisted growth of axial Ge–Si heterojunction nanowire (left). SEM image and EDX analysis line-scan profile of Ge–Si heterojunction nanowire (right). Source: Perea et al. [70].

preparation methods of 2D materials mainly consist of mechanical exfoliation [38, 39], liquid exfoliation [40, 41], and CVD [42–44]. Among these methods, CVD is a promising approach for controllable growth and scaled-up production of high-quality 2D materials [45]. And the liquid metal-assisted synthesis by the CVD process was reported in many 2D materials. Until now, the liquid metal-assisted synthesis mainly realized 2D growth and growth behavior control of 2D materials, and this chapter will discuss from these two parts in detail.

5.2.4.1 Liquid Metal for the Two-Dimensional Growth of Materials

The liquid metal-assisted 2D growth mainly achieves by three approaches (liquid metal substrate, liquid metal skin, and vapor). And then the recent progresses of as-mentioned approaches are discussed in detail.

Liquid Metal Substrate-Assisted Synthesis In the CVD process, the selection of growth substrates is critical. The influence of growth substrates mainly reflects on the decomposition efficiency and supply method of the precursors, growth rate, and final quality of 2D materials. To date, using liquid metal as substrate has developed in the 2D growth of many materials. The appropriate catalytic ability, smooth isotropic surface, fast diffusion rate, and great tolerance for precursors of liquid metal play an important role in limiting the material dimension down to 2D. Besides, the surficial metal layers of the liquid metal substrate can act as reaction media to assist the synthesis of 2D compounds due to their low-energy liquid surfaces.

The self-limited growth using liquid metal substrate has been well developed in the graphene growth for the layered control. During the graphene growth process, the growth substrates play an important role in catalytically decomposing the carbon source [87–90]. Thus far, many transition metals (such as Fe, Ni, Co, Cu, Pt, Au, Pd, Ru, Rh, and Ir) and their alloys group (Ni–Mo, Cu–Ni) have shown their outstanding abilities to grow graphene as substrates [42, 91–101]. However, owing to the difficult-to-control graphene growth kinetics on their heterogeneous solid surface, what results in an uncontrollable carbon precipitation and final causes inhomogeneous nucleation at grain boundaries, resulting in yield of graphene with uncontrollable layers and quality. This problem could be solved by employing a single-crystal metal substrate to grow uniform high-quality monolayer graphene (MLG), but this approach is also limited by its high cost and finite yield [102–105].

Many liquid metal substrates were confirmed to be suitable for growing graphene. In the early stages, the researchers have fabricated graphene thin films using liquid Ga [46, 106–108], and molten Ni and Cu [15, 109]. Amini and coworkers demonstrated that the molten phase (Ni and Cu) could grow monolayer or few-layer graphene films [109]. First, the carbon was dissolved into a molten metal, and then the dissolved carbon could nucleate and grow on the surface of the molten metal with decreasing temperature. This work provided a new approach for growing large-area graphene films, although the obtained graphene films embody nonuniform layers, the graphene growth on the liquid metal surfaces is proved to be reasonable. Two years later, Geng and coworkers first reported that liquid Cu is a particularly efficient approach for uniform single-layered hexagonal graphene crystals and continuous MLG films [15]. Figure 5.4a illustrates the approach of hexagonal graphene crystals formed, the grain boundaries of polycrystalline Cu were completely eliminated during the melting process, and leading to uniform nucleation distribution and low nucleation density of graphene. As shown in Figure 5.4b,c, the graphene crystals gradually realized a clear alignment relation when the density or coverage of graphene crystals on the Cu surface increased. With the growth time extended to two hours, the shape and edges of the hexagonal graphene crystals disappeared and then formed a large area of continuous graphene film (Figure 5.4d). Figure 5.4e shows the transferred graphene on SiO_2/Si substrate, Raman spectroscopy of transferred graphene is exhibited in Figure 5.4f, showing the single-layer and high-quality characteristics of the obtained graphene. The electrical characterization of the hexagonal graphene was also measured, and the resistance of the single-layer graphene device is approximately 87 Ω (Figure 5.4g). Besides, the author also found that the 2D resistivity showed no connection with the width of hexagonal graphene crystals, which showed that the electrical properties of large hexagonal graphene crystals are microscopically uniform.

The aforementioned works trigger the realization that the liquid Cu is effective mean for the preparation of graphene. Except for the molten metals, the liquid metal Ga, which owns super low melting point, also attracts the researchers' attention. Fujita and coworkers developed the preparation of few-layered graphene at the

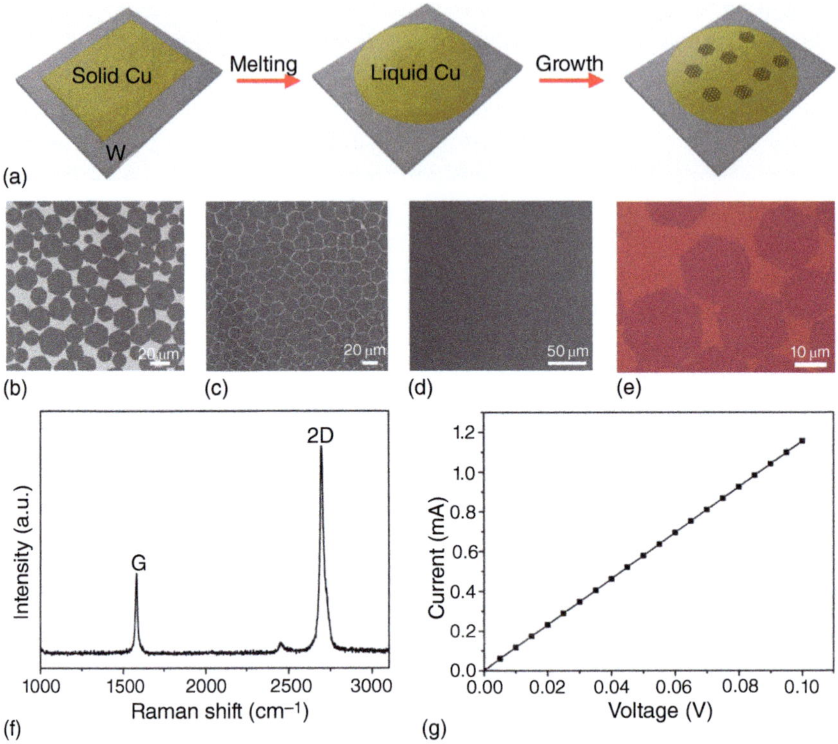

Figure 5.4 Liquid Cu-assisted self-limited growth of monolayer graphene. (a) Scheme showing CVD process for the synthesis of HGFs on liquid Cu surface. (b–d) SEM image showing partially covered and well-dispersed HGFs (c), a compact assembly HGFs (c), and continuous graphene film with uniform contrast (d). (e) Typical optical image of HGFs transferred onto 300 nm SiO_2/Si substrate. (f) Typical Raman spectroscopy of the transferred HGFs showing single-layer characteristics and no detectable D-band. (g) The corresponding I–V curve of the device with a resistance and 2D resistivity values. Source: Geng et al. [15].

solid amorphous carbon/liquid Ga interface [106–108]. An amorphous C film was pressed onto liquid Ga surface and then annealed at 1000 °C for 30 min under a background pressure of 5×10^{-4} Pa. The author presented that liquid Ga exhibits as a good graphitizing catalyst, owing to the insolubility of C in Ga, the graphitization only can occur on the surface. The obtained few-layered graphene always consisted of four to six layers even when the thickness of the C source was changed. However, the growth of multilayered graphene appeared to be strongly restricted by the remaining amorphous carbon backing layer, resulting in poor crystallinity due to the lack of continuity of the graphene layer. And its electrical performance is also strongly compromised. This work provided a new liquid metal–graphene system by utilizing the catalytic property of Ga and the insolubility between C and Ga, which gives the follow-up researchers a lot of inspiration.

The potential for single-layer graphene growth on liquid Ga is better embodied by changing the supplied approach of carbon source. Wang and coworkers reported that liquid Ga as growth substrate is an effective mean to controllably acquire

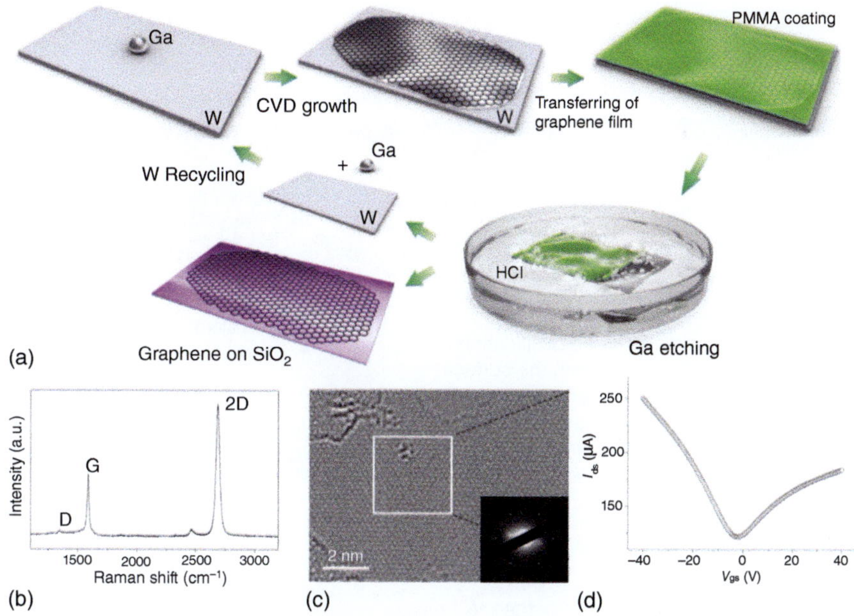

Figure 5.5 Liquid Ga-assisted self-limited growth of monolayer graphene. (a) Schematic illustration of graphene growth over liquid Ga supported on a W substrate. (b) Raman spectrum of the single-layer graphene. (c) HRTEM image and the selected area electron diffraction pattern (inset) of the single-layer graphene. (d) The typical transfer characteristics of the obtained graphene FETs under ambient conditions. Source: Wang et al. [46].

high-quality monolayer crystals by atmospheric pressure CVD (APCVD) [46]. Figure 5.5a illustrated the process of the growth and transfer of graphene. First, liquid Ga could well spread on a W support foil due to the great wettability between W and Ga, the graphene growth was realized at 1020 °C with 5–20 sccm CH_4 as C source in an APCVD system. After growth process, the graphene could be separated from Ga by poly(methyl methacrylate) (PMMA) coating and dilute hydrochloric acid (HCl) etching, the as-transferred graphene was placed on a SiO_2/Si substrate, and the W substrates could be reused as a Ga-support. The Raman spectroscopy shown in Figure 5.5b substantiated the obtained graphene was monolayer with relatively low defect density, and high-resolution transmission electron microscope (HRTEM) image and selected area electron diffraction (SAED) patterns further verified the single-crystal nature and high-quality of the obtained graphene (Figure 5.5c). The electrical properties were also representative indicator of graphene quality, as shown in Figure 5.5d, the extracted carrier mobility of electrons was as high as 7400 $cm^2/V\,s$. The author believed that this remarkable high mobility mainly benefits from the low vapor pressure at high temperature of Ga, which avoided the evaporates and redeposits during synthesis.

Whereafter, Zeng and coworkers extended the liquid metal system for the single-layer graphene growth. The author choose a series of liquid metals (Ga, In, and Cu) as the growth substrates, and strictly obtained MLG films with remarkable

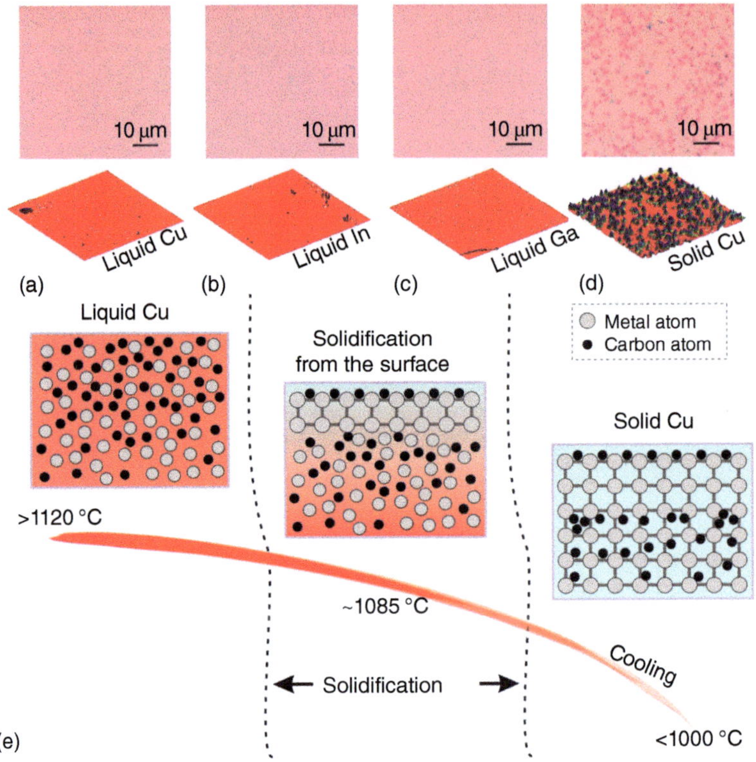

Figure 5.6 The mechanism of liquid metal-assisted self-limited growth of monolayer graphene. (a–d) Optical microscope images and the corresponding layer distribution analysis of graphene grown on liquid Cu, liquid In, liquid Ga and solid Cu, respectively. The layer thickness is represented by different colors as seen from the bottom color codes. (e) Schematic illustration of the carbon distribution in liquid Cu, solid Cu, and the solidified surface of liquid Cu during phase transition. Source: Zeng et al. [1].

tolerance of synthesis conditions [1]. Figure 5.6a–d showed the optical microscope (OM) images of the transferred graphene films, which grow over liquid Cu, In and Ga, and solid Cu, respectively. The graphene films grown on the liquid metals exhibit outstanding uniformity; however, the graphene films obtained from the solid Cu presented local nonhomogeneous domains, and the visual false 3D images further revealed the different uniformity between the graphene films obtained from liquid metals and solid Cu. To explore the advantages of the liquid metals, the author investigated the carbon distribution of Cu surfaces in three different states (liquid, solid, and the solidified surface) (Figure 5.6e). The X-ray photoelectron spectroscopy (XPS) depth profile indicated the carbon concentration in the liquid Cu remain at a quite high level from the surface to the bulk. Compared with that in the solid Cu, the liquid Cu showed a better ability to dissolve redundant C. During the cooling process, the frozen metal lattices near the surface blocked carbon precipitation from the deeper liquid Cu, and the deeper liquid Cu also can buffer the redundant carbon supply. Owing to this self-limited surface catalytic process, the grown graphene

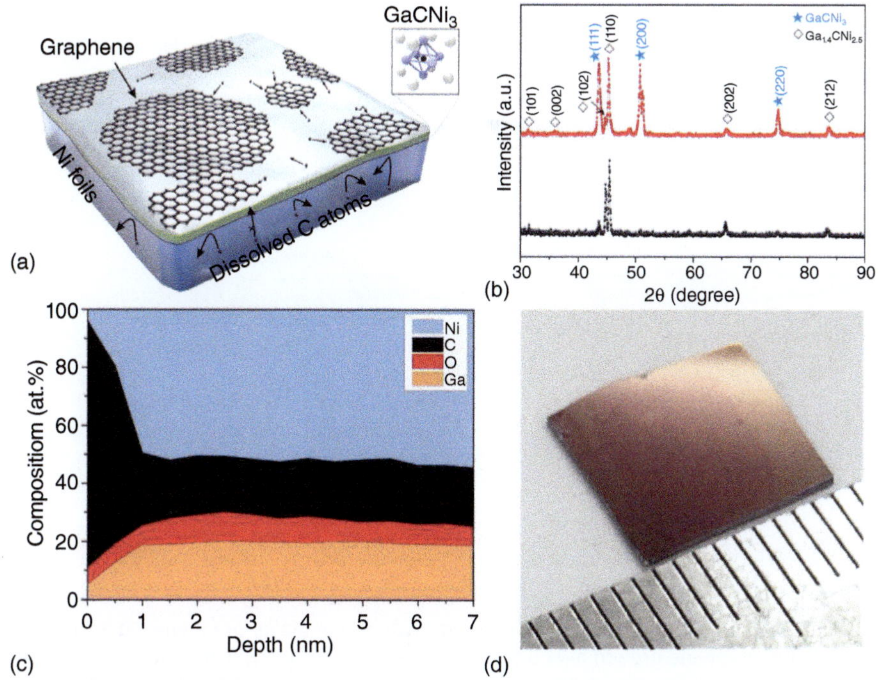

Figure 5.7 Liquid alloy-assisted self-limited growth of monolayer graphene. (a) Schematic illustration of the GaCNi3 formation and the suppressed upward precipitation of the carbons. (b) XRD spectra of the Ga–Ni foils after and before the introduction of CH_4. (c) X-ray photoelectron spectroscopic analyses of graphene on a Ga–Ni foil. XPS composition profiles of elements along the surface normal direction on the foil. (d) Graphene film transferred onto a SiO_2/Si substrate. Source: Chen et al. [111].

realized strictly single-layer and the growth conditions could tolerate robust variations. Besides, the quasi-atomically smooth surface of liquid metal could avert the formation of defects and grain boundaries. It seems that the liquid metal may offer a universal solution for the growth of high-quality, single-layer graphene.

In addition, the liquid alloy system (such as In–Cu, Sn–Ni, Sn–Ag–Cu) was also proved as appropriate substrate for the preparation of graphene [110]. Chen and coworkers utilized liquid Ga to assist iron group metals (such as Fe, Co, Ni) to form an antiperovskite layer and realized the uniform single-layer graphene films with 100% surface coverage [111]. Figure 5.7a demonstrated the process of the formation and effect of the antiperovskite layer. The structure of the antiperovskite layer was confirmed in Figure 5.7b; the X-ray diffraction (XRD) pattern appeared as obvious signal of carbides, principally $GaCNi_3$ after CVD growth, which not only possessed excellent stability throughout high temperature, but sealed the passageway of carbon transmission from the highly dissolved carbon metals during cooling process. And the XPS depth profile of Ga–Ni foil further proved the formation of carbide and the process of carbon sealing (Figure 5.7c). Figure 5.7d showed the OM image of a transferred graphene film on SiO_2/Si substrate, which proved the uniformity of the obtained graphene film on a large surface coverage (~0.4 cm^2). Similar to the single

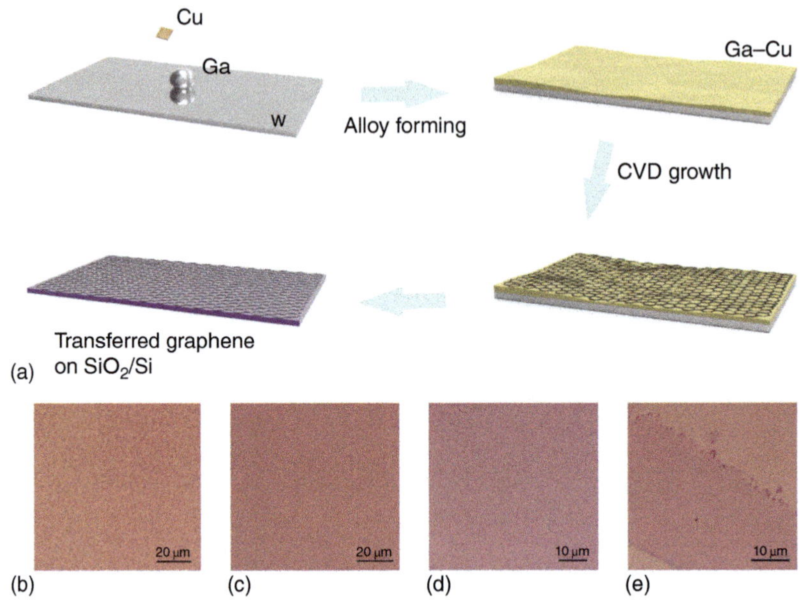

Figure 5.8 Liquid alloy-assisted self-limited growth of monolayer graphene. (a) Schematic illustration of graphene growth over liquid Ga–Cu system. (b–e) Typical optical microscope images of graphene films grown on Ga–Cu alloy with different Ga concentration. Source: Wang et al. [112].

liquid metal, the liquid alloy substrate also succeeds in effectively suppressing of the extra carbon deposition process and presents excellent tolerance in robust variations of the growth conditions.

For the design of the liquid alloy, many other transition metals also added to reform the preparation conditions or the growth behavior of graphene. Wang and coworkers realized the growth of uniform graphene at reduced temperature on Ga–Cu alloy [112]. Compared with the liquid Ga, the Cu–Ga alloy not only has much lower surface tension, but provides a better catalyst to decompose CH_4, leading to growth of large-area, high-quality graphene at lower temperature. Figure 5.8a illustrated the preparation process of graphene on liquid Ga–Cu alloy. For better understanding the Cu–Ga alloy system as growth substrate, the author investigated the influence of Cu concentration in Cu–Ga alloy (Figure 5.8b–e). When the Cu concentration decreased from 80.00 to 52.00 atom%, the graphene films gradually became uniform and realized 100% surface coverage, and this result could remain until the Cu concentration decreases to 30 atom %. With the Cu content further decreased to 0.00 atom %, the uniformity of the graphene films did not change while its surface coverage is reduced owing to the poor spreadability of Ga over W. This work designed the Ga-based liquid alloy as growth substrate to carry out the large-area, low-cost, and high-quality graphene, which might open a new approach at the field of decreased-temperature growth.

Except for single-layer graphene, designing the bilayer graphene can realize the widely tunable bandgap and lead to many interesting physical phenomena

[113, 114]. And magic-angle bilayer graphene ("magic" twist angles of around 1 degree) got more attention owing to its notable superconductivity and orbital magnets [115], these excellent properties make the bilayer graphene as a prospective material for electronic, photonic, and spintronic devices [116–119]. Synthesizing large-area, high-quality, uniform bilayer graphene by a scalable approach is critical. Although the traditional method of mechanical exfoliation provides a straightforward method to prepare high-quality bilayer graphene, the controllability, yield, and size of bilayer graphene needs to be promoted for further applications [38, 120]. Ma and coworkers designed a core–shell structured liquid Pt_3Si/solid Pt substrate to achieve the preparation of high-quality wafer-scale AB-stacked bilayer graphene (AB-BLG) by CVD, and the AB-BLG also possessed millimeter-sized grains with 100% coverage between two layers [121]. First, the author synthesized the Pt_3Si/Pt substrate by coating Si film on polycrystalline Pt foil and annealed in H_2 atmosphere at 1095 °C for 12 h, the Pt_3Si composition was further confirmed by XRD (Figure 5.9a). The uniform MLG shown in Figure 5.9b was formed in mixed gas flow of CH_4 and H_2 at 1100 °C for 10 min. The cooling process with a slowly cooling rate (12.5 °C/h) to 1025 °C contributed to the growth of the second layer domains, and these domains could expand and join together for a wafer-scale uniform BLG film (~$1.5 \times 3.5\,cm^2$) with prolonging the growth time (Figure 5.9c). The process of AB-BLG interlayer epitaxial growth on liquid Pt_3Si/solid Pt substrate was illustrated in Figure 5.9d–g. Owing to the high carbon solubility of solid Pt core and the low carbon solubility of liquid Pt_3Si shell, the liquid Pt_3Si shell could act as a diffusion channel to get rid of excess carbon, so that the excess carbon atoms could diffuse into the Pt core during the growth of MLG. Subsequently, as the temperature decreases, the solubility of carbon in the Pt core would reduce, which could cause the precipitation of carbon atoms dissolved in the solid Pt core and the formation of the second layer with interlayer epitaxy under the preformed MLG. Furthermore, the liquid Pt_3Si shell could also clear off the effect of different crystallographic lattices, defects, and grain boundaries of the normal polycrystalline Pt, which is contributed to the growth of uniform BLG.

Owing to the chemical inertness of liquid metal, the liquid metal substrate could help to grow various 2D materials (such as TMDs and transition metal carbides (TMCs)) with great tolerance for precursors and avoid corrosion of the reactants to the substrate [122–125]. Xu and coworkers reported the fabrication of high-quality ultrathin α-Mo_2C crystals on Cu/Mo substrate at the growth temperature above 1085 °C [126]. In such high temperature, the Cu was molten and the Mo–Cu alloy could form at the interface of liquid Cu and Mo foil. After that, Mo atoms transferred to the surface and reacted with the carbon atoms to form α-Mo_2C crystals (Figure 5.10a). A very low concentration of carbon source not only resulted in the ultrathin thickness of α-Mo_2C crystal shown in Figure 5.10b,c, but hampered the formation of graphene. Moreover, the 2D superconducting characteristics of ultrathin α-Mo_2C crystals were revealed, which were consistent with Berezinskii–Kosterlitz–Thouless behavior and presented strong anisotropy with magnetic field orientation. Analogously, this growth strategy also developed for fabricating the other ultrathin TMCs crystals, such as WC and TaC, which were

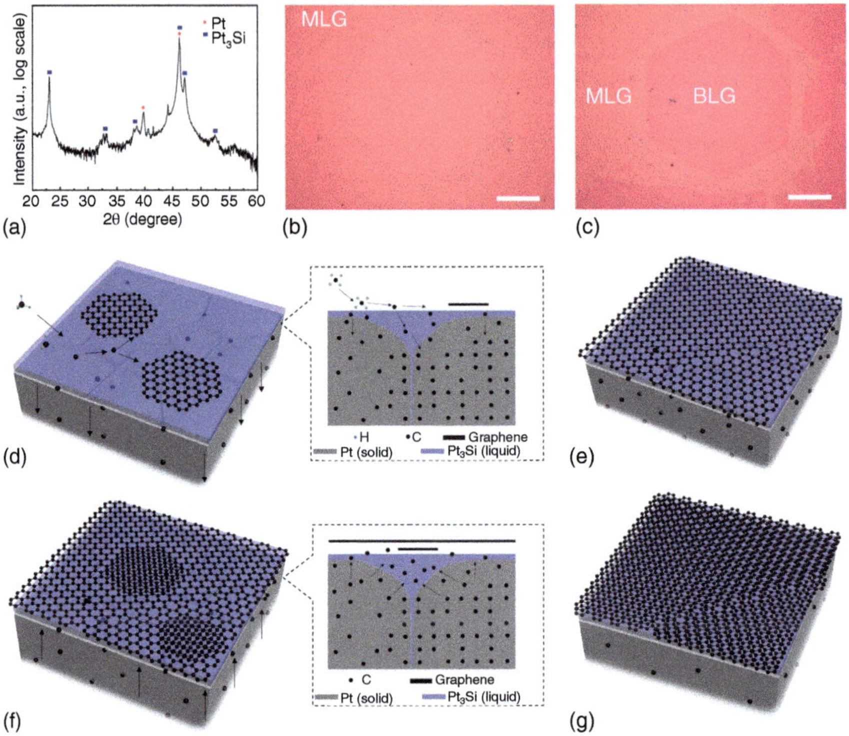

Figure 5.9 Liquid alloy-assisted self-limited growth of bilayer graphene. (a) XRD pattern of the substrate, showing the coexistence of Pt and Pt3Si. (b–c) Optical images showing the formation process of BLG. (b) A MLG film obtained after 10 min growth at 1100 °C. (c) A MLG film with the second layer domains obtained after slow cooling for 4 h. (d–g) Schematic for the interlayer epitaxial growth of AB-BLG film on liquid Pt3Si/solid Pt substrate. Source: Ma et al. [121].

obtained by using W or Ta foils to replace the Mo foils as the supporting substrate. Except for the TMCs, the substrate-induced alloy strategy has developed for many other 2D materials later. Shivayogimath and coworkers provided a universal approach for fabricating ultrathin 2D binary compounds by epitaxial growth on the Au-based alloy substrate [127]. The gold (Au) was selected to form single component solid solution with alloying metal M as the substrate. Specifically, the metal M (~20 nm) was sputtered onto a sapphire substrate first, then covered with Au (~500 nm). The preheated process above 850 °C helped to form M–Au alloy layer, and the concentration of M in M–Au alloy was limited to ≤5 at % for maintaining single-phase solid solution of Au as substrate. Au-based single-phase solid solution could also limit the solubility of the vapor-phase precursor (the elements: S, Se, Te, N) for controlling the thickness of the obtained compounds in few-atom-thick layers at the surface of the alloy. Besides, the Au didn't react with the vapor-phase precursors at the growth temperature so that Au could be used as template for the epitaxial growth of the obtained materials. The synthesis process was illustrated in Figure 5.10d, and a series of 2D materials, such as sulfides, tellurides, nitrides, and selenides, were obtained by this approach.

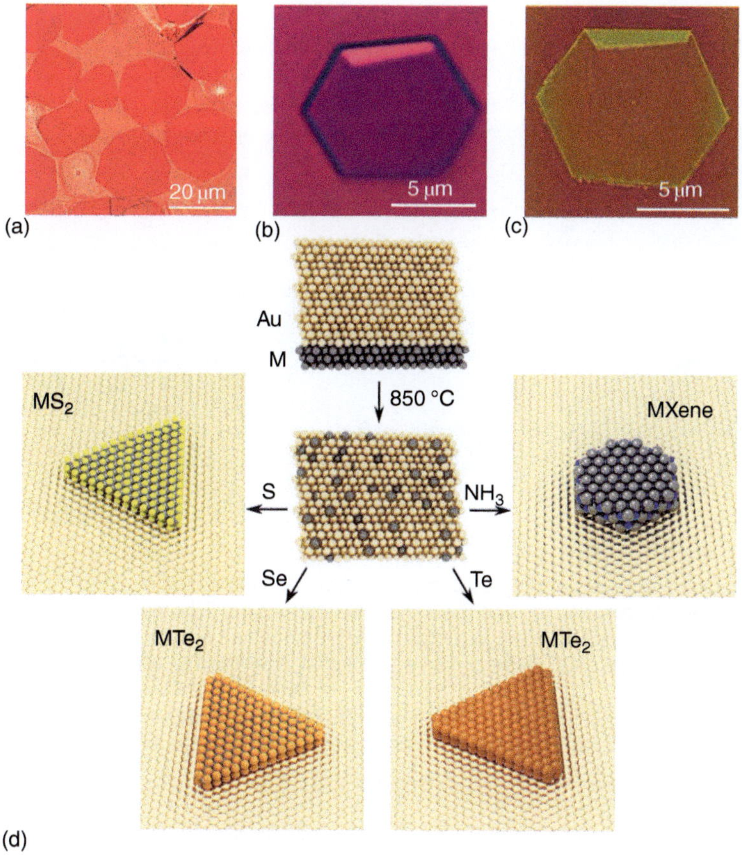

Figure 5.10 Liquid metal substrate-assisted synthesis of the 2D materials. (a) Typical optical image of ultrathin α-Mo2C crystals with different regular shapes on a Cu/Mo substrate. (b–c) Optical image and AFM image of hexagonal ultrathin α-Mo2C crystal with thicknesses of 6.7 nm. Source: (a–c) Xu e al. [126]. (d) Schematic overview of the synthesis process. A thin layer (~20 nm) of metal M and a thick layer (~500 nm) of Au are sputtered onto a c-plane sapphire substrate, respectively. The Au-M alloy is then exposed to a vapor of S, Se, Te, or more generally an elemental X vapor. Source: Shivayogimath et al. [127].

The liquid metal substrate with a low-energy liquid surface can act as reaction media to fabricate desired 2D-layered compounds. For instance, Zhou and coworkers reported the growth of 2D-layered GaSe and $GaxIn_{1-x}Se$ crystals on Ga and eutectic gallium–indium binary alloy (EGaIn), respectively, by utilizing the low-energy liquid metals with ultralow melting points (Figure 5.11a–d) [19]. Specifically, these room temperature liquid metals offered a low-energy surface for growing 2D materials. Meanwhile, these liquid metals were used as both growth substrates and reactants at the temperature about 100–150 °C, and hence the formation of GaSe and $GaxIn_{1-x}Se$ single crystals can be realized as the Ga–Se nanoclusters vaporized from the Ga_2Se_3 precursors when reacted with the Ga or In–Ga metals. This method provides the potential of low-temperature synthesis of 2D layered materials by utilizing

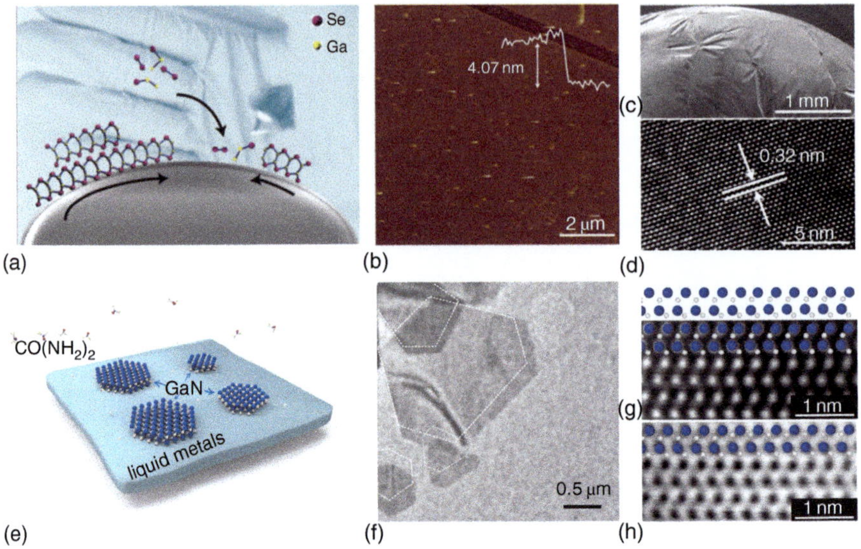

Figure 5.11 The surficial layers of the liquid metal substrate acting as the reaction precursors to fabricate 2D compounds and their characterizations. (a) Schematic of the growth of 2D GaSe single crystals on liquid metal. (b) AFM image and height profile of 2D GaSe crystal. (c) A SEM image of 2D GaSe crystal on liquid Ga after growth. (d) A HRTEM image of 2D GaSe crystal grown on liquid Ga. Source: (a–d) Zhou et al. [19]. (e) Synthetic diagram of the growth of 2D GaN single crystals via the SCNR method. (f) A low-magnification TEM image for the morphology characterization of the 2D GaN single crystals. (g) HAADF-STEM and (h) BF-STEM images of the 2D GaN crystal collected along the $[10\bar{1}0]$ zone axis. Source: (e–h) Chen et al. [128].

these ultralow melting point liquid metals, which satisfies the request of industrial procedures.

Additionally, the surficial layers of the liquid metal can also be utilized to grow 2D nonlayered GaN crystals on the liquid metal substrate through a surface-confined nitridation strategy, as shown in Figure 5.11e [128]. The growth process on the Ga metals spreading over the tungsten (W) foil was carried out in a CVD system, with urea acting as the nitrogen source. Specifically, stratified Ga/Ga–W structure induced by surface tension difference of the Ga and W was created to realize the confined nitridation process. Thereinto, the surficial ultrathin Ga layers on the liquid metal surface can act as templates to form 2D GaN single crystals, simultaneously, the inward Ga–W alloy would hinder the thickening of crystals as a result of the greater nitridation ability of W metal than Ga metal. Therefore, the 2D growth behavior can be achieved, suggesting that the liquid metals were a facile substrate to grow 2D materials by controlling the surficial nitridation reaction on the liquid surface and the reaction energy barrier in the liquid metal bulk. The micrometer-sized GaN crystals were confirmed by the HRTEM image (Figure 5.11f). The collected high-angle annular dark-field scanning transmission electron microscope (HAADF–STEM) as well as BF-STEM images along the $(10\bar{1}0)$ plane gain insight into the structure (Figure 5.11g,h), in which the Ga atoms were located at the tetrahedral sites and were consistent with the wurtzite structure. The as-obtained 2D GaN displayed obvious blue shift of the photon energy in comparison with bulk GaN, suggesting the high quality and the 2D nature of GaN achieved by this confined nitridation method.

Liquid Metal Oxide Skins-Assisted Synthesis The surface of the liquid metals, such as Ga-based alloys [129, 130], easily formed self-passivating thin oxide skins under ambient conditions or the oxide environment [131]. Particularly, these self-passivating films with atomic thickness play an important role in synthesizing 2D nanomaterials.

Actually, the atomically thin oxide skin of the liquid metal is regarded as a kind of 2D material. Zavabeti and coworkers reported liquid eutectic Ga as a reaction media to fabricate various 2D oxides [20]. Inspired by the phenomenon that self-passivating oxidation behavior of gallium alloys such as EGaIn and gallium–indium–tin ternary alloy (galinstan) were exclusively composed of gallium oxide [129, 130], this work created co-alloyed 2D oxides at the air–metal interface by selecting co-alloying reactive metals in view of the thermodynamics. Namely, the oxide layer resulting in the maximal decrease of Gibbs free energy (ΔG_f) would occupy the surface (Figure 5.12a), and then these surficial oxides would be promising to be synthesized as 2D materials after separating them from the alloyed metals. They imported ~1 wt% of elemental hafnium (Hf), aluminum (Al), and gadolinium (Gd) to the galinstan alloys to form the precursors for the synthesis of the respective oxides. These oxides satisfied the needs of a smaller ΔG_f than Ga_2O_3, and then the surficial 2D oxides on the liquid alloys can be formed (Figure 5.12a,b). The synthesis of three different 2D oxides, such as transition metal (HfO_2), a posttransition metal (Al_2O_3), and a rare earth metal (Gd_2O_3), demonstrated the versatility of the approach. In these obtained oxides, HfO_2 nanosheets displayed a promising insulating property in electronics.

To obtain these 2D oxide skins, effective method is required to separate these ultrathin films from the liquid metals. In this work, they reported two strategies for isolating the ultrathin oxide films. The first one was a vdW exfoliation approach (Figure 5.12c), which allows printing the 2D oxide films of the liquid metal onto a targeted substrate. Typically, the macroscopic forces between inside metal and its oxide films is weak, which allows a clean delamination of the liquid metal [132]. This approach can be applied for producing the highly crystalline 2D oxide sheets. The second one of forming oxides depended on injecting the air into the liquid metal (Figure 5.12d). The oxides formed quickly when air bubbles passed across the liquid metal. In addition, these air bubbles enabled the oxides to be obtained in an aqueous suspension when these air bubbles rose through the water. This approach is promising for producing high-yield 2D oxides. The surficial oxide layer of the liquid Ga metals such as the Ga_2O_3 can also be exfoliated by the printing technique to prepare the quasi-2D semiconducting β-Ga_2O_3 in constructing electronic devices [133].

The surficial oxide layer of the liquid metals can also act as templates to grow wafer-area films [49–52], whatever they are intrinsically layered materials or nonlayered materials. For instance, the large-scale 2D layered GaS semiconductors

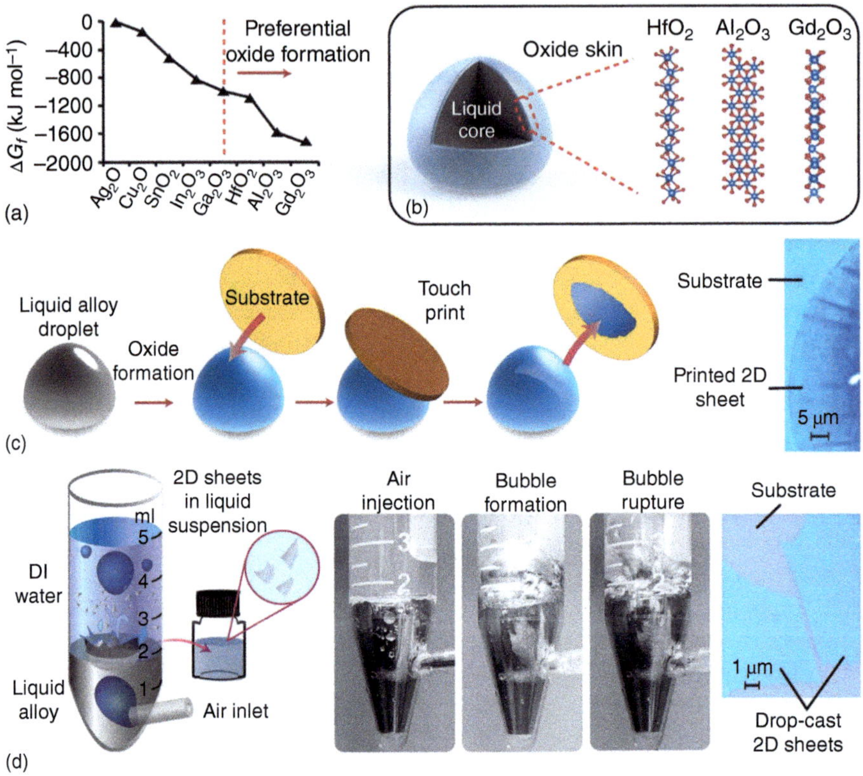

Figure 5.12 Liquid metal printing technique for exfoliating its films. (a) Gibbs free energy for the oxides. (b) The crystal structure diagram of the HfO_2, Al_2O_3, and Gd_2O_3 layers. (c) The diagrams of the van der Waals exfoliation technique. An OM image of the printed 2D sheet can be shown at the right. (d) The diagrams of the gas injection approach (left), photograph images of the gas injection through the liquid metal (center), and an OM image of the 2D materials obtained on the SiO2/Si substrates (right). Source: Zavabeti et al. [20].

can be obtained by utilizing this oxide layer as templates [49]. The first step was to transfer the oxide layer liquid Ga onto the targeted substrates. Notably, the atomically thin oxide layer increased the wettability of the Ga put on SiO_2/Si substrates, thus printing the thin oxide layer on the substrates. This ultrathin oxide film serving as templates was first exposed to HCl vapor and then was sulfurized through a low-temperature process, then the wafer-scale GaS films with unit cell thickness (~1.5 nm) were fabricated. The as-obtained GaS displayed an electron mobility of ~0.2 cm^2/V/s and an on/off ratio above 150, which were comparable to the exfoliated GaS layer in the previous reports, indicating the high quality of the GaS achieved by this liquid metal-assisted printing approach. This method provides etchless patterning of wafer-scale 2D nanomaterials, which is compatible with the industry applications.

Apart from the synthesis of layered 2D materials, this printing process could be extended to grow the 2D nonlayered materials. For example, the gallium phosphate ($GaPO_4$), which does not intrinsically crystallize in a layered structure, hence cannot

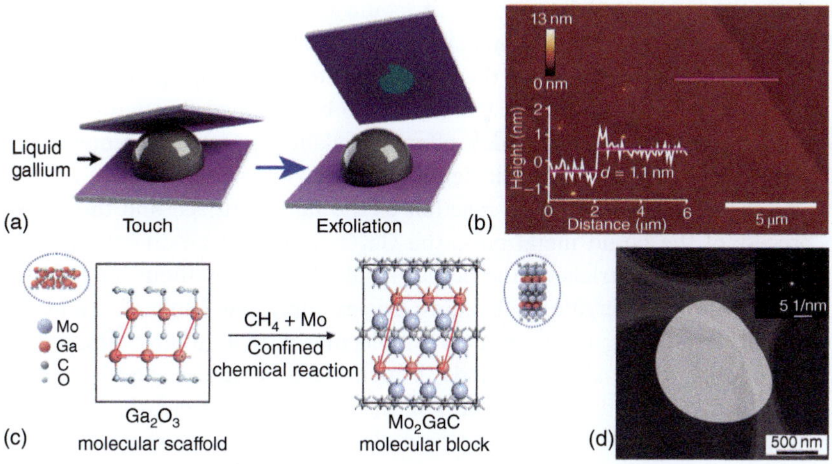

Figure 5.13 Oxide skins of the liquid metal-assisted the synthesis of 2D materials and their characterizations. (a) Synthetic diagram of the 2D printing approach of Ga_2O_3 films from liquid Ga. (b) AFM image for the morphology characterizations of the 2D $GaPO_4$ films. The inset shows the corresponding height profile. Source: (a, b) Syed et al. [50]. (c) Schematic of the growth of 2D Mo_2GaC single crystals via the design of Ga_2O_3 molecular scaffold on the liquid metal surface. (d) A low-magnification HAADF-STEM image for the morphology characterization of a Mo_2GaC single crystal. Source: (c, d) Zeng et al. [134].

be achieved by conventional exfoliation approaches. Syed and coworkers developed the growth of liquid metal-based 2D semiconductors at a low temperature by this printing technique [50]. The ultrathin Ga_2O_3 layer of liquid Ga was first exfoliated (Figure 5.13a), followed by a vapor-phase reaction with phosphoric acid at temperature of 350 °C. Then the large-area ultrathin $GaPO_4$ nanosheets with wide bandgap can be prepared. The atomic force microscopy (AFM) investigation indicated the $GaPO_4$ nanosheets with unit cell thickness (Figure 5.13b). The as-prepared $GaPO_4$ nanosheets exhibited a large piezoelectric coefficient, which was collected by using a combination of piezoresponse force microscopy (PFM). The phosphatization process conducted at relatively low temperature was compatible with device fabrication processes, offering a promising method for creating 2D piezoelectric nanomaterials.

The Ga_2O_3 skin can be extended to the synthesis of the nonlayered 2D GaN nanosheet with a wurtzite structure [52]. The liquid gallium droplet was squeezed between two SiO_2/Si substrates to deposit the Ga_2O_3 films; the exfoliated ultrathin oxide layer can be obviously confirmed by the optical image. Then the 2D GaN nanosheets can be synthesized after the ammonolysis reaction at temperature of 800 °C. The GaN film figuring a typical thickness of 1.3 nm suggested the 2D nature of the GaN. In addition, the GaN films are laterally large, with the films being translucent, demonstrating the thin nature of the 2D GaN films. The atomic resolution images of the lattice and the fast Fourier transform (FFT) patterns confirmed the wurtzite structure of GaN films. But the crystallinity of the GaN films are not very highly crystalline because of the oxygen doping in the GaN films, and as a result, no obvious band gap shift could be observable. But the carrier mobility of 21.5 $cm^2/V\,s$ of GaN films still indicates the potential of the GaN in photoelectric devices. In addition to the synthesis of the GaN, the centimeter-sized InN films featuring a thickness of 2.0 nm can also be obtained through this vdW squeeze-printing approach [52], which offers a scalable approach for integrating the 2D semiconductors into electronic devices.

Through controlling the surficial reaction of the liquid surface and the segregation process of the liquid metal bulk, the Ga_2O_3 layer can be directly used to grow 2D nonlayered materials on the liquid metal substrates without the transfer of these films onto the targeted substrates. For example, the films of the liquid Ga, atomically thin Ga_2O_3 layer, acted as a molecular scaffold to grow the strong interlayer-bonding MAX single crystals, which have mixed covalent–metallic character of the M–X bonds [134]. To design this molecular scaffold, the liquid Ga with poorly crystallized self-passivating oxide layer formed at room temperature was heated at the temperature above 300 °C, as a result, the poorly crystallized oxide layer was turned into a crystalline β-Ga_2O_3 layer, which served as a molecular scaffold to build the 2D Mo_2GaC crystals. As the schematic diagram shown in Figure 5.13c, with this ultrathin Ga_2O_3 layer acting as the molecular scaffold, the ultrathin Mo_2GaC crystals can be obtained after the supply of the CH_4 at 1000 °C. The Ga underneath the self-passivating Ga_2O_3 layer is kept in a liquid state at this temperature, subsequently, some Mo atoms originated from the Mo substrates would diffuse from the Ga bulk to its surface and reacted with the ultrathin Ga_2O_3 layers and the C atoms decomposed by CH_4 to grow Mo_2GaC crystals. The morphology of the ultrathin Mo_2GaC crystals can be confirmed by the low-magnification, HAADF-STEM image (Figure 5.13d).

The self-limiting oxide layer of the liquid metal can act as reaction media to assist the synthesis of 2D layered or nonlayered materials and can be extended to the deposition of other 2D nanomaterials in the future, which will provide a new inspiration for future optoelectronics.

Liquid Metal Vapor-Assisted Synthesis The liquid metals have a relatively high sublimation pressure [135], which allows sufficient quantity of metal particles to guarantee more collision possibilities with reactant species, thus accelerating and controlling the reaction process. Therefore, apart from the liquid substrates and their skins, the liquid metal vapor has also been employed for the synthesis of the 2D nanomaterials. The preparation of the monolayer WSe_2 crystals was reported via designing a Cu vapor-assisted growth approach [136]. Usually, layer aggregation and lateral growth of the TMDs would coexist during the reaction process, which would cause undesirable multilayer TMDs. The strategy utilized the Cu vapor to impede the undesired growth process. Specifically, the Cu vapor was imported to occupy the center site of the WSe_2 upper surface and hence suppressed the growth along the vertical directions, enabling only the lateral growth to occur, giving rise to monolayer WSe_2 (Figure 5.14a). Apart from TMDs, the liquid metal vapor could be employed to prepare other 2D materials. Generally, the nonmetal substrates such as quartz are chemically inert and hence have no catalytic reactivity for growing

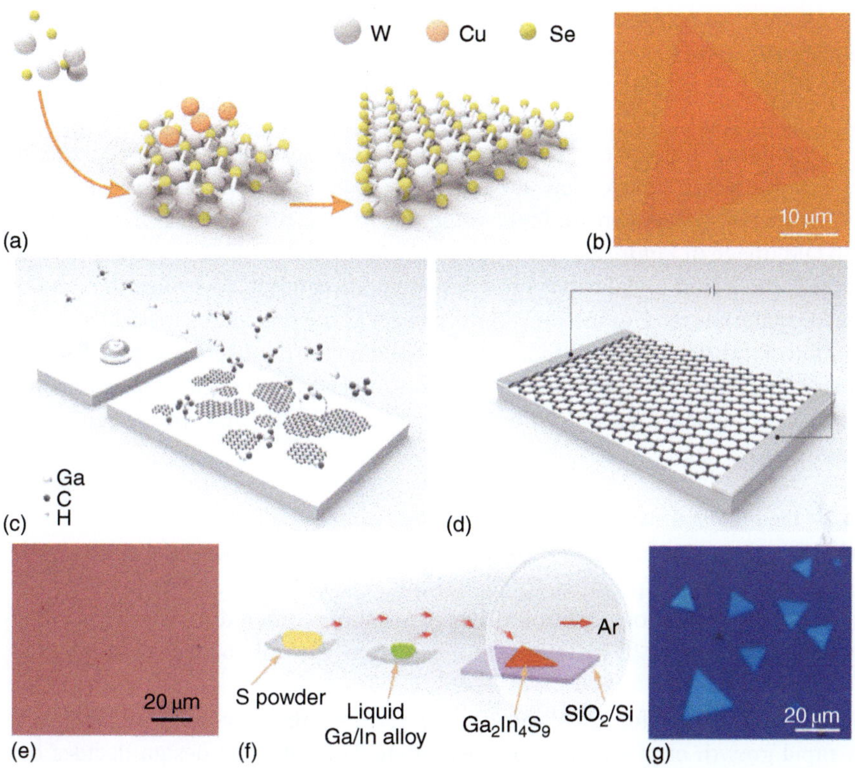

Figure 5.14 Liquid metal vapor-assisted synthesis of the 2D materials and its characterizations. (a) Schematic of the Cu vapor-assisted growth strategy. (b) OM image of a triangular WSe_2 single crystal. Source: (a, b) Liu et al. [136]. (c) Schematic of Ga vapor-assisted remote catalyzation for the directly growth of graphene on the quartz substrates. (d) Device structure diagram of the fabricated graphene defogger. (e) OM image of a graphene film transferred on 300 nm SiO_2/Si substrate. Source: (c–e) Tan et al. [137]. (f) Schematic of growing 2D Ga2In4S9 from the precursors of liquid Ga/In alloy and S powder. (g) OM image of triangular $Ga_2In_4S_9$ crystals on SiO_2/Si substrates. Source: (f, g) Wang et al. [138].

graphene, giving rise to a poor crystallinity and an uncontrollable layer number. The highly crystalline MLG synthesized on quartz by a Ga vapor-assisted growth strategy has been reported [137]. Specifically, the sublimation pressure of Ga metal is much higher than that of Cu metal at the same temperature and pressure [135]. Additionally, unlike Cu, Ga metals more easily absorb carbon than Cu [103, 139]. Therefore, sufficient quantity of Ga atoms can be provided to ensure more collision possibilities with reactant species, thus accelerating the decomposition of CH_4. Subsequently, the floating Ga atoms can prevent the accumulation of undecomposed carbon sources, and hence guarantee that self-limited growth of highly crystalline graphene growth on the quartz substrates, as the remote catalyzation illustrated in Figure 5.14c. The OM image (Figure 5.14e) confirmed the large-scale and uniform graphene. The as-obtained graphene was designed to act as highly effective defoggers (Figure 5.14d), indicating the highly crystalline graphene films. The liquid metal vapor can also be utilized for the 2D growth of the TMDs. For instance, the synthesis of 2D $Ga_2In_4S_9$ single crystals by utilizing the liquid Ga–In alloy has been reported [138]. The liquid Ga–In alloy with low melting points provides a reduced reaction temperature [19], simultaneously, it evaporates easily. In a liquid metal vapor-containing atmosphere, these metal atoms can serve as metal precursors to react with the sulfur powders, as illustrated in Figure 5.14f. The as-synthesized 2D $Ga_2In_4S_9$ single crystals were uniform triangles with micrometer-scale, as confirmed by the OM image in Figure 5.14g.

In brief, the liquid metal vapor provides adequate catalysts to realize the growth of the 2D materials, and, at the same time, can act as the precursors to accelerate the formation of the 2D crystals. Compared with solid metal, liquid counterpart displays obvious superiorities in controlling the layer number as well as the domain, which provides a promising strategy for the 2D growth during the vapor-phase deposition process.

5.2.4.2 Liquid Metal-Assisted Growth Behavior Control

Controlling the growth behavior of 2D materials has great potential for its wide applications. In recent progress, the liquid metal as reaction medium realized the rapid growth, morphology control, oriented stitching, ordered assembly, and heterogeneous stacking of 2D materials. In this section, we will discuss these aspects.

Rapid Growth Except for the layer-limited growth, the low-density nucleation with rapid growth of graphene is also essential. The substrate design decides the adsorption and diffusion during the growth process. As mentioned earlier, the quasi-atomically smooth surface of liquid metal catalyst, which offers a high diffusion rate and low nucleation density, has enormous potential for the fast growth of graphene [18, 88, 140]. Zheng and coworkers first proposed the rapid growth strategy of graphene single crystals on liquid Cu, which possessed millimeter-size and high-quality [2]. During the growth process of graphene crystals on the polycrystalline solid Cu, the occurrence of nucleation process is preferred at the places with high surface energy, including the crystal boundaries, defects, and impurities. Free electrons on the Cu surface would saturate the edges of graphene significantly, resulting in uneven nucleation density and confined crystal size. Compared with the solid Cu, the liquid Cu substrates could eliminate the places with high surface energy and provide a smooth surface with high diffusion rate. Figure 5.15a illustrated the approaches of C supply during the growth of graphene on liquid Cu, which could be attributed to both the surface diffusion and bulk segregation. The two approaches are also proved by the time-of-flight secondary ion mass spectra (TOF–SIMS) shown in Figure 5.15b. The depth profiling SIMS measurements expounded the different C dissolution behaviors in Cu substrates with two states, compared to graphene/solid Cu, the G/liquid Cu exhibited a higher saturation state of C content in the bulk phase, and the bulk segregation as extra approach of the precursor supply accelerated the growth rate of graphene crystals. The rapid growth result was shown in Figure 5.15c, when the time of CH_4 introduction to the CVD furnace is two seconds, the size of graphene grains could

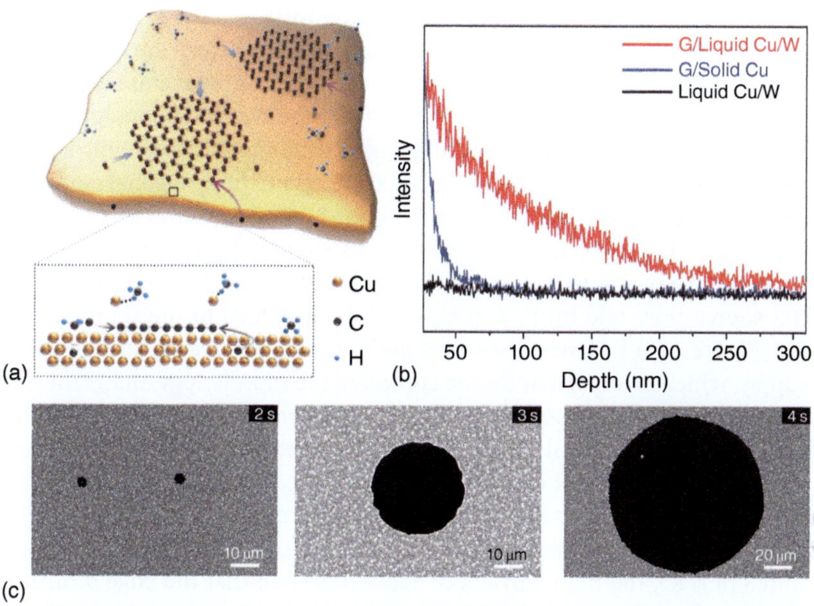

Figure 5.15 Liquid Cu-assisted rapid growth of graphene. (a) Schematic illustration of the precursor supply via the surface adsorption and the bulk segregation for the graphene growth on liquid Cu. (b) TOF–SIMS characterizations shown the intensity profile of carbon along the surface of G/liquid Cu and G/solid Cu. (c) SEM images of graphene single crystals synthesized at $t = 2$, 3, and 4 s. $t = 0$ s is defined as the moment when CH_4 is introduced to the CVD furnace. Source: Zheng et al. [2].

grow up to ~5 μm, which confirmed the accelerated nucleation of graphene on liquid Cu. Incidentally, the growth rate on liquid Cu reached up to 79 μm/s, this record was also higher than that reported on the solid Cu. This work further extends the possibility of the rapid growth behavior of graphene on the liquid metal, and this unique dual mass transfer strategy of liquid metal substrate may also develop to the rapid growth of various 2D materials.

Morphology Control The hexagonal graphene grain is the most widespread since it was obtained on solid Cu surface [141]. The precise morphology control of graphene is crucial for the research of the growth mechanism and has received intensive efforts [142]. Compared with the solid substrate, the liquid metal substrate provides more possibilities for the morphology of graphene grains. Wang and coworkers explored the relationship between the shape and size distribution of graphene domains grown on a liquid Ga surface [46]. The nucleation points are mostly less than 4 μm. When the domains grow up to 7–20 μm, the morphologies of graphene are mainly reflected in hexagons and dodecagons, and the graphene with twelve-sided symmetry may derive from the fast transfer of C atoms on a liquid Ga surface. Owing to the smooth surface of liquid Ga, the morphology of domains turns to near-circular shape so as to obtain a low surface energy as the growth of graphene. Finally, most graphene domains will grow up to a continuous film.

The morphology control of graphene is crucial to investigate the growth mechanism, it is still extremely challenging in precise control of the details during reaction process, which leads to a large effect of the morphology of as-grown graphene. Geng and coworkers reported the controllable synthesis of the twelve-pointed graphene grains (TPGGs) on liquid Cu surface and proposed the growth mechanism [16]. First, the carbon source flow rate was maintained at 1 sccm, the hexagonal graphene grains were successfully obtained as shown in Figure 5.16a. When the carbon source flow rate increased to 3–5 sccm, the interedge of six-pointed graphene grains further lead to form the seven- to twelve-pointed graphene grains (Figure 5.16b). With the carbon source flow rate up to 6 sccm, the TPGGs turned to the round shape eventually (Figure 5.16c). Hence, the stable grain growth required a relatively low carbon supply, which is important for the control of the morphology and grain size of graphene. The author proposed the edge-diffusion limited model to explain the shape formation and the critical grain size required of graphene grains. The energy barrier of edge diffusion of graphene, which was calculated through the density functional theory, was 0.56 eV (zigzag edge) and 2.02 eV (armchair edge), respectively. Compared with the energy barrier of thermal fluctuation (0.12 eV) at the growth temperature (1400 K), the high barriers of edges indicated that the edge diffusion was a rate-determining process. Then the local triangular tips would nucleate and fast grow in graphene grains with sizes beyond 12.5 μm. This work provides a deep thinking for the mechanism in morphology regulation of graphene and has inspiration in preparation of graphene with other morphologies. In addition to the TPGGs, Geng and co-workers also demonstrated the controlled growth of four-, five-, and six-lobed graphene grains on liquid Cu (311) surface [143]. It's worth noting that the Cu (311) surface is reconstructed through the mild oxygenation by Ar annealing. The Cu (311) surface with undulant texture and low surface density resulted in reducing the surface diffusivity of carbon sources. And the nonuniform diffusion of carbon sources to the edge sites would cause the instable graphene shape during the growth process. Some edges would grow faster and others disappear, which led to form multiple-lobed grains (Figure 5.16d). While decreasing the Ar flow rate and annealing time, the Cu (311) surface could turn to Cu (110), which contributed to the preparation of triangular graphene grains. This work proposes a substrate-induced growth mechanism of asymmetric graphene grains, which could be achieved by changing the lattice surface of liquid metal. And this approach might have the potential to realize the morphology control of other 2D materials.

Furthermore, the morphology control also could be realized in hierarchical graphene architectures (HGAs) by liquid metal-assisted. Luo and coworkers developed the growth of HGAs on liquid Cu surface [144]. The morphologies of HGAs, which include the simple stacking structure (stacking of three hexagonal graphene layers) and complex stacking structure (stacking of hexagonal, snowflake, and dendritic graphene layers), are mainly controlled by the flow rate ratio of Ar/H_2. In addition, the HGA morphology was not affected by the growth temperature, and a higher temperature leads to a faster growth rate. As simulated in Figure 5.16e, the model of the HGAs layer stacking consisted of three stacked layers of graphene. Except for simultaneous nucleating at the center, the nonsimultaneous nucleating

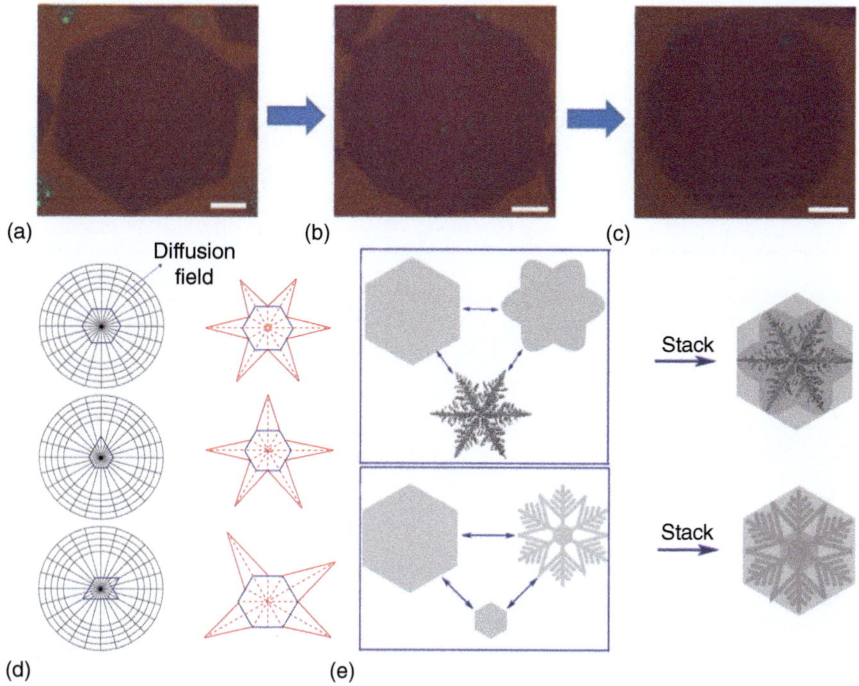

Figure 5.16 Liquid metal-assisted morphology control of graphene. (a–c) Transforming process from hexagonal to round-shaped graphene grain. All scale bars are 5 μm. Source: (a–c) Geng et al. [16]. (d) Schematics showing the mechanism of multilobe grain growth in the diffusion limited mode. Source: Geng et al. [143]. (e) Schematics showing the two different stacking models (hexagonal–star-shaped–dendritic stacking and hexagonal–snowflake–hexagonal stacking). Source: Luo et al. [144].

also occurred on the HGAs. The liquid Cu catalyzed the decomposition of C source and promoted the diffusion of C atoms under the preformed graphene. The final forms of graphene were determined by the diffusion competition of adatom along the edge of the island and surface. When the active C atoms diffused to adsorb, it is more likely to appear repeated and exhibit nonsimultaneous nucleation. So that the layer-stacking growth of graphene could occur from top to bottom. This work provided some insight into the precise design of stacking graphene based on liquid metal and indicated that the anisotropic electrical transport results could be achieved by designing the heterogeneous thickness of graphene.

Oriented Stitching Control Usually, the wafer-scale 2D materials grow on polycrystalline solid metal substrates suffer from inevitable grain boundaries problems, which result from the incommensurate dislocated stitching of the grains. Such nonnegligible grain boundaries would give rise to a polycrystalline quality as well as a declining electrical performance of the 2D materials [145, 146]. To settle these problems, a liquid Cu metal-assisted the growth of isotropic graphene grains (IGGs) was proposed to realize the commensurate-oriented stitching, which was caused

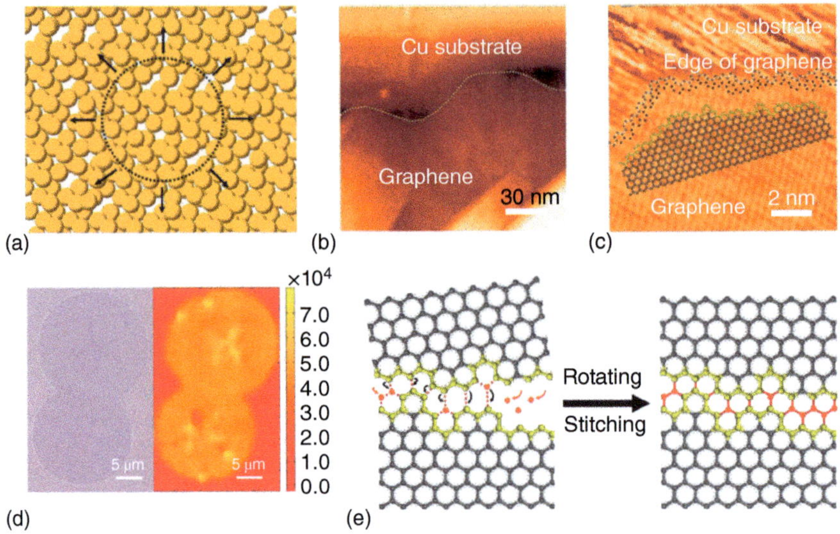

Figure 5.17 Oriented stitching control of graphene grains on the liquid metal Cu. (a) Schematic of the IGGs formed on the liquid metal Cu. (b) Scanning tunneling microscopy (STM) image demonstrating the edge atom structure of an individual graphene. (c) Atom-resolution STM image demonstrating the precise edge atom structure. (d) OM image and the corresponding Raman mapping of the 2D band. (e) Schematic of the stitching process of graphene. Source: Zeng et al. [17].

as a result of the isotropic liquid metal surface, as illustrated in Figure 5.17a [17]. Different from the solid metal substrates, which own an ordered crystal orientation, the liquid metal substrates have an amorphous and isotropic surface, which enabled an assemble ability of the carbon atoms along isotropic directions. Notably, the concentration of the provided carbon source should be introduced as low as possible, and hence the as-resulted slow growth rate allowed a fully isotropic diffusion of the carbon atoms along all directions on this liquid surface. Subsequently, the resulting round graphene grains would lead to a smaller rotational steric hindrance than that of the angular graphene grains. The edge of the IGG behave as a sawteeth arranged with periodicity (Figure 5.17b), and as a more precise edge structure information shown in Figure 5.17c, the carbon atoms arranged in both the armchair and zigzag structures can be observed. This active mixed edge structure drove the rotation of IGGs. Therefore, the low rotational steric hindrance effect as well as mixed edges contributed to a low energy to promote the stitching of the two neighboring graphene grains on liquid Cu because of its rheological characteristics. No detectable defects could be found in the interface of the two grains, as the OM and the corresponding Raman mapping images displayed in Figure 5.17d, suggesting the oriented stitching results. In brief, the self-adjustment of the orientation will drive the stitching between two neighboring IGGs, thus forming a perfect and smooth stitching, as displayed in Figure 5.17e, from which we can further understand the fundamental principles of this smooth oriented stitching process.

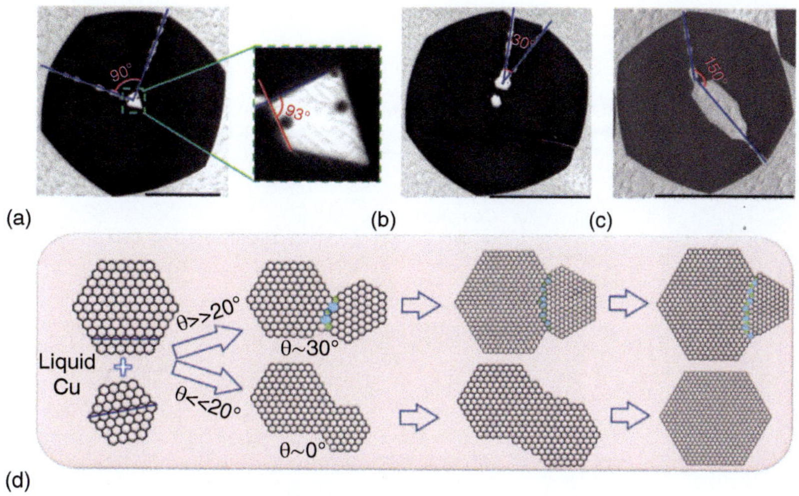

Figure 5.18 The formation of twinned graphene polycrystals on the liquid Cu substrate. (a–c) SEM images of polycrystalline graphene domains, which consist of two parts. The selected area from (a) is amplified for a better observation. The scale bar in (a) and (b, c) represents 10, 10, and 20 mm, respectively. (d) Schematic of the two different behaviors of graphene coalescence. Source: Dong et al. [147].

Except for the oriented stitching between two crystals with round shape, the oriented stitching behavior can also take place inside the crystal. Dong and coworkers reported the formation of 30 °C-twinned graphene polycrystals on the liquid Cu substrate, as the SEM images shown in Figure 5.18a–c [147]. The atomic simulations reveal two factors that lead to the formation of this unique domain alignment in graphene polycrystals. One is that formation energy of the 30°-grain boundaries in graphene is minimum, indicating the high stability of the 30°-grain boundaries in graphene. Another one is that the graphene islands can rotate freely during the coalescence process between the adjacent domains on the liquid Cu. Therefore, these would give rise to the formation of a 30 °C-twinned polycrystal or a perfect graphene single crystal, as the schematic illustrated in Figure 5.18d.

With the exception of the graphene, the fabrication of wafer-scale hexagonal boron nitride (*h*-BN) also suffers from the problems of defects and grain boundaries. Tan and coworkers reported that the *h*-BN single crystal can also form circular grains with edge configuration terminated alternately by boron (B) atoms and nitrogen (N) atoms on the liquid Cu, and the resulting isotropic homogeneity allows the adjacent domains to be orderly aligned with an accordant orientation [148].

When the growth substrates were changed into liquid Au, the synthesis of large-scale single-crystalline *h*-BN (SC-*h*-BN) films can be realized via a self-collimation process between circular *h*-BN grains through a CVD [149]. Specifically, to induce the formation of the circular *h*-BN grains, the solubility of B and N atoms in liquid Au was limited, which accelerated their high diffusion on the liquid Au surface. Afterward, the circular *h*-BN grains on the surface of the liquid Au would rotate facilely, and then the B and N edges would be self-collimated by

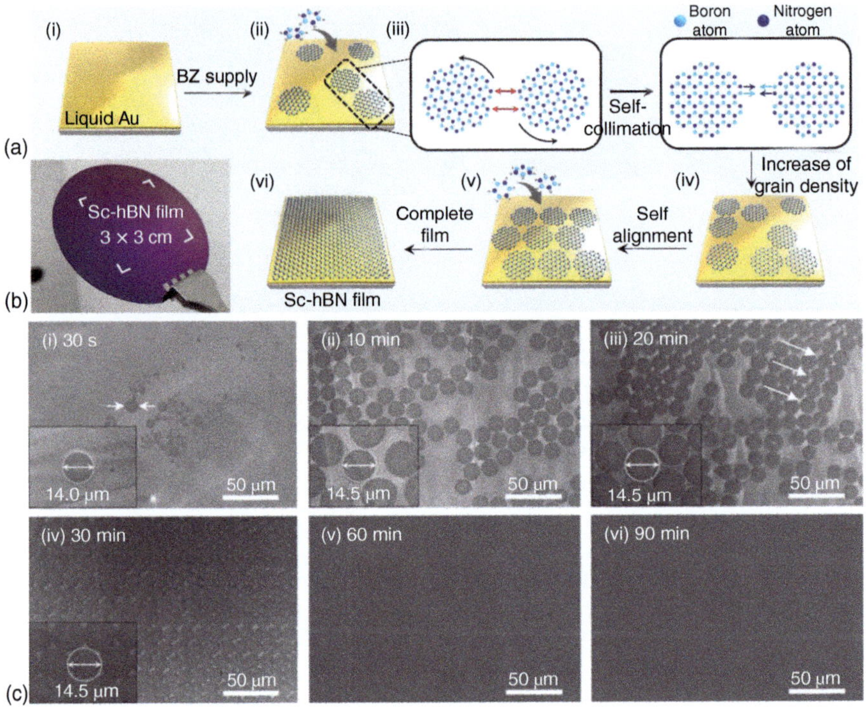

Figure 5.19 Schematic illustration for the growth of SC-*h*-BN films on liquid metal Au substrate and characterizations. (a) Schematic for the self-collimated growth of SC-*h*-BN film induced by the attractive Coulomb interaction between neighboring grains (i–vi). BZ, borazine. (b) Photograph image of a wafer-scale SC-*h*-BN film. (c) SEM images for illustrating the growth procedures of wafer-scale *h*-BN film. Source: Lee et al. [149].

means of electrostatic interaction between neighboring grains (Figure 5.19a, b), after a growth for a long time, the wafer-scale growth of SC-*h*-BN film could be realized, as indicated in Figure 5.19c. And no grain boundary can be detected in the SC-*h*-BN films.

Ordered Assembly Control The array orderly assembled by the 2D single crystals is highly anticipated because of its unique physicochemical properties, which take an important role especially in the integrated circuits. For example, the graphene super ordered structure (GSOS) can act as a conductive interconnection to realize the barrier-free contacts in integrated circuits [150]. The self-assembly growth of graphene array on solid metal substrates was undesirable because of the disorganized distribution of the nucleation sites, which were caused by the favorable absorption of carbon atoms on the defect sites or stages of the solid metal substrates. Zeng and coworkers employed the liquid metal Cu substrates, which have smooth surfaces as well as outstanding fluidity, to drive a uniform nucleation and a self-assembly [18]. Specifically, the PMMA was utilized as precursors to supply the nucleation seeds, simultaneously, the airflow was imported to provoke the assembly. During lateral growth procedure in the hydrogen atmosphere,

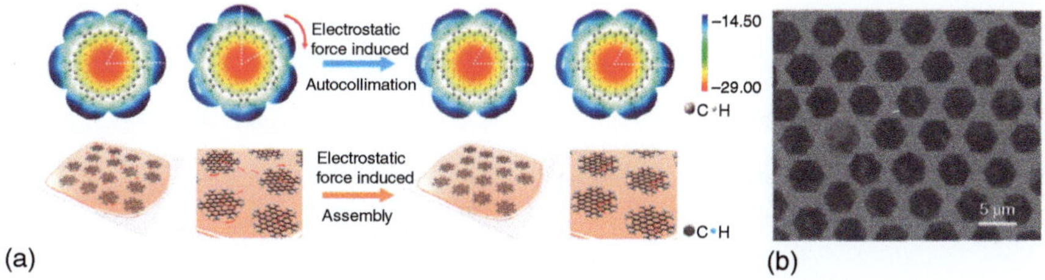

Figure 5.20 Ordered assembly control toward the 2DSOS. (a) Illustrations of the self-assembly process of GSOS. (b) SEM image of GSOS. Source: Zeng et al. [18].

the electrostatic repulsive force between adjacent crystals contributed to the self-assembly of the GSOS, as seen in Figure 5.20a. Afterward, they obtained the first 2D superordered structure (2DSOS) arrays, which is orderly assembled by the high-crystalline graphene single crystals. As indicated in Figure 5.20b, the excellent periodicity with uniform size, spacing, and orientation can be seen obviously.

Furthermore, Xue and coworkers also proposed that the gas flow can drive the assembled growth of graphene on the molten metal substrate [151]. This report also indicated that reduced viscosity property, resulting from the amorphous molten Cu under a relatively high temperature, could give rise to the drift and the rotation to align with the gas flow of graphene on the Cu surface. Noteworthily, the surface tension and viscosity of the molten Cu could cause an effect on the alignment. The change of surface tension of the molten Cu metal was weak within the growth temperature range [152], but the viscosity would decline obviously with the increase of the temperature [153]. Thus, it can be induced that the gas flow-derived graphene rotation and alignment were mainly caused by the reduced viscosity under a high temperature. However, the viscosity of the liquid Cu would increase after increasing the growth temperature, and hence the airflow-induced assembled growth of graphene on the molten Cu substrates would become irregular growth when the system temperature increases up to a critical value. Therefore, the physical property of the liquid metal that affects the growth behaviors should be concerned to achieve a better assembly or alignment control.

In addition to the gas flow-driven self-assembly on the liquid metal, the nanoparticles in molten substrates can also drive the spontaneous assembly toward a large-scale graphene array through the formation of the electrical charge of the metal oxide, that is, Al_2O_3, in molten Cu substrates [154]. Herein, the assembly of Al_2O_3 nanoparticles on molten Cu surface was achieved first to induce the synthesis of a wafer-scale graphene array. Specifically, once the solid nanoparticles, such as Al_2O_3, were introduced to contact with molten Cu the electron transfer would take place at the interface until it reached the equilibrium state, leading to a contact potential between the Al_2O_3 and Cu. The electrons would flow from the Al_2O_3 into the molten Cu because of the lower work function of the Al_2O_3 than that of the Cu, afterward, the molten Cu obtain a negative charge, while the Al_2O_3 achieves a positive charge. Then the potential difference was created. Figure 5.21a indicates the underlying mechanism of the formation of the ordered Al_2O_3 arrangement. Initially, Al_2O_3 particles were randomly distributed inside

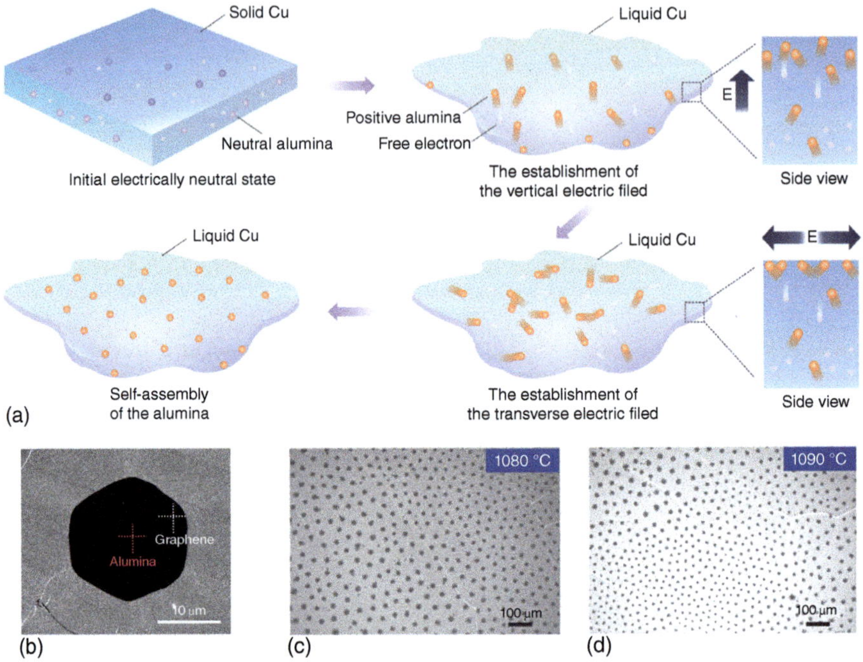

Figure 5.21 Self-Assembly of metal oxide nanoparticles in liquid metal toward the nucleation control of graphene single-crystal arrays. (a) The mechanism for the spontaneous construction of ordered alumina (Al_2O_3) array structures in liquid metal via the formation of electric fields. (b) SEM image of a graphene crystal with a Al_2O_3 particle in its center. (c) 1080 °C, and (d) 1090 °C. Source: Zeng et al. [154].

the solid Cu, but when the solid Cu was heated to a molten state, the potential difference would occur and then it influenced the arrangement of the Al_2O_3. The Al_2O_3 would be enriched on the molten metal surface because of the existence of oxygen in the growth atmosphere [155], which caused the formation of the vertical electric field. Meanwhile, a transverse electric field would be formed due to the migration of the positively charged Al_2O_3 or free electrons between the neighboring nanoparticles. Therefore, the Al_2O_3 particles suffered from the vertical electric field as well as the transverse field force on the molten Cu surface, therefore, the ordered Al_2O_3 nanoparticle arrays could be formed and displayed a near-hexagonal arrangement. The ordered Al_2O_3 islands triggered by electrical charge could precisely control the graphene nucleation, yielding the formation of wafer-scale graphene arrays.

Figure 5.21b exhibited a SEM image of the Al_2O_3 particle located in the center of one graphene crystal. SEM images of graphene arrays at 1080 and 1090 °C were shown in Figure 5.21c,d, which indicated that the graphene array achieved at 1080 °C has a better regularity. The thermal movements of the Al_2O_3 particles would be promoted as the increase of the growth temperature, as a result, the induced electrical field was improved, which could aid in the self-assembly of the ordered Al_2O_3 array. But if the growth temperature was too high, the drastic

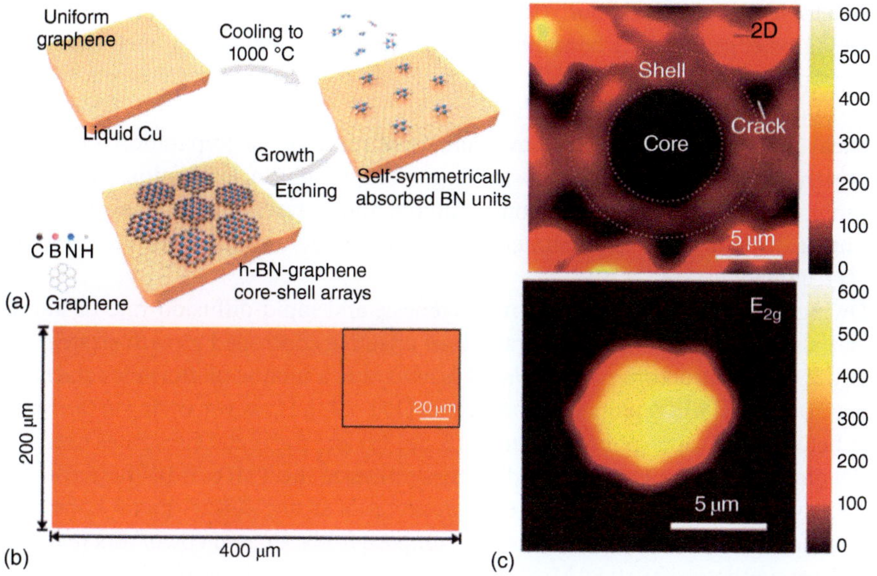

Figure 5.22 The self-symmetrical etching growth of the *h*-BN-G CSA on the liquid Cu substrate. (a) Schematic of the formation process toward the *h*-BN-G CSA. (b) OM image of large-area CSA. (c) Raman mapping to verify the core–shell structure, including the intensity mapping of the 2D peak of graphene and the E2g peak of *h*-BN. Source: Wang et al. [156].

thermal movements of the Al_2O_3 particles would break the balanced state, thus the regularity would be affected. In brief, the ordered arrangement of the nanoparticles in the molten metal can provide a more robust self-assembly system to obtain wafer-scale single arrays.

In addition to the graphene, the *h*-BN single crystals can also be assembled into arrays on the liquid Cu surfaces [148]. Besides, Wang and coworkers prepared a *h*-BN–G core–shell array (CSA) for the first time, which is a highly desirable heterostructure in integrated devices [156]. The formation procedures of the *h*-BN–G CSA were exhibited in Figure 5.22a. The crucial prerequisite was the growth of highly crystalline graphene films with uniformity on the molten Cu surface. The graphene acted as an excellent platform for the anisotropic electrostatic interaction of the *h*-BN units, thus the *h*-BN islands adsorbed on the smooth surface of graphene would tend to self-assembly toward an ordered array. Subsequently, the hydrogen-rich *h*-BN units could cause the etching of the graphene, and then the etched graphene would be substituted by the *h*-BN to grow the *h*-BN–G CSA. As the OM image displayed in Figure 5.22b, the size of the *h*-BN–G CSA could be up to millimeter-scale. The composition and structure of the core (*h*-BN)–shell (graphene) components were determined by Raman mapping displayed in Figure 5.22c. This self-symmetrical etching growth strategy is instrumental in precisely controlling the 2D ordered heterostructure arrays, which offered a new avenue in promoting the fabrication of the 2D integrated devices.

Heterogeneous Stacking Control Combining the different 2D materials with vdW stacking or edge covalent bonding construct, 2D heterostructures have extended the properties of 2D materials in various fields (such as electronics, photonics, and catalysis) [85, 157–166]. For the later practical applications of both lateral and vertical 2D heterostructures, the development of controllable preparation is critical. Compared with the mechanical stacking or liquid-phase exfoliation, the CVD technologies have shown higher controllability and universality in the synthesis of 2D heterostructures. To date, numerous 2D heterostructures have been realized by CVD technologies [167–170].

Owing to the great corrosivity, inclusiveness, and rapid diffusion rate of liquid metal, using liquid metal as substrate is an effective controlled growth strategy in many 2D heterostructure systems. *h*-BN is always as a unique platform for stacking other 2D materials to explore their properties [171–173]. However, owing to the reaction between the vapor-phase precursor (S, Se, Te) and the metal substrates used for *h*-BN growth (such as Cu and Ni), directly synthesize the TMDs/*h*-BN heterostructures is very challenging. Besides, the formation of strong sulfur–metal bond can break the metal surface structure and decompose *h*-BN. Fu and coworkers realized the direct synthesis of vertical MoS_2/*h*-BN heterostructures via liquid Ni–Ga alloy as sulfide-resistant substrate, and the liquid Ni–Ga alloy also could catalyze the growth of *h*-BN [174]. The Mo source for MoS_2 growth was provided by the diffusion of supporting Mo foil under the Ni–Ga alloy. Figure 5.23a illustrated the preparation process of MoS_2/*h*-BN heterostructures, the *h*-BN film was synthesized by the catalytic decomposition of ammonia borane first, then the vapor-phase precursor (H_2S) was introduced and reacted with the diffused Mo, the MoS_2 was grown on *h*-BN film finally. The crystal size reached 200 μm^2, which is larger than the grain size of

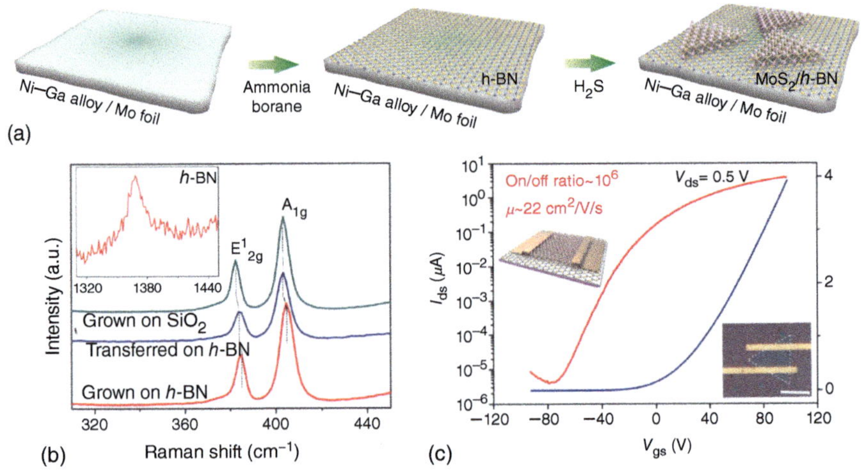

Figure 5.23 Liquid metal-assisted heterogeneous stacking control. (a) Schematic showing the preparation of TMDs/h-BN heterostructures. (b) The Raman spectra of different MoS2 domains grown on three types of substrates. (c) Electrical properties of Ids–Vgs curves for the FET devices made on MoS_2/h-BN heterostructure. Inset: Optical image of the device; the scale bar is 5 μm. Source: Fu et al. [174].

concurrent reports. The further TEM and SAED measurements were confirmed by the coexistence of MoS_2 and *h*-BN. As shown in Figure 5.23b, compared with the Raman spectra of MoS_2 domains grown on SiO_2 surface (green) and transferred on *h*-BN film (blue), the peaks of MoS_2 directly grown (red) on *h*-BN film exhibited a blue shift, which indicated that directly grown *h*-BN film by CVD process as substrate could help release lattice strain and reduce the doping levels, and the inset of Figure 5.23b provided the Raman peak of *h*-BN film under MoS_2 crystals. The electrical properties of MoS_2/*h*-BN heterostructures were measured in Figure 5.23c, which presented n-type conduction with an on/off current ratio of $\sim 10^6$ and a high mobility of MoS_2 up to 22 $cm^2/V/s$, indicating an enhanced device performance of TMD materials. Furthermore, the WS_2/*h*-BN heterostructures were also fabricated via liquid Ni–Ga alloy as substrate, which confirmed the universality of this liquid metal-assisted approach for creating novel 2D heterostructures.

In addition, prepared vertical heterostructures of MXenes and other 2D materials have attracted extensive interest in recent years owing to its significance of charge transfer or proximity effects in the energy storage or catalysis [175–177]. For such applications, the large-size synthesis of MXenes is also significative. Geng and coworkers reported a direct synthesis strategy of uniform 2D Mo_2C/graphene heterostructures by liquid Cu-catalyzed CVD with large size [178]. Owing to the good wettability, the liquid Cu could cover the Mo substrate entirely, then the Mo diffused to the surface and reacted with the C source (CH_4) to form Mo_2C crystals. The different growth results have been performed by tuning the flow rate of CH_4. The low flow rate (0.3 sccm) of CH_4 led to thick Mo_2C crystals. With the increased flow rate (3 sccm) of CH_4, which reached the surface saturation limit of liquid Cu, the large MLG formed and covered the Cu surface first, and reduced the transfer of Mo atoms to the surface. Therefore, the growth rate was reduced and limited the thickness of the obtained Mo_2C crystals. Besides, compared with the Mo_2C catalyst, this vertically stacked Mo_2C/graphene heterostructures showed an improved catalytic activity in hydrogen evolution reaction (HER) owing to its high-efficiency interfacial charge-transfer kinetics between graphene and Mo_2C. Analogously, the liquid Au substrate also was confirmed for the controlled synthesis of 2D Mo_2C/graphene heterostructure.

The stacking of two different 2D TMDs extends the tunable electronic structure [179–181] and has great potential in optoelectronic devices [182–184]. Increasing the stacking area significantly and controlling the growth behavior precisely (especially in the control of the overlap percentage) are difficult to realize in vertically stacked TMD heterostructures. Zhang and coworkers demonstrated the twinned growth of 2D ReS_2 and WS_2 vertically stacked TMDs heterostructures, which achieved 100% overlap with large crystal sizes over 600 μm^2 [47]. The strategy for the twinned growth of ReS_2/WS_2 vertical heterostructures was illustrated in Figure 5.24a. The liquid Au was chosen as the substrate to dissolve Re and W sources for the twinned growth of ReS_2/WS_2 vertical heterostructures. And the ReS_2/WS_2 vertical heterostructures possessed a high uniformity. The Raman spectra further confirmed the vertically stacked structure of ReS_2/WS_2, since Figure 5.24b showed the simple superposition of peaks from both ReS_2 and WS_2 rather than the peaks with their positions between those of pure ReS_2 or WS_2. The later HRTEM image

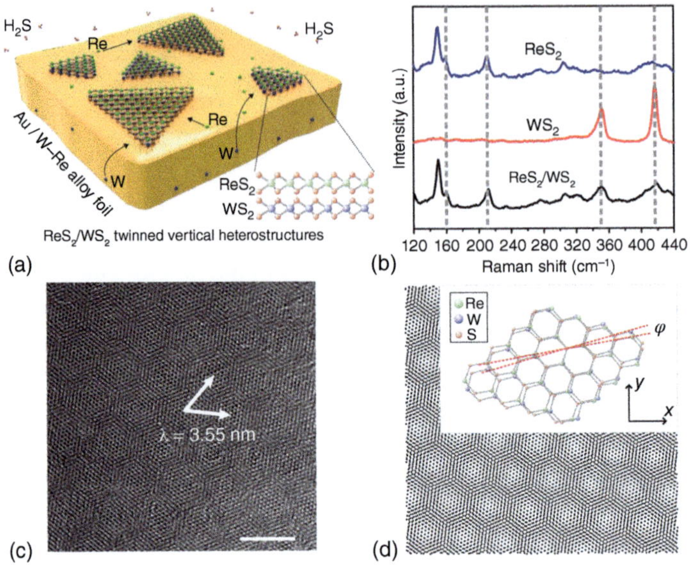

Figure 5.24 Liquid metal-assisted heterogeneous stacking control. (a) Schematic for the twinned growth of ReS_2/WS_2 heterostructures. (b) Comparison of the Raman spectra of ReS_2, WS_2, and ReS_2/WS_2. (c) HRTEM image of ReS_2/WS_2 heterostructure showing the resulting Moiré pattern. Scale bar, 4 nm. (d) Tentative orientation model of rotating the upper ReS_2 by an angle of 5.6° with respect to the ground WS_2. Insert is the atomistic illustration of the ReS_2/WS_2 heterostructure. Source: Zhang et al. [47].

of the stacked structures was shown in Figure 5.24c, which had displayed the clear Moiré pattern with periodicity about 3.55 nm. And the corresponding FFT revealed that the two hexagonal reciprocal lattices were rotated with angle of $\Phi = 5.6°$. Figure 5.24d further simulated the atomic structure of the upper ReS_2 by a rotating angle of 5.6° with the ground WS_2 layer, which exhibited the similar Moiré pattern with that shown in Figure 5.21f and proved the ReS_2/WS_2 stacked structures. Furthermore, the presented strategy also has potential as a general method for the fabrication of other TMD vertical heterostructures such as MoS_2/WS_2, which could be synthesized by simply replacing the Mo–W alloy with W–Re alloy during the growth process. Besides, the reaction temperature enables us to realize the selection of growth order.

Direct growth of lateral 2D heterostructures is quite hard to achieve owing to the uncontrollability of the nucleation process and layer stacking. Liquid metals as the substrates are equipped to control the segregation process effectively, which is potential to realize the direct growth of lateral 2D heterostructures. Zeng and coworkers reported a liquid metal solvent-based co-segregation approach to obtain the in-plane WC/graphene heterostructure via one-step direct CVD process [48]. The liquid metal Ga could accelerate the co-segregation process of W and C atoms, and the growth process of the in-plane WC/graphene heterostructure was illustrated in Figure 5.25a. During elevated temperature, the fluctuated atomic clusters of liquid Ga served as heteroatom containers, so that W atoms could diffuse into liquid Ga following the

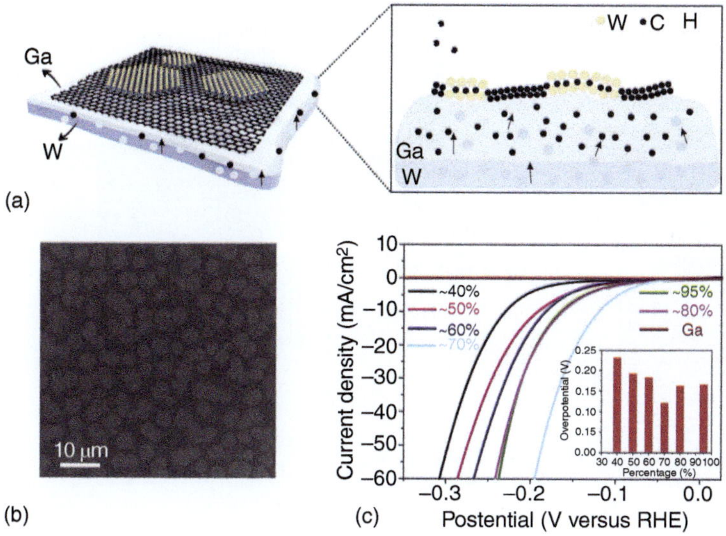

Figure 5.25 Liquid metal-assisted heterogeneous stacking control. (a) Schematic of the formation of 2D in-plane WC/graphene heterostructures on a Ga–W substrate. (b) The SEM image of the WC single crystals embedded in graphene. (c) The polarization curves for 2D in-plane WC/graphene heterostructures with different WC% coverages (40–95%). Source: Zeng et al. [48].

atomic thermal motion, and the C atoms after decomposed were also "dissolved" into liquid Ga. Owing to the competitive relationship between WC and graphene, the author found that growth temperature and cooling rate determined the coverage and thickness of WC. For further effective HER, the author employed a high growth temperature and rapid cooling rate to obtain the in-plane WC/graphene heterostructures with larger coverage and thinner thickness. The ultrathin WC single crystals embedded in graphene with high density was shown in Figure 5.25b. The polarization curves for a series of 2D in-plane WC/graphene heterostructures with different WC% coverages (40–95%) were exhibited in Figure 5.25c, and the sample with 70% WC coverage had the lowest overpotential value of 120 mV at 10 mA/cm^2 in 0.5 M H_2SO_4. This enhanced HER performance mainly attributed to the facilitation of electron transfer at the WC/graphene interface. This research put forward a co-segregation strategy, which controlled the unique solvent effect to adjust the segregation process, had the potential to realize the direct growth for other lateral 2D heterostructures.

5.2.4.3 Liquid Metal-Assisted Sliding Transfer

In the device fabrication process, in order to construct the devices, the 2D materials were required to be transferred onto the target substrates, however, current PMMA-assisted transfer approach caused unavoidable residues. Lu and coworkers proposed a liquid Ga metal-assisted sliding transfer for ultraclean 2D materials [185], as the sliding transfer installation displayed in Figure 5.26a,b. To exfoliate the graphene from the Ga surface, the liquid Ga metal with graphene on the surface was

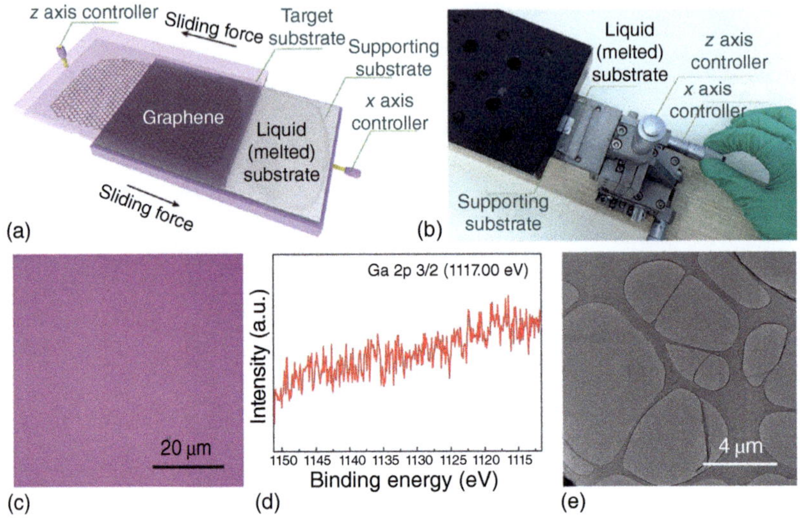

Figure 5.26 Liquid metal-assisted sliding transfer process. (a) Synthetic diagram of the sliding transfer method for obtaining uniform graphene. (b) A photograph of the device for transferring the graphene. (c) OM image of the transferred graphene on the targeted SiO_2/Si substrate. (d) High-resolution XPS spectrum of Ga 2p. (e) A low-magnification TEM image of the transferred graphene. Source: Lu et al. [185].

contacted face to face with the SiO_2/Si substrate. Specifically, the contact strength was manipulated by the Z-axis controller, afterward, the Ga would slide by rotating the x-axis controller. After a few seconds, then the large-area, uniform graphene would be separated from the liquid Ga surface and transferred onto the targeted substrate (Figure 5.26c). Compared with traditional PMMA-assisted transfer, where there existed residues of the solvents and chloride etchants, the Ga-assisted sliding transfer could not detect observable Ga content, as the high-resolution XPS confirmed in Figure 5.26d. The low-magnification TEM image indicated wrinkle-free and unclean graphene (Figure 5.26e). The liquid metal enabled the ability to slide transfer ultraclean, large-area 2D materials onto arbitrary substrates, which has good prospects in future industrial application.

5.3 Alloy Preparation

Alloy materials are one of the most widely used materials in society today. Among metal materials, liquid metal is a kind of material that can remain liquid at room temperature, which is different from traditional metal materials. The exploration of liquid metal is an opportunity and revolution for people to study metal materials. Liquid metal can not only assist the preparation of nanomaterials such as 2D materials, but also have an important effect in the process of assisted alloy structure and morphology formation during alloy preparation.

From the perspective of phase structure, alloys can be generally divided into solid-solution phase, amorphous phase, and ordered phase. The solute atoms dissolve into the lattice of solvent, so it is called solid-solution phase. And the atoms of various elements are arranged in the lattice in disorder. In general, solid-solution phase includes some simple phases of face-centered cubic (FCC) phase, body-centered cubic (BCC) phase, and hexagonally closed packed (HCP) phase. Amorphous phase alloy refers to the alloy of which the atom arrangement characteristics is "short-range ordered, long-range disordered" in the solid state, also known as metallic glass. The atoms of the ordered phase alloys occupy different structural positions in the lattice, having ordered atomic arrangement.

Here, we will discuss the status of alloy preparation and the research progress of liquid metal-assisted alloy preparation. Among them, the current situation of alloy preparation will be discussed from the perspective of single-principal component alloy and multiprincipal component. The preparation of liquid metal-assisted alloy preparation is mainly divided into the formation of alloy structure and alloy morphology.

5.3.1 Current Progress of Alloy Preparation

The preparation method of the alloy largely determines the properties and application fields of the alloy. In this section, we will demonstrate the status of alloy preparation.

Magnesium (Mg) alloy and iron (Fe) alloy are two kinds of single-principal component bulk alloys with broad application fields and superior performance. In the automotive industry, Mg alloy is regarded as the lightest metal structural material because of its excellent strength and specific stiffness, which is considered to be the first choice for lightweight automotive structures in the future [186, 187]. Common preparation methods for Mg alloy include casting methods such as pressure casting and sand casting [188]. Pressure casting, supplemented by a high cooling rate, is suitable for magnesium–aluminum (Mg–Al) series alloys, which can improve the uneven structure and poor creep performance of the alloys. Sand casting has high fire resistance, which is suitable for magnesium–rare earth (Mg–RE) series alloys that require higher forming temperature. The casting method to prepare Mg alloys makes use of the fluidity of molten metal, which is easy to form and can be filled in the mold with complicated shapes to obtain the alloy with complicated shapes.

Fe alloy is the most widely used alloy in the building industry and daily life. Compared with the crystalline alloy, Fe-based amorphous alloy has no defects such as grain boundary, dislocation, and slip, showing much higher strength and yield stress than the corresponding crystalline alloy. At the same time, Fe-based amorphous alloy is an important material for transformer core in power field. There are two main types of Fe-based amorphous alloys preparation: direct rapid solidification and powder metallurgy [189]. The direct rapid solidification method means that the liquid alloy is directly cooled to form a bulk amorphous alloy under a certain cooling rate. The powder metallurgy method is to fill the amorphous powder into a mold with a certain shape according to the preparation needs [190]. The methods not only guarantee excellent amorphous forming ability of the Fe alloy, but also endow it with magnetic properties.

In addition, high-entropy alloy, which possesses multiprincipal elements, also attracts people's attention on account of its remarkable wear resistance, hardness, and other physical properties. The preparation methods of high-entropy alloy samples are different. Vacuum melting and powder metallurgy are usually used to prepare bulk samples [191]. The vacuum smelting method makes use of the fluidity of the molten metal to facilitate the forming of high-entropy alloy. At the same time, its smelting temperature is high, it can smelt alloys with higher melting points, and it has a good effect on the removal of volatile impurities and certain gases. The design of high-entropy alloys is generally based on the idea of "required performance-oriented design." High-entropy alloy elements could be selected and chosen based on empirical criteria such as mechanical parameters, atomic radius parameters, and electronegativity differences [192]. Utilizing the different characteristics of the atoms of each element, it can give full play to form a solid-solution phase with a simple crystal structure, so that the alloy has the required properties. The research on the preparation of high-entropy alloy is beneficial to further broaden its application fields.

Through the above summary of the current alloy preparation methods, it is found that alloy preparation methods are usually inefficient. The product quality still needs to be improved and the preparation cost also needs to be reduced. Therefore, it is important to find a universal and inclusive reaction medium to assist in the preparation of alloys.

5.3.2 Liquid Metal-Assisted Preparation of Alloy

Liquid metal, as a metal that can exist in liquid form at low temperatures, has an atomic structure and physicochemical properties different from traditional solid metals. Liquid metal has good tolerance and can provide low mixing enthalpy with various elements, so that liquid metal can form alloy with various metal elements as reaction medium. The good fluidity and formability of liquid metal as well as the interaction between liquid metal atoms and other metal atoms can promote the formation of alloy morphology. Therefore, when the liquid metal assists the formation of the alloy, the liquid metal can affect the alloy in terms of both structure and morphology. Structurally, liquid metal, as a component element, can form alloy with solid-solution phase, amorphous phase, and ordered phase alloy by giving rise to its short-range ordered and long-range disordered structure. In terms of morphology, the alloy-assisted preparation of liquid metal can be used for the introduction of pores, shape formation, and the process of dealloying.

5.3.2.1 Liquid Metal-Assisted Structure Formation

Alloy structures mainly include solid-solution phase, amorphous phase as well as ordered phase. Solid-solution phase alloys have excellent strength and hardness due to solid-solution strengthening. Usually, the toughness and forming ability of amorphous alloy are better than that of crystal alloys. Ordered phase alloys usually have excellent high-temperature strength and rigidity. As constituent elements, liquid metal elements can participate in the preparation of solid-solution phase such as high-entropy alloy, amorphous phase such as Fe-based amorphous alloy, and ordered phase such as $AuGa_2$ alloy.

Liquid metal Ga can affect the lattice structure during preparation of high-entropy alloy, so as to assist the formation of solid-solution alloy. The addition of Ga results in increasing lattice occupancy and lattice strain, which will change the lattice parameters of the alloy. Li and coworkers prepared as-cast $FeCoNi(CuAl)_{0.8}Gax$ ($0 \leq x \leq 0.08$) high-entropy alloy by vacuum arc melting method protected by argon (Ar) atmosphere [193]. By XRD analysis, only BCC and FCC phase structures exist in the sample (Figure 5.27a). The addition of Ga element increases the relative proportion of BCC structure. As for the atomic radius of constituent elements, Ga is larger than other elements. Accompanied by the addition of Ga, the strain effect at the atomic level will result in improved lattice occupancy and lattice strain [194]. As a result, the FCC phase transformed into BCC phase (Figure 5.27b). The addition of liquid metal Ga accelerates the formation of the BCC structure and further affects the properties of the alloy. The magnetic and mechanical properties of the solid-solution alloy are related to its crystal structure to a great extent. The saturation magnetization (M_s), maximum magnetic flux density (B_m), remanence (B_r), coercive force (H_c), and hysteresis loss (Pu) increased with the relative proportion of the BCC structure [195]. While the maximum permeability (μ_{max}) changes with increasing FCC volume fraction. FCC phase high-entropy alloy has good ductility but lower strength and BCC phase solid-solution high-entropy alloy has high strength but poor ductility. The addition of Ga increases the relative proportion of the BCC phase. Meanwhile, the physical properties of the alloy are relatively increased.

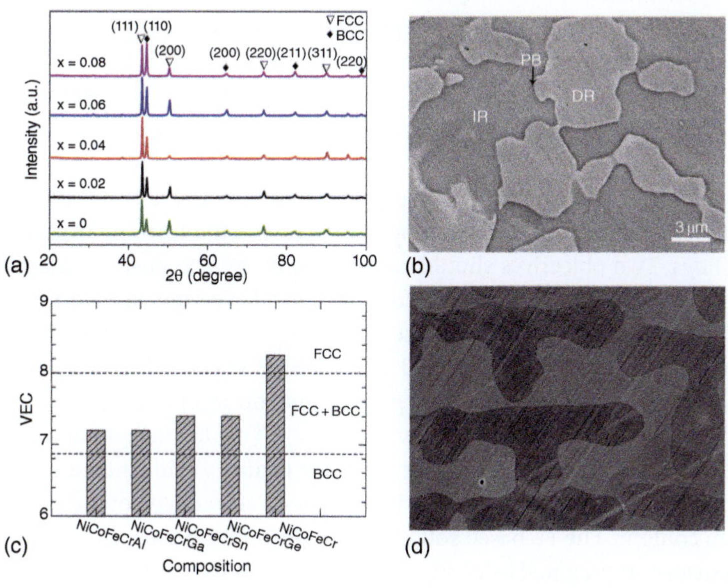

Figure 5.27 Microstructure characterization of solid-solution alloys formed with the assist of liquid metal. (a) XRD patterns of as-cast FeCoNi(CuAl)0.8Gax ($0 \leq x \leq 0.08$) HEAs. (b) SEM-BSE morphologies of the as-cast FeCoNi(CuAl)0.8Gax HEAs ($x = 0.02$). Source: (a, b) Li et al. [193]. (c) VEC of the investigated NiCoFeCr⁻ based samples. (d) SEM images of the microstructures of NiCoFeCrGa. Source: (c, d) Vida et al. [23].

Liquid metal can also assist the formation of solid-solution alloy by affecting the electronic structure of the solid-solution alloy. Ga is a p-block element, which can regulate the valence electron concentration for the alloy. Vida and coworkers prepared NiCoFeCrGa high-entropy alloy by induction smelting with water-cooled Cu mold in Ar atmosphere [23]. According to the research on the average valence electron concentration of the high-entropy alloy system, it can be seen that the average valence electron concentration of NiCoFeCr samples is higher than 8. With the addition of Ga content, the average valence electron concentration is reduced to between 7 and 7.5 (Figure 5.27c). When the average valence electron concentration is greater than 8, the alloy shows a single-phase FCC structure. (Figure 5.27d). When it is less than 6.87, the alloy is found to have a BCC structure [196]. By adding Ga, the average valence electron concentration of the alloy is changed, then crystal structure is also affected [197]. The properties of alloys formed with the assistance of liquid metal can be improved. The hardness of FeCrCoNiGa was found to be improved by the mechanical property test of the above samples by nano-indentation technology.

In addition, the thermal stability of high-entropy alloys containing Ga is related to paramagnetic and diamagnetic properties, and it is a promising magnetocaloric material. Huang and coworkers found that the high-entropy alloy FeCrCoNiGa exhibited abnormal thermal expansion behavior [198]. The engineering design of magnetic and structural transformation provides a wealth of opportunities for the design and optimization of new high-entropy alloy materials with interesting thermophysical properties, so that high-entropy alloys can become the focus of magnetocaloric applications.

Except for the solid-solution phase alloys represented by high-entropy alloys, liquid metal can assist in the formation of amorphous phase structural alloys as well. The amorphous alloy containing liquid metal not only has extremely high strength, toughness, and abrasion resistance, but also exhibits excellent soft magnetic and corrosion resistance.

Liquid metal can also assist to prepare amorphous alloys, because it can utilize the intrinsic fluidity and viscosity characteristics to enhance the forming capacity of amorphous alloys. Two objectives should be pursued in the composition design of Fe-based soft magnetic bulk amorphous alloys: one is to have high glass-forming capacity (GFA); the second is a good soft magnetic property. Generally, GFA of amorphous alloys is characterized by supercooled liquid-phase region ($\Delta T_x = T_x - T_g$, T_x as crystallization temperature, T_g of the glass transition temperature). In the $Fe_{77.5\text{-}x\text{-}y\text{-}z}Cr_xMn_yGa_zP_{12}C_5B_{5.5}$ ($2 \leq x \leq 4$, $2 \leq y \leq 4$, $0 \leq z \leq 4$) bulk amorphous alloy system, the GFA and thermal stability of the alloy are changed with the addition of different elements. And, further changes will occur with the change of the alloying element atom content. The Fe-based soft magnetic bulk amorphous alloy of this system of different sizes is prepared by copper mold casting. As the content of Ga element increases, its GFA also increases. This is because the addition of an appropriate amount of elements makes the amorphous structure in a denser random stacking structure [199]. As a result, liquid metal is of great significance for amorphous alloy preparation and performance improvement.

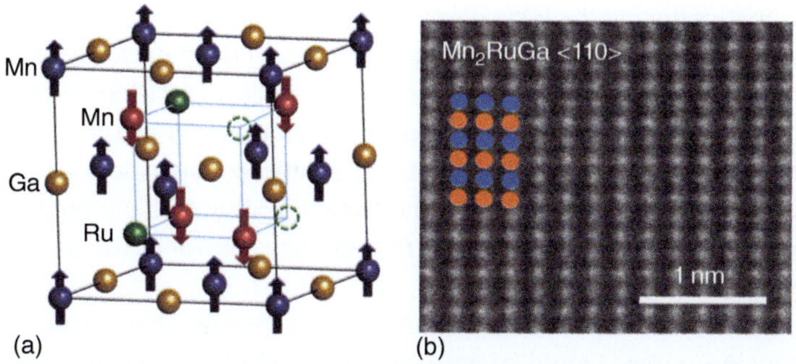

Figure 5.28 Microstructure characterization of ordered phase alloys formed with the assist of liquid metal. (a) Crystal structure of Heusler alloy $Mn_2Ru_{0.5}Ga$. (b) STEM image of Heusler alloy $Mn_2Ru_{0.5}Ga$. Source: (a,b) Finley et al. [21].

Unlike amorphous alloys, ordered phase alloys, also called intermetallic compounds, are alloys with definite chemical composition. The liquid metal atoms can occupy a certain position in the lattice of the ordered phase alloy and participate in the formation of the ordered phase alloy structurally. At the same time, the properties of ordered phase alloys are affected. Finley and coworkers found that effective current-induced magnetic switching can be realized in compensation samples with strong vertical anisotropy by studying the spin orbit torque-induced switch in $Mn_2Ru_{1-x}Ga_x$ [21]. The magnetization of the intermetallic compound could be regulated by adjusting the number of Ga atoms as well as valence electrons in each cell [200]. This work demonstrated the possibility that the $Mn_2Ru_{1-x}Ga_x$ alloy can be used in fast, high-energy spintronic devices (Figure 5.28). The binary intermetallic compound $AuGa_2$ has a fluorite (CaF_2) structure under normal pressure. Fluorite belongs to the cubic system. Au is located at the apex of the cubic unit cell and the face-centered position, forming a FCC stack. Ga is filled in the body centers of the eight small cubes. The intermetallic compounds $AuGa_2$ have superconducting properties at low temperatures, and the superconducting transition temperature is relatively low. Their properties have also been applied to low-temperature measurements. In addition, $AuGa_2$ is also a good material for solar energy absorption and research on the transition metal 5d energy band. Because of its important application in industry and basic scientific research, it has been widely concerned with experimental and theoretical workers. The first-principles calculation method of density functional was used to study the relationship between the electronic, lattice vibration, and transport properties of $AuGa_2$ as a function of pressure [201].

Ordered phase alloys have excellent high temperature strength and rigidity, having broad application prospects in high-tech fields such as aviation, aerospace, and automotive. The preparation of ordered phase alloys with the assistance of liquid metal brings development and breakthrough for its further preparation and development.

5.3.2.2 Liquid Metal-Assisted Morphology Formation

Liquid metal has excellent fluidity and deformability. Therefore, liquid metals can be used for pore introduction and macroscopic shape control of alloy by giving full play to their characteristics during the formation of alloys. In addition, liquid metal can be used to assist the formation of porous alloy morphology at the nanoscale. Liquid metal can be used as an excellent solvent due to its wide range of melting points, strong reducing property, and strong electric property. The attraction or repulsive force between the liquid metal and other alloy elements enables the specific components to be dissolved from the solid alloy into the melt, and the alloy with nanopores morphology can be obtained, so as to realize the morphology control of the liquid metal assisted alloy formation.

As shown subsequently, the elastic cooling stability of Ni–Fe–Ga alloy has been improved by introducing pores. Liquid metal-assisted morphology formation of alloys can be realized by liquid metal infiltration technology, in which the reinforced prefabricated parts are directly put into the mold and poured into the liquid metal. With the help of liquid metal pressure and capillary siphon action, alloy penetrates into the enhanced body gap. Then the porous morphology of the alloy can be obtained by acid etching [202]. Imran and coworkers prepared Ni–Fe–Ga alloy foam with 53% porosity by using vacuum induction smelting complex casting and liquid metal infiltration technology, which showed a completely reversible superelastic response [203]. Since $NaAlO_2$ powder has a high melting point, a good chemical stability with molten metal and a high solubility in acid, this paper used the powder as a space container in the preparation of alloy foam. The Ni–Fe–Ga alloy foam prepared by the introduction of porosity has good elastic cooling stability. Elastic cooling stability is an important factor in solid-state refrigeration. It is expected to replace the traditional steam compression technology due to its high efficiency, low cost, and environmental protection. Ferromagnetic shape memory alloys including Ni–Mn–Ga and Ni–Fe–Ga can exhibit large adiabatic temperature changes, and have cyclic stability and good fatigue resistance, which are important materials for elastic thermal effect [204]. Apart from the introduction of pores, liquid metal-assisted alloy morphology formation includes the control of shape owing to the excellent fluidity and good moldability of liquid metal. The liquid metal can assist to realize the shape control of the alloy by impressed current. Stoica and coworkers developed a strategy of liquid metal casting by applying current to achieve the formation of bulk amorphous alloys [26] (Figure 5.29a). By applying a direct current between the molten metal and the mold, the heterogeneous nucleation can be suppressed by changing the surface state of the fluid alloy [205]. In addition, as a result of reducing surface tension, the unformed alloy is more fluid and can be filled with complex molds, avoiding solidification before the cavity is fully filled. By changing the contact state, the overall state of the alloy changes greatly. Therefore, the molten alloy has better fluidity in the mold. The cooling rate is lower in the shorter casting time. Using this new technology, the casting process can be used to prepare amorphous parts with complex geometry and meet the requirements of industrial production (Figure 5.29b).

In addition to the assisted synthesis of liquid metal mentioned above in terms of alloy shape, the liquid metal dealloying strategy is another way to assist the synthesis of alloy morphology from a microscopic perspective.

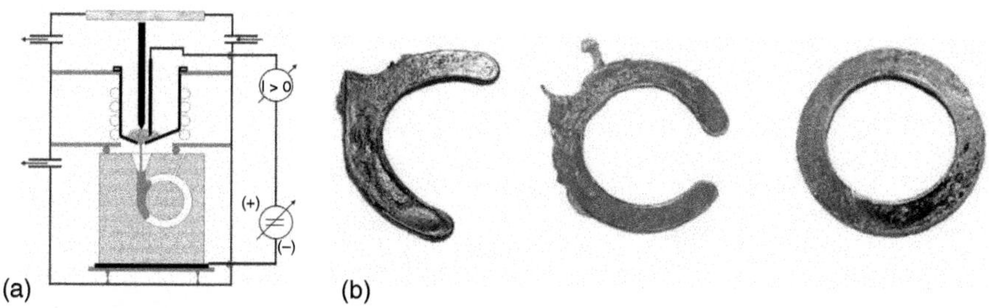

Figure 5.29 Liquid metal-assisted alloy shape formation with DC current impressed. (a) Schematic of the current-assisted liquid metal forming experimental device. The device consists of two cavities: an induction coil and a working crucible are placed in the upper cavity, and a copper mold is placed in the lower cavity. (b) Forming process of molten metal with and without DC current. Source: Stoica et al. [26].

Dealloying refers to the selective dissolution of elements in alloys, which is an important process of corrosion and also a simple process for the preparation of a large number of nanostructured metals [206]. Usually, it is an undesired corrosion mechanism. Some components are dissolved away, resulting in brittle failure occurring, stress corrosion failure, and other undesired material failures [207]. With the deepening of people's research, dealloying has developed from a destructive process to the manufacture of nanoporous metals approach. This process can be used to dissolve the lower-level elements in the binary alloy in the electrolyte, resulting in the nanopore structure of the alloy [208]. Wang and coworkers proposed a liquid metal that participated in the dealloying method to convert whole metals into nanoporous metals [209]. By dealloying the alloy in which Ga is involved in the formation, nanoporous metals were prepared successfully. By using the low-temperature characteristic of liquid metal, the precursor alloy was formed with other metals at a certain temperature. A variety of nanoporous metals with different shapes, scales, and dimensions have been successfully prepared by dealloying the precursor alloy. In addition, by recovering liquid Ga and GaOOH with a recovery rate of 94.5%, the process could realize effective recovery of Ga. This method solves the problem of recovery of active components in the process of traditional dealloying and opens a new way for the preparation of nanoporous metal materials from block metals (Figure 5.30a).

In addition to the liquid metal-assisted alloying-dealloying process described earlier, liquid metal can also be directly used in the dealloying process. In a related research, Wada and coworkers extended the dealloying strategy to a wider range of alloy, which formed similar nanostructures as well [25]. This emerging dealloying strategy — liquid metal dealloying — is beginning to attract attention. This strategy relies on the melt to selectively dissolve. Atomic interactions between alloy components and metal melts cause specific components to be dissolved into the melt from alloy solids. The remaining components have a self-organized nanopore structure. At the same time, liquid metal dealloying expands the range of applicable alloys, including almost the entire element of the periodic table. Liquid metal dealloying is a new method for preparing nanoporous metal by using liquid metal instead of electrolyte as the demetallizing liquid. Liquid metal dealloying does not selectively

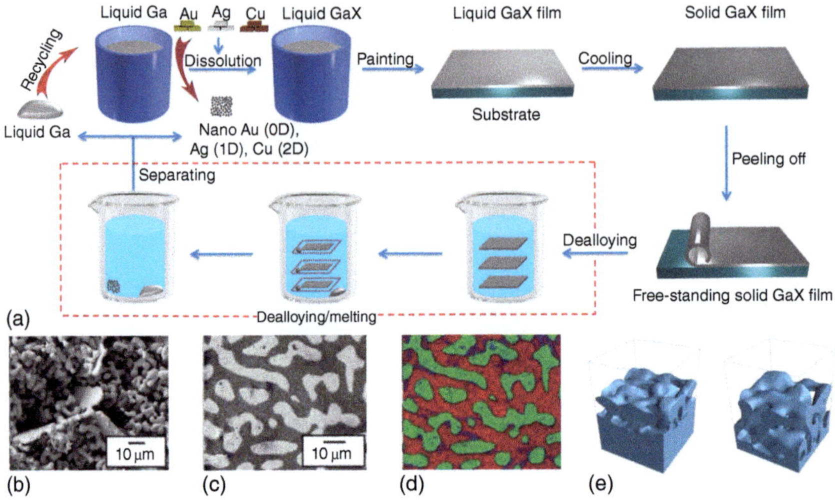

Figure 5.30 Description of dealloying process and characterization of nanoporous alloys structure. (a) Schematic illustration showing the fabrication process of nanostructured metals (0D, 1D, 2D) from bulk counterparts (Au, Ag, and Cu) through the liquid Ga-assisted alloying-dealloying strategy. Source: Wang et al. [209]. (b) SEM micrograph of a Ta-Cu composite with the Cu matrix phase dissolved away to reveal the nanoporous refractory dealloyed phase and a Ta-rich grain boundary. (c) SEM micrograph of a polished Ta-Cu composite for elemental mapping, with a 5 mm ligament size. (d) Elemental mapping of (c): green is Ta, red is Cu, and blue is residual Ti in the Cu-rich matrix phase. Source: (b, c, d) McCue et al. [210]. (e) Phase field simulation illustrates the formation of topologically connected bicontinuous nanostructures during liquid metal dealloying. Source: Geslin et al. [24].

remove alloy components through the corrosion process, but through the attraction or repulsion between alloy constituent elements. This technology enables even rare metals such as Ti to form nanopore structures [211], which can be used to develop multipurpose materials including filter units, gas assimilation devices, and biological applications. This technology can also be used to prepare other new materials with excellent properties, such as nanoporous silicon for battery electrodes with good circulating quality [212], ultrahigh surface area niobium for capacitors [213], and Cu–Tantalum (Ta) nanostructure metals with excellent application prospects [210]. With this method, nanoporous alloys can be prepared for many alloys. The formation of nanoporous Ti alloys can be achieved by partial removal of components in typical Ti alloys. Binary Ti–Cu alloy of $Ti_{30}Cu_{70}$ was obtained via arc melting method with high purity Ti and Cu as raw materials. Nanoporous Ti alloys can be prepared by soaking the Ti–Cu alloy with liquid pure Mg (mass: 99.9%) in a dealloying process [25]. McCue and coworkers prepared a TiTa alloy by vacuum arc smelting with molten Cu. Ti has good solubility in liquid Cu (up to 70% at 1100 °C), while Ta is undissolved in any state Cu [210]. After immersion, Ti is dissolved from the alloy, leaving porous Ta surrounded by molten Cu (Figure 5.30b–d). Upon cooling, a completely dense metal composite is formed. If copper is subsequently etched out, porous Ta can be extracted. In addition, although porous Ta has excellent mechanical properties, we can also add other elements to form bicontinuous metal nanoalloys. In addition, this liquid metal dealloying method is also applicable to multicomponent alloys. Nanoporous high-entropy alloy TiVNbMoTa can be prepared by dealloying $(TiVNbMoTa)_{25}Ni_{75}$ alloy in molten Mg [214].

Liquid metal dealloying is a new technology for preparing nanopores and nanocomposite structures with complex topological structures. It has an ultrahigh interface area and a wide range of applications for various materials. This process requires the dissolution of sectional constituents from its multicomponent original state alloy into the liquid metal to achieve the desired structure. However, there is still no clear understanding of the pattern formation in the liquid metal dealloying process. The mesoscale phase field model and experiments prove that the formation of the nanostructure pattern during the liquid metal dealloying process is regulated by the diffusion-coupled growth of the solid phase enriched with incompatible elements and there are more phase-contained elements in the liquid system (Figure 5.30e) [24]. Due to the interaction of interface spin state decomposition, a component domain structure enriched in immiscible elements and an enriched solid-phase and liquid-phase diffusion-coupled growth into the alloy are formed. These two simultaneous mechanisms interact to produce diverse topological disconnection and connection structures. The liquid metal dealloying technology depends on the liquid metal element, which requires a high mixing enthalpy with one of the elements of the original alloy. This is helpful for the element to dissolve in liquid metal aiming at dispersing the distribution of alloy elements. One of the phases is etched to prepare a nanopore structure with a high specific surface area. By comparing and contrasting the microphysically and chemically controlled microstructural evolution in liquid metal dealloying, the core phenomena of pore evolution – the competition between dissolution and interfacial diffusion exist in the processes [215]. In liquid metal dealloying, the dealloying rate depended on the diffusion of dissolved components in liquid metal. The liquid metal dealloying strategy is beneficial to further expand the use of dealloying as a method to control the structure of nanoscale materials with ultrahigh interface regions.

In conclusion, through the discussion of the structure and morphology of the liquid metal-assisted alloy formation, the liquid metal as a reaction medium has a good application in the alloy preparation.

5.4 Conclusion and Prospect

As described in the previous chapter, liquid metal can play an important role in the synthesis of nanomaterials and alloys as a new reaction medium. Based on its unique properties, liquid metal can be used to prepare nanomaterials and alloys from multiple angles.

Liquid metal has the characteristics of long-range disordered atomic structure. The appropriate interaction between atoms facilitates the fabrication of liquid metal droplet with 0D sizes. Then the liquid metal droplet can control the size, morphology, density, and element composition of other 0D materials. During the assisted preparation of 1D nanomaterial, liquid metal can be used as an effective catalyst to control the growth behavior of 1D nanomaterial based on the growth mechanism of VLS and SLS. This enables the preparation of 1D nanomaterial with specific components, diameters, and growth directions. Liquid metal-assisted preparation of 2D material mainly includes the control of 2D material growth and 2D material growth behavior. Liquid metal-assisted 2D material growth is mainly realized by three ways: liquid metal substrate, liquid metal skin, and vapor. Liquid metal has the advantages of smooth surface, isotropy, and fast mass transfer, so it is suitable as the substrate to assist 2D material growth. Liquid metal surface is easy to form a self-passivating thin oxide layer. Such self-passivating skin with atomic thickness is not only a 2D oxide, but also plays an important role as a template in the synthesis of other 2D nanomaterials. Liquid metals have a relatively high sublimation pressure. The liquid metal vapor provides sufficient catalyst and more possibility of collision with reactants to control the preparation process of 2D materials. As for the control of 2D materials growth behavior, liquid metal realizes rapid growth, morphology control, oriented stitching, ordered assembly, and heterogeneous stacking of 2D materials based on the characteristics of loose atomic arrangement, smooth surface, isotropy, and liquid fluidity.

With regard to the assisted preparation of alloy materials, the liquid metal can control alloy formation from the perspectives of structure and morphology. Liquid metal has excellent tolerance and can provide a low mixing enthalpy with various metal elements, so that liquid metal can form alloys with various elements. Structurally, liquid metal as a component element can participate in the formation of solid-solution phase, amorphous phase, and ordered phase alloy. Liquid metal has outstanding fluidity and formability. At the same time, liquid metal is a good solvent with wide melting point range and strong reducing property. In terms of alloy morphology formation, the introduction of pores, shape formation, and dealloying can be realized by liquid metal.

Liquid metals are a kind of deformable and flowing metal, which has the characteristics of excellent molten plasticity and fluidity. Based on the previous research, exploration of liquid metals as reaction medium can open up a new avenue of making use of metallic materials. Liquid metals as reaction medium have broad application prospects in semiconductor materials, structural materials, catalysis, and so on.

In terms of nanomaterials preparation, taking advantage of its surfaces that have no grain boundaries and are isotropic, liquid metals are hopeful to assist the growth of wafer-scale 2D materials. Besides, liquid metals are also expected to achieve the preparation of some emerging 2D materials, such as III–V group semiconductors and transition metal phosphides. Based on the ultra-smooth surface and easily shearing features of liquid metals, developing a strategy of floating production and transfer is in prospect, in order to fulfill precise fabrication and contamination-free transfer of 2D materials on a super-large scale. By utilizing its structural features of long-range disorder, unique interatomic interactions, outstanding temperature fluidity, and surface tension controllability, liquid metal has the potential to achieve formation of ultra-multiple elements alloying. What's more, it is possible for alloys of liquid metals and other metals to adjust the composition in an arbitrary proportion. Combined with high-throughput machine learning, it is hoped that alloys can be customized preparation according to performance requirements, thus expanding the application of alloys. Liquid metals also have the potential to control the morphology of alloys, which is expected to obtain patterned alloys for custom circuits and other applications.

In addition, liquid metals are expected to possess characteristics to adjust the size of the alloy, thereby achieving the preparation of nanosized alloys. And it is expected to control the morphology at the microscale, such as adjusting and designing the pore size of porous alloy through dealloying process and minimizing the pore size to obtain the mesoporous or microporous structure. Additionally, it is promising to efficiently disperse and stabilize catalysts by means of excellent fluidity and super solvent of liquid metals, aiming at obtaining monatomic catalysts with high stability and activity.

Liquid metal as a new reaction medium is going out a way of innovation and exploration and carrying out a comprehensive technological breakthrough, which shows that liquid metal havs an inestimable value.

References

1 Zeng, M., Tan, L., Wang, J. et al. (2014). Liquid metal: an innovative solution to uniform graphene films. *Chem. Mater.* 26: 3637.

2 Zheng, S., Zeng, M., Cao, H. et al. (2019). Insight into the rapid growth of graphene single crystals on liquid metal via chemical vapor deposition. *Sci. China Mater.* 62: 1087.

3 Gherasimova, M., Su, J., Cui, G. et al. (2005). A nanocluster route to zero- and one-dimensional quantum structures by MOCVD. *Phys. Status Solidi (C)* 2: 2361.

4 Qi, Z., Fang, Y., Chen, C., and Dai, J. (2016). Fabrication of GaN quantum dots using Ga droplet epitaxy and thermal annealing by metal organic chemical vapor deposition. *Phys. Status Solidi (C)* 13: 225.

5 Kumar, M., Bhat, T., Rajpalke, M. et al. (2010). Self-assembled flower-like nanostructures of InN and GaN grown by plasma-assisted molecular beam epitaxy. *Bull. Mater. Sci.* 33: 221.

6 Tian, B., Xie, P., Kempa, T.J. et al. (2009). Single-crystalline kinked semiconductor nanowire superstructures. *Nat. Nanotechnol.* 4: 824.

7 Adhikari, H., Marshall, A.F., Chidsey, C.E., and McIntyre, P.C. (2006). Germanium nanowire epitaxy: shape and orientation control. *Nano Lett.* 6: 318.

8 Dick, K.A., Deppert, K., Larsson, M.W. et al. (2004). Synthesis of branched 'nanotrees' by controlled seeding of multiple branching events. *Nat. Mater.* 3: 380.

9 Burgess, T., Breuer, S., Caroff, P. et al. (2013). Twinning superlattice formation in GaAs nanowires. *ACS Nano* 7: 8105.

10 Lehmann, S., Jacobsson, D., Deppert, K., and Dick, K.A. (2012). High crystal quality wurtzite-zinc blende heterostructures in metal-organic vapor phase epitaxy-grown GaAs nanowires. *Nano Res.* 5: 470.

11 Wen, C.Y., Reuter, M.C., Bruley, J. et al. (2009). Formation of compositionally abrupt axial heterojunctions in silicon-germanium nanowires. *Science* 326: 1247.

12 Caroff, P., Dick, K.A., Johansson, J. et al. (2009). Controlled polytypic and twin-plane superlattices in III–V nanowires. *Nat. Nanotechnol.* 4: 50.

13 Panciera, F., Chou, Y.C., Reuter, M.C. et al. (2015). Synthesis of nanostructures in nanowires using sequential catalyst reactions. *Nat. Mater.* 14: 820.

14 Laocharoensuk, R., Palaniappan, K., Smith, N.A. et al. (2013). Flow-based solution-liquid-solid nanowire synthesis. *Nat. Nanotechnol.* 8: 660.

15 Geng, D., Wu, B., Guo, Y. et al. (2012). Uniform hexagonal graphene flakes and films grown on liquid copper surface. *Proc. Natl. Acad. Sci. U. S. A.* 109: 7992.

16 Geng, D., Meng, L., Chen, B. et al. (2014). Controlled growth of single-crystal twelve-pointed graphene grains on a liquid Cu surface. *Adv. Mater.* 26: 6423.

17 Zeng, M., Tan, L., Wang, L. et al. (2016). Isotropic growth of graphene toward smoothing stitching. *ACS Nano* 10: 7189.

18 Zeng, M., Wang, L., Liu, J. et al. (2016). Self-assembly of graphene single crystals with uniform size and orientation: the first 2D super-ordered structure. *J. Am. Chem. Soc.* 138: 7812.

19 Zhou, Y., Deng, B., Zhou, Y. et al. (2016). Low-temperature growth of two-dimensional layered chalcogenide crystals on liquid. *Nano Lett.* 16: 2103.

20 Zavabeti, A., Ou, J.Z., Carey, B.J. et al. (2017). A liquid metal reaction environment for the room-temperature synthesis of atomically thin metal oxides. *Science* 358: 332.

21 Finley, J., Lee, C.H., Huang, P.Y., and Liu, L. (2019). Spin-orbit torque switching in a nearly compensated heusler ferrimagnet. *Adv. Mater.* 31: 1805361.

22 Inoue, A., Murakami, A., Zhang, T., and Takeuchi, A. (1997). Thermal stability and magnetic properties of bulk amorphous Fe-Al-Ga-P-C-B-Si alloys. *Mater. Trans, JIM* 38: 189.

23 Vida, A., Varga, L.K., Chinh, N.Q. et al. (2016). Effects of the sp element additions on the microstructure and mechanical properties of NiCoFeCr based high entropy alloys. *Mater. Sci. Eng. A* 669: 14.

24 Geslin, P.A., McCue, I., Gaskey, B. et al. (2015). Topology-generating interfacial pattern formation during liquid metal dealloying. *Nat. Commun.* 6: 8887.

25 Wada, T., Yubuta, K., Inoue, A., and Kato, H. (2011). Dealloying by metallic melt. *Mater. Lett.* 65: 1076.

26 Stoica, M., Bàrdos, A., Roth, S. et al. (2011). Improved synthesis of bulk metallic glasses by current-assisted copper mold casting. *Adv. Eng. Mater.* 13: 38.

27 Roduner, E. (2006). Size matters: why nanomaterials are different. *Chem. Soc. Rev.* 35: 583.

28 Radisavljevic, B., Radenovic, A., Brivio, J. et al. (2011). Single-layer MoS_2 transistors. *Nat. Nanotechnol.* 6: 147.

29 Han, T.-H., Lee, Y., Choi, M.-R. et al. (2012). Extremely efficient flexible organic light-emitting diodes with modified graphene anode. *Nat. Photonics* 6: 105.

30 Lin, H., Sturmberg, B.C.P., Lin, K.-T. et al. (2019). A 90-nm-thick graphene metamaterial for strong and extremely broadband absorption of unpolarized light. *Nat. Photonics* 13: 270.

31 Voiry, D., Yamaguchi, H., Li, J. et al. (2013). Enhanced catalytic activity in strained chemically exfoliated WS_2 nanosheets for hydrogen evolution. *Nat. Mater.* 12: 850.

32 Du, G., Guo, Z., Wang, S. et al. (2010). Superior stability and high capacity of restacked molybdenum disulfide as anode material for lithium ion batteries. *Chem. Commun.* 46: 1106.

33 Xie, Y., Qian, Y., Wang, W. et al. (1996). A benzene-thermal synthetic route to nanocrystalline GaN. *Science* 272: 1926.

34 Esken, D., Turner, S., Wiktor, C. et al. (2011). GaN@ZIF-8: selective formation of gallium nitride quantum dots inside a zinc methylimidazolate framework. *J. Am. Chem. Soc.* 133: 16370.

35 Kim, J.S., Mohanty, B.C., Han, C.S. et al. (2014). In situ magnetic field-assisted low temperature atmospheric growth of GaN nanowires via the vapor–liquid–solid mechanism. *ACS Appl. Mater. Interfaces* 6: 116.

36 Kuykendall, T.R., Altoe, M.V.P., Ogletree, D.F., and Aloni, S. (2014). Catalyst-directed crystallographic orientation control of GaN nanowire growth. *Nano Lett.* 14: 6767.

37 Wu, S., Wang, L., Liu, Z. et al. (2018). Ultrafast growth of horizontal GaN nanowires by HVPE through flipping the substrate. *Nanoscale* 10: 5888.

38 Novoselov, K.S., Geim, A.K., Morozov, S.V. et al. (2004). Electric field effect in atomically thin carbon films. *Science* 306: 666.

39 Gusmão, R., Sofer, Z., Bouša, D., and Pumera, M. (2017). Pnictogen (As, Sb, Bi) nanosheets for electrochemical applications are produced by shear exfoliation using kitchen blenders. *Angew. Chem. Int. Ed.* 56: 14417.

40 Abellán, G., Ares, P., Wild, S. et al. (2017). Noncovalent functionalization and charge transfer in antimonene. *Angew. Chem. Int. Ed.* 56: 14389.

41 Naguib, M., Kurtoglu, M., Presser, V. et al. (2011). Two-dimensional nanocrystals produced by exfoliation of Ti_3AlC_2. *Adv. Mater.* 23: 4248.

42 Li, X., Cai, W., An, J. et al. (2009). Large-area synthesis of high-quality and uniform graphene films on copper foils. *Science* 324: 1312.

43 Lee, Y.-H., Zhang, X.-Q., Zhang, W. et al. (2012). Synthesis of large-area MoS_2 atomic layers with chemical vapor deposition. *Adv. Mater.* 24: 2320.

44 Bae, S., Kim, H., Lee, Y. et al. (2010). Roll-to-roll production of 30-inch graphene films for transparent electrodes. *Nat. Nanotechnol.* 5: 574.

45 Novoselov, K.S., Fal'ko, V.I., Colombo, L. et al. (2012). A roadmap for graphene. *Nature* 490: 192.

46 Wang, J., Zeng, M., Tan, L. et al. (2013). High-mobility graphene on liquid p-block elements by ultra-low-loss CVD growth. *Sci. Rep.* 3: 2670.

47 Zhang, T., Jiang, B., Xu, Z. et al. (2016). Twinned growth behaviour of two-dimensional materials. *Nat. Commun.* 7: 13911.

48 Zeng, M., Chen, Y., Li, J. et al. (2017). 2D WC single crystal embedded in graphene for enhancing hydrogen evolution reaction. *Nano Energy* 33: 356.

49 Carey, B.J., Ou, J.Z., Clark, R.M. et al. (2017). Wafer-scale two-dimensional semiconductors from printed oxide skin of liquid metals. *Nat. Commun.* 8: 14482.

50 Syed, N., Zavabeti, A., Ou, J.Z. et al. (2018). Printing two-dimensional gallium phosphate out of liquid metal. *Nat. Commun.* 9: 3618.

51 Alsaif, M.M.Y.A., Pillai, N., Kuriakose, S. et al. (2019). Atomically thin Ga2S3 from skin of liquid metals for electrical, optical, and sensing applications. *ACS Appl. Nano Mater.* 2: 4665.

52 Syed, N., Zavabeti, A., Messalea, K.A. et al. (2019). Wafer-sized ultrathin gallium and indium nitride nanosheets through the ammonolysis of liquid metal derived oxides. *J. Am. Chem. Soc.* 141: 104.

53 Yu, I.S., Chang, C.P., Yang, C.P. et al. (2014). Characterization and density control of GaN nanodots on Si (111) by droplet epitaxy using plasma-assisted molecular beam epitaxy. *Nanoscale Res. Lett.* 9: 682.

54 Choi, I., Lee, H., Lee, C.R. et al. (2018). Formation of spherical-shaped GaN and InN quantum dots on curved SiN/Si surface. *Nanotechnology* 29: 315603.

55 Zheng, R., Peng, Z., Fu, Y. et al. (2020). A novel conductive core–shell particle based on liquid metal for fabricating real-time self-repairing flexible circuits. *Adv. Funct. Mater.* 30: 1910524.

56 Wagner, R.S. and Ellis, W.C. (1964). Vapor-liquid-solid mechanism of single crystal growth. *Appl. Phys. Lett.* 4: 89.

57 Hiruma, K., Yazawa, M., Katsuyama, T. et al. (1995). Growth and optical properties of nanometer-scale GaAs and InAs whiskers. *J. Appl. Phys.* 77: 447.

58 Lauhon, L.J., Gudiksen, M.S., Wang, D., and Lieber, C.M. (2002). Epitaxial core–shell and core–multishell nanowire heterostructures. *Nature* 420: 57.

59 Schwalbach, E.J. and Voorhees, P.W. (2008). Phase equilibrium and nucleation in VLS-grown nanowires. *Nano Lett.* 8: 3739.

60 Algra, R.E., Verheijen, M.A., Borgström, M.T. et al. (2008). Twinning superlattices in indium phosphide nanowires. *Nature* 456: 369.

61 Yan, C., Singh, N., and Lee, P.S. (2010). Kinking-induced structural evolution of metal oxide nanowires into single-crystalline nanorings. *ACS Nano* 4: 5350.

62 Wang, F., Dong, A., Sun, J. et al. (2006). Solution–liquid–solid growth of semiconductor nanowires. *Inorg. Chem.* 45: 7511.

63 Sun, J., Liu, C., and Yang, P. (2011). Surfactant-free, large-scale, solution–liquid–solid growth of gallium phosphide nanowires and their use for visible-light-driven hydrogen production from water reduction. *J. Am. Chem. Soc.* 133: 19306.

64 Dong, A., Yu, H., Wang, F., and Buhro, W.E. (2008). Colloidal GaAs quantum wires: solution–liquid–solid synthesis and quantum-confinement studies. *J. Am. Chem. Soc.* 130: 5954.

65 Dong, A., Wang, F., Daulton, T.L., and Buhro, W.E. (2007). Solution–liquid–solid (SLS) growth of ZnSe–ZnTe quantum wires having axial heterojunctions. *Nano Lett.* 7: 1308.

66 Sun, J. and Buhro, W.E. (2008). The use of single-source precursors for the solution–liquid–solid growth of metal sulfide semiconductor nanowires. *Angew. Chem. Int. Ed.* 47: 3215.

67 Lu, X., Fanfair, D.D., Johnston, K.P., and Korgel, B.A. (2005). High yield solution–liquid–solid synthesis of germanium nanowires. *J. Am. Chem. Soc.* 127: 15718.

68 Heitsch, A.T., Fanfair, D.D., Tuan, H.-Y., and Korgel, B.A. (2008). Solution–liquid–solid (SLS) growth of silicon nanowires. *J. Am. Chem. Soc.* 130: 5436.

69 Misra, S., Yu, L., Chen, W., and Roca i Cabarrocas, P. (2013). Wetting layer: the key player in plasma-assisted silicon nanowire growth mediated by tin. *J. Phys. Chem. C* 117: 17786.

70 Perea, D.E., Li, N., Dickerson, R.M. et al. (2011). Controlling heterojunction abruptness in VLS-grown semiconductor nanowires via in situ catalyst alloying. *Nano Lett.* 11: 3117.

71 Jeong, H., Park, T.E., Seong, H.K. et al. (2009). Growth kinetics of silicon nanowires by platinum assisted vapour–liquid–solid mechanism. *Chem. Phys. Lett.* 467: 331.

72 Wang, Y., Schmidt, V., Senz, S., and Gösele, U. (2006). Epitaxial growth of silicon nanowires using an aluminium catalyst. *Nat. Nanotechnol.* 1: 186.

73 Toko, K., Nakata, M., Jevasuwan, W. et al. (2015). Vertically aligned Ge nanowires on flexible plastic films synthesized by (111)-oriented Ge seeded vapor–liquid–solid growth. *ACS Appl. Mater. Interfaces* 7: 18120.

74 Hudak, B.M., Chang, Y.-J., Yu, L. et al. (2014). Real-time observation of the solid–liquid–vapor dissolution of individual tin(IV) oxide nanowires. *ACS Nano* 8: 5441.

75 Klamchuen, A., Suzuki, M., Nagashima, K. et al. (2015). Rational concept for designing vapor–liquid–solid growth of single crystalline metal oxide nanowires. *Nano Lett.* 15: 6406.

76 Zhu, Z., Suzuki, M., Nagashima, K. et al. (2016). Rational concept for reducing growth temperature in vapor–liquid–solid process of metal oxide nanowires. *Nano Lett.* 16: 7495.

77 Meng, G., Yanagida, T., Yoshida, H. et al. (2014). A flux induced crystal phase transition in the vapor–liquid–solid growth of indium-tin oxide nanowires. *Nanoscale* 6: 7033.

78 Kang, T.-T., Liu, X., Zhang, R.Q. et al. (2006). InN nanoflowers grown by metal organic chemical vapor deposition. *Appl. Phys. Lett.* 89: 071113.

79 Johnsson, M., Carlsson, M., and Nygren, M. (2003). Synthesis of transition metal carbide, carbonitride and boride whiskers. *Key Eng. Mater.* 247: 145.

80 Chen, Y., Liu, J., Liu, K. et al. (2019). GaN in different dimensionalities: properties, synthesis, and applications. *Mater. Sci. Eng., R* 138: 60.

81 Hu, Y.-L., Zhu, Y., Ji, H. et al. (2018). Fabrication of InxGa$_{1-x}$N nanowires on tantalum substrates by vapor–liquid–solid chemical vapor deposition. *Nanomaterials* 8: 990.

82 Ogino, T., Yamauchi, M., Yamamoto, Y. et al. (2015). Preheating temperature and growth temperature dependence of InP nanowires grown by self-catalytic VLS mode on InP substrate. *J. Cryst. Growth* 414: 161.

83 Bauer, J., Paetzelt, H., Gottschalch, V., and Wagner, G. (2010). GaAs nanowires grown by MOVPE. *Phys. Status Solidi (B)* 247: 1294.

84 Tan, C. and Zhang, H. (2015). Wet-chemical synthesis and applications of non-layer structured two-dimensional nanomaterials. *Nat. Commun.* 6: 7873.

85 Tan, C., Cao, X., Wu, X.-J. et al. (2017). Recent advances in ultrathin two-dimensional nanomaterials. *Chem. Rev.* 117: 6225.

86 Zhang, H. (2015). Ultrathin two-dimensional nanomaterials. *ACS Nano* 9: 9451.

87 Liu, J. and Fu, L. (2019). Graphene: controllable growth of graphene on liquid surfaces. *Adv. Mater.* 31: 1970060.

88 Zeng, M. and Fu, L. (2018). Controllable fabrication of graphene and related two-dimensional materials on liquid metals via chemical vapor deposition. *Acc. Chem. Res.* 51: 2839.

89 Tan, L., Zeng, M., Zhang, T., and Fu, L. (2015). Design of catalytic substrates for uniform graphene films: from solid-metal to liquid-metal. *Nanoscale* 7: 9105.

90 Mattevi, C., Kim, H., and Chhowalla, M. (2011). A review of chemical vapour deposition of graphene on copper. *J. Mater. Chem.* 21: 3324.

91 An, H., Lee, W.-J., and Jung, J. (2011). Graphene synthesis on Fe foil using thermal CVD. *Curr. Appl. Phys.* 11: S81.

92 Sutter, E., Albrecht, P., and Sutter, P. (2009). Graphene growth on polycrystalline Ru thin films. *Appl. Phys. Lett.* 95: 133109.

93 Ago, H., Ito, Y., Mizuta, N. et al. (2010). Epitaxial chemical vapor deposition growth of single-layer graphene over cobalt film crystallized on sapphire. *ACS Nano* 4: 7407.

94 Liu, M., Zhang, Y., Chen, Y. et al. (2012). Thinning segregated graphene layers on high carbon solubility substrates of rhodium foils by tuning the quenching process. *ACS Nano* 6: 10581.

95 Vo-Van, C., Kimouche, A., Reserbat-Plantey, A. et al. (2011). Epitaxial graphene prepared by chemical vapor deposition on single crystal thin iridium films on sapphire. *Appl. Phys. Lett.* 98: 181903.

96 Reina, A., Jia, X., Ho, J. et al. (2009). Large area, few-layer graphene films on arbitrary substrates by chemical vapor deposition. *Nano Lett.* 9: 30.

97 Kwon, S.-Y., Ciobanu, C.V., Petrova, V. et al. (2009). Growth of semiconducting graphene on palladium. *Nano Lett.* 9: 3985.

98 Gao, T., Xie, S., Gao, Y. et al. (2011). Growth and atomic-scale characterizations of graphene on multifaceted textured Pt foils prepared by chemical vapor deposition. *ACS Nano* 5: 9194.

99 Oznuluer, T., Pince, E., Polat, E.O. et al. (2011). Synthesis of graphene on gold. *Appl. Phys. Lett.* 98: 183101.

100 Dai, B., Fu, L., Zou, Z. et al. (2011). Rational design of a binary metal alloy for chemical vapour deposition growth of uniform single-layer graphene. *Nat. Commun.* 2: 522.

101 Liu, X., Fu, L., Liu, N. et al. (2011). Segregation growth of graphene on Cu–Ni alloy for precise layer control. *J. Phys. Chem. C* 115: 11976.

102 Bhaviripudi, S., Jia, X., Dresselhaus, M.S., and Kong, J. (2010). Role of kinetic factors in chemical vapor deposition synthesis of uniform large area graphene using copper catalyst. *Nano Lett.* 10: 4128.

103 Li, X., Cai, W., Colombo, L., and Ruoff, R.S. (2009). Evolution of graphene growth on Ni and Cu by carbon isotope labeling. *Nano Lett.* 9: 4268.

104 Ogawa, Y., Hu, B., Orofeo, C.M. et al. (2012). Domain structure and boundary in single-layer graphene grown on Cu(111) and Cu(100) films. *J. Phys. Chem. Lett.* 3: 219.

105 Zhang, Y., Gomez, L., Ishikawa, F.N. et al. (2010). Comparison of graphene growth on single-crystalline and polycrystalline Ni by chemical vapor deposition. *J. Phys. Chem. Lett.* 1: 3101.

106 Fujita, J., Miyazawa, Y., Ueki, R. et al. (2010). Fabrication of large-area graphene using liquid gallium and its electrical properties. *Jpn. J. Appl. Phys.* 49: 06GC01.

107 Ueki, R., Nishijima, T., Hikata, T. et al. (2012). In-situ observation of surface graphitization of gallium droplet and concentration of carbon in liquid gallium. *Jpn. J. Appl. Phys.* 51: 06FD28.

108 Fujita, J.-i., Ueki, R., Miyazawa, Y., and Ichihashi, T. (2009). Graphitization at interface between amorphous carbon and liquid gallium for fabricating large area graphene sheets. *J. Vac. Sci. Technol., B* 27: 3063.

109 Amini, S., Garay, J., Liu, G. et al. (2010). Growth of large-area graphene films from metal-carbon melts. *J. Appl. Phys.* 108: 094321.

110 Ding, G., Zhu, Y., Wang, S. et al. (2013). Chemical vapor deposition of graphene on liquid metal catalysts. *Carbon* 53: 321.

111 Chen, L., Kong, Z., Yue, S. et al. (2015). Growth of uniform monolayer graphene using iron-group metals via the formation of an antiperovskite layer. *Chem. Mater.* 27: 8230.

112 Wang, J., Chen, L., Wu, N. et al. (2016). Uniform graphene on liquid metal by chemical vapour deposition at reduced temperature. *Carbon* 96: 799.

113 Zhang, Y., Tang, T.-T., Girit, C. et al. (2009). Direct observation of a widely tunable bandgap in bilayer graphene. *Nature* 459: 820.

114 Xia, F., Farmer, D.B., Lin, Y.-m., and Avouris, P. (2010). Graphene field-effect transistors with high on/off current ratio and large transport band gap at room temperature. *Nano Lett.* 10: 715.

115 Lu, X., Stepanov, P., Yang, W. et al. (2019). Superconductors, orbital magnets and correlated states in magic-angle bilayer graphene. *Nature* 574: 653.

116 Ju, L., Wang, L., Cao, T. et al. (2017). Tunable excitons in bilayer graphene. *Science* 358: 907.

117 Zibrov, A.A., Kometter, C., Zhou, H. et al. (2017). Tunable interacting composite fermion phases in a half-filled bilayer-graphene Landau level. *Nature* 549: 360.

118 Yan, J., Kim, M.H., Elle, J.A. et al. (2012). Dual-gated bilayer graphene hot-electron bolometer. *Nat. Nanotechnol.* 7: 472.

119 Ju, L., Shi, Z., Nair, N. et al. (2015). Topological valley transport at bilayer graphene domain walls. *Nature* 520: 650.

120 Meyer, J.C., Geim, A.K., Katsnelson, M.I. et al. (2007). The structure of suspended graphene sheets. *Nature* 446: 60.

121 Ma, W., Chen, M.-L., Yin, L. et al. (2019). Interlayer epitaxy of wafer-scale high-quality uniform AB-stacked bilayer graphene films on liquid Pt3Si/solid Pt. *Nat. Commun.* 10: 2809.

122 Chen, Y., Sun, J., Gao, J. et al. (2015). Growing uniform graphene disks and films on molten glass for heating devices and cell culture. *Adv. Mater.* 27: 7839.

123 Ju, M., Liang, X., Liu, J. et al. (2017). Universal substrate-trapping strategy to grow strictly monolayer transition metal dichalcogenides crystals. *Chem. Mater.* 29: 6095.

124 Xu, D., Chen, W., Zeng, M. et al. (2018). Crystal-field tuning of photoluminescence in two-dimensional materials with embedded lanthanide ions. *Angew. Chem. Int. Ed.* 57: 755.

125 Chen, J., Zhao, X., Tan, S.J.R. et al. (2017). Chemical vapor deposition of large-size monolayer $MoSe_2$ crystals on molten glass. *J. Am. Chem. Soc.* 139: 1073.

126 Xu, C., Wang, L., Liu, Z. et al. (2015). Large-area high-quality 2D ultrathin Mo_2C superconducting crystals. *Nat. Mater.* 14: 1135.

127 Shivayogimath, A., Thomsen, J.D., Mackenzie, D.M.A. et al. (2019). A universal approach for the synthesis of two-dimensional binary compounds. *Nat. Commun.* 10: 2957.

128 Chen, Y., Liu, K., Liu, J. et al. (2018). Growth of 2D GaN single crystals on liquid metals. *J. Am. Chem. Soc.* 140: 16392.

129 Dickey, M.D., Chiechi, R.C., Larsen, R.J. et al. (2008). Eutectic gallium-indium (EGaIn): a liquid metal alloy for the formation of stable structures in microchannels at room temperature. *Adv. Funct. Mater.* 18: 1097.

130 Kim, D., Thissen, P., Viner, G. et al. (2013). Recovery of nonwetting characteristics by surface modification of gallium-based liquid metal droplets using hydrochloric acid vapor. *ACS Appl. Mater. Interfaces* 5: 179.

131 Cabrera, N. and Mott, N.F. (1949). Theory of the oxidation of metals. *Rep. Prog. Phys.* 12: 163.

132 Chandler, D., Weeks, J.D., and Andersen, H.C. (1983). Van der Waals picture of liquids, solids, and phase transformations. *Science* 220: 787.

133 Lin, J., Li, Q., Liu, T.-Y. et al. (2019). Printing of quasi-2D semiconducting β-Ga_2O_3 in constructing electronic devices via room-temperature liquid metal oxide skin. *Phys. Status Solidi (RRL) – Rap. Res. Lett.* 13: 1900271.

134 Zeng, M., Chen, Y., Zhang, E. et al. (2019). Molecular scaffold growth of two-dimensional, strong interlayer-bonding-layered materials. *CCS Chemistry* 1: 117.

135 Honig, R.E. (1969). Vapor pressure data for the solid and liquid elements. *RCA Rev.* 30: 285.

136 Liu, J., Zeng, M., Wang, L. et al. (2016). Ultrafast self-limited growth of strictly monolayer WSe_2 crystals. *Small* 12: 5741.

137 Tan, L., Zeng, M., Wu, Q. et al. (2015). Direct growth of ultrafast transparent single-layer graphene defoggers. *Small* 11: 1840.

138 Wang, F., Gao, T., Zhang, Q. et al. (2019). Liquid-alloy-assisted growth of 2D ternary $Ga_2In_4S_9$ toward high-performance UV photodetection. *Adv. Mater.* 31: 1806306.

139 Pan, Z.W., Dai, S., Beach, D.B. et al. (2003). Gallium-mediated growth of multi-wall carbon nanotubes. *Appl. Phys. Lett.* 82: 1947.

140 Liu, J. and Fu, L. (2019). Controllable growth of graphene on liquid surfaces. *Adv. Mater.* 31: 1800690.

141 Wu, B., Geng, D., Guo, Y. et al. (2011). Equiangular hexagon-shape-controlled synthesis of graphene on copper surface. *Adv. Mater.* 23: 3522.

142 Jacobberger, R.M. and Arnold, M.S. (2013). Graphene growth dynamics on epi-taxial copper thin films. *Chem. Mater.* 25: 871.

143 Geng, D., Gao, E., Wang, H. et al. (2016). Large-area growth of five-lobed and triangular graphene grains on textured Cu substrate. *Adv. Mater. Interfaces* 3: 1600347.

144 Luo, B., Chen, B., Meng, L. et al. (2014). Layer-stacking growth and electrical transport of hierarchical graphene architectures. *Adv. Mater.* 26: 3218.

145 Liu, Y. and Yakobson, B.I. (2010). Cones, pringles, and grain boundary land-scapes in graphene topology. *Nano Lett.* 10: 2178.

146 Yazyev, O.V. and Louie, S.G. (2010). Electronic transport in polycrystalline graphene. *Nat. Mater.* 9: 806.

147 Dong, J., Geng, D., Liu, F., and Ding, F. (2019). Formation of twinned graphene polycrystals. *Angew. Chem. Int. Ed.* 58: 7723.

148 Tan, L., Han, J., Mendes, R.G. et al. (2015). Self-aligned single-crystalline hexagonal boron nitride arrays: toward higher integrated electronic devices. *Adv. Electron. Mater.* 1: 1500223.

149 Lee, J.S., Choi, S.H., Yun, S.J. et al. (2018). Wafer-scale single-crystal hexagonal boron nitride film via self-collimated grain formation. *Science* 362: 817.

150 Liu, Y., Wu, H., Cheng, H.-C. et al. (2015). Toward barrier free contact to molybdenum disulfide using graphene electrodes. *Nano Lett.* 15: 3030.

151 Xue, X., Xu, Q., Wang, H. et al. (2019). Gas-flow-driven aligned growth of graphene on liquid copper. *Chem. Mater.* 31: 1231.

152 Harrison, D.A., Yan, D., and Blairs, S. (1977). The surface tension of liquid copper. *J. Chem. Thermodyn.* 9: 1111.

153 Assael, M.J., Kalyva, A.E., Antoniadis, K.D. et al. (2010). Reference data for the density and viscosity of liquid copper and liquid tin. *J. Phys. Chem. Ref. Data* 39: 033105.

154 Zeng, M., Cao, H., Zhang, Q. et al. (2018). Self-assembly of metal oxide nanoparticles in liquid metal toward nucleation control for graphene single-crystal arrays. *Chem* 4: 626.

155 Gan, L. and Luo, Z. (2013). Turning off hydrogen to realize seeded growth of subcentimeter single-crystal graphene grains on copper. *ACS Nano* 7: 9480.

156 Wang, C., Zuo, J., Tan, L. et al. (2017). Hexagonal boron nitride-graphene core-shell arrays formed by self-symmetrical etching growth. *J. Am. Chem. Soc.* 139: 13997.

157 Zhang, T. and Fu, L. (2018). Controllable chemical vapor deposition growth of two-dimensional heterostructures. *Chem* 4: 671.

158 Liu, Y., Weiss, N.O., Duan, X. et al. (2016). Van der Waals heterostructures and devices. *Nat. Rev. Mater.* 1: 16042.

159 Georgiou, T., Jalil, R., Belle, B.D. et al. (2013). Vertical field-effect transistor based on graphene–WS_2 heterostructures for flexible and transparent electronics. *Nat. Nanotechno.* 8: 100.

160 Britnell, L., Gorbachev, R.V., Jalil, R. et al. (2012). Field-effect tunneling transistor based on vertical graphene heterostructures. *Science* 335: 947.

161 Hong, X., Kim, J., Shi, S.-F. et al. (2014). Ultrafast charge transfer in atomically thin MoS_2/WS_2 heterostructures. *Nat. Nanotechnol.* 9: 682.

162 Deng, D., Novoselov, K.S., Fu, Q. et al. (2016). Catalysis with two-dimensional materials and their heterostructures. *Nat. Nanotechno.* 11: 218.

163 Chiu, M.-H., Zhang, C., Shiu, H.-W. et al. (2015). Determination of band alignment in the single-layer MoS_2/WSe_2 heterojunction. *Nat. Commun.* 6: 7666.

164 Massicotte, M., Schmidt, P., Vialla, F. et al. (2016). Picosecond photoresponse in van der Waals heterostructures. *Nat. Nanotechnol.* 11: 42.

165 Dean, C.R., Wang, L., Maher, P. et al. (2013). Hofstadter's butterfly and the fractal quantum Hall effect in moiré superlattices. *Nature* 497: 598.

166 Roy, K., Padmanabhan, M., Goswami, S. et al. (2013). Graphene–MoS_2 hybrid structures for multifunctional photoresponsive memory devices. *Nat. Nanotechno.* 8: 826.

167 Shi, J., Liu, M., Wen, J. et al. (2015). All chemical vapor deposition synthesis and intrinsic bandgap observation of MoS_2/graphene heterostructures. *Adv. Mater.* 27: 7086.

168 Behura, S., Nguyen, P., Che, S. et al. (2015). Large-area, transfer-free, oxide-assisted synthesis of hexagonal boron nitride films and their heterostructures with MoS_2 and WS_2. *J. Am. Chem. Soc.* 137: 13060.

169 Zhang, C., Chuu, C.-P., Ren, X. et al. (2017). Interlayer couplings, Moiré patterns, and 2D electronic superlattices in MoS_2/WSe_2 hetero-bilayers. *Sci. Adv.* 3: e1601459.

170 Naylor, C.H., Parkin, W.M., Gao, Z. et al. (2017). Synthesis and physical properties of phase-engineered transition metal dichalcogenide monolayer heterostructures. *ACS Nano* 11: 8619.

171 Banszerus, L., Schmitz, M., Engels, S. et al. (2015). Ultrahigh-mobility graphene devices from chemical vapor deposition on reusable copper. *Sci. Adv.* 1: e1500222.

172 Yang, W., Chen, G., Shi, Z. et al. (2013). Epitaxial growth of single-domain graphene on hexagonal boron nitride. *Nat. Mater.* 12: 792.

173 Zhang, M., Zhu, Y., Wang, X. et al. (2015). Controlled synthesis of ZrS2 monolayer and few layers on hexagonal boron nitride. *J. Am. Chem. Soc.* 137: 7051.

174 Fu, L., Sun, Y., Wu, N. et al. (2016). Direct growth of MoS2/h-BN heterostructures via a sulfide-resistant alloy. *ACS Nano* 10: 2063.

175 Lukatskaya, M.R., Mashtalir, O., Ren, C.E. et al. (2013). Cation intercalation and high volumetric capacitance of two-dimensional titanium carbide. *Science* 341: 1502.

176 Ghidiu, M., Lukatskaya, M.R., Zhao, M.-Q. et al. (2014). Conductive two-dimensional titanium carbide 'clay' with high volumetric capacitance. *Nature* 516: 78.

177 Shahzad, F., Alhabeb, M., Hatter, C.B. et al. (2016). Electromagnetic interference shielding with 2D transition metal carbides (MXenes). *Science* 353: 1137.

178 Geng, D., Zhao, X., Chen, Z. et al. (2017). Direct synthesis of large-area 2D Mo_2C on in situ grown graphene. *Adv. Mater.* 29: 1700072.

179 Xu, M., Liang, T., Shi, M., and Chen, H. (2013). Graphene-like two-dimensional materials. *Chem. Rev.* 113: 3766.

180 Splendiani, A., Sun, L., Zhang, Y. et al. (2010). Emerging photoluminescence in monolayer MoS_2. *Nano Lett.* 10: 1271.

181 Wang, X., Feng, H., Wu, Y., and Jiao, L. (2013). Controlled synthesis of highly crystalline MoS_2 flakes by chemical vapor deposition. *J. Am. Chem. Soc.* 135: 5304.

182 Wang, Q.H., Kalantar-Zadeh, K., Kis, A. et al. (2012). Electronics and optoelectronics of two-dimensional transition metal dichalcogenides. *Nat. Nanotechno.* 7: 699.

183 Baugher, B.W.H., Churchill, H.O.H., Yang, Y., and Jarillo-Herrero, P. (2014). Optoelectronic devices based on electrically tunable p–n diodes in a monolayer dichalcogenide. *Nat. Nanotechno.* 9: 262.

184 Cheng, R., Jiang, S., Chen, Y. et al. (2014). Few-layer molybdenum disulfide transistors and circuits for high-speed flexible electronics. *Nat. Commun.* 5: 5143.

185 Lu, W., Zeng, M., Li, X. et al. (2016). Controllable sliding transfer of wafer-size graphene. *Adv. Sci.* 3: 1600006.

186 Kulekci, M.K. (2007). Magnesium and its alloys applications in automotive industry. *J. Adv. Manuf. Tech.* 39: 851.

187 Yang, M.B., Pan, F.S., Zhang, J., and Zhang, J. (2005). An analysis of the development and applications of current and new Mg-Al based elevated temperature magnesium alloys. *Mater. Sci. Forum* 488-489: 923.

188 Pan, F., Yang, M., and Chen, X. (2016). A review on casting magnesium alloys: modification of commercial alloys and development of new alloys. *J. Mater. Sci. Techno.* 32: 1211.

189 Yao, K.F. and Zhang, C.Q. (2007). Fe-based bulk metallic glass with high plasticity. *Appl. Phys. Lett.* 90: 061901.

190 Sha, Y.H., Zhang, F., Zhou, S.C. et al. (2008). Improvement of recrystallization texture and magnetic property in non-oriented silicon steel by asymmetric rolling. *J. Magn. Magn. Mater.* 320: 393.

191 Borkar, T., Gwalani, B., Choudhuri, D. et al. (2016). A combinatorial assessment of $AlxCrCuFeNi_2$ (0<x<1.5) complex concentrated alloys: microstructure, microhardness, and magnetic properties. *Acta Mater.* 116: 63.

192 Zhang, Y., Zhou, Y.J., Lin, J.P. et al. (2008). Solid-solution phase formation rules for multi-component alloys. *Adv. Eng. Mater.* 10: 534.

193 Li, Z., Xu, H., Gu, Y. et al. (2018). Correlation between the magnetic properties and phase constitution of $FeCoNi(CuAl)_{0.8}Ga$ ($0 \leq x \leq 0.08$) high-entropy alloys. *J. Alloys Compd.* 746: 285.

194 Zhang, S., Wu, C.L., and Zhang, C.H. (2015). Phase evolution characteristics of $FeCoCrAlCuV_xNi$ high entropy alloy coatings by laser high-entropy alloying. *Mater. Lett.* 141: 7.

195 Kao, Y.-F., Chen, S.-K., Chen, T.-J. et al. (2011). Electrical, magnetic, and Hall properties of $Al_xCoCrFeNi$ high-entropy alloys. *J. Alloys Compd.* 509: 1607.

196 Guo, S., Ng, C., Lu, J., and Liu, C.T. (2011). Effect of valence electron concentration on stability of fcc or bcc phase in high entropy alloys. *J. Appl. Phys.* 109: 103505.

197 Tian, F., Varga, L.K., Chen, N. et al. (2015). Empirical design of single phase high-entropy alloys with high hardness. *Intermetallics* 58: 1.

198 Huang, S., Vida, Á., Li, W. et al. (2017). Thermal expansion in FeCrCoNiGa high-entropy alloy from theory and experiment. *Appl. Phys. Lett.* 110: 241902.

199 Stoica, M., Eckert, J., Roth, S., and Schultz, L. (2004). Preparation of bulk amorphous Fe-Cr-Mo-Ga-P-C-B alloys by copper mold. *Mater. Sci. Eng. A* 375–377: 399.

200 Galanakis, I., Dederichs, P.H., and Papanikolaou, N. (2002). Slater–Pauling behavior and origin of the half-metallicity of the full-Heusler alloys. *Phys. Rev. B* 66: 174429.

201 Schirber, J.E. (1972). Pressure-induced electron transition in $AuGa_2$. *Phys. Rev. Lett.* 28: 1127.

202 Boonyongmaneerat, Y. and Dunand, D.C. (2008). Ni-Mo-Cr foams processed by casting replication of sodium aluminate preforms. *Adv. Eng. Mater.* 10: 379.

203 Imran, M., Zhang, X., Qian, M., and Geng, L. (2020). Enhancing the elastocaloric cooling stability of Ni–Fe–Ga alloys via introducing pores. *Adv. Eng. Mater.* 22: 1901140.

204 Wu, Y., Ertekin, E., and Sehitoglu, H. (2017). Elastocaloric cooling capacity of shape memory alloys – role of deformation temperatures, mechanical cycling, stress hysteresis and inhomogeneity of transformation. *Acta Mater.* 135: 158.

205 Turnbull, D. (1969). Under what conditions can a glass be formed? *Contemp. Phys.* 10: 473.

206 Erlebacher, J., Aziz, M.J., Karma, A. et al. (2001). Evolution of nanoporosity in dealloying. *Nature* 410: 450.

207 Chen, Q. and Sieradzki, K. (2013). Spontaneous evolution of bicontinuous nanostructures in dealloyed Li-based systems. *Nat Mater* 12: 1102.

208 Li, W.C. and Balk, T.J. (2010). Achieving finer pores and ligaments in nanoporous palladium–nickel thin films. *Scripta Mater.* 62: 167.

209 Wang, Z., Gao, H., Niu, J. et al. (2019). Transforming bulk metals into metallic nanostructures: a liquid-metal-assisted top-down dealloying strategy with sustainability. *ACS Sustainable Chem. Eng.* 7: 3274.

210 McCue, I., Ryan, S., Hemker, K. et al. (2016). Size effects in the mechanical properties of bulk bicontinuous Ta/Cu nanocomposites made by liquid metal dealloying. *Adv. Eng. Mater.* 18: 46.

211 Abe, H., Sato, K., Nishikawa, H. et al. (2009). Dealloying of Cu-Zr-Ti bulk metallic glass in hydrofluoric acid solution. *Mater. Trans.* 50: 1255.

212 Wada, T., Ichitsubo, T., Yubuta, K. et al. (2014). Bulk-nanoporous-silicon negative electrode with extremely high cyclability for lithium-ion batteries prepared using a top-down process. *Nano Lett.* 14: 4505.

213 Kim, J.W., Tsuda, M., Wada, T. et al. (2015). Optimizing niobium dealloying with metallic melt to fabricate porous structure for electrolytic capacitors. *Acta Mater.* 84: 497.

214 Joo, S.H., Bae, J.W., Park, W.Y. et al. (2020). Beating thermal coarsening in nanoporous materials via high-entropy design. *Adv. Mater.* 32: 1906160.

215 McCue, I., Karma, A., and Erlebacher, J. (2018). Pattern formation during electrochemical and liquid metal dealloying. *MRS Bull.* 43: 27.

6

Constructing Techniques of Liquid Metal-Based Architectures

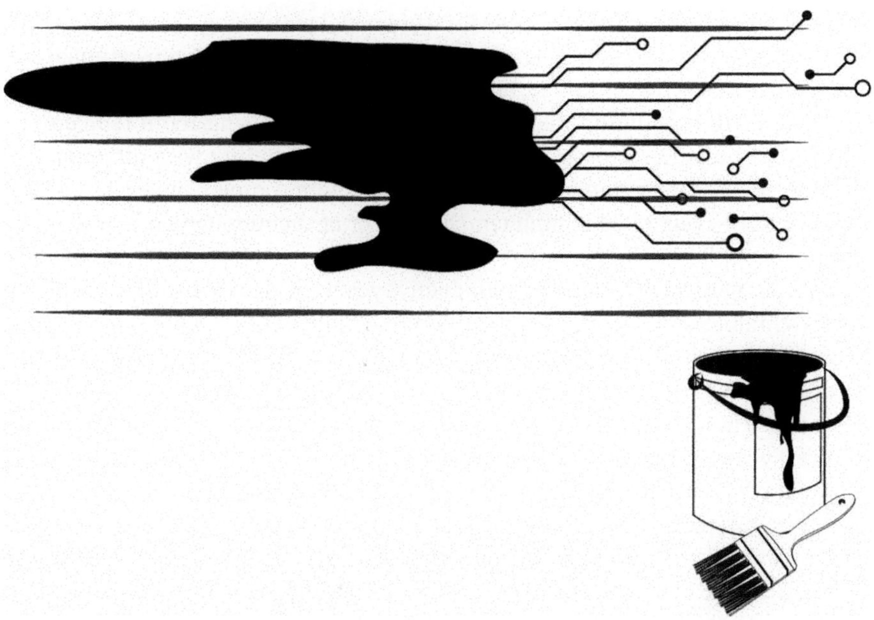

In the final analysis, the fantasy of science is the bold creation of science and technology.
Source: Feiding (1892–1977).

6.1 Introduction

Liquid metals possess many intrinsic special properties, including low melting point, low vapor pressure, and excellent combination of conductivity and deformability [1]. These remarkable advantages offer liquid metals promising potential in various important fields, such as flexible and stretchable electronics, biomedicine, and soft robots [2–4]. To utilize promising liquid metals for these applications, it is of great

Liquid Metals: Properties, Mechanisms, and Applications, First Edition. Lei Fu and Mengqi Zeng.

importance to develop techniques for constructing liquid metal-based architectures. In this chapter, we mainly focus on Ga-based liquid metals-based architectures due to its low melting, low evaporation pressure, safety, fluidity, and moldable properties.

Patterning liquid metals on a target substrate is considered to be a major method to construct liquid metals-based architectures [5]. Ga-based liquid metals spontaneously form surface oxide, which could decrease the effective surface tension [6]. Therefore, liquid metals exhibit good adhesion on various substrates, including paper, silica gel plate, and rubber sheet [7]. Besides, the presence of oxide layer also provides mechanical stability to patterned liquid metal with different geometries [8, 9]. These properties make liquid metals difficult to use conventional photolithography method, which employs light to irradiate photosensitive polymers on target substrate and selectively removes photoresist to obtain desired topography [10]. Although photolithography of liquid metals has not been realized, this method can be used to make molds or stencils in liquid metal patterning.

Due to the special properties, liquid metals allow easier and unconventional pattern techniques. Owing to fluidity, liquid metals can be injected into microchannels [11]. Also, patterning liquid metals could be conducted using room temperature and ambient pressures process that exhibit high compatibility with different substrates. On the contrary, patterning solid metals is more complex since it often need to be melted or evaporated at high temperature or at vacuum, which is destructive and not compatible with thermosensitive and volatile substrates [12]. Meanwhile, liquid metals with intrinsically high conductivity make it ideal conductive ink. However, the conductivity of patterning traditional solid metals is usually obtained after removing polymer, which is employed to promote dispersion and avoid agglomeration of solid metals [13, 14]. Additionally, the moldable properties allow the liquid metals to use a number of unconventional patterning methods, such as imprinting and direct writing [15].

The simplest method to pattern liquid metals is injecting it into predefined features by surpassing the pressure demanded to rupture surface oxide layer, such as microchannels and hollow fibers [11, 16]. Fortunately, the demanded pressure is modest and can be reached by pushing a syringe with hand [16]. Besides, well-designed, flexible, and complex liquid metals pattern can be accomplished by using imprinting, mask-assisted deposition, direct writing, laser ablation, and the like. Meanwhile, the incorporation of magnetic and electric field has an impact on the constructed liquid metal architecture. Apart from those patterning techniques, some techniques to construct liquid metals-based architectures that do not need pattering is also discussed in this chapter, such as directly mixing liquid metal with elastomer, polymer, and hydrogen. Here, we organize the constructing techniques into seven categories, which would be discussed in the next sections:

(i) *Injection:* It takes advantages of required force to fill liquid metals into predefined structure, which can be divided into direct injection and vacuum filling.
(ii) *Imprinting:* Liquid metal can adhere to the surface of elastomeric mold (stamp) by manual pressing. Imprinting liquid metals on the recesses and protrusions can obtain patterned liquid metals.

(iii) *Mask-assisted deposition:* This method fills and flow the liquid metals on a mask. After removing the mask, liquid metal pattern with the same mask feature can be created.
(iv) *Direct writing:* By moving the equipped liquid metals ink deposition nozzle, controlled composition and architecture can be obtained by direct writing. According to properties of the conductive inks and the manner of the delivery process, the techniques can be mainly classified into two categories, including inkjet printing and microextrusion printing.
(v) *Laser ablation:* Laser ablation belongs to typically subtractive manufacturing. In this process, laser is applied to selectively remove unnecessary areas to obtain desired patterns. It has been well-documented that CO_2 and UV laser can be used to ablate liquid metals.
(vi) *External field guiding:* This process utilizes external field, such as electrical field and magnetic field to control the liquid metal patterning.
(vii) *Others:* Different from the methods mentioned earlier, other methods that do not need pattering are discussed here, such as shearing mixing and sedimentation method.

6.2 Injection

Injection, using stress to fill the fluid material (e.g., ionic liquids, metal–salt mixtures, molten solders, and liquid metal) into predefined features, is a versatile and well-developed constructing technique [12]. Although the injection of molten solders has been used for creating metallic microstructures, it requires heating and cooling steps, which increases the fabrication time, makes it incompatible with many organic materials, and even results in structural cracking and mechanical fatigue. While the injection of liquid metal addresses these limitations of solders. This simple and convenient technique of molding liquid metal into functional architectures has been widely used in the fabrication of low-cost, soft, and stretchable electronics, such as wires, circuit elements, electrodes, antennas, and sensors [17]. Since liquid metal possesses a high surface tension and surface oxide, pressure is required to inject liquid metal into structures with void space. Based on the applied pressure (relative to atmospheric pressure) during the filling process, we sort injection methods into two categories: direct injection (positive pressure) and vacuum filling (negative pressure).

6.2.1 Direct Injection

In the direct injection process, it is usually required to punch holes in channels as inlets and outlets, where inlets are used for injection and outlets allow air to escape from the channel during filling. And a syringe or pump is used to provide the required pressure. This method usually creates liquid metal structures encased in an elastomer. Similar to the conventional replica molding, it is possible to remove the solidified liquid metal from the soft elastomeric channel. Therefore, independent liquid metal patterns can be produced by freezing the injected liquid metal to solid state and peeling off the channel [18, 19]. According to the study of Dickey and coworkers, liquid metal can flow rapidly and fill the microchannel as long as the applied stress exceeds the stress necessary to rupture the gallium oxide skin [11]. Then the injected liquid metal within the channel can form stable microstructure with the assistance of the newly formed oxide skin. This elastic oxide skin distinguishes liquid metal from Hg, which spontaneously retracts from the channels and forms a shape with minimum surface energy once we relieve the inlet pressure. Later, this group found that the pressure required for injecting EGaIn–ethanol suspension was one-third of that for pure EGaIn in the microchannel with a cross-sectional area of $50 \times 300\,\mu m^2$, which was due to the fact that fluid properties of suspension are dominated by the solvent [20]. Moreover, a gold coating layer can reactively wet liquid metal and improve the filling process by decreasing interfacial free energy of the liquid metal channel [21].

Based on Young–LaPlace's equation, the critical pressure P_C (minimum pressure necessary to induce flow) for injection varies inversely with the cross-sectional width and height of the microfluidic conduit [11]. Therefore, liquid metal can be selectively injected into the desired channel by designing the microchannel structure. For example, Ota and coworkers fabricated a liquid–liquid heterojunction device by introducing a high flow resistance structure, where liquid metal serves as metal electrodes and ionic liquid is used as the sensing element [22]. There are two kinds of pressure during injection: the total pressure P_T and capillary pressure P_C (critical pressure), where P_T depends on the applied pressure and injection velocity. Given that liquid metal has a high contact angle with PDMS, the direction of P_C is opposite to P_T. When P_T is less than P_C, liquid metal is confined to its territory, thus preventing liquid metal from penetrating the position assigned for the ionic liquid (Figure 6.1a). This can be realized by the structure composed of a series of junction channels with 30 μm wide and 31 μm high. While for a 300 μm wide channel, liquid metal enters into the area for ionic liquid. Moreover, inherently aligned microposts (so-called Laplace barriers) can also block liquid metal and guide it only to the desired area in microchannels, as long as P_C is not exceeded [26, 27].

Although the passivating surface oxide of liquid metal provides mechanical stability within the microchannels, it can also adhere to the inner walls, thus preventing the achievement of liquid metal reconfigurable electronics. To reversibly actuate liquid metal in microchannels, the oxide skin can be removed by treating liquid metal with acid, alkaline solution, acidified silicone oils, and an external electric field [11, 28–30]. Alternatively, it is possible to minimize liquid metal residues by prefilling channels with a water slip layer [31]. Joshipura and coworkers sprayed the channel with commercial NeverWet before injection [23]. The NeverWet, composed of silica nanoparticles grafted with polysiloxanes, has a dual roughness scale from micro to nano, which indicates its superhydrophobic characteristic and can prevent the adhesion of oxide. Therefore, the loaded liquid metal can withdraw from the channel similar to Hg without oxide residue (Figure 6.1b). By adjusting the length of liquid metal within the NeverWet-coated channel, they established a reconfigurable monopole antenna in the gigahertz range. They also fabricated

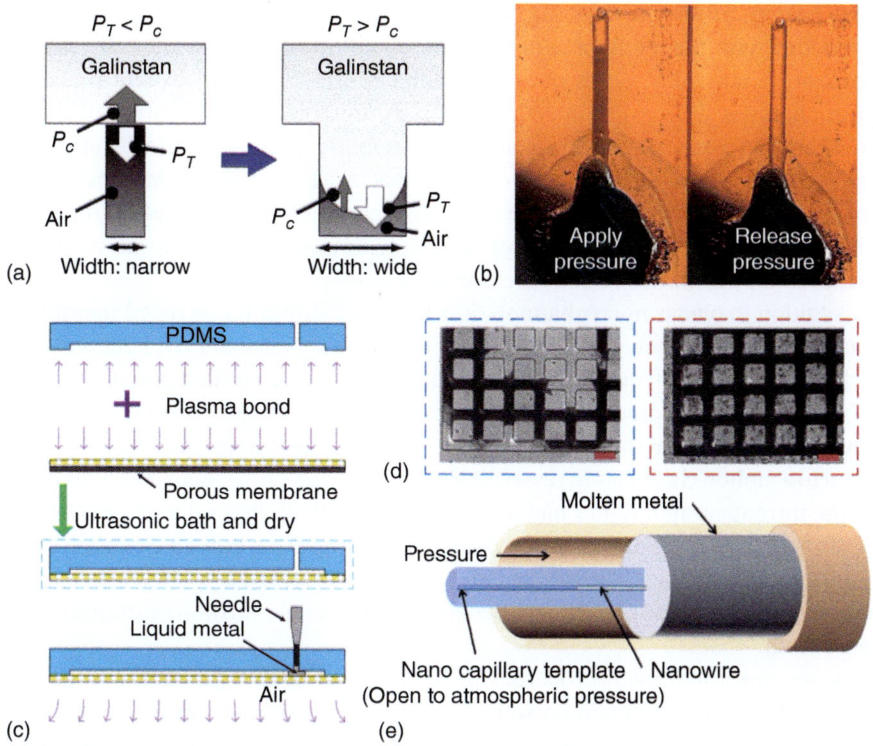

Figure 6.1 (a) Schematic illustrations of narrow and wide junction channels during Galinstan injection, depicting the relationship between the capillary pressure (P_C) and total pressure (P_T). Source: Ota et al. [22]. (b) Reversible motion of liquid metal in a closed NeverWet-coated channel. Source: Joshipura et al. [23]. (c) The Schematic process of porous membrane-assisted injection. (d) Comparison of injection without a porous membrane (left) and with membrane (right), respectively. Scale bar is 100 μm. Source: (c, d) Wang et al. [24]. (e) The schematic of modified high-pressure filling technique. Source: Zhao et al. [25].

surface-exposed liquid metal traces on the uncoated channel path by stripping the substrate coated with NeverWet.

For the direct injection, the requirements for both inlets and outlets make it difficult to fill the blind-ended (without outlets) microchannels. Guo and coworkers realized a complete filling of liquid metal into a closed-end microchannel, which was driven by a constant pressure imposed at the inlet [32]. The distinctive gas permeability of PDMS guaranteed the release of air from the channel during the pressure-driven filling process. However, the accumulation of high injection pressure may also cause delamination of microchannels. Recently, Wang and coworkers have developed a novel porous membrane-assisted injection to achieve fast liquid metal filling in the blind-ended microchannel without any damage or air residues [24]. The key to produce a porous membrane is using granular sugar as a sacrificial structure. The schematic processes are shown in Figure 6.1c. First, a PDMS slice with the desired feature was fabricated via replica molding. Then, the mixture of PDMS and granular sugar powder was spin-coated on the silicon wafer, followed by the baking, ultrasonic cleaning, and drying. Finally, the patterned PDMS slice was bonded with the membrane to form the microfluidics system after air plasma treatment. Remarkably, pores not only facilitated the release of air and pressure within channels during the injection, but also promoted the formation of oxide skin after the injection, which enhanced the stability of the structure. Therefore, it showed complete filling in branched structures (Figure 6.1d). Moreover, the minimum feature size that could be injected successfully was around 12 μm wide and 20 high. This novel injection is promising in advancing the community of fast fabrication for ultrathin stretchable electronics.

In addition to planar microchannels, it is feasible to inject liquid metal into other structures such as hollow fibers structures or three-dimensional (3D) helical channels [33, 34]. Zhao and coworkers utilized the modified spliced-fiber pressure-filling technique to deliver liquid metal into hollow silica fiber (Figure 6.1e) [25]. Specifically, the nanocapillary template with one side open to atmospheric pressure was spliced to a molten Ga or Ga–In reservoir under high pressure for injecting. This realized the introduction of liquid metal into the fiber with an inner diameter as small as 150 nm at a pressure of ~10 MPa, which demonstrated the available breadth during the injection process. The injected liquid metal exhibited super cooling behavior inside the nanowire and could be used to produce the single-fluxon-controlled resistance switch.

In summary, rheological properties of liquid metal enable the ability of moldability, which makes injection possible. For direct injection, as long as the imposed pressure exceeds the critical pressure, liquid metal can be successfully filled into the structures using a syringe or pump. Spontaneously formed surface oxide is crucial to hold the injected liquid metal into place. Meanwhile, the principle of oxide adhesion minimum may be utilized to construct reconfigurable electronics. Direct injection is usually suitable for simple microchannels. For complex channels, it usually obtains incomplete filling of liquid metal because it may only flow into the main branch of channels. When the size is quite small, the applied pressure will be very high due to the large surface tension of liquid metal. And the injection pressure may cause liquid metal leakage and delamination of the channels. The novel porous membrane-assisted injection offers a potential method for the complex channels and has realized the fast filling of thin blind-ended microchannel.

6.2.2 Vacuum Filling

Vacuum filling is the complementary method to achieve injection of liquid metal, which can completely introduce liquid metal into every trench of patterned mold by applying vacuum to create negative pressure. Unlike direct injection, this method does not require outlets to expel air, and it is very suitable for complex structures, such as blind-ended microchannels or 3D nanonetwork structures. It can also reduce liquid metal leakage and damage the channels, which is common in pressurized injection. Note that the liquid metal can be processed in vacuum environment without considering boiling or evaporation under reduced pressure, due to its negligible vapor pressure.

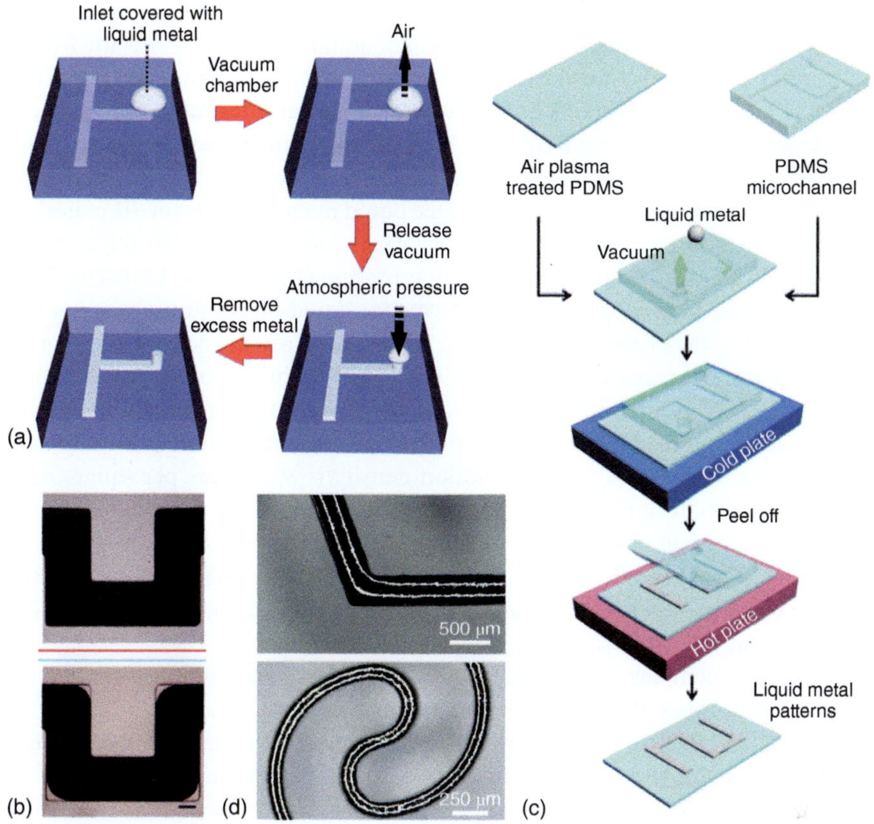

Figure 6.2 (a) Schematic illustration of vacuum filling. (b) Optical micrographs of vacuum filling (top) and injection (bottom). The scale bar is 200 μm. Source: (a, b) Lin et al. [35]. (c) Scheme of a defect-free, high-resolution liquid metal patterning process. (d) The independent defect-free patterns [36]. Source: (c, d) Yang et al. [36].

Lin and coworkers placed the PDMS microfluidic device with a drop of liquid metal upon the inlet in a vacuum chamber to expel air and reduce the injection resistance (Figure 6.2a) [35]. When exposed to the atmosphere, the channel had a lower pressure than environment (atmospheric pressure) due to the seal of liquid metal at the inlet. Therefore, the generated pressure difference enabled liquid metal to be sucked into the channel. They demonstrated that this vacuum filling can pull liquid metal into complex and branched PDMS microfluidic channels within seconds. The comparison between this method and direct injection is illustrated in Figure 6.2b. The channels with branches and dead ends can be completely filled by vacuum filling. If the microchannel is built from dissolvable poly(vinyl alcohol) (PVA), this technique can fabricate flexible transient circuits beyond liquid metal patterns [37]. Moreover, similar to direct injection, vacuum filling can also achieve complex pattern independence by solidifying liquid metal and stripping channels off (Figure 6.2c, d) [36]. Notably, the air plasma-treated PDMS substrate has better adhesion to liquid metal compared with the untreated PDMS microchannel. This originates that air plasma treatment can generate an oxidized PDMS layer, which has a strong interaction with the oxidized skin of liquid metal. Therefore, the frozen EGaIn can be completely transferred from the untreated microchannel and form the defect-free independent patterns on the air plasma-treated substrate.

For direct injection, the requirement of multiple inlets and outlets limits its ability to produce dense and distinct liquid metal patterns. Vacuum filling provides a comparatively easy and robust method to produce liquid metal contacts for 2D materials in a high-integration density. Chen and coworkers introduced a filter paper and a PDMS sacrificial layer during vacuum filling process (Figure 6.3a) [38]. The filter paper can release air and exhibit poor wetting for liquid metal, thus confining liquid metal within the channel. After filling, liquid metal residue at the inlets can be removed by stripping the PDMS sacrificial layer. Both of these contribute to creating a well-defined clean interface conforming to the shape of the channel. Therefore, they produced patterned liquid metal contacts with high resolution and quality for graphene. The obtained highest integration density is 36 devices per square centimeter. Furthermore, Andrews and coworkers discovered that vacuum filling can enhance the adhesion of liquid metal, which is due to the fact that pressure forces liquid metal to form a conformal coverage (Figure 6.3b) [39]. Therefore, it decreases the contact resistance between the liquid metal contacts and carbon nanotube (CNT) transistor, compared to direct writing.

In addition to complex planar microfluidic structures, vacuum filling can also be applied in infiltrating 3D deep features. Park and coworkers vacuum-filled liquid metal into 3D PDMS net-shaped nanostructures to create a conductor with high levels of stretchability and conductivity, which can be used as interconnects in LED [41]. However, it requires a multi-step lithography process to engineer 3D PDMS nanonetworks, which is quite tedious. Liang and coworkers put PDMS

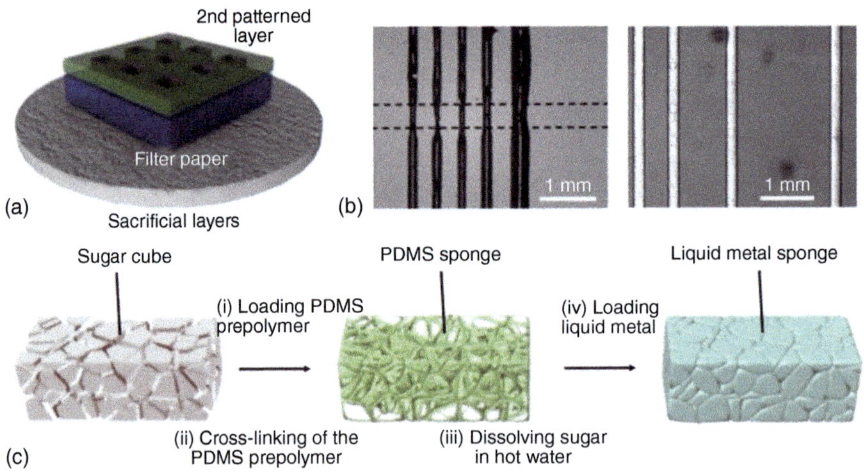

Figure 6.3 (a) Schematic illustration of the device with a filter paper and a PDMS sacrificial layer. Source: Chen et al. [38]. (b) Optical micrographs of liquid metal patterns: direct writing (left) and vacuum filling (right). Source: Andrews et al. [39]. (c) Schematic illustration of the fabrication of liquid metal PDMS sponges. Source: (c) Liang et al. [40].

sponges in a bath containing GaInSn under vacuum to fabricate liquid metal and PDMS soft electric conductors [40]. The PDMS sponges were obtained via the sugar-templating method (Figure 6.3c). First, PDMS prepolymer filled sugar cubes, and then the sugar was dissolved away, leading to a sponge with random pores that can be loaded with liquid metal. The vacuum was utilized to ensure the complete filling of liquid metal in the voids of PDMS sponges. The presence of liquid metal does not significantly disturb the deformation of the PDMS sponge under compression. The elasticity of 3D sponge-like structure and the fluidity of liquid metal endow the all-soft conductor with high mechanical flexibility and electrical conductivity.

In summary, a pressure differential created by vacuum enables liquid metal to spontaneously flow into microchannels. This filling technique is especially suitable for creating deep features and infiltrating the microchannels with blind ends or branches. It has been demonstrated to fabricate high-integration liquid metal contacts and 3D stretchable conductors. However, the vacuum treatment usually requires several hours to create a full pressure drop. And it needs sufficient chip stiffness and thickness for the recovery of channel after filling.

6.3 Imprinting

Imprinting liquid metal with an elastomeric mold (stamp) is a conceptually very simple method to pattern liquid metal, which mainly exploits the adhesion of liquid metal surface oxide to elastomer. Both the recesses and protrusions of the stamp can be used to form liquid metal pattern.

As reported by Gozen and coworkers, liquid metal was dispersed on a flat substrate using a roller and then flattened by compressing with a flat elastomer [42]. Then a lithography-defined elastomer mold with micron-patterned features was pressed against the flattened liquid metal film. Under the applied normal force, liquid metal conformably entered into the recesses of the mold. Finally, the desired liquid metal was created after separating the mold from the substrate. Surface oxide assists to keep liquid metal intact in the mold during the separation. If needed, an additional elastomer layer was employed to encapsulate the EGaIn-filled mold as a circuit element for a co-planar capacitor. The whole process is illustrated in Figure 6.4a. Using this method, they created liquid metal trace with minimum line width and spacing of 2 and 1 μm, which is almost impossible for injection of elastomer microchannel. This is due to the fact that the injection pressure for single-micron-scale structures is at least 1 MPa based on the Young–Laplace's equation, which is beyond the failure limit of elastomers. The obtained liquid metal trace with micron-scale line width enables the fabrication of the high-density stretchable microelectronics. Liquid metal also exhibits wettability on the outside areas of channel. Kim and coworkers presented a uniform and residue-free liquid metal thin-line imprinting method. They used toluene to selectively modify the PDMS mold surface [43]. Compared with untreated PDMS, the chemically modified PDMS has a lower surface energy, thus increasing the hydrophobicity. This could efficiently transfer the redundant liquid

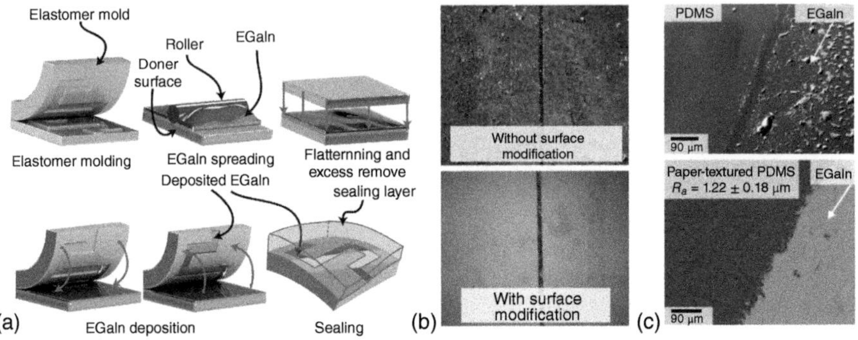

Figure 6.4 (a) Schematic description of the liquid metal imprinting process. Source: Gozen et al. [42]. (b) Optical micrographs of patterned liquid metal line and surrounding PDMS surfaces without and with toluene modification and residue transfer step. Source: Kim et al. [43]. (c) Patterned liquid metal films on normal and paper-textured PDMS surfaces. Source: Kim et al. [15].

metal on the outside areas of channel to a nonmodified sacrificial PDMS substrate. To completely remove liquid metal residue, they repeated this transfer process several times. Figure 6.4b shows a comparison of liquid metal pattern between this method and imprinting without modification. This advanced technique can obtain high-quality liquid metal patterns with a spectrum from single micrometers to a couple of millimeters.

In addition, it is also possible to create liquid metal pattern with the protruding surfaces of stamp. Tabatabai and coworkers fabricated a stamp with geometry protruding 2 mm from the base, which effectively prevented liquid metal from invading into any recesses during gently pressing the stamp onto liquid metal [44]. The finest feature with 0.75 mm width was created after transferring liquid metal onto the target substrate. Since the surface oxide has interference in uniform wetting on receiving substrate, the produced liquid metal features may have nonuniform thickness and missing areas. Kim and coworkers enhanced the wetting property of the substrate for liquid metal by physically modifying the target substrate surface. They poured liquid PDMS on the paper and peeled it off after solidifying PDMS, thus obtaining paper-textured PDMS surface [15]. Notably, the micro-cellulose fiber structures of paper are completely transferred to the PDMS, and most of the nano-sized structures disappear in the transferring process. It has been reported that paper with a micro–nano texture (dual-scale fibers) exhibits super-lyophobic characteristics. Liquid metal forms Cassie state and shows superior nonwetting property on paper [45]. But for the resulting PDMS surface only with microsized bumpy surface morphology of paper, liquid metal can fully contact and wet with it, which can be explained by the Wenzel model [45]. Therefore, uniform and smooth liquid metal thin film can form on the modified PDMS after multiple transfer processes (Figure 6.4c). Although it creates liquid metal pattern with a poor lateral resolution, this method is suitable for large-scale (millimeter–centimeter) thin-film patterning. In addition to the target substrate, it is also possible to tune wetting behavior by designing the stamp. Zhang and coworkers obtained a super liquid metal-phobic

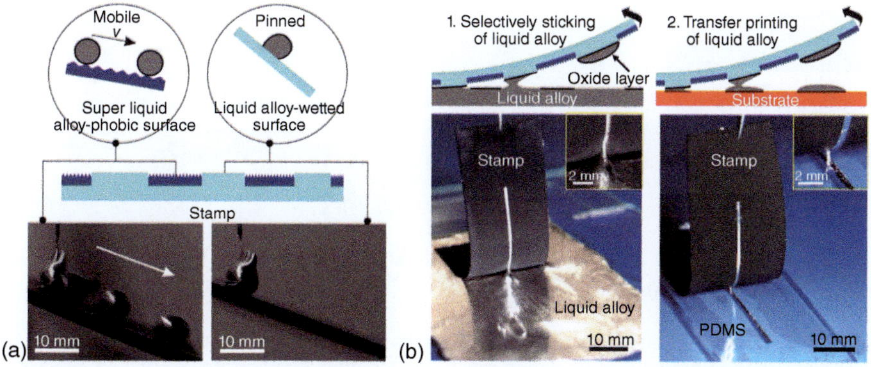

Figure 6.5 (a) A selectively amphiphilic stamp. (b) Illustration and photographs of the main processes: selectively sticking (left) and transfer printing (right). Source: (a, b) Zhang et al [46].

surface in desired areas of the carbon-dosed PDMS (cPDMS) surface by creating micro/nanoscale roughness under UV treatment [46]. Liquid metal was repelled on the UV-treated region, while pinned on the untreated region (Figure 6.5a). Therefore, it selectively adhered to the modified cPDMS, forming an inked stamp (Figure 6.5b). Due to the excellent mechanical compliance of cPDMS, this method can transfer liquid metal pattern onto curved surfaces conformably without any deformation.

In conclusion, liquid metal can adhere to every surface of elastomeric stamp by manually pressing, including concave and raised areas. The oxide can preserve the shape of liquid metal during the separating process. For imprinting with recesses, it creates liquid metal pattern within the channel of stamp. But there is usually liquid metal residue on the outside of channel. While imprinting with protrusion it produces patterns on the surface of stamp, which can be further transferred to the target substrate (a flat or curved surface) by using physical contact. But liquid metal has uneven wetting on the target substrate. There have been some strategies to tune wetting behavior and thus improve pattern quality and resolution, such as chemical or physical modification of substrate and design of stamp.

6.4 Mask-Assisted Deposition

Liquid metal can go through and fill the gaps if a mask is placed on the receiving substrate. While redundant liquid metal remains on the surface of mask. After removing the mask, liquid metal pattern with the same feature of mask can be created. This technique is called mask-assisted deposition. There are mainly three deposition ways by utilizing masks, including pressing manually, spraying, and evaporation.

In the simplest embodiment for the mask-assisted deposition, Jeong and coworkers adapted a roller to press liquid metal through the mesh of a stiff copper stencil onto the semi-cured PDMS substrate [47]. Liquid metal pattern with line width as small as 200 μm and spacing of 100 μm could be obtained after removing stencil mask

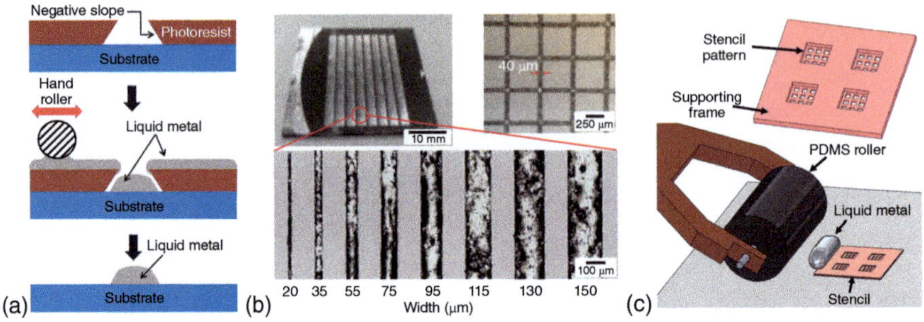

Figure 6.6 (a) The scheme of photolithography-based liquid metal patterning process. (b) The deposited liquid metal lines and mesh pattern on a Si substrate. Source: (a, b) Park et al. [48]. (c) Multilayer stencil printing process. Source: Lazarus et al. [49].

from substrate. However, it produced rough line edges. Later, Park and coworkers introduced the lift-off technique. The whole process is presented in Figure 6.6a [48]. First, liquid metal was overlaid onto the surface of a patterned photoresist (a sacrificial mask) and adhered to the openings by using a hand-roller. The sidewalls of the photoresist pattern were designed with a negative slope, which could suppress liquid metal residue of the sidewalls efficiently. Afterward, the sample was submerged in acetone to dissolve the resist, thus leaving liquid metal feature on substrate. Notably, the formed patterns have sharp edges with a line width as small as 20 μm and a height of 10 μm (Figure 6.6b). Although it produces nonuniform thickness, this photolithographic approach reduces the width of patterns by an order of magnitude. Moreover, it requires a specialized, expensive facility and chemicals, which is not suitable for large-area fabrication. Lazarus and coworkers employed a microfabricated multilevel stencil as the mask to print liquid metal, as exhibited in Figure 6.6c [49]. The electroplated Cu multilevel stencils consisted of 3 μm thin regions and 30 μm thicker regions. The thin layer enabled finer patterning because of the reduced sidewall pinning during the mask deposition. While the thicker layer was robust enough as supporting frame for printing. The pattern with a feature of 20 μm width was obtained here.

Although depositing liquid metal with pressing is quite simple to operate, it is labor-intensive and usually produces features with relatively rough edges. Zhang and coworkers proposed the atomized spraying of liquid metal by using an airbrush for the first time (Figure 6.7a) [50]. Based on the Bernoulli's principle, airflow force at the nozzles can disintegrate liquid metal into microdroplets. The discretized microdroplets with a high specific surface area can quickly oxidize, which significantly increases the adhesion of liquid metal on the substrate. Then the atomized microdroplets hit the substrate at high density and velocity and are firmly deposited onto the surface. Due to the hit of high velocity, the oxide can rupture and liquid metal microdroplets merged. Finally, the desired pattern was produced after removing the mask. This method results in smooth line edges and can deposit liquid metal pattern on diverse substrates and even curve surfaces. The resolution of spray deposition is related to the microdroplet size and mask. Moreover, Sahlberg

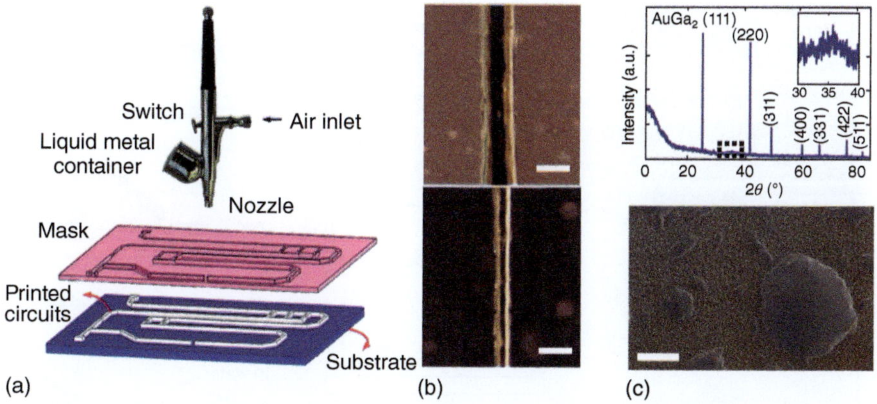

Figure 6.7 (a) Schematic illustration of atomized spray deposition of liquid metal. Source: Zhang et al. [50]. (b) Liquid metal lines with the 1D SSP before (top) and after (bottom) shrink. Scale bars indicate 500 μm. Source: Sahlberg et al. [51]. (c) Top: X-ray diffraction (XRD) pattern of biphasic metal film on PDMS substrate. Bottom: False-color SEM image of the biphasic metal film. The regions with gray and yellow colors represent the liquid Gallium and $AuGa_2$/Ga film, respectively. Scale bar is 5 μm. Source: Hirsch et al. [52].

and coworkers discovered that a stretch–shrink patterning (SSP) process can reduce line width (Figure 6.7b) [51]. They sprayed liquid metal onto the pre-strained masked substrate and stripped the mask. Then the substrate was gently relaxed, shrunk into its original condition, and cured to produce the pattern. Liquid metal can deform to accommodate the interface geometry and force conditions during the substrate-shrinking. Thus, line width with 100 μm remarkably decreased to 25 μm when the 1D shrinkage occurred in the perpendicular direction to the lines. They pointed out that a higher resolution would be obtained by employing a mask with improved resolution or a substrate with enhanced plasticity.

In addition to spray deposition, thermal evaporation is also possible to apply liquid metal onto the masked substrate. Different from the evaporation deposition of solid films, the deposited liquid metal nanoparticles still remain in liquid form because of their low melting point. Moreover, a high vacuum level prevents liquid metal from oxidation. These two factors ensure that the contact nanoparticles can merge to produce liquid metal pattern. Hirsch and coworkers first sputtered gold on the masked PDMS substrate, and then evaporated liquid metal onto substrate [52]. After stripping the mask, the desired pattern was formed. The thin biphasic metal film composed of solid $AuGa_2$ and liquid gallium is formed because liquid metal can alloy with gold (Figure 6.7c). It has a high intrinsical stretchability (400% uniaxial strain), which can be used for wafer-scale fabrication of stretchable conductors.

Moreover, if wetting metal or sacrificial polymer coating is preliminarily deposited onto the substrate as mask, liquid metal pattern can be created without the process of stripping mask. Guo and coworkers fabricated polymethacrylates (PMA) pattern on a paper substrate, which was used as a mask for depositing liquid metal [53]. As shown in Figure 6.8a, the paper has a quite rough surface microstructure, resulting in the incomplete permeation of oxides in the gap of paper fibers. While the surface

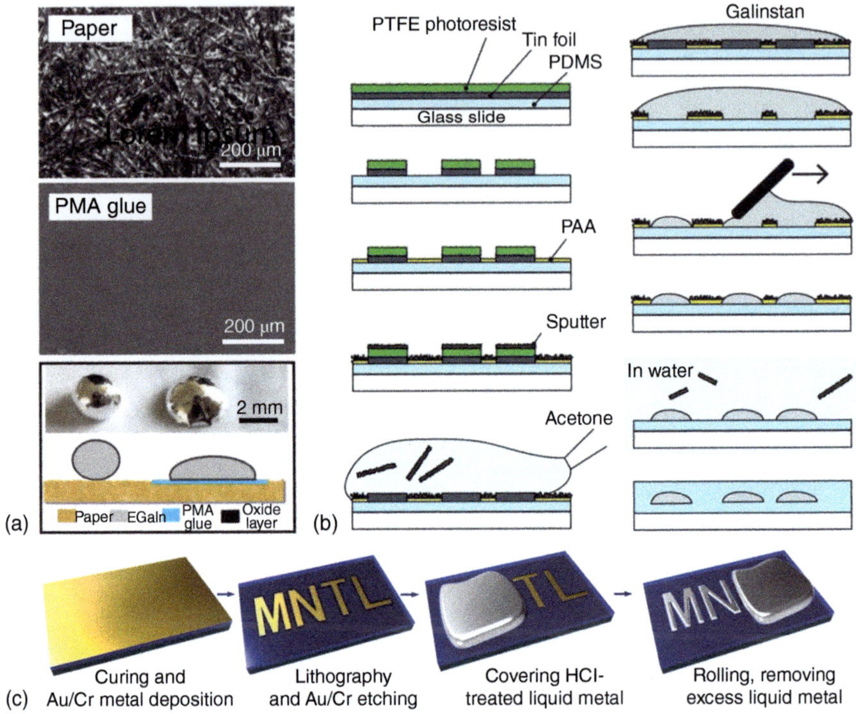

Figure 6.8 (a) SEM images of paper surface and PMA glue, and schematic illustration of EGaIn droplets on the surface of paper and PMA glue. Source: Guo et al. [53]. (b) Conceptual drawing of liquid metal-embedded elastomer created by selective wetting. Source: Kramer et al. [54]. (c) The scheme of selective liquid metal plating on Au/Cr substrate. Source: Li et al. [55].

of PMA glue printed on the paper is smooth, which promotes the attachment of liquid metal. More importantly, the methyl functional group of PMA can interact with the Ga_2O_3 by the hydrogen bond. These indicate that PMA glue has a higher affinity to liquid metal and thus liquid metal can stick firmly to the PMA glue rather than paper. A series of liquid metal patterns and paper electronics were produced. Moreover, it has also been reported to use superhydrophobic coating as sacrificial mask [56].

Since liquid metal can reactive wet with some metal, the patterned metal layer (e.g., Sn foil, Au, or Cu adhesive layer) can be used as mask. Kramer and coworkers specifically depicted the whole process (Figure 6.8b) [54]. It is based on the fact that Ga can wet reactively with Sn foil rather than the sputter-coated In [57]. First, the desired pattern was customized on the PDMS substrate via sputtering In and photolithography. It consisted of two parts: Sn film as a seed layer and In/poly(acrylic acid) (PAA) as a shield. Then, liquid metal could preferentially deposit onto the Sn area and formed the pattern after releasing In/PAA layer in water. To guarantee complete wetting on metal surface, it needs a treatment of at least 24 hours. The obtained minimum liquid metal line has a planar size of

200 μm. Later, Li and coworkers employed electron beam lithography technique to fabricate the Au/Cr adhesive layer with the desired pattern on PDMS substrate [55]. The reduced liquid metal via hydrochloric acid (HCl) treatment exhibited super-lyophobic and super-lyophilic characteristics for PDMS substrate and metal pattern, respectively. Figure 6.8c shows the whole process. In this approach, they produced the feature with uniform edges in a high resolution of less than 10 μm. Notably, the full wetting process only required tens of seconds. Recently, by combining the hybrid nanofabrication technique, submicron-scale liquid metal pattern was produced on Au wetting pattern [58]. The obtained liquid metal line has the width of 180 nm with a spacing of 1 μm, which is the highest resolution up to date. Moreover, Liquid metal can also be plated on Cu wetting pattern [59, 60].

In conclusion, mask-assisted deposition is a generalized and high throughput technique for constructing liquid metal patterns. To gain a clear boundary of pattern, the mask should be tightly located close to the substrate. Manually pressing, spraying, and evaporation can push liquid metal through the openings of the masked substrate. When the mask is removed, liquid metal is deposited onto the exposed regions of the substrate. Noted there are some disadvantages, such as excessive liquid metal loss during the process of removing mask, poor line edges for manually pressing and cumbersome operation for both thermal evaporation and spray deposition. The use of photoresist, multilevel mask or pre-strained substrate can improve deposition quality and resolution. Moreover, a sacrificial layer (polymer wetting layer or superhydrophobic coating, metal adhesive layer) can be used as a mask to selectively deposit liquid metal, thus directly producing the desired pattern without the process of removing mask. However, it needs to pre-deposit sacrificial layer and this kind of mask cannot be recycled.

6.5 Direct Writing

Direct writing is a versatile and advanced technique to construct materials with controlled composition and architecture by moving the equipped ink deposition nozzle [61]. Various materials can be used as inks, including polymeric, colloidal, semiconductor materials, and so on [62–66].

Initially, Russo and coworkers proposed that conductive silver ink could be directly written for a variety of conductive constructions by a roller-ball pen [67]. Such direct writing conductive ink technology provides a simple and easy method for the manufacture of flexible printing electronic products, and undoubtedly sheds light on direct conductive electronics constructions. As mentioned earlier, the room temperature liquid metals are promising as electric printing inks due to their low melting point and toxicity, admirable stretchable properties, relatively high metallic conductivity, and environmentally friendly, applying to microfluidic systems and soft flexible conductive devices [8, 16, 68]. Among the flexible patterning techniques of liquid metal, direct writing technique, belonging to additive manufacturing, can realize rapidly fabricating or constructing 2D patterning on various substrate surfaces and complex 3D architectures and does not need expensive toolings,

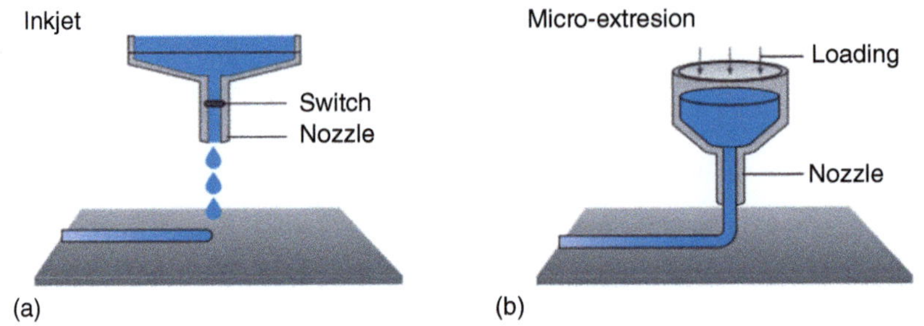

Figure 6.9 (a) Schematic of inkjet printing techniques. (b) Schematic of micro-extrusion printing techniques. Source: Zhang et al. [69].

troublesome dies, or pre-prepared masks. These techniques can be mainly classified into two categories, including inkjet printing and micro-extrusion printing, which are defined on the basis of the properties of the conductive inks and the manner of the delivery process (Figure 6.9a,b). Inkjet printing exploits nozzles to deliver liquid metal droplets to the target substrate using thermal or acoustic processes. By contrast, micro-extrusion printing forms continuous filaments of liquid metals with the help of applying extra pressure. In this section, we will summarize the progress made so far in these two techniques [69].

6.5.1 Inkjet Printing

As for inkjet printing, liquid metal droplets are used as inks. Micro- or nanodroplets are usually needed in proper solvents to stabilize the liquid metal [20, 70–74]. After printing, a subsequent sintering process, such as mechanical sintering and laser sintering, is needed to reconnect making it transform into a conductor. This step is necessary because the surface of the liquid metal nanoparticles spontaneously forms the gallium oxide, which blocks the conduction of the nanoparticles liquid metals [20, 70, 71]. For example, Ren and coworkers obtained the EGaInSn nanodroplets in the presence of thiols via a simple ultra-sonication process [71]. The TEM images combine with energy dispersive X-ray (EDX) data of EGaInSn nanodroplets shown in Figure 6.10a–c distinctly indicate that liquid metal oxide layer surrounds each nanodroplets, which is needed for the subsequent sintering process to conduct when it is prepared into an electronic device. To reveal the applicability of this technique, they fabricated stretchable and flexible superconducting liquid metal-based circuits on common polyethylene terephthalate (PET) and PDMS substrate by inkjet printing with EGaInSn nanodroplets (Figure 6.10d). The realization of novel superconducting system promotes the advanced flexible superconductor-devices utilization and provides a new insight for expanding commercial applications of superconductors.

In contrast, liquid metal droplets in aqueous solvents are expected to show more potential in biomedical applications [76]. Here, Li and coworkers obtained a biocompatible liquid metal ink, through encapsulating EGaIn nanodroplets into marine polysaccharides or alginate microgels via ultrasonication (Figure 6.10e) [75]. During the ultrasonication process, on one hand, alginate could accelerate the downsizing process of EGaIn due to the coordination effect of their carboxyl

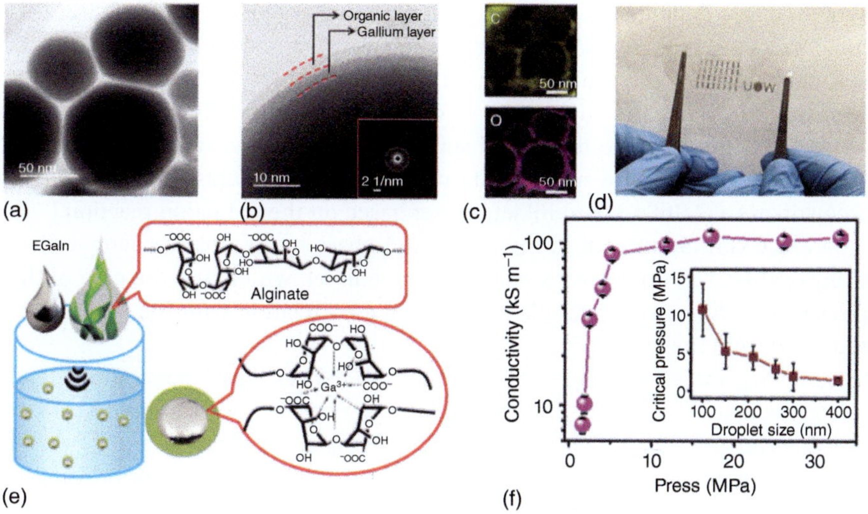

Figure 6.10 (a, b) TEM images of EGaInSn nanodroplets via a simple ultrasonication process; (c) A representative EDX elements spectrum of EGaInSn nanodroplets. (d) Image of sintered flexible circles of the EGaInSn nanodroplet by inkjet printing techniques. Source: (a–d) Ren et al. [71]. (e) Schematic of encapsulating EGaIn nanodroplets into alginate microgels via ultrasonication. (f) Conductivity of EGaIn nanodroplets under different sintering pressures. The inset gives the dependence of critical sintering pressure on droplet diameter. Source: (e, f) Li et al. [75].

groups with Ga^{3+}. On the other hand, chelating Ga^{3+} into structural "egg-box" crosslinkers could form around the EGaIn droplets. At the same time, EGaIn aqueous nanodroplets ink could maintain long stable period, such as 7 days in air and >60 days after N_2 purge, no coalescing and oxidizing occurred because of decreasing oxygen permeability of the microgel shells and admirable mechanical robustness. Remarkable, cross-linked alginate gel possessed tunable thickness affording high biocompatibility owing to retarding the generation and release of toxic cations. In addition, owing to alginate shells with the intrinsic soft feature as coating layers, the EGaIn droplets could achieve high electrical conductivity (the largest value of 4.8×10^5 S m^{-1}) through sintering and pattern on various substrates (Figure 6.10f). This EGaIn aqueous nanodroplets ink provides great opportunity for applications in electric skins and artificial limbs.

At present, inkjet printing is regarded as a flexible and promising patterned direct-write method for controllably constructing various 2D liquid metal electronic components. About constructing 3D liquid metal electronic components, science researchers have also made unremitting efforts and achieving commendable results. For example, Zhou and coworkers revealed a totally automated 3D inkjet printing method, which could directly print liquid metal-silicone (LMS) inks to generate soft, stretchable electronics [77]. What's more, Yu and coworkers also proposed a new suspension 3D printing method, which could prepare macroscopic liquid metal structures with the help of self-healing hydrogel [78]. They also manufactured the stereo electronic systems to reveal the capability and potential application of this method in constructing more complicated liquid metal-based architectures.

For inkjet printing, the challenges faced by this technique are avoiding the clogging of nozzles. As mentioned many times earlier, gallium-based liquid metal is easy to oxidize spontaneously to form solid-state oxides layer in the atmospheric environment, which stack at the nozzle and cause clogging. To address this problem, Li and coworkers a GaInSn-based inkjet printer based on the reduction reaction [79]. To guarantee GaInSn remain in the true liquid phase in the nozzle channel, they first injected HCl solution into a microfluidic channel, which was the core of the approach. The diameter of GaInSn droplet had a correlation with the size of proposed nozzle and flow rate.

In summary, inkjet printing exploits nozzles to deliver liquid metal droplets to the target substrate using thermal or acoustic processes. After printing, a subsequent sintering process, such as mechanical sintering and laser sintering, is needed to reconnect making it transform into a conductor. At present, inkjet printing is regarded as a flexible and promising patterned direct-write method for controllably constructing various 2D liquid metal electronic components. In addition, progress is also being made in the preparation of 3D liquid metal-based architectures. Note that the resolution of inkjet printing is about 100 μm and needed further improved.

6.5.2 Micro-extrusion Printing

By comparison, micro-extrusion printing can offer more possibilities in 2D structure and 3D structure fabrication due to the flexible ink designs and feature sizes. Generally, loaded inks are extruded onto the target substrate by a printing head control. The implementation of this process depends on a handling and dispensing system in virtue of extra pressure, such as aeration and machinery with a piston or screw. Hereon, writing liquid metal directly with a brush instead of a nozzle also boils down to micro-extrusion writing/printing, considering that the mechanism of writing is similar. Micro-extrusion printing is regarded as the most flexible and promising patterned direct-write method for controllably constructing various 2D and 3D liquid metal electronic components. Generally, pursuing the large-area rapid printing, improving the adhesion between liquid metal and substrate and resolution, and the realization of 3D stereoscopic printing is the key to accelerate its technological development, and a lot of efforts have also been performed to focus on these factors.

In 2014, rapid development of micro-extrusion printing had caught the eye of researchers to study liquid metal direct writing techniques. For example, Boley and coworkers reported a novel micro-extrusion printing direct writing system, which was achieved by flowing EGaIn through a syringe needle [80]. The schematic of the system is revealed in Figure 6.11a. The trace geometry of the constructing architectures depends on the jet flow rate, the inner diameter of the nozzle, and the intercepting distance. Representative optical microscope (OM) images of traces resulting from nozzles with various inner diameters on the surface of glass and PDMS are shown in Figure 6.11b. The average normalized raggedness of trace was 3% verifying the creditable quality of the written architectures. In addition, they also fabricated

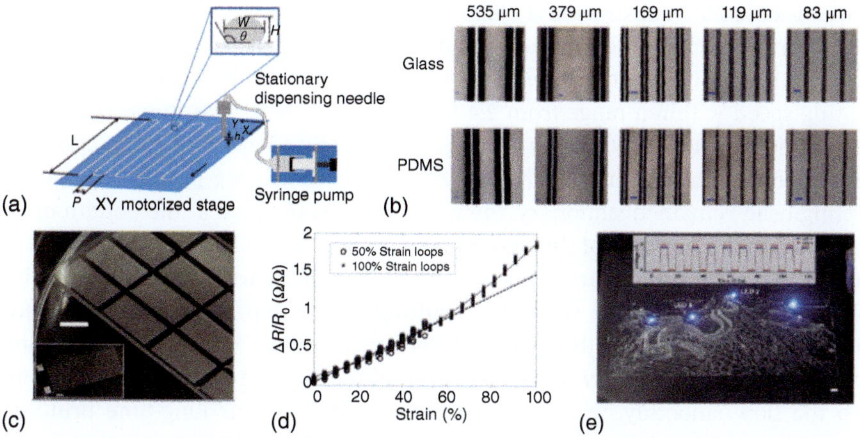

Figure 6.11 (a) Schematic of direct writing system. (b) OM of traces from nozzles of with different nozzle inner on glass and PDMS, Scale bars are 100 μm in length. (c) (main image) Photograph of multiple strain gauge devices on glass. Scale bar is 5 mm in length. Patterns were written using the 119 μm inner diameter nozzle. (inset) Photograph of a larger scale directly written strain gauge on glass. Scale bar is 5 mm in length. Device and pads were written using a 379 μm inner diameter nozzle. Source: (b, c) Boley et al. [80]. (d) Performance characterization of directly written strain gauges encapsulated in PDMS. Source: (a, d) Boley et al. [80]. (e) The LED with connected by a complex pattern of liquid metal with "Great Wall" appearance through the tapping mode printer system, the scale bars is 10 mm. Source: Zheng et al. [81].

an elastomer-encapsulated strain gauge by this novel direct writing system, which represented up to 50% strain and an approximately linear relationship accompanied (Figure 6.11c,d). Compared with filling microchannels by injection, which encountered geometry limitation and alleviating pressure requirement, this one-step direct writing system is automated, possessing the ability to prepare high-volume applications. Moreover, Boley and coworkers clearly revealed that the attachment of oxide skin on the substrate is by contact angle characterization. The spontaneous formation of gallium oxide layer on the surface of the EGaIn brought a strong adhesive force to the receiving substrates and facilitated the traces successfully written. This result is critically important and provides a theoretical basis for understanding the printing behavior of liquid metals.

In the same year, Zheng and coworkers presented a micro-extrusion printer possessing a driver with built-in monitoring sensor, which realized the widespread applications of the printed electronics with large area and low cost. This printer could facile delivery of the liquid metal ink by tapping motion of the cartridge and automatically and directly print out target liquid metal-based electronics [81]. The main components of this printer included a printing driver, an ink cartridge, two pulleys, and a computer with controlling software. This printer could quickly and automatically print out liquid metal-based electronics with desired features such as size, form, and resistance. Except for various basic figurate electrical elements, such as straight line, polyline, circle, and complex curve, a more sophisticated electronic drawing manufacture of the world-famous "Great Wall" can also be easily printed on the flexible PVC substrate and connected to the LEDs (Figure 6.11e). The wire resolution of this method for printing liquid metal-based ink can reach 100 mm. To guarantee the quality of created conductive electronic, it is suggested to set the printing speed within a range from 2.4 to 12 m min^{-1}. Note that this is the fastest high-quality printing method up to date. The printing head with tapping mode was the core of this method in the electronics manufacturing process. When the printing head remained stationary, the roller beads were fixed at the bottom of its seat owing to the pressure of the liquid metal, so that the roller beads fitted closely with the gap to prevent ink from flowing out. While, in the working state, the roller beads rhythmically hit the substrate resulting in the gap opened continuously, allowing ink outflow with the help of roller-bead rotation. The rhythmic tapping motion allowed the raceway to oscillate around the gap to remove impurities and keep the flow smoothly, which is a necessary condition for long-time printing. This tapping mode of the roller-bead possesses self-clearance function that it can compatibly print the liquid metal ink. In addition, the opening amplitude of the gap can be automatically adjusted by continuous tapping motion and evenly control the ink outflow. This easily controllable and quickly automatic micro extrusion print system sheds light on the large-scale application in the printed electronics and provides a promising tool for electronics manufacture in the future.

Here, the polymer substrate is generally used due to the wettability limitation of liquid metal (common EGaIn or GaInSn alloy) on paper or other substrates. Enormous efforts have also been performed to enhance the adhesion between liquid metal ink and the target substrate [82, 83]. For example, Zheng and coworkers adopted the dispensing machine and designed a brush-like porous pinhead to realize modifying the adhesion of the liquid metal-based inks [83]. Figure 6.12a reveals schematic of the brush-like porous pinhead. And the representative directly printed curved and regular circuits and annulus wire with LED are presented in Figure 6.12b,c, which illustrate the excellent adhesion between liquid metal-based inks and paper. Except for optimizing the pinhead, it was confirmed by Wang and coworkers that pre-treatment paper surface through microdroplet could help smooth writing of liquid metal [82]. Furthermore, Tang and coworkers discovered liquid metal mixtures (or hybrids) by adding suitable ingredients that would improve its adhesion and performance. A series of stable transitional state metallic mixtures named as TransM2ixes with a mixture of GaIn alloy and Cu particles were developed by a two-stage route consisting of the first-step wet-processing stage and the second stage of vacuum-drying, which not only revealed excellent adhesion, but also enhanced electrical conductivity κE, and thermal conductivity κT [84]. The adhesion improvement was attributed to the fact that dispersed solid particles in the liquid metal roughen the surface of the liquid metal and form a large number of oxide layer wrinkles. Generally, almost all substrates are also rugged and wrinkled, which will significantly increase the contact area between the liquid metal surface and substrate, bringing about improvement of adhesion (Figure 6.12e). As shown in Figure 6.12d, true liquid metal could hardly stick to a brush, while large amounts of TransM2ix remainder will be firmly attached after a single dip in the brush head. This TransM2ix ink can also be suitable for a wide variety of writing instruments,

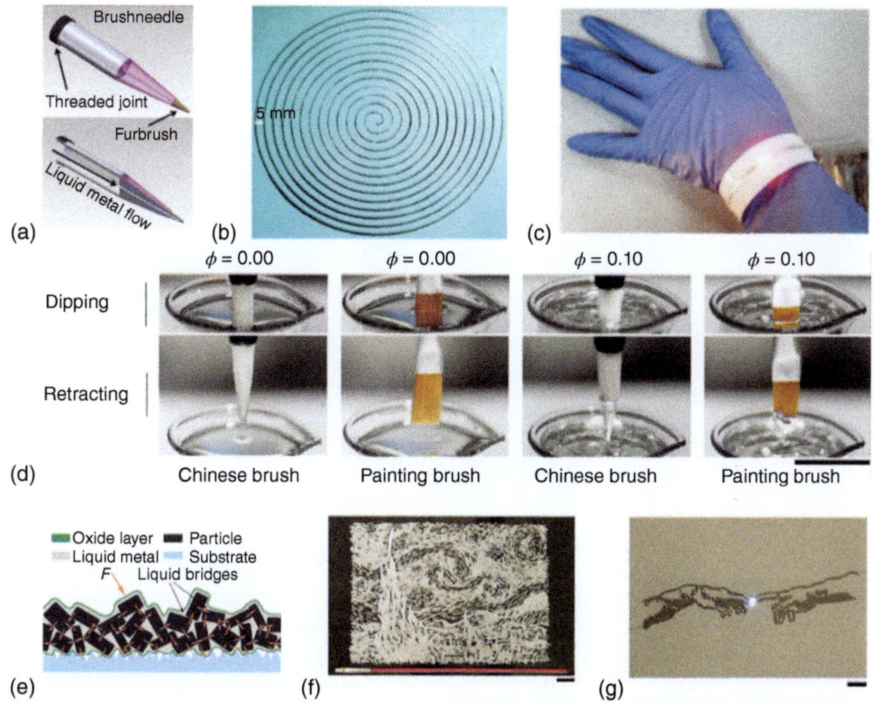

Figure 6.12 (a) Schematic of the brush-like porous pinhead. (b) OM of printed circular components on coated paper. (c) OM of LED connected by wearable annulus wire. Source: (a–c) Zheng et al. [83]. (d) Adhesion conditions of TransM2ix with different packing ratio ϕ using two types of brushes of a single dip-in test. (e) Schematic of dispersed solid particles in the liquid metals roughen the surface of the TransM2ixes and form a large number of oxide layer wrinkles. Insets: bumpy surface of TransM2ixes sue to dispersed solid particles (top view) and liquid bridges formed between dispersed solid particles (cross-sectional view). Source: Tang et al. [84]. (f) The graphic work of TransM2ixes patterns using a painting brush on plastic substrate (g) Lighted LED connected by TransM2ixes patterns painted on flexible transparent substrate. Scale bars: 20 mm. Source: (d, f, g) Tang et al. [84].

such as the Chinese brush. TransM2ixes ($\phi = 0.10$ and 0.15) was painted directly for the complex patterns using the common painting brush, as shown in Figure 6.12f. Pure gallium-based liquid metals can't write and paint directly like this. Notably, the obtained TransM2ixes patterns could be further fabricated to flexible circuits given the TransM2ixes intrinsic conductivity (Figure 6.12g). These easy-to-handle liquid metal mixtures are excellent choices for highly conductive architecture, while providing flexibility and comfort.

Except for flat substrates, micro-extrusion direct writing/printing can also realize the construction of liquid metal on curved or stretched substrates [85]. For example, Yoon and coworkers developed a novel micro-extrusion direct writing system that could make out EGaIn patterns on uneven surfaces, such as inclined and curved substrates, relying on four degrees-of-freedom (4-DOF, XYZ, and θ) motion. This 4-DOF motion could control the pressurized EGaIn dispenser and accurately senses the distance between the tip and the substrate. The conceptual schematic for the

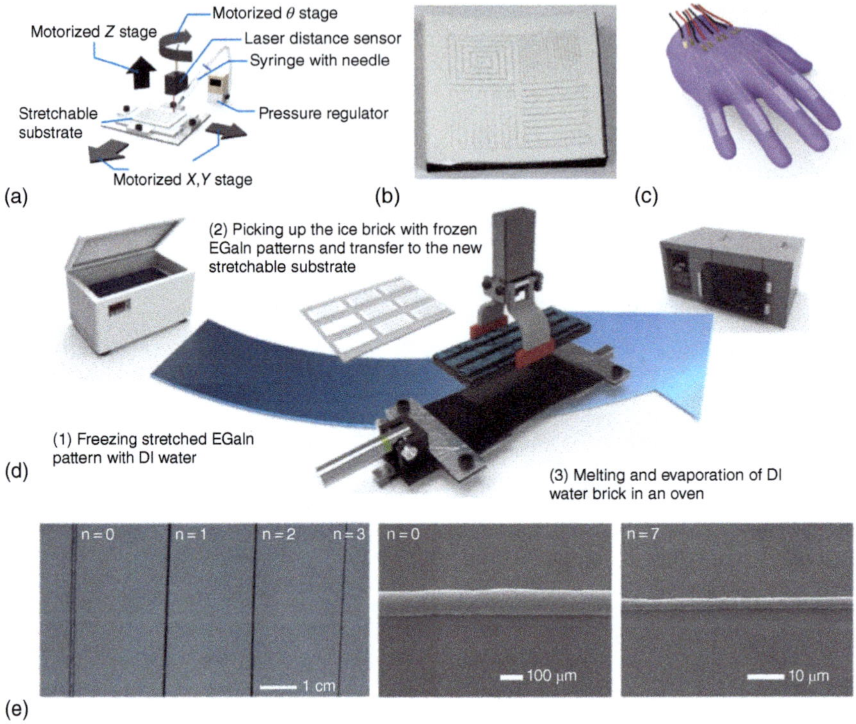

Figure 6.13 (a) Schematic of the 4-DOF direct writing system. (b) Complicated liquid metals patterned on an uneven surface by 4-DOF direct writing system. (c) 3D model of glove-type wearable liquid metals device with four meander resistors. Source: (a–c) Yoon et al. [85]. (d) Fabrication process of transferring EGaIn patterned on a stretchable substrate. Source: Kim et al. [86]. (e) The OM and SEM images of the change for the width of linear EGaIn patterns [86]. Source: Kim et al. [86].

proposed 4-DOF motion-direct writing system is revealed in Figure 6.13a. The main components of this experimental hardware were made up of a rotation stage three linear stages, two electronic pressure regulators, a laser distance sensor, and operating programs. Extremely complicated EGaIn circuits could be easily obtained by using this direct writing system, as shown in Figure 6.13b. The 3D model of a wearable device is showed in Figure 6.13c. Here, the obtained EGaIn patterns had a width of 70–80 μm, the minimum width between two parallel EGaIn lines is 200 μm, and the local slope allows for a maximum of 20° of reliable direct writing. The successful preparation of curved patterns for wearable device avoids weak extra links and paves the way for the preparation of high-performance electronic components.

Directly writing liquid metals on stretched substrates is commonly reported to improve the resolution of liquid metal architectures [86, 87]. Here, Liang and coworkers claimed that the width of liquid metal patterns could be reduced below 10 μm by writing liquid metal patterns on or in stretchable elastic substrate [87]. The realization of the designed pattern mainly included two steps: shadow mask printing or directly handwriting liquid metal on a pre-stretched elastic substrate surface and subsequent releasing the substrate. Admirably, the resolution of written resulting liquid metal pattern could be further improved nearly 50 times with the help of the transfer printing method. In addition, the resistance of the resulting liquid metal patterns remained almost unchanged even after the substrate was stretched multiple times. Here, Kim and coworkers employed direct writing coupled with pick-*n*-place transfer technique, realizing the resolution down to 2 μm [86]. The preparation process was as follows: they first patterned liquid metal with a line width of 120 μm on a stretchable elastomer Ecoflex. Subsequently, the substrate was stretched uniaxially and then frozen in DI water. Finally, the frozen liquid metal pattern was transferred to another unstretched elastomer, which results in the molting and evaporation of frozen water immediately. The schematic of fabrication process is showed in Figure 6.13d. The desired resolution could be realized by repeating this process over and over again and the tolerable repeating times determined by the applied tensile strain. From Figure 6.13e, which shows the SEM images of initial direct liquid metal printing and transferred liquid metal patterns, we can clearly see that the change for the width of linear liquid metal patterns with the applied strain of 2.0 after the sequential transfer, where n represents transfer times. Note that the width of the liquid metal pattern becomes ≈2 μm after seven sequential transfers ($n = 7$), deriving from an initial liquid metal pattern with the width of 120 μm. This combination of direct printing and posttransfer method achieves controlled preparation of different resolution patterns for a variety of applications.

In addition to the above involved in controllably constructing various 2D liquid metal electronic components, micro-extrusion printing is also regarded as the most flexible and promising patterned direct-write method for 3D liquid metal electronic components. Here, based on the immediately formed ultrathin (~1 nm) gallium oxide layer once exposed to atmosphere resulting in having a solid nonflowing property, Ladd and coworkers used four direct-write patterning methods to obtain free-standing liquid metal structures, such as filaments, interconnects, and the stacking of droplets [88]. Here, at modest pressures (<5 kPa), free-standing liquid metal filament structures could be extruded to form. The upper limit of the wire length was determined by the movement of the flat stage, which could reach several centimeters when the wire extended from the matrix to the tip of the syringe (Figure 6.14a). Conveniently, employing suitable amount of vacuum to the syringe could stop the wires at desired time. The obtained tallest free-standing wire was 1 cm high when its diameter was ~200 μm, and the diameter of the narrowest wire was 30 μm. Here, free-standing wires were formed originating from skin oxidation crossing the nozzle to the substrate. Thus, increased distance between the nozzle and the substrate could generate stretching force along the axis of the wire, producing skin and allowing the elongation of wire. It's worth noting that the pressure of the liquid metal delayed any unstable capillary forces allowing new skin to form, which improved the mechanical stability of the wire. Except for extruding wires, free-standing liquid metal microstructures were also easily obtained using three additional methods, including forming stable liquid metal filaments by fast discharge, stacking of droplets, injecting the metal into the microchannels, and then removing utilized channels (Figure 6.14b–d). This four direct-write patterning methods have tremendous potential for true 3D-printing

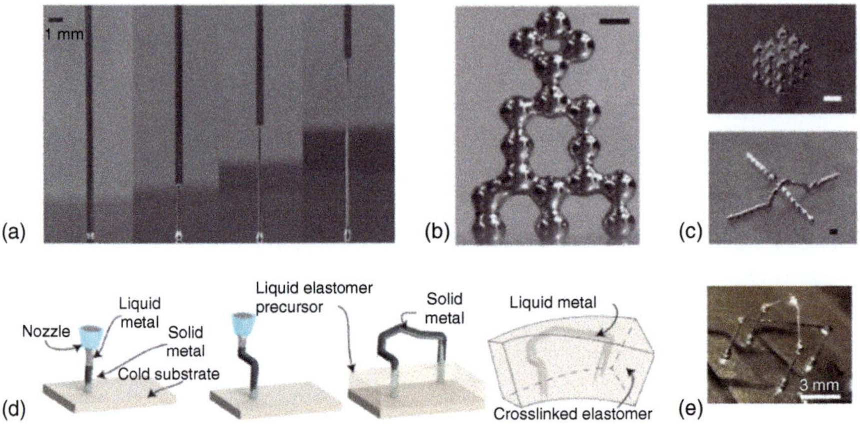

Figure 6.14 (a) Sequential images of direct-writing free-standing liquid metals wires. (b) A tower of stacking liquid metals droplets. (c) 3D cubic array and wire made up of liquid metal droplets. Source: (a–c) Ladd et al. [88]. (d) Schematic of 3D freeze-printing process. (e) A complex 3D liquid metal wire fabricated by 3D freeze-printing process. Source: (d, e) Gannarapu et al. [89].

liquid metal-based architectures. Soon afterward, Gozen and coworker presented that mechanically flexible true 3D conductive networks could be obtained by 3D freeze-printing [89]. The 3D freeze-printing process is schematically shown in Figure 6.14e. Briefly, they first dispensed EGaIn on a substrate by a nozzle, and the temperature used was below the melting point of the liquid metal. When the liquid metal structure was formed in 3D orientations, more liquid metals were added to the print structure along the conversion path and remain cylindrical owing to the spontaneous formation and deformation of the oxide skin. At the same time, the printed structure begun to freeze from the contiguous substrate, creating a moving freeze-front and making the structure more stable compared to oxide skin alone. The obtained structures were then encapsulated in liquid elastomer precursors, and the encapsulated structure was then heated so that the alloy melt and the elastomer were cross-linked at the same time. Finally, a flexible composite was obtained at room temperature attributing to the cross-linking elastomer, which mechanically supported the melting structure. The typically complex 3D network by this 3D freeze-printing technique is also shown in Figure 6.11f. This method had made a breakthrough in liquid metal 3D printing that allows for a variety of any 3D direction flexible liquid metal-based architectures preparations.

It's worth noting that the resolution in those 3D micro-extrusion printing approaches mentioned earlier is at the scale of sub-millimeter, which can't meet the demand of high-integrity electronic devices. The blockages for the improvement of the resolution are the poor controllability of the liquid metal base materials and the relatively large droplet diameter. To address the problem, Park and coworkers reported a high-resolution liquid metal printing technique [90]. This technique could produce continuous, uniform, and high-resolution liquid metal filaments and directly reconfigure them into diverse 3D structure just by using fine nozzles facilitating ejection of narrow metallic filaments under ambient conditions.

Schematic illustration of the 3D printing system through a nozzle is illustrated in Figure 6.15a. Here, EGaIn (75.5% Ga and 24.5% In alloy) was used as the ink. Figure 6.15b displays a SEM image of printed EGaIn structures with complex 2D and 3D geometries. According to atomic force microscope (AFM) test results, we can see that the printed minimum line width of EGaIn structures was 1.9 mm, which is almost the highest resolution reported so far (Figure 6.15c) [88]. The pre-printed liquid metals patterning could be raised by the tip of the nozzle. Also, it could be precisely reconfigured to other desired position to prepare different 3D structures. To verify its availability, they represented a reconfigurable antenna whose resonance frequency and radiation characteristics could be changed by changing their geometry, and also showed the reversible movement of 3D interconnects in the form of mechanical switches. Colorized SEM image of the micro-LED array and EGaIn interconnects through this 3D printing system are shown in Figure 6.15d. In addition, it is also due to the spontaneous and instantaneous formation of oxide thin layers during the lifting and movement of liquid metal filaments, which effectively prevents the crucible from penetrating from the liquid metal filament to the substrate. Compared with the traditional 3D printing method, this reconfigurable 3D printing permits the interconnections possessing minimizing number and space for complex 3D integrations.

Notably, similar system mentioned by Park and coworkers can also print composites as stretchable electrical interconnections [91]. The SEM image of printed 3D structures of carbon nanotube/liquid metal (CNT/LM) composites is shown in Figure 6.15e. The mechanical strength of liquid metals can be improved by adding CNT. Figure 6.15f shows the elastic modulus and strength increased linearly with the CNT loading, which is three times at 15 wt% higher than the elastic original modulus of liquid metals. However, the composite materials condensed at the nozzle and then could not be printed when the content of CNT was increased to 20 wt%. In addition, the printed 3D structure still retained when it was placed at 340 °C for 70 minutes (Figure 6.15g). And after heat treatment, from current–voltage characteristics, we can see that the resistance of the printed CNT/LM composite also had a negligible change (Figure 6.15h). Meanwhile, the printed structure exhibited high conductivity ($3\ \mathrm{S\,\mu m^{-1}}$) and excellent current density ($3.5 \times 10^{10}\ \mathrm{A\,m^{-2}}$), which is at the same order with traditional metal interconnections of $10^{10}\ \mathrm{A\,m^{-2}}$. These CNT/LM composites structures have high resolutions with 5 μm of minimum diameter.

In summary, micro-extrusion printing can controllably construct various 2D and 3D free-standing liquid metal electronic components with arbitrary complex shapes and structures at room temperature, such as assemble 3D flexible, stretchable, and shape reconfigurable devices, including antennas, electrical interconnects, and so forth. In addition, researchers have made remarkable progress in large-area rapid printing, improving the adhesion between liquid metal and substrate and resolution, and the realization of 3D stereoscopic printing. These researches offer promising candidates toward constructing the complex and variable with excellent performance liquid metal-based architectures of practical application in future. The disadvantages of the micro extrusion printing are relatively complex and requiring pressure control.

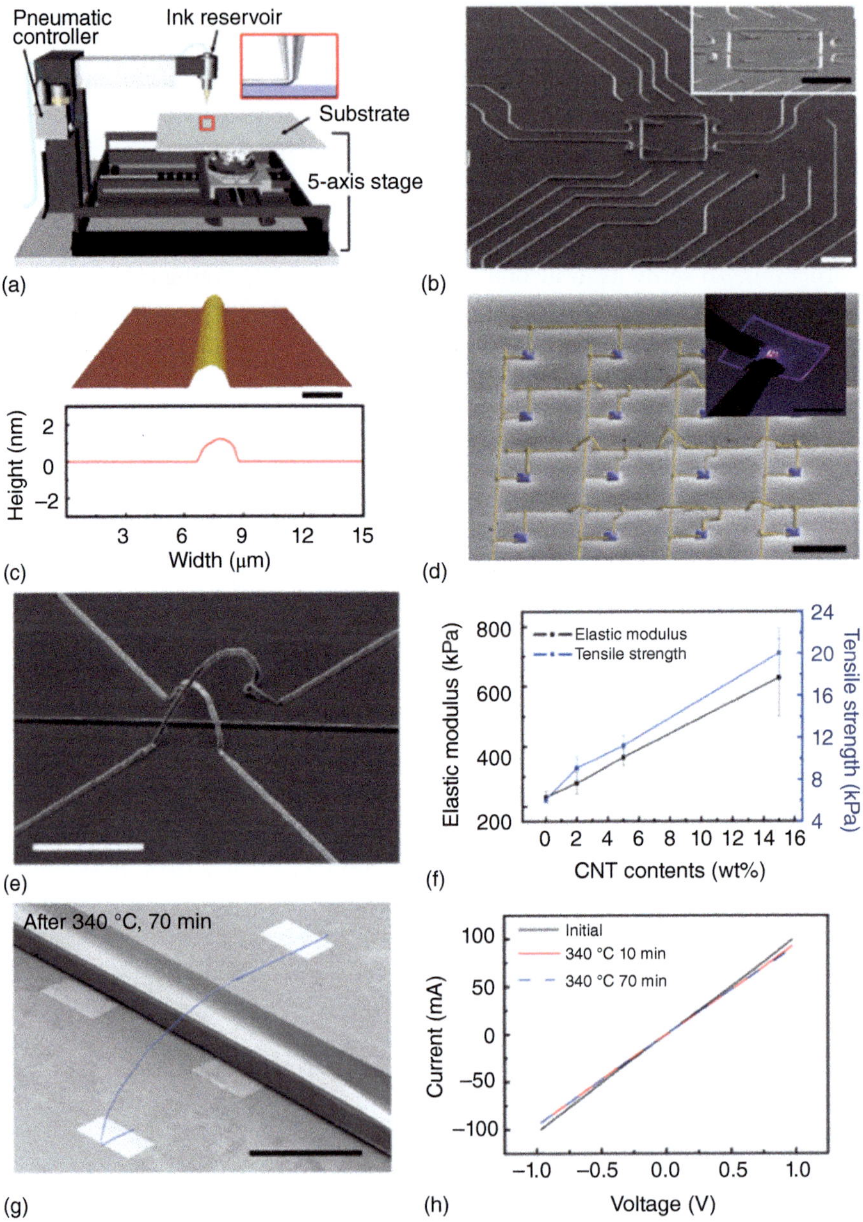

Figure 6.15 (a) Schematic of a 3D printing apparatus. (b) SEM images of printed EGaIn structures with complex 2D and 3D geometries. Scale bar, 100 mm. Inset is a magnified SEM image of the 3D geometries. Scale bar, 100 mm. (c) AFM image and cross-sectional profile of printed EGaIn line. Scale bar, 2 mm. (d) Colorized SEM image of the micro-LED array and EGaIn interconnects. Blue and yellow colors correspond to micro-LED and EGaIn, respectively. Scale bar, 300 mm. Inset is photograph of light emission of the micro-LED array. Scale bars, 1 cm [90]. Source: (a–d) Park et al. [90]. (e) SEM image of printed 3D CNT/LM composites structures. (f) Elastic modulus and tensile strength of the composite as a function of CNT contents from 0 to 15 wt%. (g) Colorized SEM image of a 3D CNT/LM composite structure after the heat treatment. The blue color corresponds to liquid metals. Scale bars, 500 μm (h) Current–voltage characteristics of CNT/LM composites before and after heat treatment. Source: (f, h) Park et al. [91].

6.6 Laser Ablation

Direct or indirect laser ablates liquid metal selectively leaving only the necessary areas that belong to typically subtractive manufacturing is a fairly quick and easy approach to realizing rapid prototyping liquid metal-based materials. It has been well documented that CO_2 and UV laser can be used to pattern liquid-phase EGaIn alloy [92–94].

Lu and coworkers rapidly patterned liquid metal embedded in the thin PDMS film by using a CO_2 laser (30 W, 10.6 μm wavelength) [93]. The process is depicted in Figure 6.16a. Laser irradiation vaporizes PDMS due to the local heating, when the vapor pressure exceeded surface tension of liquid metal and PDMS, both materials are displaced and removed. The planar size of the electronic material obtained was >150 μm through this method, and the minimum size of the produced feature was about 100 μm (Figure 6.16b).

Notably, liquid metals can be ablated directly using low wavelength of a UV laser. For example, using a 1W UV(Nd : YAG) laser with a wavelength of 355 nm, the planar dimensions liquid metal pattern can be obtained was about 50 μm [92]. The electrically conductive films constructed by liquid metal lines array on the elastomer were obtained by UV laser-based microfabrication technique, as shown in Figure 6.16c [94]. To enhance adherence between EGaIn and PDMS, a "bi-phasic" Cr/Cu architecture was deposited on the PDMS substrate before EGaIn deposition and patterning. This laser ablation writing method can generate a conductive grid with line widths and pitches of 4.5 and 96 μm, respectively (Figure 6.16c). This conductive film has extremely high transparency and conductivity when it is stretched to 100%, which has enormous potential for sensing applications.

In conclusion, direct laser ablation, which belongs to typically subtractive manufacturing is a quick and facile approach to realizing rapid prototyping liquid metals and other conductive materials. This technique is higher wastage of liquid metal. In addition, the width (resolution) of obtained pattern strongly depends on the wavelength of the adopted laser. This rapid prototyping liquid metals method has a promising prospect for industrial application of liquid metal construction.

6.7 External Field Guiding

In addition to the pattern liquid metal methods mentioned earlier, the external field, including magnetic field and electric field, can also guide the pattern of liquid metal for obtaining various complex structures.

6.7.1 Magnetic Field-Guided Patterning

The mixture of liquid metals and magnetic particles respond to magnetic field. Therefore, employing magnetic field can control patterning process. Recently, Ma and coworkers accidently noticed that when moving magnetic field was utilized for liquid metals and magnetic particles composite, liquid metals could be deformed to

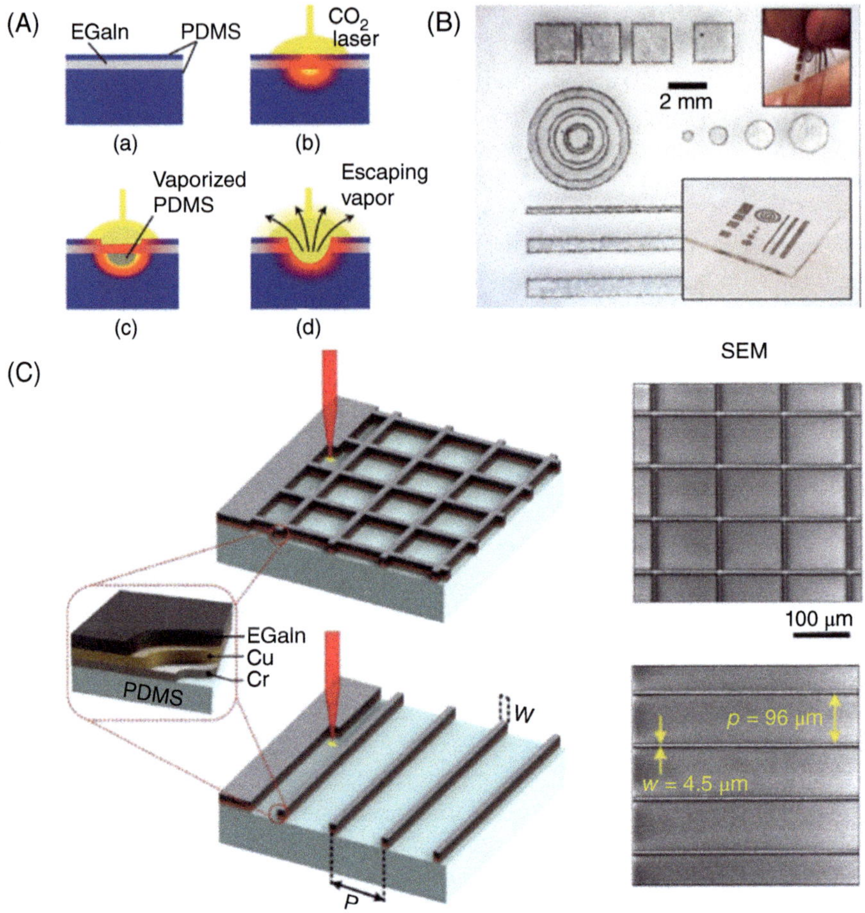

Figure 6.16 (a) Schematic of direct CO_2 laser ablation process for preparing liquid metals patterns; (b) liquid metals patterns by direct CO_2 laser ablation. Source: (a, b) Lu et al. [93]. (c) Schematic of direct UV laser ablation process and SEM images of the obtained square grid and parallel line of liquid metals patterns. Source: Pan et al. [94].

obtain a continuous line for patterning [95]. Figure 6.17a shows the direct patterning of liquid metals with magnetic field. Further information about the process is shown in Figure 6.17b. In this simple method, only a low-cost magnet was required, and it could be adopted on not only stretchable films (e.g., PDMS) and wet substrates (e.g., hydrogel) but also supermetallophobic substrates (e.g., paper) because the magnetic field enhanced the adhesion of the pattern. Besides, magnetic field can be remote controlled, leading to direct patterning on nonplanar surfaces (e.g., eggshell and the inner wall of the glass vial, Figure 6.17c), which cannot be realized by almost the reported patterning strategies. Compared with traditional strategies, this approach does not need modified substrates or complex fabrication facilities. Figure 6.17d shows the fabricated liquid metals and PDMS thin film with embedded LED array. The device with mechanical flexibility enabled it to bear different strains including stretching and twisting. What's more, the liquid metals could be directly

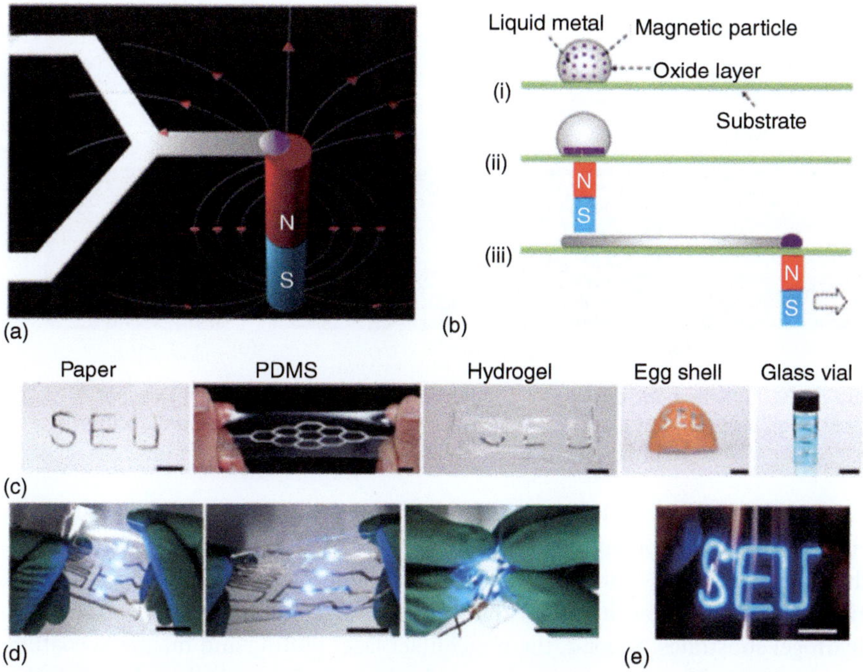

Figure 6.17 (a) Schematic of the direct patterning of liquid metals using external magnetic. (b) Schematic of detailed preparation process. (c) Photographs of the obtained liquid metals patterns on various receiving substrates; (d) Photographs of flexibility device of liquid metals. (e) Photograph of an electroluminescentdevice with patterned liquid metals electrode. Scale bars: 2 cm for (d) and 1 cm for (f). Source: Ma et al. [95].

fabricated on an active phosphor layer (Cu-doped ZnS embedded in PDMS) of an electroluminescent device, which could work under bending (Figure 6.17e). This functional method provide potential for soft robot, flexible electronics, and the like.

However, the aforementioned technique remains challenging to fabricate high-resolution liquid metal patterning. Xu and coworkers further used a magnet combined with digital laser cutting to prepare high-resolution liquid metal patterning on PVA hydrogels [96]. The fabrication process utilizing magnetic field is shown in Figure 6.18a. First, a thin PET film was deposited on PVA substrate, which was then subjected in situ digital laser patterning of PET film to fabricate the shadow mask. And, EGaIn and Ni microparticle dispersion was induced into the obtained PET film [95]. A permanent magnet was utilized to attract and aggregate the Ni microparticles. The aggregated Ni microparticles and liquid metals moved through the PET mask via actuating the magnet, leading to desired laser cut patterns. Then, they tilted the glass substrate and absorbed liquid metals to remove excess liquid metals. Meanwhile, the soft squeegee was applied to thoroughly clean liquid metals, bringing about a smooth liquid metal surface. Then PET mask was peeled off to form liquid metal patterns with very low line widths of 85 μm. Figure 6.18a shows different liquid metal patterns including a logo of Southeast University (Figure 6.18b), a quick response code (Figure 6.18c), spiral lines (Figure 6.18d), and a circuit pattern (Figure 6.18e). The prepared liquid metal patterns on PVA hydrogel

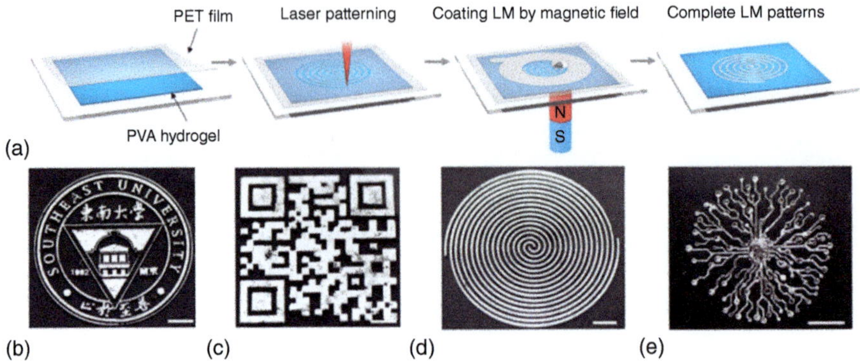

Figure 6.18 (a) Schematic of the direct patterning of liquid metals using external magnetic combined with the laser. (b–e) Photographs of various liquid metals patterns on PVA hydrogels. Source: Xu et al. [96].

showed mechanical and electrical self-healing, providing promising potential for robust electronic equipment. The strategy can also pattern liquid metals on other hydrogels substrates. Meanwhile, the reported high-resolution liquid metal patterning strategy owns fascinating features: First, it does not require the retreatment of hydrogel substrates. Second, the noncontact laser cutting and magnetic coating steps have no influence on deformations of the soft, fragile hydrogel. Third, the cost-effective and useful method can be applied to large-scale productions.

In summary, magnetic field-guided patterning is a simple and fast patterned method for controllably constructing various liquid metal architectures. It can achieve a resolution as small as 4.5 μm via the combination with digital laser cutting. Moreover, both magnetic coating and laser cutting do not deform hydrogel, which is promising for the cost-effective and large-scale fabrication. Note that the homogeneity of pattern is still a challenge.

6.7.2 Electric Field-Guided Patterning

As in Chapter 4, electrochemistry can induce liquid metal droplets to drive, deform, and shift, which provides a possibility for the controllable construction of electrochemical components [97–99]. The special approach utilized electrical bias to enable depositing or eliminating surface skin of liquid metals. The oxide skin of liquid metals can effectively decrease the surface tension and enable it to flow into capillaries. However, removing the oxide displays opposite performance. Reconfigurable wires can be achieved by using electrical field to move liquid metals in the capillary [100, 101]. Figure 6.19a,b shows the movement of liquid metals through capillaries via using continuous electrowetting (CEW). However, there are some disadvantages: (i) Generally, the metal plugs display constant length. (ii) It remains challenging to directly connect metal plugs with the fixed external RF feed because metal plugs can move in channel. (iii) Injecting metals from the reservoir into capillary require sufficient pressure differential, which is hard to produce by CEW. To address the limitations of CEW mentioned

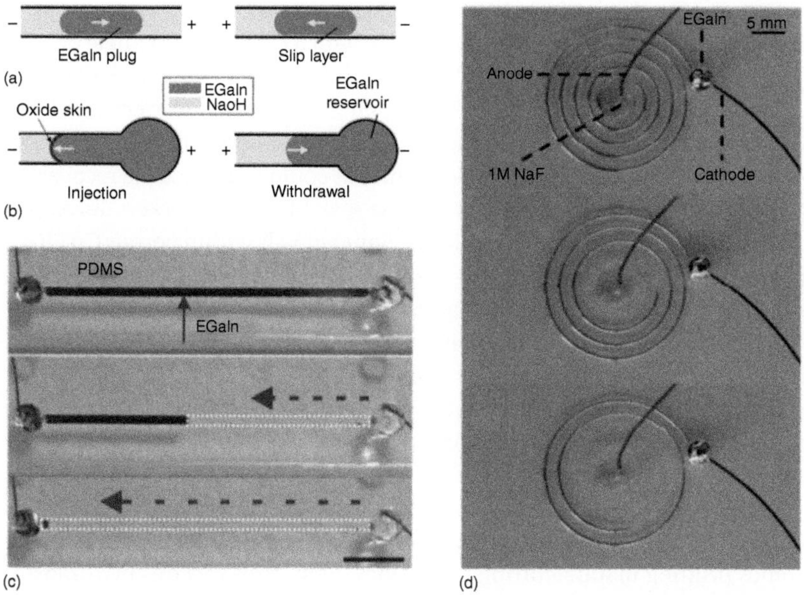

Figure 6.19 (a) Schematic of continuous electrowetting using an EGaIn plug. (b) Schematic of ECC of an EGaIn filament [100]. Source: (a, b) Wang et al. [100]. (c) Reduction-driven liquid metals withdrawal in a PDMS microchannel. Source: Khan et al. [101]. (d) Electrochemical withdrawal of liquid metals from the microchannels. Source: Parekh et al. [102].

earlier, Rashed Khan and coworkers developed a "electrochemically controlled capillarity" (ECC) approach. The length of liquid metals can be changed by ECC. The ECC method can use voltage to actuate liquid metals between the capillary and reservoir [27]. When liquid metal within the reservoir has a higher Laplace pressure than that within the capillary, it can easily flow into the capillary. DC potential with reversed polarity can electrochemically reduce the oxide and change liquid metals into the state with large tension, which is called recapillarity because it is an important reduction reaction to induce capillarity [101]. Without oxide skin, Laplace pressure can move the EGaIn to external metal reservoir, shorten the metal filament, and fill the vacated space with electrolyte. Recapillarity was found to be an effective method for manipulating fluids to produce microfluidic channels [101, 102]. Figure 6.19c shows the reduction-induced liquid metal withdrawal in the PDMS microchannel. The liquid metals were written and drawn to different patterns. The electric field can be employed in NaF electrolyte by encapsulation. And liquid metals exited in channel result from capillary action (Figure 6.19d).

The aforementioned strategy can prepare solid 3D structure. However, it remains challenging to control aspect ratios and sizes. Yan and coworkers reported a unique strategy for fabrication liquid metal channels by amalgamating liquid metals with patterned Cu tape [103]. This strategy can create complex liquid metal structures with various dimensions and sizes, such as trees, serpentine channel, and spiral channel. Cu surface was covered by a mask of laser-engraved adhesive sheet. When

EGaln was put on Cu via applying 5 V DC, it spread and wetted Cu surface due to removing the oxide layer, leading to a cylinder cap with a small contact angle (<90°), instead of forming a spherical droplet with large contact angle (>90°). The fabrication process is revealed in Figure 6.20a. The liquid metal pattern was encapsulated when the solution was dried. Because the encapsulated material did not change its structure, the pipe restored shape of the liquid metals could be formed via recovering the liquid metals by alkaline solution. In the same year, Ma and coworkers developed a novel method of manipulating gallium-based liquid metals within porous Cu [104]. The mechanism of this electrochemical method, which enabled reactive-wetting, coating, and spreading of the liquid metals on the Cu surface, is completely different from the aforementioned electrochemical method. Figure 6.20b illustrates the designed experiment test principle. Directional locomotion of liquid metals on Cu surface was realized by a mechanism that electric potential produced the local flow near the contact line, therefore producing $CuCa_2$ precursor film on Cu. The different wet properties of liquid metals on Cu and $CuGa_2$@Cu induced the dynamic and continuous spreading of liquid metals on Cu surface. Based on this, complex patterns of liquid metals and flexible devices can be fabricated by a unique transfer electronics printing manufacturing technology. First, PVC was thermal-transfer printed onto Cu foil. Then it was cut into desired patterning. The exposed Cu area was coated by $CuGa_2$ and selectively wetted by liquid metals using the above method. PET was finally printed with a circuit pattern via freezing and heating transfer process. The sharp and clear pattern boundary was formed in the method, which is an obvious strength, and important to print microsized conductive lines. Various shapes of large-area $CuGa_2$ film could also be fabricated using this strategy (Figure 6.20c) [105, 106].

In addition, voltage-induced liquid metals lithography has been reported, which also has the ability to print multiple crosses or complex patterns by using a low conductivity anode [9]. Tang and coworkers reported a novel strategy to prepare 3D Ga microstructures utilizing dielectrophoresis. This method has capability of creating 3D Ga microstructures with different aspect ratios and sizes [107].

In conclusion, the varied liquid metal-based architectures can be derived by applying an electrical field. For example, reconfigurable wires can be achieved by applying electrical potential to actuate a liquid metal in a capillary. Recapillarity is an effective method for manipulating fluids to produce microfluidic channels. In addition, compared with recapillarity, amalgamating liquid metals with patterned copper tape and voltage-induced electrochemical lithography offer more possibilities for constructing different aspect ratios and sizes of the liquid metal-based architectures. However, the resolution of the electric field-guided patterning methods needs to be improved.

6.8 Others

As mentioned earlier, constructing liquid metal-based architectures has been realized by various patterning techniques that usually require the forces to inject liquid

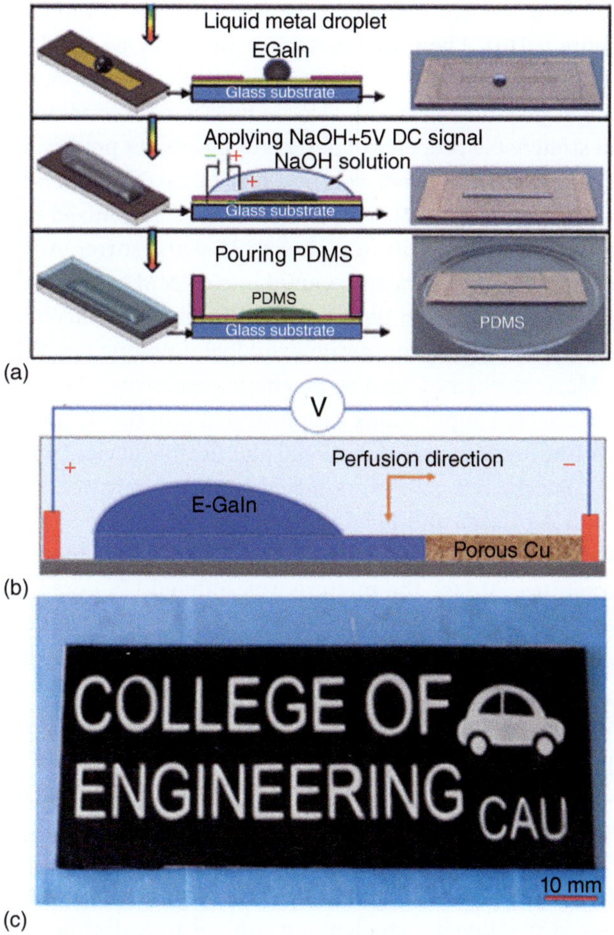

Figure 6.20 (a) Schematic of preparing liquid metals channels. Source: Yan et al. [103]. (b) Schematic of the experimental set-up of voltage-induced liquid metals perfusion within porous Cu. (c) Digital images of the $CuGa_2$ conductive film with complicated patterns. Source: (b, c) Ma et al. [104].

metal to the microchannels, or different injection, direct writing, and laser ablation processes to prepare desired patterns. Generally, these methods require the preparation of microchannels or sacrificial substrates, which are time-consuming and complicated. Developing easier methods that do not require patterning and can be directly utilized for different fields have attracted research attention. Direct engineering of liquid metals on elastomer, polymer, and hydrogel is considered as a promising strategy. Owing to the good adhesion on various surfaces, liquid metals can be directly implemented with elastomer, polymer, hydrogel, and the like. The property makes it promising for directly constructing liquid metals-related materials. For now, several synthesis techniques have been devoted to directly mix liquid metal with elastomer, polymer, and hydrogel, such as stirring, shear mixing, and sedimentation method.

Lin and coworkers directly dispensed EGaIn lines on the putty substrate composed of viscoelastic polymer (Figure 6.21a). Then, they stretched the putty by hand to obtain an elongated structure for metallic wires without any specialized equipment (Figure 6.21b). The conductive metallic wires can also be fabricated into various shapes owing to the soft and stretchable property of putty [108]. Another promising method to fabricate liquid metal and elastomer composite is mixing and stirring the raw material for a period of time [115–117]. Yun and coworkers also mixed the raw materials and then stirred the mixture using a high-speed electric stirrer. After removing air bubbles of the mixture in vacuum and poured into a PMMA mold, the mixture was then placed in the oven to obtain liquid metal-filled magnetorheological elastomer (Figure 6.21c). The prepared composite exhibited a unique positive piezoconductive response. More researches on the magnetic responsive thermal property and several proof-of-concept applications demonstrated that the obtained composite holds prospect for future novel multifunctional and flexible devices with excellent performance [109]. Parida and coworkers mixed healable polyurethane acrylate (PUA) elastomer with Ag flakes and EGaIn to prepare conductive slurry, which was then drop-casted and placed between two PUA filled liquid metals, resulting in sandwiched structures. At last, the mixture was suffered UV cured to obtain stretchable and healing triboelectric nanogenerator (SH-TENG) [110], which is shown in Figure 6.21d. A machining tool assembled with a brush head can also be used for high-speed mixing of liquid metals alloy and PDMS [118]. In this process, the time of stirring and curing play a vital role in obtaining the final composite [119].

Shear mixing usually disperses liquid metals in elastomers via mortar and pestle or mechanical mixers [112, 120–123]. Wang and coworkers employed shear mixing of the uncured liquid metal elastomer in the presence of liquid ethanol (Figure 6.21e). During heating condition, liquid metal elastomer swelled and changed shape freely and then actuated in response to the increasing pressure caused by ethanol evaporation. Programed and desired swelling are challenging abilities to realize in soft actuation. Changing heat source location is a convenient way to selectively heat liquid metal composites (LMC), resulting in local deformation or controlled swelling of LMC. Based on this, Wang and coworkers observed that the planar LMC swelled and uplifted in a selective heating area, and finally obtained a desired 3D structure. The obtained 3D structures could return to the initial planar shapes when cooled. Therefore, it is possible to change the shape of selected area of LMC and obtain the programmable desired 3D structures, which broaden its application in soft robots [111]. Kazem and coworkers also embedded liquid metals in elastomers by shear mixing, the obtained liquid metal-embedded elastomer (LMME) film is shown in Figure 6.21f [112]. However, shear-mixing approach cannot independently control the size and loading of liquid metal particles, likely because the viscosity of obtained composite changed with liquid metals loadings [124]. Tutika and coworkers combined probe sonicator with the shear mixer to systematically control liquid metal particle size and loading. This fabrication approach is not only suitable for silicone elastomers (Ecoflex and PDMS), but also be employed to various elastomers, such as thermoplastic elastomers and so forth [125].

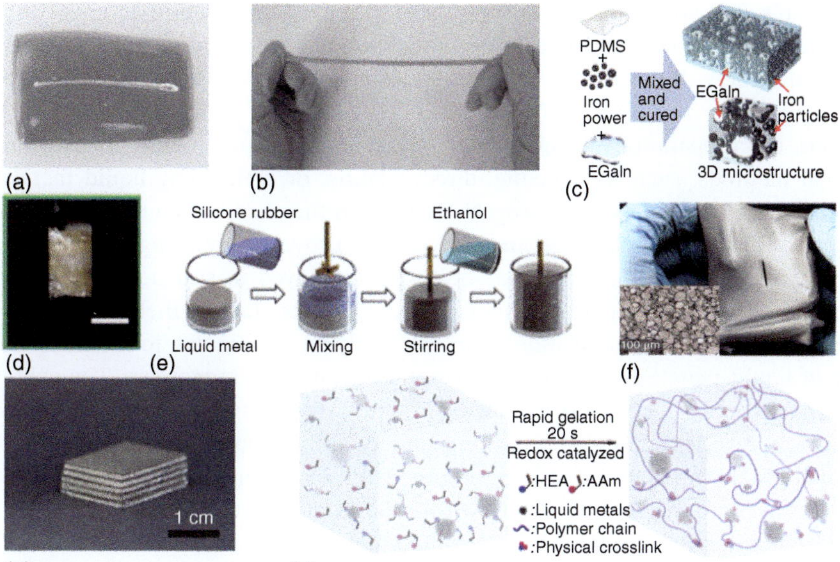

Figure 6.21 Direct mixing of liquid metals with elastomer, polymer, hydrogels, and the like. (a) EGaIn lines dispensed on the putty substrate. (b) Stretched putty with elongated liquid metal lines. Source: (a, b) Lin et al. [108]. (c) Schematic of the preparation of liquid metal-filled magnetorheological elastomer. Source: Yun et al. [109]. (d) Digital photo of the SH-TENG. Scale bar is 5 mm. Source: Parida et al. [110]. (e) Preparation of the liquid metal and elastomer composite. Source: Wang et al. [111]. (f) A notched liquid metal-embedded elastomer can deform without crack propagation. The inset shows the optical micrography of LMEE without deformation. Source: Kazem et al. [112]. (g) An anisotropic composite with five layers prepared by sedimentation method. The dark and bright layers represent Ga-rich and PDMS-rich layers, respectively. Source: Zhu et al. [113]. (h) Fast preparation of LMHs. Source: Peng et al. [114].

Sedimentation methods employ gravity to allow sedimentation of the liquid metals. It is a universal, reliable, and easy method to obtain shape morphing material with anisotropic mechanical, thermal, and electrical properties. Zhu and coworkers utilized the sedimentation method to fabricate anisotropic liquid metals and elastomer composite. They first mixed PDMS and Ga and transferred to the mold, which was then sustained for 10 minutes to allow the sedimentation of liquid metals. After that, the mixture was degassed and cured. Multilayer composites with Janus structure can be obtained by repeating these steps (Figure 6.21g). Due to the Janus structure and different properties, the obtained composite displayed anisotropic mechanical, electrical, and heat transport behavior. Meanwhile, the geometry of this anisotropic composite can be changed into rectangular, copious shapes, and the like [113]. Park and coworkers utilized sedimentation-induced EGaIn in poly(ethylene glycol) diacrylate (PEGDA) to prepare liquid metal hydrogels (LMHs). During this progress, one-pot UV irradiation was employed to synthesis PEGDA hydrogel with vertically phase-segregated EGaIn. The subsequent UV irradiation made LMHs solid-like film with light gray color, mechanically robust, and flexible [7]. Wang and coworkers also utilized UV irradiation to cross-link liquid metal and polymer network together. The obtained suspension was mixed and immediately drop-casted onto a 2-hydroxyethyl acrylate (HEA) substrate to fabricate polymerized liquid metal networks [119].

So far, the preparation of liquid metal-based architectures mainly focuses on physical methods, such as injecting, direct writing, or embedding liquid metals into elastomers, polymers, or hydrogels. Reports on the chemical methods are still scarce. Exploring new chemical methods would prove a promising concept for future applications [126]. Peng and coworkers employed in situ polymerization method to prepare liquid metal–hydrogels composites. Liquid metal–hydrogels were fabricated via in situ gelation in an extremely rapid time due to the catalytic effect of liquid metals and unique surface interaction with monomers. The oxide layer enabled polar monomers anchoring at liquid metal surface, and this surface interaction resulted in enhanced homogeneity of composite. Meanwhile, the obtained LMHs can stabilize at least one hour in polar monomer solution. Then, gelation was performed absolutely fast within 20 seconds (Figure 6.21h) after adding a radical initiator. In this rapid progress, liquid metals acted as "reactive" fillers that promoted ammonium persulfate (APS) for rapid gelation [114]. Similar to in situ polymerization method, Yan and coworkers demonstrated a universal tetherable initiator attached on the liquid metal surface, which could allow surface-initiated atom transfer radical polymerization (SI-ATRP). By employing the tetherable initiators to modify the liquid metal surface, numerous EGaIn–polymer hybrid materials were fabricated by SI-ATRP, such as silicone-based elastomer and thermoplastic elastomers. Meanwhile, the obtained material revealed well-defined microstructure. The research provides a promising prospect for developing new materials with liquid metals for future soft electronics [127]. Thrasher and coworkers also synthesized polymerized functionalized liquid metals networks with surface-bound acrylate ligands. The obtained polymer functionalized liquid metals exhibited outstanding electromechanical property, responsive behavior, making it an ideal material for stretchable devices, sensing, and reconfigurable electronics and soft robotics [128].

In summary, some techniques that do not need pattering are also discussed here, directly mixing liquid metal with polymer or hydrogel, elastomer is another method to prepare liquid metal-based architectures. Utilizing chemical methods is emerged as a new strategy and still in its infancy. Compared with patterning methods mentioned above, directly mixing method is easier. However, this method is not very suitable for complex and well-designed patterning.

6.9 Conclusion and Prospect

Due to the remarkable properties of liquid metals and the development of advanced technology, liquid metals could be engineered by many strategies to fabricate various architectures for different devices and applications, such as stretchable, flexible devices, lightweight, self-healing devices, antennas, and biomedicine. This chapter

Table 6.1 The comparison of different patterning techniques of liquid metal-based architectures.

Constructing techniques		Minimum feature size	Advantages	Disadvantages	Scope of techniques	References
Injection	Direct injection	12 μm (elastomer channel), 150 nm (silicon fiber)	Simple, fast, reliable and low-cost process, channels consistency	Incomplete filling for complex structure; high pressure	Simple channel or structures with void space	[11, 24, 25]
	Vacuum filling	10 μm	Channels consistency, no need of outlet; complete filling	Vacuum treatment; sufficient chip stiffness and thickness	Complex structures with deep feature	[35]
Imprinting	Imprinting with recesses	2 μm	Simple and fast process; embedded patterns	Liquid metal residue, less reliable	Small feature size (microscale line patterning)	[42]
	Imprinting with protrusions	500 μm	Simple and fast process; independent patterns	Uneven wetting, less reliable	Large feature size (large-area patterning: millimeter–centimeter)	[15, 44]
Mask-assisted deposition		180 nm	High throughput process	Mask fabrication or pre-deposited sacrificial layer	Arbitrary independent patterns	[47, 50, 52–54, 58]
Direct writing	Inkjet printing	100 μm	various 2D liquid metal electronic components	Relatively complex technique	Microscale droplets derived various 2D liquid metal electronic components	[75, 77]
	Microextrusion printing	2 μm	3D free-standing structures		Antennas; electrical interconnects; 3D free-standing structures	[80, 88, 90]
Laser ablation		4.5 μm	Fast and reliable; Relatively simple	Higher wastage of liquid metal	Patterned circuit	[92–94]
External field guiding	Magnetic field guided	85 μm	Simple and fast	High-resolution, homogeneity of pattern	Flexible electronics NFC liquid metal coil	[95, 129]
	Electric field guided	500 μm	Controlling surface tension of the liquid metal	High resolution	Complex channels with diverse structures	[102, 104]

presents the recently advanced achievements on constructing techniques of liquid metal-based architectures and their fabrication processes. Various fabrication strategies can be generally categorized into seven classes, which are further divided into numerous techniques involving different methods for preparing related structures. Patterning liquid metals is a major technique to construct liquid metal-based architectures, different patterning process has been discussed, each of them has their own advantages and disadvantages and different scopes. We compare the different patterning techniques of liquid metal-based architectures shown in Table 6.1. Despite the extensive efforts made so far, the development of liquid metal-based architectures still poses challenges.

1. Liquid metals might display corrosiveness. For instance, Ga can penetrate into Al and contribute to embrittlement. Most equipment demands the connection of metal with other components. Yet, Ga could alloy with many metals including Al, Ni, Pt, and Au [130]. Controlling the destructive behavior plays a pivotal role in practical applications. Liquid metals' thermal interface material is proposed to alleviate the corrosive behavior on Al by combining liquid metals with methyl silicone oil [131]. Graphene can be used as a diffusion barrier to effectively protect metals from the attack of Ga [132]. Besides, encapsulated liquid metals with semiconductors and insulated coating can significantly hinder corrosion [68].
2. The oxide skin can be a disadvantage in fabricating liquid metal structures. This oxide skin can effectively reduce surface tension and improve wettability on various substrates [30], which is beneficial for stabilizing patterned liquid metals in position, and maintaining nonspherical morphology [133]. However, the wettability originating from the oxide layer provides liquid metals the tendency to adsorb to arbitrary solid substrates even if unwanted, making it hard to handle liquid metals. Therefore, this special character can be a strength or weakness depending on the applications.
3. Mechanical stability of liquid metal-based architectures should be further improved. The oxide skin provides mechanical stability to patterned liquid metal with different geometries. However, the very thin oxide skin is readily destroyed by acid, alkaline solution, and mechanical stress. Although this property enables liquid metals to be easily manipulated, the patterned liquid metals still need to stay stable for different applications. Three-dimensional freeze-printing is an effective method to further improve the stability compared to oxide skin alone. This technique utilizes the substrate with low temperature to freeze the printed liquid metals structure and create a moving freeze-front to make the structure more stable [89].
4. There is limitations to the minimum size of patterned liquid metals: resolution limitations. The patterning strategies shown in this chapter exhibit relatively low resolution of micrometers size compared with traditional methods, due to the high surface tension of liquid metals [86]. Despite the fact that liquid metals have been induced inside CNTs [39], and they also can be injected into $\approx$150 nm capillaries under high pressure [25]. Most fabrication technology for patterning the metals cannot reach such small length scales. High precision is still difficult to

achieve during the liquid metal patterning process due to the high surface tension. And liquid metals can hardly be directly painted onto flexible substrates with high precision due to its high surface tension [134]. Nevertheless, the current resolution limits can satisfy most stretchable and soft electronic devices [8].

5. Large-scale production is hard to achieve since most of the techniques reported now are realized on laboratory techniques. It is possible to scale the liquid metal patterning by adapting existing manufacturing techniques. However, the high-throughput preparation of these manufacturing methods remains to be explored.

References

1 Wang, X.L., Guo, R., and Liu, J. (2019). Liquid metal based soft robotics: materials, designs, and applications. *Adv. Mater. Technol.* 4: 1800549.

2 Fassler, A. and Majidi, C. (2015). Liquid-phase metal inclusions for a conductive polymer composite. *Adv. Mater.* 27: 1928.

3 Wang, D., Xie, W., Gao, Q. et al. (2019). Non-magnetic injectable implant for magnetic field-driven thermochemotherapy and dual stimuli-responsive drug delivery: transformable liquid metal hybrid platform for cancer theranostics. *Small* 15: 1900511.

4 Guo, X., Ding, Y., Xue, L. et al. (2018). A self-healing room-temperature liquid-metal anode for alkali-ion batteries. *Adv. Funct. Mater.* 28: 1804649.

5 Kim, T., Kim, D.M., Lee, B.J., and Lee, J. (2019). Soft and deformable sensors based on liquid metals. *Sensors (Basel)* 19: 4250.

6 Khondoker, M.A.H. and Sameoto, D. (2016). Fabrication methods and applications of microstructured gallium based liquid metal alloys. *Smart Mater. Struct.* 25: 093001.

7 Park, J.E., Kang, H.S., Baek, J. et al. (2019). Rewritable, printable conducting liquid metal hydrogel. *ACS Nano* 13: 9122.

8 Dickey, M.D. (2017). Stretchable and soft electronics using liquid metals. *Adv. Mater.* 29: 1606425.

9 Yu, Z., Yun, F.F., and Wang, X. (2018). A novel liquid metal patterning technique: voltage induced non-contact electrochemical lithography at room temperature. *Mater. Horiz.* 5: 36.

10 Stoykovich, M.P. and Nealey, P.F. (2006). Block copolymers and conventional lithography. *Mater. Today* 9: 20.

11 Dickey, M.D., Chiechi, R.C., Larsen, R.J. et al. (2008). Eutectic gallium-indium (EGaIn): a liquid metal alloy for the formation of stable structures in microchannels at room temperature. *Adv. Funct. Mater.* 18: 1097.

12 Joshipura, I.D., Ayers, H.R., Majidi, C., and Dickey, M.D. (2015). Methods to pattern liquid metals. *J. Mater. Chem. C* 3: 3834.

13 Wang, X. and Liu, J. (2016). Recent advancements in liquid metal flexible printed electronics: properties, technologies, and applications. *Micromachines (Basel)* 7: 206.

14 Cano-Raya, C., Denchev, Z.Z., Cruz, S.F., and Viana, J.C. (2019). Chemistry of solid metal-based inks and pastes for printed electronics – a review. *Appl. Mater. Today* 15: 416.

15 Kim, M.G., Kim, C., Alrowais, H., and Brand, O. (2018). Multiscale and uniform liquid metal thin-film patterning based on soft lithography for 3D heterogeneous integrated soft microsystems: additive stamping and subtractive reverse stamping. *Adv. Mater. Technol.* 3: 1800061.

16 Dickey, M.D. (2014). Emerging applications of liquid metals featuring surface oxides. *ACS. Appl. Mater. Interfaces* 6: 18369.

17 Zhu, L.F., Wang, B., Handschuh-Wang, S., and Zhou, X.C. (2020). Liquid metal-based soft microfluidics. *Small.* 16: 1903841.

18 Wang, J., Liu, S., Guruswamy, S., and Nahata, A. (2014). Injection molding of free-standing, three-dimensional, all-metal terahertz metamaterials. *Adv. Opt. Mater.* 2: 663.

19 Fassler, A. and Majidi, C. (2013). 3D structures of liquid-phase GaIn alloy embedded in PDMS with freeze casting. *Lab. Chip.* 13: 4442.

20 Lin, Y.L., Cooper, C., Wang, M. et al. (2015). Handwritten, soft circuit boards and antennas using liquid metal nanoparticles. *Small* 11: 6397.

21 Kim, H.-J., Son, C., and Ziaie, B. (2008). A multiaxial stretchable interconnect using liquid-alloy-filled elastomeric microchannels. *Appl. Phys. Lett.* 92: 011904.

22 Ota, H., Chen, K., Lin, Y. et al. (2014). Highly deformable liquid-state heterojunction sensors. *Nat. Commun.* 5: 5032.

23 Joshipura, I.D., Ayers, H.R., Castillo, G.A. et al. (2018). Patterning and reversible actuation of liquid gallium alloys by preventing adhesion on rough surfaces. *ACS Appl. Mater. Interfaces* 10: 44686.

24 Wang, R., Gui, L., Zhang, L. et al. (2019). Porous membrane-enabled fast liquid metal patterning in thin blind-ended microchannels. *Adv. Mater. Technol.* 4: 1900256.

25 Zhao, W., Bischof, J.L., Hutasoit, J. et al. (2015). Single-fluxon controlled resistance switching in centimeter-long superconducting gallium-indium eutectic nanowires. *Nano Lett.* 15: 153.

26 So, J.-H. and Dickey, M.D. (2011). Inherently aligned microfluidic electrodes composed of liquid metal. *Lab on a Chip* 11: 905.

27 Rashed Khan, M., Hayes, G.J., So, J.-H. et al. (2011). A frequency shifting liquid metal antenna with pressure responsiveness. *Appl. Phys. Lett.* 99: 013501.

28 Zavabeti, A., Daeneke, T., Chrimes, A.F. et al. (2016). Ionic imbalance induced self-propulsion of liquid metals. *Nat. Commun.* 7: 12402.

29 Holcomb, S., Brothers, M., Diebold, A. et al. (2016). Oxide-free actuation of gallium liquid metal alloys enabled by novel acidified siloxane oils. *Langmuir* 32: 12656.

30 Khan, M.R., Eaker, C.B., Bowden, E.F., and Dickey, M.D. (2014). Giant and switchable surface activity of liquid metal via surface oxidation. *Proc. Natl. Acad. Sci. U. S. A.* 111: 14047.

31 Khan, M.R., Trlica, C., So, J.H. et al. (2014). Influence of water on the interfacial behavior of gallium liquid metal alloys. *ACS Appl. Mater. Interfaces* 6: 22467.
32 Guo, W., Teo, A.J.T., Gañán-Calvo, A.M. et al. (2018). Pressure-driven filling of closed-end microchannel: realization of comb-shaped transducers for acoustofluidics. *Phys. Rev. Appl.* 10: 054045.
33 Cooper, C.B., Joshipura, I.D., Parekh, D.P. et al. (2019). Toughening stretchable fibers via serial fracturing of a metallic core. *Sci. Adv.* 5: eaat4600.
34 Liu, H., Li, M., Ouyang, C. et al. (2018). Biofriendly, stretchable, and reusable hydrogel electronics as wearable force sensors. *Small* 14: 1801711.
35 Lin, Y., Gordon, O., Khan, M.R. et al. (2017). Vacuum filling of complex microchannels with liquid metal. *Lab Chip.* 17: 3043.
36 Yang, J., Zhou, T., Zhang, L. et al. (2017). Defect-free, high resolution patterning of liquid metals using reversibly sealed, reusable polydimethylsiloxane microchannels for flexible electronic applications. *J. Mater. Chem. C* 5: 6790.
37 Teng, L., Ye, S., Handschuh-Wang, S. et al. (2019). Liquid metal-based transient circuits for flexible and recyclable electronics. *Adv. Funct. Mater.* 29: 1808739.
38 Chen, Y.L., Nguyen, Y., Wu, S.E. et al. (2018). Patterned liquid metal contacts for high density, stick-and-peel 2D material device arrays. *Nanoscale* 10: 14510.
39 Andrews, J.B., Mondal, K., Neumann, T.V. et al. (2018). Patterned liquid metal contacts for printed carbon nanotube transistors. *ACS Nano* 12: 5482.
40 Liang, S.Q., Li, Y.Y., Chen, Y.Z. et al. (2017). Liquid metal sponges for mechanically durable, all-soft, electrical conductors. *J. Mater. Chem. C* 5: 1586.
41 Park, J., Wang, S., Li, M. et al. (2012). Three-dimensional nanonetworks for giant stretchability in dielectrics and conductors. *Nat. Commun.* 3: 916.
42 Gozen, B.A., Tabatabai, A., Ozdoganlar, O.B., and Majidi, C. (2014). High-density soft-matter electronics with micron-scale line width. *Adv. Mater.* 26: 5211.
43 Kim, M.G., Alrowais, H., Pavlidis, S., and Brand, O. (2017). Size-scalable and high-density liquid-metal-based soft electronic passive components and circuits using soft lithography. *Adv. Funct. Mater.* 27: 1604466.
44 Tabatabai, A., Fassler, A., Usiak, C., and Majidi, C. (2013). Liquid-phase gallium-indium alloy electronics with microcontact printing. *Langmuir* 29: 6194.
45 Li, G., Parmar, M., and Lee, D.W. (2015). An oxidized liquid metal-based microfluidic platform for tunable electronic device applications. *Lab Chip* 15: 766.
46 Zhang, S., Wang, B., Jiang, J. et al. (2019). High-fidelity conformal printing of 3D liquid alloy circuits for soft electronics. *ACS Appl. Mater. Interfaces* 11: 7148.
47 Jeong, S.H., Hagman, A., Hjort, K. et al. (2012). Liquid alloy printing of microfluidic stretchable electronics. *Lab Chip* 12: 4657.
48 Park, C.W., Moon, Y.G., Seong, H. et al. (2016). Photolithography-based patterning of liquid metal interconnects for monolithically integrated stretchable circuits. *ACS Appl. Mater. Interfaces* 8: 15459.

49 Lazarus, N., Bedair, S.S., and Kierzewski, I.M. (2017). Ultrafine pitch stencil printing of liquid metal alloys. *ACS Appl. Mater. Interfaces* 9: 1178.

50 Zhang, Q., Gao, Y., and Liu, J. (2014). Atomized spraying of liquid metal droplets on desired substrate surfaces as a generalized way for ubiquitous printed electronics. *Appl. Phys. A.* 116: 1091.

51 Sahlberg, A., Nilsson, F., Berglund, A. et al. (2018). High-resolution liquid alloy patterning for small stretchable strain sensor arrays. *Adv. Mater. Technol.* 3: 1700330.

52 Hirsch, A., Michaud, H.O., Gerratt, A.P. et al. (2016). Intrinsically stretchable biphasic (solid-liquid) thin metal films. *Adv. Mater.* 28: 4507.

53 Guo, R., Tang, J., Dong, S. et al. (2018). One-step liquid metal transfer printing: toward fabrication of flexible electronics on wide range of substrates. *Adv. Mater. Technol.* 3: 1800265.

54 Kramer, R.K., Majidi, C., and Wood, R.J. (2013). Masked deposition of gallium-indium alloys for liquid-embedded elastomer conductors. *Adv. Funct. Mater.* 23: 5292.

55 Li, G., Wu, X., and Lee, D.-W. (2015). Selectively plated stretchable liquid metal wires for transparent electronics. *Sens. Actuators B Chem.* 221: 1114.

56 Yao, Y.Y., Ding, Y.J., Li, H.P. et al. (2019). Multi-substrate liquid metal circuits printing via superhydrophobic coating and adhesive patterning. *Adv. Eng. Mater.* 21: 1801363.

57 Kramer, R.K., Boley, J.W., Stone, H.A. et al. (2014). Effect of microtextured surface topography on the wetting behavior of eutectic gallium-indium alloys. *Langmuir* 30: 533.

58 Kim, M.G., Brown, D.K., and Brand, O. (2020). Nanofabrication for all-soft and high-density electronic devices based on liquid metal. *Nat. Commun.* 11: 1002.

59 Ozutemiz, K.B., Wissman, J., Ozdoganlar, O.B., and Majidi, C. (2018). EGaIn-metal interfacing for liquid metal circuitry and microelectronics integration. *Adv. Mater. Interfaces* 5: 1701596.

60 Li, G. and Lee, D.W. (2017). An advanced selective liquid-metal plating technique for stretchable biosensor applications. *Lab Chip* 17: 3415.

61 Hon, K.K.B., Li, L., and Hutchings, I.M. (2008). Direct writing technology—advances and developments. *CIRP Ann.* 57: 601.

62 Cumpston, B.H., Ananthavel, S.P., Barlow, S. et al. (1999). Two-photon polymerization initiators for three-dimensional optical data storage and microfabrication. *Nature.* 398: 51.

63 Campbell, M., Sharp, D.N., Harrison, M.T. et al. (2000). Fabrication of photonic crystals for the visible spectrum by holographic lithography. *Nature* 404: 53.

64 Ellis, J., Mavromatos, N.E., Nanopoulos, D.V., and Sakharov, A.S. (2004). Cosmology: synchrotron radiation and quantum gravity. *Nature* 428: 386.

65 van Blaaderen, A., Ruel, R., and Wiltzius, P. (1997). Template-directed colloidal crystallization. *Nature* 385: 321.

66 Lin, S.Y., Fleming, J.G., Hetherington, D.L. et al. (1998). A three-dimensional photonic crystal operating at infrared wavelengths. *Nature* 394: 251.

67 Russo, A., Ahn, B.Y., Adams, J.J. et al. (2011). Pen-on-paper flexible electronics. *Adv. Mater.* 23: 3426.

68 Wang, Q., Yu, Y., and Liu, J. (2018). Preparations, characteristics and applications of the functional liquid metal materials. *Adv. Eng. Mater.* 20: 1700781.

69 Zhang, Y., Zhang, F., Yan, Z. et al. (2017). Printing, folding and assembly methods for forming 3D mesostructures in advanced materials. *Nat. Rev. Mater.* 2: 17019.

70 Boley, J.W., White, E.L., and Kramer, R.K. (2015). Mechanically sintered gallium-indium nanoparticles. *Adv. Mater.* 27: 2355.

71 Ren, L., Zhuang, J., Casillas, G. et al. (2016). Nanodroplets for stretchable superconducting circuits. *Adv. Funct. Mater.* 26: 8111.

72 Tevis, I.D., Newcomb, L.B., and Thuo, M. (2014). Synthesis of liquid core-shell particles and solid patchy multicomponent particles by shearing liquids into complex particles (SLICE). *Langmuir.* 30: 14308.

73 Hohman, J.N., Kim, M., Wadsworth, G.A. et al. (2011). Directing substrate morphology via self-assembly: ligand-mediated scission of gallium-indium microspheres to the nanoscale. *Nano Lett.* 11: 5104.

74 Yamaguchi, A., Mashima, Y., and Iyoda, T. (2015). Reversible size control of liquid-metal nanoparticles under ultrasonication. *Angew. Chem. Int. Ed.* 54: 12809.

75 Li, X., Li, M., Zong, L. et al. (2018). Liquid metal droplets wrapped with polysaccharide microgel as biocompatible aqueous ink for flexible conductive devices. *Adv. Funct. Mater.* 28: 1804197.

76 Tang, S.Y., Qiao, R., Yan, S. et al. (2018). Microfluidic mass production of stabilized and stealthy liquid metal nanoparticles. *Small* 14: 1800118.

77 Zhou, L.y., Fu, J.z., Gao, Q. et al. (2019). All-printed flexible and stretchable electronics with pressing or freezing activatable liquid-metal–silicone inks. *Adv. Funct. Mater.* 30: 1906683.

78 Yu, Y., Liu, F., Zhang, R., and Liu, J. (2017). Suspension 3D printing of liquid metal into self-healing hydrogel. *Adv. Mater. Technol.* 2: 1700173.

79 Li, G.Y., Wu, X., and Lee, D.W. (2015). A novel liquid metal-based inkjet nozzle for flexible electronics. *Actuators Microsyst. (Transducers). IEEE* 339: 15362810.

80 Boley, J.W., White, E.L., Chiu, G.T.C., and Kramer, R.K. (2014). Direct writing of gallium-indium alloy for stretchable electronics. *Adv. Funct. Mater.* 24: 3501.

81 Zheng, Y., He, Z.Z., Yang, J., and Liu, J. (2014). Personal electronics printing via tapping mode composite liquid metal ink delivery and adhesion mechanism. *Sci. Rep.* 4: 4588.

82 Wang, L., Wang, M., Lu, J. et al. (2019). Enhanced adhesion between liquid metal ink and the wetted printer paper for direct writing electronic circuits. *J. Taiwan Inst. Chem. Eng.* 95: 202.

83 Zheng, Y., He, Z., Gao, Y., and Liu, J. (2013). Direct desktop printed-circuits-on-paper flexible electronics. *Sci. Rep.* 3: 1786.

84 Tang, J., Zhao, X., Li, J. et al. (2017). Gallium-based liquid metal amalgams: transitional-state metallic mixtures (TransM (2)ixes) with enhanced and tunable electrical, thermal, and mechanical properties. *ACS Appl. Mater. Interfaces* 9: 35977.

85 Yoon, Y., Kim, S., Kim, D. et al. (2019). Four degrees-of-freedom direct writing of liquid metal patterns on uneven surfaces. *Adv. Mater. Technol.* 4: 1800379.

86 Kim, D., Yoon, Y., Kauh, S.K., and Lee, J. (2018). Towards sub-microscale liquid metal patterns: cascade phase change mediated pick-n-place transfer of liquid metals printed and stretched over a flexible substrate. *Adv. Funct. Mater.* 28: 1800380.

87 Liang, B., Wei, J., Fang, L. et al. (2019). High-resolution rapid prototyping of liquid metal electronics by direct writing on highly prestretched substrates. *ACS Omega.* 4: 21072.

88 Ladd, C., So, J.H., Muth, J., and Dickey, M.D. (2013). 3D printing of free standing liquid metal microstructures. *Adv. Mater.* 25: 5081.

89 Gannarapu, A. and Gozen, B.A. (2016). Freeze-printing of liquid metal alloys for manufacturing of 3D, conductive, and flexible networks. *Adv. Mater. Technol.* 1: 1600047.

90 Park, Y.-G., An, H.S., Kim, J.-Y., and Park, J.-U. (2019). High-resolution, reconfigurable printing of liquid metals with three-dimensional structures. *Sci. Adv.* 5: 2844.

91 Park, Y.G., Min, H., Kim, H. et al. (2019). Three-dimensional, high-resolution printing of carbon nanotube/liquid metal composites with mechanical and electrical reinforcement. *Nano Lett.* 19: 4866.

92 Lu, T., Markvicka, E.J., Jin, Y., and Majidi, C. (2017). Soft-matter printed circuit board with UV laser micropatterning. *ACS Appl. Mater. Interfaces* 9: 22055.

93 Lu, T., Finkenauer, L., Wissman, J., and Majidi, C. (2014). Rapid prototyping for soft-matter electronics. *Adv. Funct. Mater.* 24: 3351.

94 Pan, C., Kumar, K., Li, J. et al. (2018). Visually imperceptible liquid-metal circuits for transparent, stretchable electronics with direct laser writing. *Adv. Mater.* 30: 1706937.

95 Ma, B., Xu, C., Chi, J. et al. (2019). A versatile approach for direct patterning of liquid metal using magnetic field. *Adv. Funct. Mater.* 29: 1901370.

96 Xu, C., Ma, B., Yuan, S. et al. (2019). High-resolution patterning of liquid metal on hydrogel for flexible, stretchable, and self-healing electronics. *Adv. Electr. Mater.* 6: 1900721.

97 Tang, S.Y., Sivan, V., Khoshmanesh, K. et al. (2013). Electrochemically induced actuation of liquid metal marbles. *Nanoscale* 5: 5949.

98 Sheng, L., Zhang, J., and Liu, J. (2014). Diverse transformations of liquid metals between different morphologies. *Adv. Mater.* 26: 6036.

99 Sheng, L., He, Z., Yao, Y., and Liu, J. (2015). Transient state machine enabled from the colliding and coalescence of a swarm of autonomously running liquid metal motors. *Small.* 11: 5253.

100 Wang, M., Trlica, C., Khan, M.R. et al. (2015). A reconfigurable liquid metal antenna driven by electrochemically controlled capillarity. *J. Appl. Phys.* 117: 194901.

101 Khan, M.R., Trlica, C., and Dickey, M.D. (2015). Recapillarity: electrochemically controlled capillary withdrawal of a liquid metal alloy from microchannels. *Adv. Funct. Mater.* 25: 671.

102 Parekh, D.P., Ladd, C., Panich, L. et al. (2016). 3D printing of liquid metals as fugitive inks for fabrication of 3D microfluidic channels. *Lab Chip* 16: 1812.

103 Yan, S., Li, Y., Zhao, Q. et al. (2018). Liquid metal-based amalgamation-assisted lithography for fabrication of complex channels with diverse structures and configurations. *Lab Chip* 18: 785.

104 Ma, J.-L., Dong, H.-X., and He, Z.-Z. (2018). Electrochemically enabled manipulation of gallium-based liquid metals within porous copper. *Mater. Horiz.* 5: 675.

105 Lee, K.T., Jung, Y.S., Kwon, J.Y. et al. (2008). Role of electrochemically driven cu nanograins in cuga2electrode. *Chem. Mater.* 20: 447.

106 Steichen, M., Thomassey, M., Siebentritt, S., and Dale, P.J. (2011). Controlled electrodeposition of Cu-Ga from a deep eutectic solvent for low cost fabrication of $CuGaSe_2$ thin film solar cells. *Phys. Chem. Chem. Phys.* 13: 4292.

107 Tang, S.-Y., Zhu, J., Sivan, V. et al. (2015). Creation of liquid metal 3D microstructures using dielectrophoresis. *Adv. Funct. Mater.* 25: 4445.

108 Lin, Y., Ladd, C., Wang, S. et al. (2016). Drawing liquid metal wires at room temperature. *Extreme Mech. Lett.* 7: 55.

109 Yun, G., Tang, S.Y., Sun, S. et al. (2019). Liquid metal-filled magnetorheological elastomer with positive piezoconductivity. *Nat. Commun.* 10: 1300.

110 Parida, K., Thangavel, G., Cai, G. et al. (2019). Extremely stretchable and self-healing conductor based on thermoplastic elastomer for all-three-dimensional printed triboelectric nanogenerator. *Nat. Commun.* 10: 2158.

111 Wang, H., Yao, Y., Wang, X. et al. (2019). Large-magnitude transformable liquid-metal composites. *ACS Omega* 4: 2311.

112 Kazem, N., Bartlett, M.D., and Majidi, C. (2018). Extreme toughening of soft materials with liquid metal. *Adv. Mater.* 30: 1706594.

113 Zhu, L., Chen, Y., Shang, W. et al. (2019). Anisotropic liquid metal–elastomer composites. *J. Mater. Chem. C* 7: 10166.

114 Peng, H., Xin, Y., Xu, J. et al. (2019). Ultra-stretchable hydrogels with reactive liquid metals as asymmetric force-sensors. *Mater. Horiz.* 6: 618.

115 Mei, S., Gao, Y., Deng, Z., and Liu, J. (2014). Thermally conductive and highly electrically resistive grease through homogeneously dispersing liquid metal droplets inside methyl silicone oil. *J. Electron. Packag.* 136: 011009.

116 Markvicka, E.J., Bartlett, M.D., Huang, X., and Majidi, C. (2018). An autonomously electrically self-healing liquid metal-elastomer composite for robust soft-matter robotics and electronics. *Nat. Mater.* 17: 618.

117 Hayashi, Y., Saneie, N., Yip, G. et al. (2016). Metallic nanoemulsion with galinstan for high heat-flux thermal management. *Int. J. Heat Mass Transfer.* 101: 1204.

118 Jeong, S.H., Chen, S., Huo, J. et al. (2015). Mechanically stretchable and electrically insulating thermal elastomer composite by liquid alloy droplet embedment. *Sci. Rep.* 5: 18257.

119 Wang, H., Yao, Y., He, Z. et al. (2019). A highly stretchable liquid metal polymer as reversible transitional insulator and conductor. *Adv. Mater.* 31: 1901337.

120 Bartlett, M.D., Fassler, A., Kazem, N. et al. (2016). Stretchable, high-k dielectric elastomers through liquid-metal inclusions. *Adv. Mater.* 28: 3726.

121 Bartlett, M.D., Kazem, N., Powell-Palm, M.J. et al. (2017). High thermal conductivity in soft elastomers with elongated liquid metal inclusions. *Proc. Natl. Acad. Sci. U. S. A.* 114: 2143.

122 Ralphs, M.I., Kemme, N., Vartak, P.B. et al. (2018). In situ alloying of thermally conductive polymer composites by combining liquid and solid metal microadditives. *ACS Appl. Mater. Inter.* 10: 2083.

123 Tutika, R., Zhou, S.H., Napolitano, R.E., and Bartlett, M.D. (2018). Mechanical and functional tradeoffs in multiphase liquid metal, solid particle soft composites. *Adv. Funct. Mater.* 28: 1804336.

124 Koh, A., Sietins, J., Slipher, G., and Mrozek, R. (2018). Deformable liquid metal polymer composites with tunable electronic and mechanical properties. *J. Mater. Res.* 33: 2443.

125 Tutika, R., Kmiec, S., Haque, A. et al. (2019). Liquid metal-elastomer soft composites with independently controllable and highly tunable droplet size and volume loading. *ACS Appl. Mater. Inter.* 11: 17873.

126 Xin, Y., Peng, H., Xu, J., and Zhang, J. (2019). Ultrauniform embedded liquid metal in sulfur polymers for recyclable, conductive, and self-healable materials. *Adv. Funct. Mater.* 29: 1804336.

127 Yan, J., Malakooti, M.H., Lu, Z. et al. (et al., 2019). Solution processable liquid metal nanodroplets by surface-initiated atom transfer radical polymerization. *Nat. Nanotechnol.* 14: 684.

128 Thrasher, C.J., Farrell, Z.J., Morris, N.J. et al. (2019). Mechanoresponsive polymerized liquid metal networks. *Adv. Mater.* 31: 1903864.

129 Xu, C., Ma, B., Yuan, S. et al. (2020). High-resolution patterning of liquid metal on hydrogel for flexible, stretchable, and self-healing electronics. *Adv. Electron. Mater.* 6: 1900721.

130 Yatsenko, S.P., Sabirzyanov, N.A., and Yatsenko, A.S. (2008). Dissolution rates and solubility of some metals in liquid gallium and aluminum. *J. Phys.: Conf. Ser.* 98: 062032.

131 Fan, P., Sun, Z., Wang, Y. et al. (2018). Nano liquid metal for the preparation of a thermally conductive and electrically insulating material with high stability. *RSC Adv.* 8: 16232.

132 Ahlberg, P., Jeong, S.H., Jiao, M. et al. (2014). Graphene as a diffusion barrier in galinstan-solid metal contacts. *IEEE Trans. Electron Dev.* 61: 2996.

133 Chiechi, R.C., Weiss, E.A., Dickey, M.D., and Whitesides, G.M. (2008). Eutectic gallium-indium (EGaIn): a moldable liquid metal for electrical characterization of self-assembled monolayers. *Angew. Chem. Int. Ed. Engl.* 47: 142.

134 Guo, R., Wang, X., Chang, H. et al. (2018). Ni-GaIn amalgams enabled rapid and customizable fabrication of wearable and wireless healthcare electronics. *Adv. Eng. Mater.* 20: 1800054.

7

Liquid Metal Toward Flexible Electronics

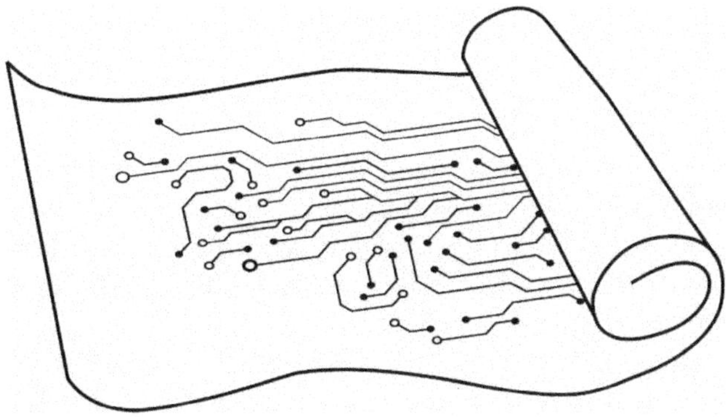

The soft and weak can overcome the hard and strong.
Source: Lao Tzu. Public Domain.

7.1 Introduction

Early in the 1950s, the construction of flexible electronics had come into being, which surprisingly had an extremely long history [1]. In 1948, a comprehensive treatment of the whole field of printed circuits had been presented [2]. Printed circuits were defined as the circuits on any insulated substrates. In a sense, some printed devices can be regarded as flexible electronics. Along with the development of material science and microelectronics, the variety of flexible electronics has improved in the aspects of material species, device structures, and functions.

Flexible electronics can be bendable or stretchable, but different device structures should be designed to achieve different forms of flexibility. Bendable and stretchable devices are under different strains [3]. In a bendable device, tensile and compressive strains are applied on the top and bottom surfaces, respectively. In a stretchable device, the strain is applied along the stretching direction, and each part of the device should tolerate the stretching strain. Therefore, the mechanical property of the whole device is essential to protect the device structure and keep the electrical

Liquid Metals: Properties, Mechanisms, and Applications, First Edition. Lei Fu and Mengqi Zeng.

performances [4]. Compared to bendable devices, special structures are expressly designed to relieve spatially uneven strains. Toward wearable devices [5], it is necessary to take the strain tolerance, substrate compatibility, and electronic responsibility into consideration during the construction of those flexible electronics. Despite that those traditional materials have been widely used for flexible electronics, the mechanical failure can appear under large strains, and device performances may not meet the expectation. Owing to the aforementioned reasons, new materials and novel construction strategies need to be discovered and developed. This chapter introduces the developing status of flexible electronics based on traditional and classical materials at first, and then expounds why and how liquid metals can be applied to electronics including the advantages and the constructing strategies. Finally, the chapter exhibits the advances of novel liquid metal-based flexible electronics and their performances.

7.2 Traditional Materials for Flexible Electronics

For flexible electronics, a diversity of materials has been utilized for electrodes, channels, and substrates. In detail, polymers, papers, or similar films are usually chosen for the substrates of all the flexible electronics, and metals or conductive compounds always serve as electrodes. In this section, traditional materials [6–8] including elementary substances, inorganic compounds, organic materials, and composite materials are discussed, which mainly act as channels of flexible electronics. The structure, design [7], and working mechanism are also analyzed in detail.

7.2.1 Elementary Substances and Inorganic Materials

In the development of electronics, numerous materials have served as milestones. Semiconducting elementary substances and inorganic compounds were used in diodes, transistors, and even more complex integrated circuits. Apart from the development of chips, the increasing requirements of flexible electronics also promote the improvement of these materials in the aspects of structures and species.

Silicon (Si) is one of the most traditional semiconductors for transistors, which have been widely used in computers and other instruments. Although Si industry and Si-based electronics developed rapidly during the period between late twentieth century and early twenty-first century, traditional fabrication of polycrystalline Si can hardly be implemented on the flexible substrates for the high process temperatures [9, 10]. Moreover, owing to the poor mechanical performances, polycrystalline Si can hardly meet with the requirements of flexible devices. Novel Si nanowire (NW) thin film transistors (TFTs) were developed to adapt to the flexible substrates [11]. Such NW-TFTs are bendable and exhibit a threshold voltage of ~3.0 V, an on–off current ratio $> 10^5$, and a high subthreshold swing of 500–800 mV dec^{-1}. NW-TFT technology has given much inspiration on Si-based flexible devices [12]. Furthermore, Ge/Si core–shell NW arrays can also be fabricated for a low-voltage

(~5 V) macroscale artificial skin with excellent flexibility and stability [13]. NW-based technology offers excellent mechanical robustness and reliability to flexible electronics. Except for Si and Ge, other semiconducting inorganic compounds can be also applied to flexible electronics such a NW-based strategy, such as CdS [11], GaAs [14–16], and metal oxides (e.g., In_2O_3, ZnO) [17]. These inorganic material NWs can show distinct bandgap values with the bulks on account of the quantum confinement, which may improve the optoelectronic performances of flexible devices. As one of the most famous one-dimensional (1D) materials with controllable metallic or semiconducting properties, carbon nanotube (CNT)-based flexible electronics have also been developed [18]. With a high average on/off current ratio of 10^4 when channel length is 100 μm, flexible CNT integrated circuits, including inverter, NOR and NAND circuits, all exhibit excellent logic performances. In short, semiconducting 1D materials (NWs, nanotubes, or nanofibers) of elementary substances and inorganic compounds help scientists overcome the dilemma of crystal quality and electrical performances.

Along with the development of nanoscience and the discovery of novel nanomaterials, some innovative semiconductors have been applied to flexible electronics. Apart from nanowires and nanotubes of semiconductors, thin semiconducting films or crystals can be compatible with flexible substrates. Flexible TFTs using metal oxide semiconductors, including amorphous In–Ga–Zn–O (a-IGZO), IZO, and In_2O_3, can exhibit good performances for transistors and oscillators [19, 20]. Actually, amorphous metal oxide films have played a crucial part in flexible electronics, which not only show comparable carrier mobility to polycrystalline Si but have advantages for low cost and large scale [21, 22]. When the film or crystal thickness decreases to a few nanometers, they can present obvious quantum confinement to induce special physical phenomena, such as two-dimensional (2D) materials. Since the discovery of graphene in 2004, 2D materials with atomic-scale thickness have gained much attention in many fields including flexible electronics [23, 24]. Two-dimensional elementary substances (such as graphene and black phosphorus) and inorganic materials (such as transition metal dichalcogenides (TMDs), hexagonal boron nitride, and metal carbides) have various band structures so that they can be used in flexible integrated circuits as electrodes or channels [25–27]. Nowadays, 2D materials become more and more important in the application of flexible electronics by virtue of their outstanding mechanical, electrical, and optical performances.

In consideration of the mechanical properties of channels and electrodes, microstructures of the electronics have been designed for flexible electronics especially for stretchable ones, because it is aforementioned that each part is under strain when the device is stretched. The microstructures often include 1D wavy structures, 1D buckled ribbons, 2D wavy structures, and 2D buckled meshes, which are designed for those that cannot allow too much strain (such as Si and metal oxides) (Figure 7.1a) [28]. Curved structure may be the rudiment of 1D wavy structure and 1D buckled ribbon. In 2004, Gray and coworkers reported the high-conductivity elastomeric electronics using wavy narrow silver wires [30]. The wire conductivity was not dramatically changed by subfailure strain of ~50%.

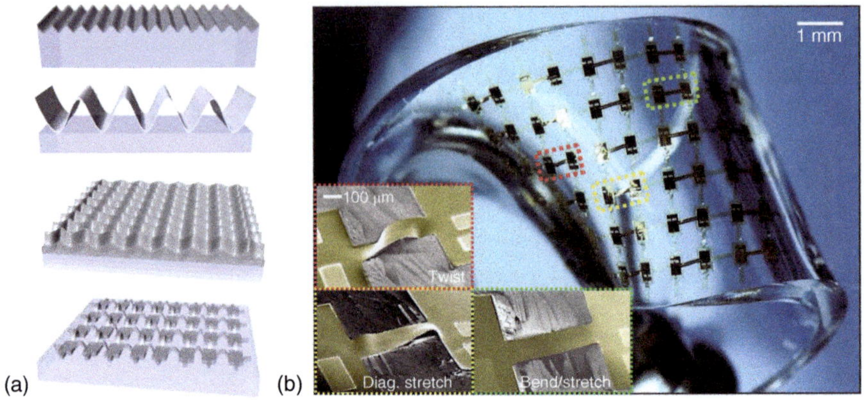

Figure 7.1 Microstructures designed for stretchable devices. (a) Schematic illustrations of several microstructures designed for Si-based stretchable devices including 1D and 2D wavy structures, 1D buckled ribbons, and 2D buckled meshes [28]. Source: Kim et al. [28]. (b) Schematic illustration of the fabrication processes of the stretchable Si-based MOSFET arrays and the optical images of stretched device and structural transformation under stretching [29]. Source: Kim et al. [29].

Stretchable complementary metal-oxide-semiconductor (CMOS) integrated circuits based on single-crystalline Si ribbons can keep the device structures and electrical performances after diagonal stretching, twisting, and bending to high strain levels (even over 100%) (Figure 7.1b) [29]. Interconnected Si islands with spiral springs can also be used to improve the mechanical properties of flexible electronics [31]. Except for Si-based devices, similar strategies are also adopted to metal-oxide-based flexible electronics [32]. Such designed structures have been widely used for large-scale Si- or metal oxides-based flexible electronics to achieve both excellent electrical performances and elastic deformation under large strain.

To sum up, elementary substances and inorganic materials play an important role in flexible electronics on account of their excellent electrical performances and the low dimensional structures [33–35]. For stretchable electronics, specific microstructures have been designed to overcome the large strain of stretching. Although flexible electronics based on elementary substances and inorganic compounds exhibit promising prospects, the complicated constructing method and structure design may hinder the practical application due to the high cost. Even so, it is noteworthy to explore manufacturing technologies to achieve low-cost fabrication of flexible devices based on those materials with outstanding physical properties.

7.2.2 Organic Materials

Elementary substances and inorganic materials always show good electrical conductivity or carrier mobility, but mechanical failure happens when there are some lattice defects. On the contrary, semiconducting organic materials exhibit excellent flexibility and can be industrially manufactured with relatively low cost. For the compatibility with the flexible substances, organic materials have been utilized in flexible electronics since the end of the twentieth century [36]. Moreover, on account of the usage of light-emitting diode (LED) and optoelectronics with low cost and large scale, organic materials have played a crucial role in both fundamental research and practical application.

To obtain high-performance flexible electronics, researchers tend to choose compounds owning large conjugated systems to exhibit higher charge density and carrier mobility than those without conjugated structures [37, 38]. Crone and coworkers designed α-sexithiophene (α-6T) and hexadecafluorocopper phthalocyanine (F-CuPc) as n- and p-type semiconductors, respectively, and used them in TFTs on plastic substrates [39]. They constructed flexible integrated circuits by connecting n- or p-channel TFTs including inverters, oscillators, and other complicated ones, and achieved high operation speed of $\sim$1 kHz in clocked sequential complementary circuits. Actually, organic TFTs can be convenient for paper-like devices and organic displays [9, 40]. Except for organic films, organic single crystals and polymers were also been utilized for flexible organic electronics [41, 42]. In a word, scientists have taken effort to explore suitable organic materials containing small molecules and polymers to enrich the choices for flexible electronics for several years [43, 44]. The breakthrough of material species indeed promotes the development of flexible organic electronics. In other aspects, researchers also have tried to change the morphology of both organic materials and device constructions, such as NW FETs and organic LEDs (OLEDs) with highly flexible patterned elastomeric substrates [45–47].

To sum up, organic materials indeed exhibit superb compatibility with flexible substrates and have been widely used in flexible electronics and optoelectronics. The construction of organic electronics is relatively convenient because organic channels can be easily fabricated on random substrates even papers by drop-casting, spin coating, and printing [46, 48–51]. Nevertheless, challenges still exist that the unexpected defects can be expended under strain and finally result in the mechanical failure [52]. Some morphological microstructures and crosslinking compounds have been designed to solve the problem, which help organic materials be applied to pressure sensors, electronic muscles, and skins, and better [53, 54].

7.2.3 Composite Materials

As mentioned in Sections 7.2.1 and 7.2.2, elementary substances and inorganic materials mainly exhibit outstanding carrier mobility and conductivity, while organic materials show good compatibility with flexible substrates. Therefore, designing composite materials is efficient to improve the performances of flexible electronics. Because of the unique advantages, constructing inorganic/organic composites may achieve high-performance flexible electronics. As a demonstration, high-mobility semiconductors and/or high-capacitance gate dielectric can help to achieve low-power consumption [55, 56]. Besides, inorganic/organic high-κ dielectrics were also involved in flexible CNT-based TFTs [57]. For 2D material-based composites, organic compounds can attach new charge-transfer routes to 2D inorganic materials to improve their performances and function diversity [58]. As a matter of fact, 2D material-based composites are more and more popular in flexible electronic applications because of the excellent synergistic effects and compatibility with flexible substrates. Whatever, the fully printed flexible electronics would be inspirable for industrial manufacture of these devices.

Except for inorganic/organic hybrid flexible electronics, metals also play a crucial role in composite material-based flexible electronics for their eminent malleability and conductivity [59, 60]. Rogers and co-workers designed Si CMOS circuits on polyimide (PI) substrate with etched hole arrays that can exhibit comparable performances to full Si circuits and excellent flexibility by releasing the strain through hole arrays [61]. Recently, researchers have paid much attention to improve the flexibility and conductivity for more practical applications such as flexible functional devices by combining metals with organic materials [62]. Except for embedding metal nanoscale structures, elastic conductor ink consisting of Ag flakes and elastomeric fluorine copolymers can be conveniently used for constructing flexible electronics such as active matrix organic TFT arrays, electromyogram detectors, flexible OLEDs, and flexible nanogenerators [63–66]. In a word, metal-based composite materials can help to improve the mechanical and electrical performances of flexible electronics.

In summary, traditional elementary substances, inorganic compounds, organic materials, and corresponding composite materials have been widely used for flexible electronics for their intrinsic properties. However, traditional flexible electronics may have a diversity of problems for the practical applications including mechanical failure at low-level strain or poor electrical performances. With the discovery of innovative 1D and 2D materials, the functional integration and mechanical performances of the flexible electronics have been enormously improved [23, 67–70]. Composite materials provide other access to high-performance flexible electronics on account of synergistic effects of different materials. Nevertheless, the flexibility of electronics is still restricted to the mechanical properties of electrode and channel materials, as well as the construction strategies. Therefore, it is vital to explore novel materials and new techniques for flexible device construction.

7.3 Advantages of Liquid Metals

Despite the fact that the development of flexible electronics based on traditional materials is thriving in the recent years for the expansion of material choices, researchers are still struggling with the dilemma of improving the electrical performances and mechanical properties. The appearance of composite materials and 2D materials indeed remits the predicament, but the complicated processes and high costs of the material fabrication and device construction may hinder the industrial manufacture and practical applications. In addition, flexible electronics based on traditional materials generally work as diodes and transistors with complex device structures, which also provide numerous opportunities for flexible integrated electronics.

As for aforementioned liquid metals, excellent mechanical properties and high conductivity make them become promising candidates. First, liquid metals can exhibit a zero Young's module, which is lower than almost all the traditional materials applied to flexible electronics and biological tissues [71–73]. It means that liquid metals are compatible with the elastomer substrates unlike solid metals. Traditional metal electrodes or films deposited on elastomer substrates allow larger strain about 30% until rupture than freestanding metal films [74], while liquid metals can exhibit excellent conductivity under large strain (even >200% [75] or much larger [76]) as previously reported. It means that liquid metals can exhibit excellent electrical performances even at high-level strains, while solid metal films may face the mechanical failure. For the good combination of mechanical performance and electrical conductivity, liquid metals can become good substitutes for traditional metal electrodes in flexible electronics. Thus, second, liquid metal-based electronics can be modulated to improve the functional integration. For instance, liquid metals can be introduced to elastomer substrates and attach self-healing behavior to the composites [77]. Magnetism can also be attached to liquid metals by introducing magnetic metal particles [78]. Third, fluidity of liquid metals can benefit the convenient construction of flexible electronics, including injection and printing instead of lithography and deposition processes [71, 79]. Therefore, liquid metals can be crucial for flexible electronics and be superior in the combination of excellent mechanical and electrical properties.

Liquid metal-based electronics cannot serve as diodes and transistors to achieve signal modulation, but they can work as resistance-dependent sensors and switchable devices, when only liquid metals are used in the devices. Composite materials based on liquid metals and other materials are possibly applied to diode- and transistor-based integrated circuits.

7.4 Construction Strategies for Liquid Metal-Based Flexible Electronics

The construction strategies for liquid -metal-based flexible electronics actually have been mentioned in Chapter 6. Herein, commonly used strategies for liquid metal-based flexible electronics are just briefly introduced.

As mentioned in Section 7.3, liquid metals are utilized to connect circuit elements or as resistors or resistance sensors. By virtue of the fluidity, liquid metals can be injected or printed to construct circuits. For example, liquid metals can be dropped on porous polymer substrate surface and squeezed into the substrate [75]. For microchannels, liquid metals can be directly injected into channels via a syringe [80, 81]. Another commonly used method is microfluidic injection technique [82, 83], which mainly adapt to liquid metal electrodes and interconnection circuits at a micrometer scale. Taking the melting points and viscosities into concern, liquid metals and their alloys with melting points above the room temperature and large viscosities are difficult to be injected into microchannels with small scales. Printing provides an access for these metals and their alloys. Initially, templates were used in liquid metal printing to ensure the patterns [84]. Along with the development of three-dimensional (3D) printing techniques, liquid metals can be involved in printing inks so that they can be directly printed onto the target substrates without any templates [85, 86]. However, the resolution of printing still needs improvement for highly integrated circuits.

7.5 Typical Flexible Electronics Based on Liquid Metals

Recently, typical flexible electronics based on liquid metals have been widely discussed, which is motivated by the mechanical properties of liquid metals suitable for human experience, such as tissue, skin, textiles, and clothing. Such types of electronics just like wearable or implantable electronics and sensors for soft robotics need to operate during deformation. Liquid metals are reliable materials to construct these devices because they can deform infinitely while keeping metallic conductivity. In contrast to traditional materials, gallium-based liquid metals have outstanding electrical conductivity, stretchability, and few defects, leading to popular and promising materials for stretchable conductors, self-healing circuits, reconfigurable antennas, soft sensors, memory devices, diodes, and so on. Although most liquids tend to bead up to reduce surface energy, the surface oxide on these metals can help to pattern them into targeted shapes by different techniques, including 3D printing and fluidic injection. In this section, we will introduce these flexible electronics including stretchable wires and interconnects, stretchable circuit elements, stretchable antennas, and soft sensors in detail, as well as further opportunities and challenges.

Flexible electronics have two significant characteristics: bendability and stretchability. Bendable products can keep their function when being bent. Some stiff materials made enough thin, such as copper, aluminum foil, and silicon [87], can be bent at will. Such a concept is widely utilized commercially to establish bendable conductors and transistors on polymers [88]. It has been reported that some researchers took advantage of this concept to deposit thin electronic materials on polyethylene naphthalate (PEN) to produce transistors, sensors, and solar cells, which can retain function in the process of folding, crumpling, and flexing [89]. Although these electronics can be bent, they cannot be elongated. Compared with bendable electronics, stretchable electronics have a wider application space as well as increased durability [90]. In principle, stretchable electronics can fit the skin [91] (and other tissue), or be included in different form factors, such as wearable electronic and textiles [92]. Soft electronics are the further development of stretchable electronics, which are motivated by the brain, the nervous system, and the sensory organs composed of many "circuits" and "sensors." Stretchable and soft electronics play a vital role as logic and communication devices, various sensors [93, 94], electronic skin [95], implantable devices, and prosthetics.

Nowadays, in contrast to traditional materials described in Section 7.2, gallium-based liquid metals have excellent fluidity on account of the high mobility of the individual atoms and have access to flexible electronics owing to the great combination of conductivity and deformability. There are various ways

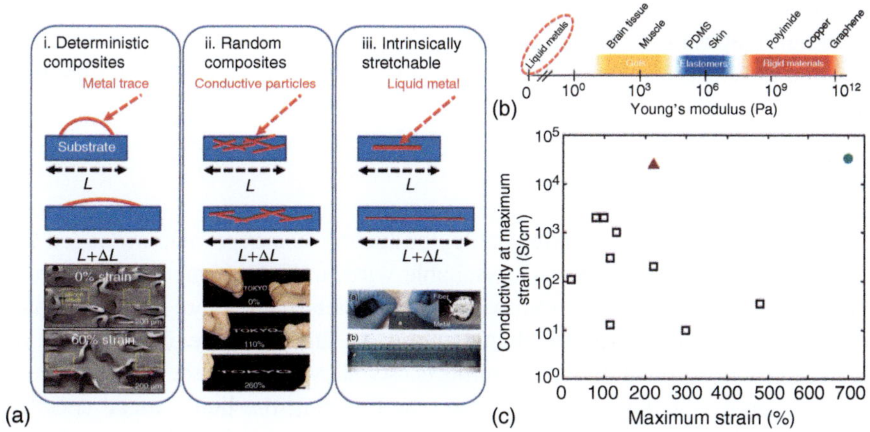

Figure 7.2 Strategies for stretchable and soft conductors. (a) (i) Solid yet thin metal traces in deterministic geometries that can flex when stretched [29]. (ii) Composites of conductive particles in elastomers. Source: Matsuhisa et al. [65]. (iii) Intrinsically stretchable materials such as liquid metal. Source: Zhu et al. [76]. (b) Liquid metals are softer (in tension) than materials found in the body and common materials used for flexible electronics. Values from [72, 73]. (c) To date, no conductive "inks" offer the combination of stretchability and conductivity as liquid metal. Source: Matsuhisa et al. [65]. The open squares represent composite approaches, whereas the triangle [75] and the circle [76] are select examples that use liquid metal embedded in elastomer [71]. Source: Dickey [71].

to fabricate stretchable and bendable conductors. As shown in Figure 7.2a, three general strategies are highlighted, which are relevant to three material geometries. Figure 7.2a-i presents that thin films with deterministic geometries allow otherwise rigid materials to elongate [4]. These traces will deform out of plane under stress and result in different shapes, such as wavy structures [30, 96], intentionally fractured thin films [97, 98], filamentary and serpentine meshes [29, 99], and fractal and kirigami structures [100–102]. This method is generally compatible with traditional microfabrication techniques to design varieties of materials, and can achieve interconnectedness and integration of all the components. However, the ultimate strain of rigid materials at failure is limited. Also, wavy and serpentine geometries, which need to take indirect paths may lead to the areal footprint and the low component densities. In addition, the resulting thin filaments will limit the electrical conduction. Figure 7.2a-ii offers another way of creating stretchable conductors, that is, conductive particles can be dispersed in elastomer to create electrical applications [103–107]. Different with bulk metals, such conductors are optically transparent. However, due to the presence of nonconductive materials in composites, these structures are not as conductive as the bulk metals. Figure 7.2a-iii shows that using intrinsically stretchable conductors is also a usual method for constructing electronics. Conductive materials that are intrinsically conductive can retain bulk conductivity during elastic deformation. Unfortunately, there are few types of materials meeting the requirements, including ionic liquids/ionogels [108], organic conductors and semiconductors, and so on, which are outside the scope of these chapters and cannot provide the combination of conductivity and fluidity as liquid metals (Figure 7.2b). As shown in Figure 7.2c, the conductivity of different materials at maximum strain is quite different, and the ideal material should be extremely stretchable and conductive, just like the upper right quadrant of that plot.

7.5.1 Stretchable Conductors

Liquid metal used to create stretchable conductors has been acknowledged for a long time. Perhaps the earliest stretchable wires used toxic mercury for strain sensing [109]. Nowadays, gallium-based liquid metals are more popular materials to be implanted into stretchable tubing to fabricate stretchable wires or self-healing wires and liquid interconnects. As shown in Figure 7.3a, Zhu and coworkers produced the conductive fibers by injecting the liquid metal (EGaIn) into hollow fibers for earphones and for a battery charger [76]. EGaIn is liquid with high conductivity and low viscosity at room temperature, and tends to form a thin oxide skin at room temperature. Researchers harness this property to form ultrastretchable conductive wires in the fibers, whose mechanical properties are also defined by the encasing material. In order to improve the mechanical property, they chose poly[styrene-*b*-(ethylene-*co*-butylene)-*b*-styrene] (SEBS) as elastomeric fibers on account of low modulus, high tensile strength, and good resistance to most chemicals. In Figure 7.3a, it can be found that such hollow fibers with gallium-based liquid metal core could be stretched to around 800% with metallic conductivity until it was broken. Park and coworkers reported high-resolution stretchable wires used as antennas by 3D printing using liquid metals. As presented in Figure 7.3b, researchers combined the advantages of EGaIn and 3D printing technique, and as-fabricated wires exhibit high-resolution according to the image of the atomic force microscopy (AFM) and magnified scanning electron microscopy (SEM) images. Compared with traditional inorganic and organic materials, of which the electrical conductivity is influenced by the defects, the structure of liquid metals is stable and less-defect, contributing to greater conductivity. Although composite materials offer different access to flexible conductors, the performances of these electronics are usually limited by the complex construction strategies and the mechanical properties of electrode and channel materials. Liquid metal has excellent mechanical property, high electrical conductivity, and great compatibility with substrates, which makes it more suitable and convenient for flexible wires.

Besides, liquid metals also have great self-healing performance, which is related to their oxidized layers. The liquid metal is extremely easy to oxidize and form a thin film on the surface, which can prevent further oxidization. When liquid metal is cut, the new oxide film forms so quickly that the liquid metal neither retreats into nor leaks from the microchannel. And, contacting two cut surfaces gives the chance of self-healing electrically just like the human body. Owing to such a characteristic, liquid metals can be used to realize self-healing conductors [80, 110, 111]. So and coworkers fabricated mechanically tunable and highly deformable conductors by injecting liquid metal into polydimethylsiloxane (PDMS) microchannels [110].

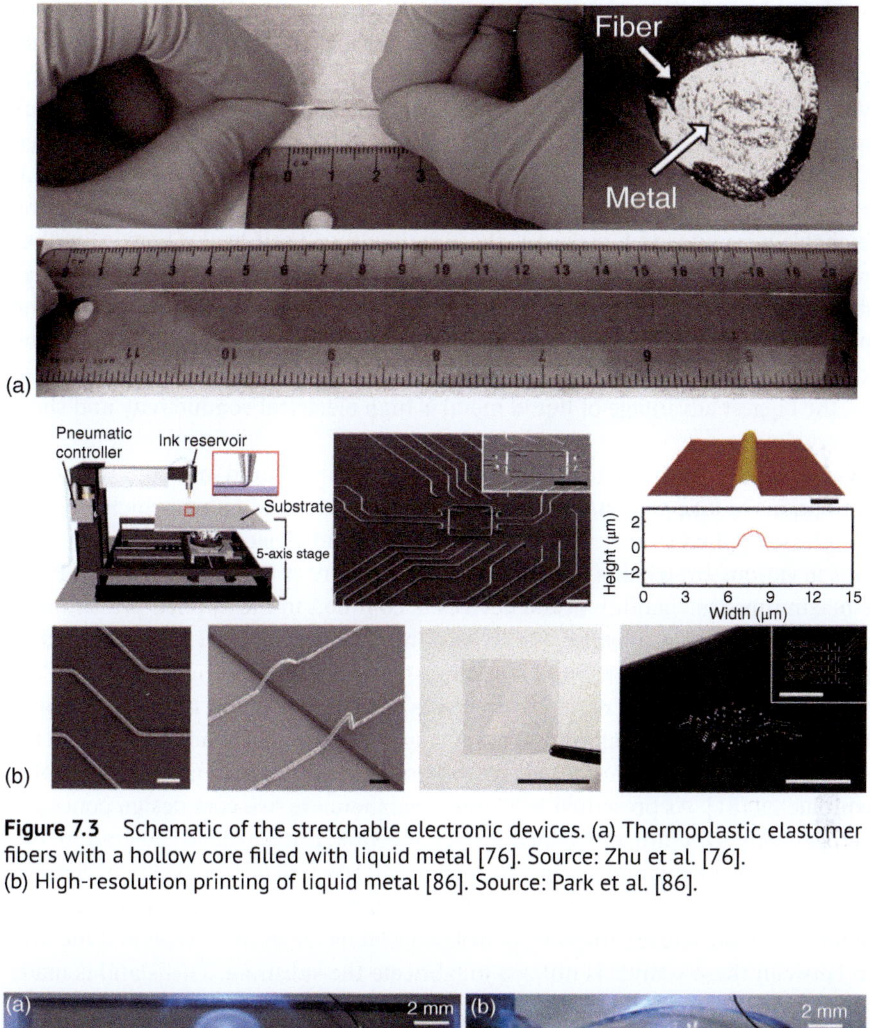

Figure 7.3 Schematic of the stretchable electronic devices. (a) Thermoplastic elastomer fibers with a hollow core filled with liquid metal [76]. Source: Zhu et al. [76]. (b) High-resolution printing of liquid metal [86]. Source: Park et al. [86].

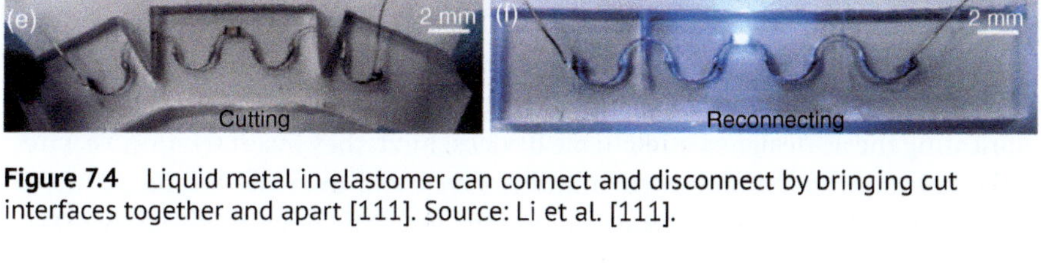

Figure 7.4 Liquid metal in elastomer can connect and disconnect by bringing cut interfaces together and apart [111]. Source: Li et al. [111].

The fluidic conductors can deform reversibly after stretching or rolling without additional help. At the bottom right of Figure 7.4b, the device was entirely cut with a razor blade with some of the PDMS complete. When the razor blade was removed, the electrical resistance of the conductor was the same as the pre-cut value, which means the device based on liquid metal can self-heal and retain conductivity. Li and coworkers achieved highly stretchable and self-healing electronics by using a galinstan-based inkjet printing system [111]. A serpentine-shaped wire integrated with an LED stably retained properties for a bend of 60°, a twist of 180°, or a strain of 60%. Even if the stretchable wire was cut by scissors, the circuit will still self-heal. There is another approach for self-healing, that is, scattering liquid metal droplets upon a polymer above a conductive Au trace. When the polymer is cut, these droplets will across the gap between the Au trace [112]. Unfortunately, this method is not available for the stretchable electronics at present. Generally, different with other self-healing materials, the biggest advantage of liquid metal is high electrical conductivity and short recovery time.

In addition to stretchable or self-healing wires, forming liquid interconnects offers more options to take good use of the characteristics of liquid metal, which can also help release applied strain. Although liquid metal and elastomeric substrates have access to various stretchable working devices, there are still some limitations. First, the maximum stretchability of the devices is confined to the applied pre-strain of the entire deformable substrate [113]. And the channel material is also limited to the repeat cycles of stretching and releasing [7]. What's more, some stretching electronics require additional complex processes for the protection of the electronics [114, 115]. Therefore, developing simpler ways to construct liquid interconnects is the focus of researchers. Yoon and coworkers reported a novel approach to solve the problems [81]. As presented in Figure 7.5a, there are five core design concepts: heterogeneous structure, embedded interconnection, double-sided integration, dry transfer of device array, and use of a Ag-nanowire (NW) sticker for electrical contact. In order to reduce the strain of the substrate brought by mechanical deformation, a heterogeneous structure composed of isolated islands for active devices and the thin film between these islands is utilized to fabricate the substrate. The island is made of a relatively stiff material (PDMS) with a high Young's modulus. And the thin film is made of a soft polymer, which is the mixture of Ecoflex and polydimethylsiloxane (MEP). Because of high electrical conductivity of the liquid metal and the small volume change during the process of stretching, EGaIn is embedded in the thin film to construct the interconnection [116]. To increase areal densities of the integrated electronics per unit area of the substrate, researchers design the double-sided active devices. Dry transfer of device array achieves the independent fabrication of the deformable substrate, which does not rely on the active sites. At last, using a Ag-NW sticker to connect the liquid–metal interconnect and the active sites do not require the metallization process for metallic electrodes [117]. Figure 7.5b is the key process of fabricating the as-designed stretchable devices. First, they insert the thin Fe wires into the holes of the Teflon-coated stainless-steel mold to create the liquid–metal channels. Second, the wires are inserted into the top mold holes followed by PDMS poured into the trench areas between the two molds. Third, they put the PDMS into a dry oven at 65 °C until it is half-curing to provide the chance of bonding with MEP film. Fourth, the MEP is poured through a space between the top and bottom PDMS islands and is annealed at 65 °C for 90 minutes after curing. Finally, the Fe wires can be extracted and leave the channels for EGaIn, which is injected with a microsyringe. In general, there are two key points when designing novel stretchable substrates including the high stretchability of the entire device array and the high areal density of the active devices, which exhibits mechanically stable device performance even if it suffers twisting, bending, and uniaxial stretching.

Apart from the methods described earlier, introducing the perforations in full 3D nanostructured forms also provides the opportunity to improve the range of design choices, as well as the stretchability of electronics. When we stretch the device, the elastic bridging elements of the structure will rotate to release applied strains. For example, compared with solid films made of PDMS, the stretchability of such 3D net-shaped nanostructures in the same material is increased by 62% and the fracture strain is also increased (more than ~200%). Park and coworkers form stretchable conductors by filling the EGaIn alloy into 3D interconnected channels in PDMS [75], which leads to highly electrical conductivity, great cyclic properties, and current-carrying capacities (Figure 7.6). As presented in Figure 7.6a, the conductivity remains almost unchanged during the stretching process even at 220% strain, which is thanks to the ability of EGaIn to change as the shape of the stretched 3D PDMS changes and remain conduction pathways through the porous network. Figure 7.6b shows that the conductivity after 10 000 cycles of 90% strain is still consistent with the original, which suggests the good cyclic properties and stability. The schematic illustration of a prototype that composed of 3D stretchable conductor and commercial LED lamps is presented in Figure 7.6c, where the LED lamps can remain bright even at ~220% strain (Figure 7.6d). It has also been proved that such a device can also survive on nonplanar substrates in Figure 7.6e, f, which is relevant to easy conformal wrapping caused by the low-bending modulus and highly elastic properties. Besides, this approach can be applied to LED arrays with various colors on hemispherical substrates (Figure 7.6g). What's more, forming liquid interconnects can penetrate the elastomer and form electrical contacts with substrates without any soldering.

In a word, stretchable conductors based on liquid metals have received much attention because of good deformability, high electrical conductivity, excellent mechanical properties, and even self-healing. As mentioned earlier, there are two approaches to stretchable conductors: the one is injecting liquid metal into a soft elastomer and the other is creating interconnections to endure significant in-plane and out-of-plane deformations. However, there are still some disadvantages restricting wider application. More clever and simple approaches to build flexible conductors need further exploration though some concepts have been proposed and applied including ingeniously designing the device structure or choosing suitable substrates. Besides, how to reduce the hurt to the channel materials after repeated stretching need be solved. That is, taking advantages of liquid metals and constructing ingenious structures help to open the door of high-performance electronics.

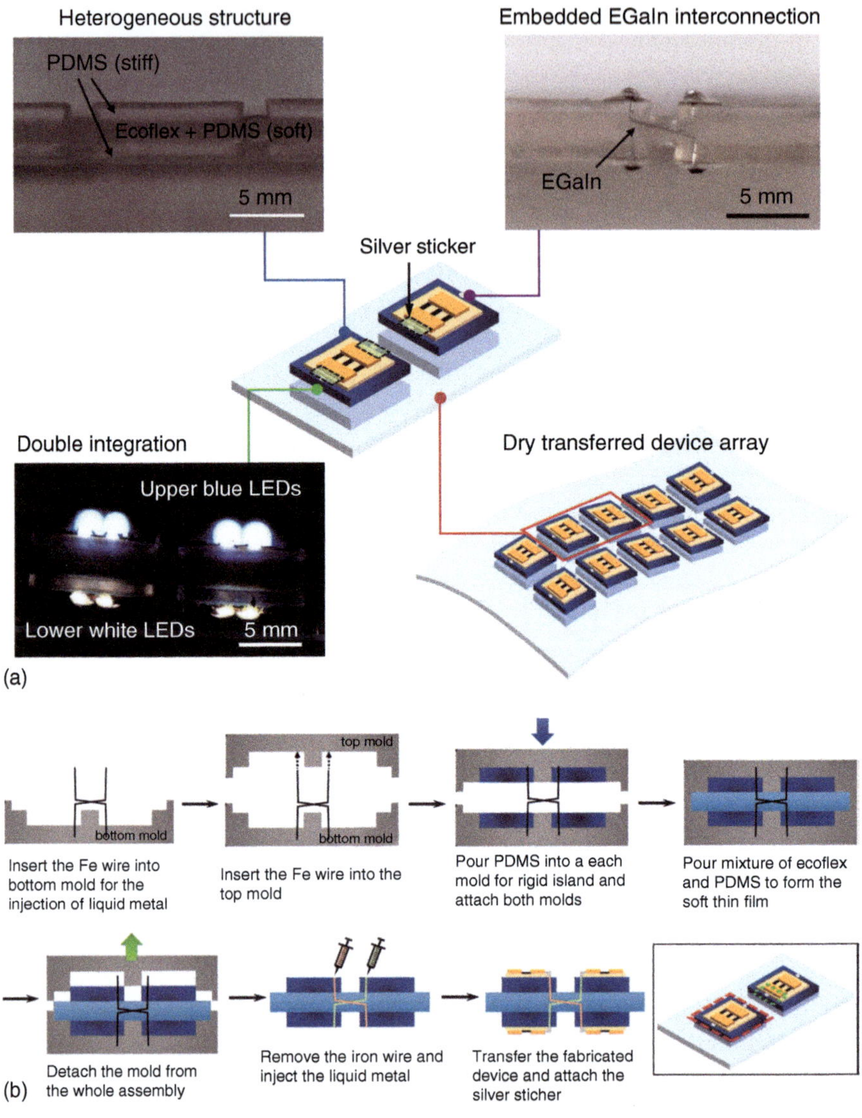

Figure 7.5 Schematic of the process of liquid interconnects. (a) The idea for suggested device on the deformable substrate. Source: Yoon et al. [81]. (b) Fabrication of the deformable substrate, embedded interconnection, and dry transfer of the active devices [81]. Reproduced with permission of Copyright © 2014, Wiley-VCH.

7.5.2 Stretchable Circuit Elements

Stretchable circuit elements based on liquid metals are also research hotspots in the field of flexible electronics, which is mainly made of four major applications: resistors, memristors, capacitors, and inductors. There are several ways to make circuit elements. One approach is patterning the liquid metal and arranging the circuit elements at the appropriate location with the help of "pick-and-place" techniques

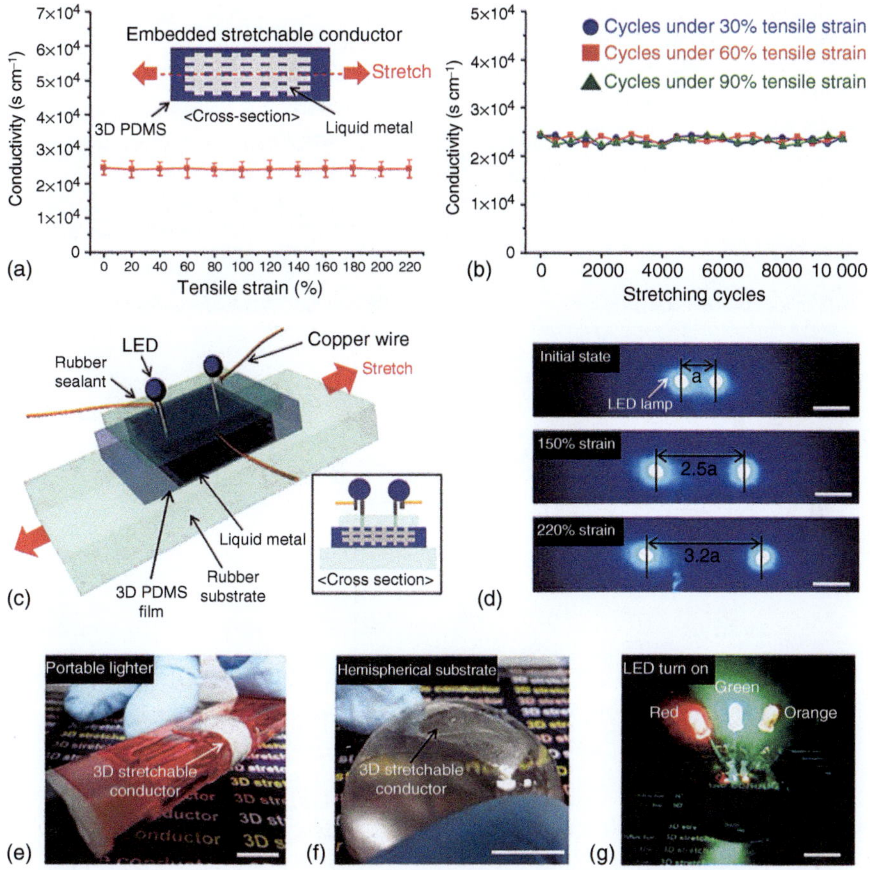

Figure 7.6 Electrical stability of 3D stretchable conductor and its performance through LED. (a) Electrical conductivity of the conductor stretched up to ~220%; (b) electrical conductivity after cyclic stretching-releasing of different strains; (c) schematic of LED devices on stretchable conductors; (d) successful operation of the device even stretched up to ~220%; (e) 3D stretchable conductor attached on the portable lighter; (f) 3D stretchable conductor attached on a hemispherical substrate; (g) realization of LED arrays (red, green, orange colors) utilizing 3D stretchable conductors on the hemispherical substrate conductor [75]. Source: Park et al. [75].

(Figure 7.7a) [82, 84]. The top two photos in Figure 7.7a are the same LED performance in different cases where the left is not applied strain and the right suffers 10% strain, suggesting that the conducted LED exhibits great electrical performance during the stretching process [84]. And, up to five LEDs could be connected in series and fully functional in parallel and serial line configurations to connect LED pads. Another approach is beginning with circuit elements on a substrate followed by patterning or injecting the metal [118]. There exist six procedures to make wireless power system. First, mold was prepared by 3D printing. Then silicone precursors were poured on the mold followed by cured silicone removed from mold in few minutes. Next, rigid components were added and bonded to partially cured silicone.

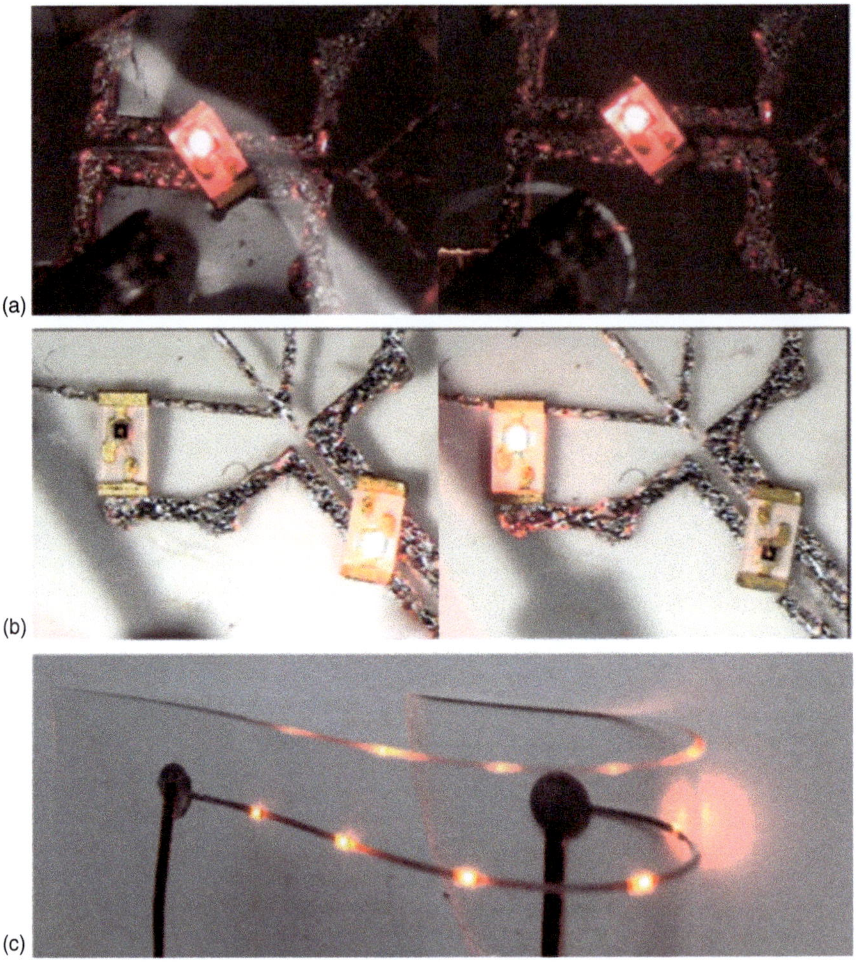

Figure 7.7 Integrated LEDs are on under different conditions [84].

Finally, galinstan was injected. The injection of liquid metal can utilize direct-write techniques [119] or inserting the legs of circuit elements into an elastomer before liquid metal patterning [120]. Actually, in contrast to some traditional contacting ways such as electrode plating, which costs relatively large and performs poorly, the most attractive point of circuit elements is that wires can penetrate the elastomer and form great electrical contacts with liquid metal without any soldering.

Circuit elements based on liquid metals can be used to form resistor-like devices [121]. Kim and coworkers presented a simple circuit with embedded resistors (Figure 7.8a, b), which consists of three series resistors with 5 μm width and 50 μm high conductor lines. Initially, they studied the direct current (DC) and alternating current (AC) electrical behaviors and found that it exhibited an ohmic behavior (Figure 7.8c) and there was no phase shift between current and voltage as expected (Figure 7.8d). As shown in Figure 7.8e, the resistance of the conductor was the

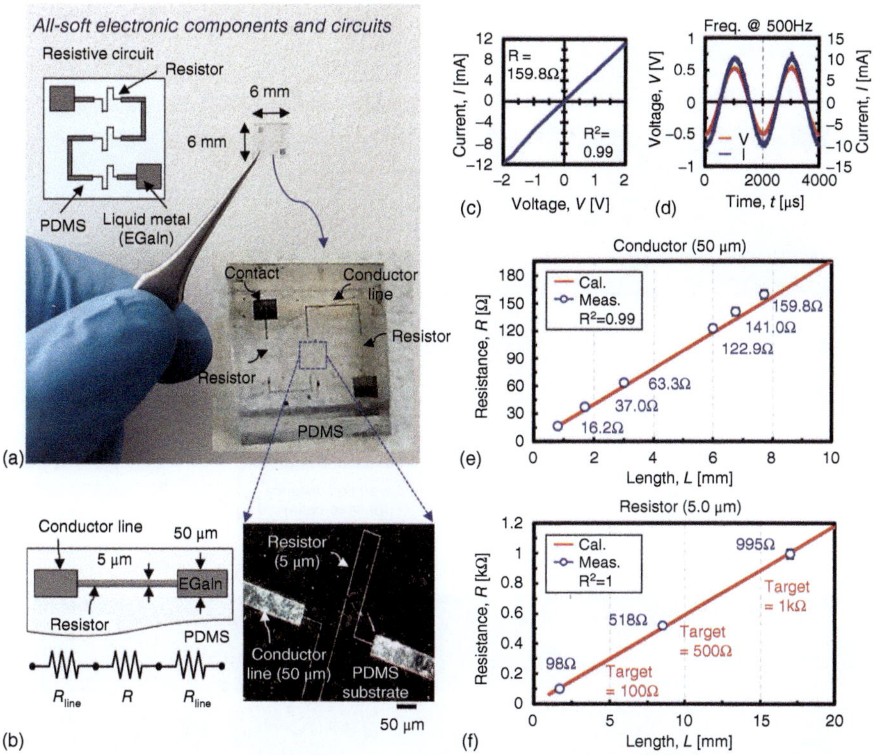

Figure 7.8 Liquid metals can form stretchable circuit elements for resistors. (a, b) Three embedded resistors (5 μm width) are connected with soft circuits; the resistors were connected with conductor lines (50 μm width); (c) DC characterization and (d) AC characterization of conductor lines; calculated and measured resistance of (e) conductor lines and (f) resistors as a function of the channel length [121]. Source: Kim et al. [121].

function of the length of channel, which was also consistent with the theoretical calculation (Figure 7.8f).

Circuit elements based on liquid metals give a novel chance to form memristor-like devices (Figure 7.9a) [122], which store binary information by either being in a resistive or conductive state. This type of switching relies on the ability to electrochemically deposit or remove the surface metal oxides. The interfacial resistance of a thick oxide layer is large, whereas that of the surface without the oxide layer is low. According to this, we can make the memristors composed of nearly 100% liquid-like materials that is fabricated by contacting EGaIn electrodes on both sides of a stack of two layers of hydrogel. Hydrogels are flexible polymer networks filled with water serving as an electrolyte for electrochemical reactions. A memristor requires asymmetry to work properly, and the two layers of hydrogel meet the requirements by creating different pH values in contact with the metal electrodes to make asymmetry. That is, the oxide film is affected by the environment pH values. When the pH is high, the oxides will remove and realize the transition from "off" (resistive) to "on" (conductive), on the contrary, it can be deposited or removed depending on the

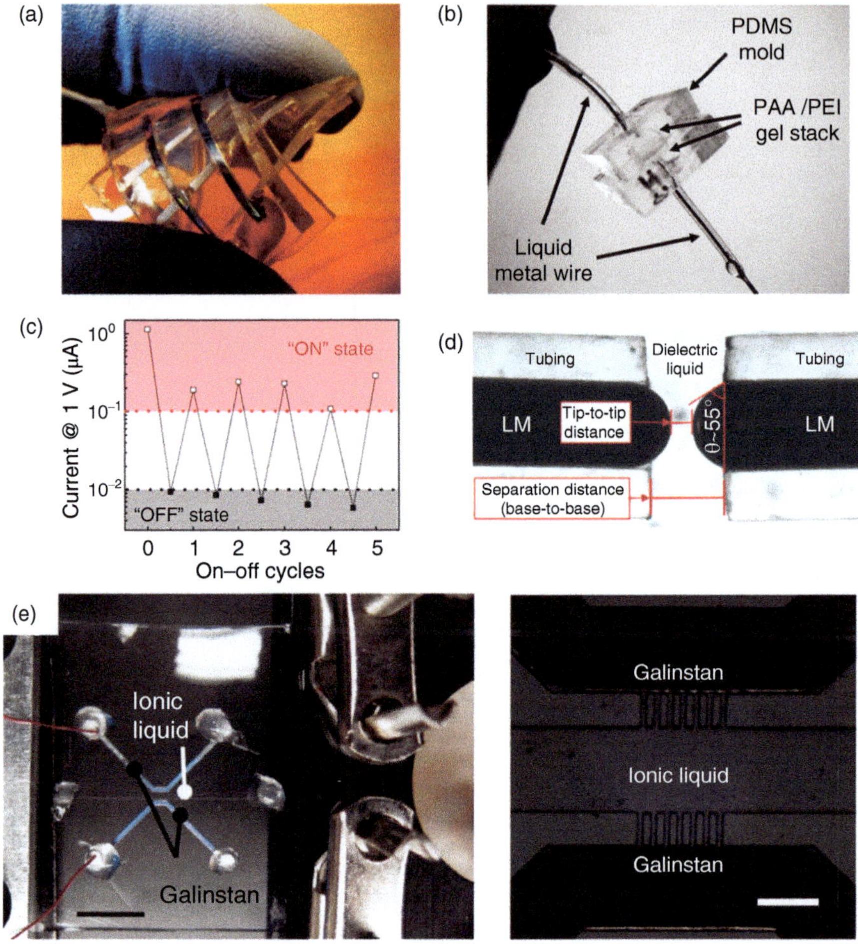

Figure 7.9 Soft, functional circuit elements using liquid metal. (a) A memristor device composed of hydrogels between liquid-metal crossbar junctions [122]. Source: Koo et al. [122]. (b) A soft diode composed of elastomer, gel, and liquid metal. Source: So et al. [123]. (c) The device can be switched from "on" (conductive) and "off" (resistive) states by varying the oxide thickness electrochemically at one interface [123]. Source: Based on So et al. [123]. (d) A capacitor composed of liquid metal plugs in an elastomeric channel separated by a dielectric liquid [124]. Source: Liu et al. [124]. (e) Liquid metal electrodes separated by ionic liquid. The conductivity of the ionic liquid varies with temperature, oxygen levels, and humidity and is therefore a soft sensor [125]. Source: Ota et al. [125].

polarity of the electrode and turn "on" and "off," as shown in Figure 7.9c. Although the switching time of the device is long compared with normal memristors, it is composed entirely of soft liquid-like materials, it utilizes ionic principles, and is hysteretic; these are all properties found in the human brain. Similar principles can be implemented to create diode devices (Figure 7.9b). Because of relatively high resistivity of the interface, the current is rectified depending on the polarity of the electric potential [123]. Encasing these hydrogel–metal devices in elastomer results in a truly soft diode device. Liquid metal can also serve as an active capacitor by injecting two slugs of the metal into microchannels separated by a gap filled with an insulating liquid (Figure 7.9d) [124]. Adjusting the distance between the slugs – for example by compressing the device – changes the capacitance. This principle can be used for sensing touch or deformation.

Besides, circuit elements can help make reconfigurable circuit wiring, which is related to the characteristics of liquid metal. Palleau and coworkers injected liquid metal into microchannel consisting of self-healing polymer to fabricate reconfigurable circuit wiring. As illustrated in Figure 7.10a-i, a self-healing stretchable wire (SHS wire), a LED, and a voltage source form an electronic circuit. When the wire is cut into two pieces (Figure 7.10a-ii), the circuit is disconnecting. Owing to the rapid formation of a thin oxide skin, the liquid metal does not exit from or retreat into the microchannels, and remains flush with the interface of the channel [110]. What's more, only the EGaIn channel needs to be aligned to restore electrical conductivity (Figure 7.10a-iii and 7.10b). Figure 7.10c presents that a SHS wire can suffer a strain up to 150%. Although the mechanical properties of repaired SHS wire are nearly identical with that of a pristine SHS wire, it failed at 60% strain as expected. Then the researchers use a straight SHS wire as part of electronic circuit in Figure 7.10d. Two-dimensional "U" shape (Figure 7.10f) or an out-of-plane 3D "bridge" (Figure 7.10g) can be created with the assistance of scissors (Figure 7.10e), where the reconfigurable circuits keep electrical continuity proven by the LED light. In general, converting planar channels into 3D channels is a great concept for designing relatively complicated functional systems, such as microvasculature networks [126], and out-of-plane microcomponents. There are other examples for circuits as conductive elements. Lazarus and coworkers used liquid galinstan in fluidic channels successfully to prove good stretchability of two multilayer inductor topologies and the double planar coil (Figure 7.11a, b) [127]. Fassler and coworkers introduced some novel inductors and capacitors by expanding the principles in soft-matter electronics. They believed that the circuit elements is in response to the planar (including x and y directions) tensile stretch and the z direction compression [128].

In summary, with the help of high conductivity and the oxide film on the surface, circuit elements based on the liquid metals have broad applications. However, there are some problems that need to be solved. For instance, as for memristors, the mechanism is dependent on the deposition and removing of the oxides, which needs some time and causes the response backward. Developing an ingenious method to improve the performance hence becomes the concerns of researchers.

7.5.3 Stretchable Antennas

Liquid metal injected into microstructure channels is able to fabricate various stretchable antennas, which is mainly studied in the electromagnetic field. Traditional conductive composites are not conductive enough to make antennas, and the intrinsic traces of thin film may lead to extra design complexity and influence

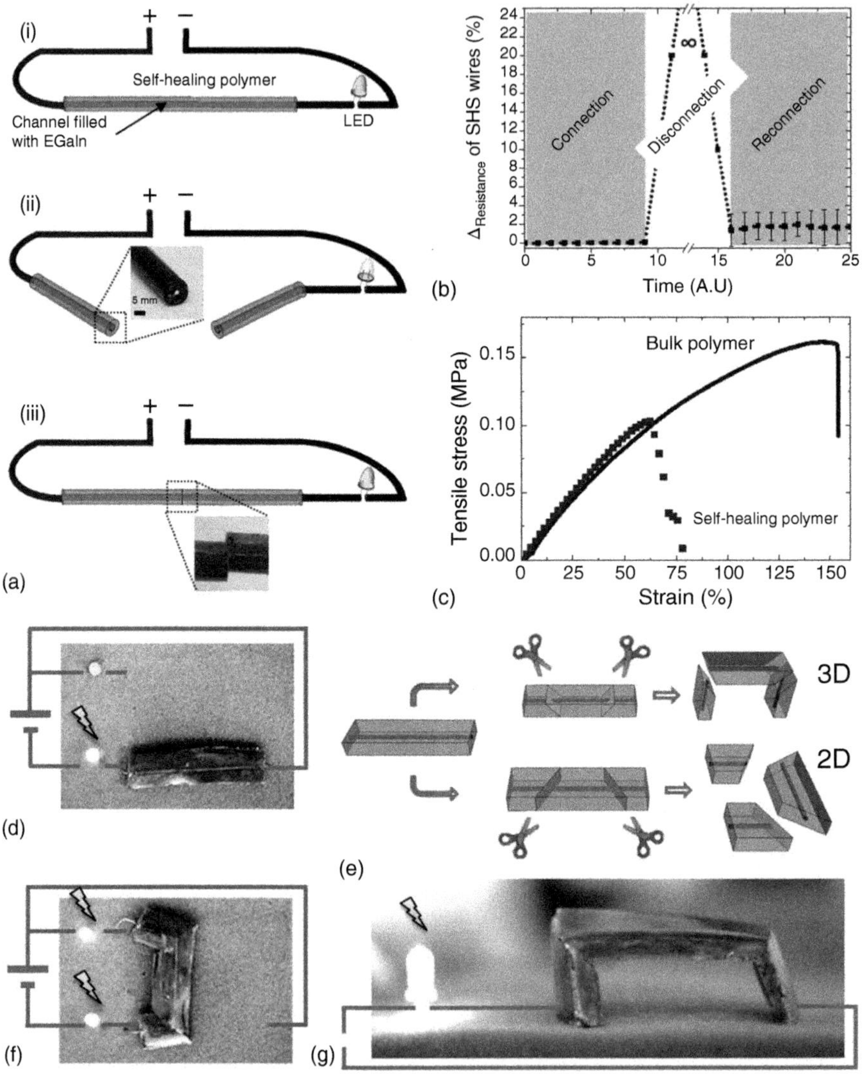

Figure 7.10 Self-healing stretchable wires for reconfigurable circuit wiring. (a) Schematics illustrating the disconnection and reconnection of a simple electronic circuit using a self-healing wire. (b) Variation of the resistance of SHS wires during connection/disconnection/reconnection experiments. The error bars indicate the variability across three sets of measurements. (c) Mechanical characterization of a SHS wire (a slab of Reverlink polymer with an embedded microchannel) before cut (black line) and after self-healing (red squares). (d) Top-down photograph of a circuit connected with a straight SHS wire. (e) Schematics explaining how cutting a straight wire can create 2D or 3D connectors. (f) Top-down photograph of the circuit in (a) after reshaping the straight wire into a "U" wire. (g) Side picture illustrating the reshaping of a SHS wire into a 3D "bridge" [80]. Source: Palleau et al. [80].

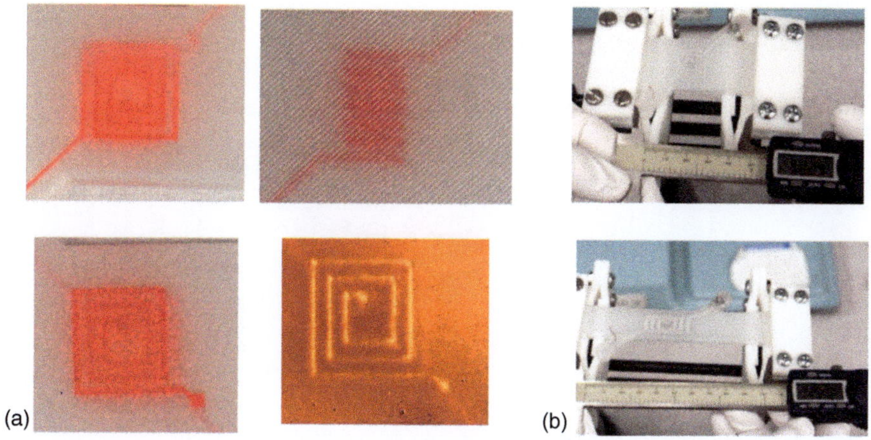

Figure 7.11 Liquid metals can form excellent stretchable electronics for circuits: (a) bottom and top side of double planar coil, bottom side of solenoid filled with water and food coloring, and top of the double planar coil after galinstan fill, respectively; (b) measured double planar coil inductor at 0% strain and 100% strain [127]. Source: Lazarus et al. [127].

antenna geometries. Therefore, liquid antennas have attracted more and more attentions [129, 130]. The spectral properties of an antenna depend on its shape, liquid metal that can physically deform and modulate its shape provides some opportunities for configurable and tunable antennas. There are various types of liquid antennas, including dipoles [110, 131], loops [132], patch antennas [133], coils and inductors [134], phase shifting coaxial transmission lines [135], monopoles [136], radiofrequency antennas [137], reconfigurable antennas [138, 139], and filters [140]. From Figure 7.12a–c, we can find that these antennas have great deformability when suffering bending and tensing. As illustrated in Figure 7.12d, antennas can change their shape by mechanical elongation, resulting in different spectral properties [131], which can be used for sensing strain wirelessly. However, the change of antenna frequency is not always ready to see, on this occasion, making vital portions of the antenna perpendicular to the strain direction is able to solve the problem [142].

Actually, compared with gallium alloys, copper is a more popular material for antennas because it has higher conductivity. For different antenna geometries, there are different choices. Such as a dipole, the radiation efficiency between Ga alloys and Cu has no significant difference, both of which are greater than 90% [110]. As for a patch antenna, the efficiency with liquid metals is lower than that of copper [141]. In addition to the antenna efficiency, other conditions including reconfigurability, stretchability, and the ability to pattern also need to be taken into consideration. That is, when fabricating liquid metal antennas, we can choose different metals and different shapes within the appropriate range on the basis of our requirements. What's more, in the field of electromagnetic devices, liquid metal can also be utilized for diffraction gratings [143], frequency selective surfaces [144], radiation shielding [145], and so on.

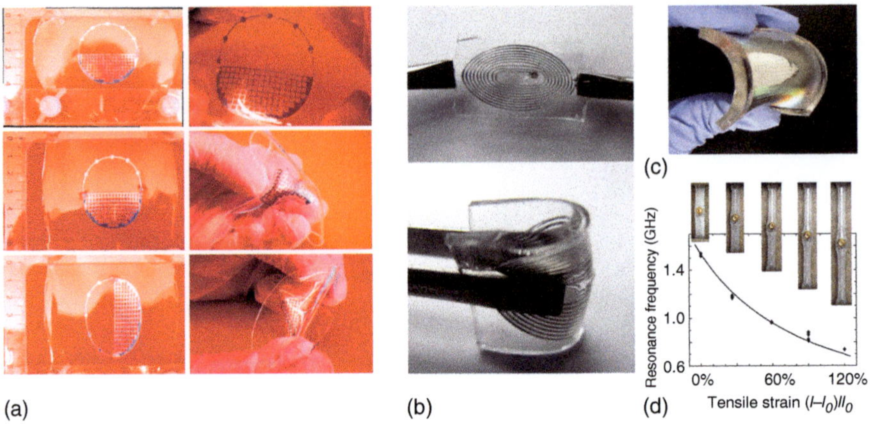

Figure 7.12 Liquid metals build excellent antennas owing to metallic conductivity and being patterned in the desired geometry. (a) Photographs of the stretchable unbalanced loop antenna: nonstretched antenna, stretched antenna with 40% *x*-axis elongation, stretched antenna with 40% *y*-axis elongation, stretched antenna with biaxial elongation, folded antenna, twisted antenna, respectively [132]. Source: Cheng et al. [132]. (b) Soft and flexible coil antenna composed of liquid metal alloy encased in biocompatible elastomeric PDMS. Metal tweezers deform the coils [134]. Source: Qusba et al. [134]. (c) Photographs of a multilayer, flexible microstrip patch antenna composed of liquid metal encased in elastomer [141]. Source: Hayes et al. [141]. (d) The resonance frequency (GHz) of the antenna as a function of the tensile strain [131]. Source: Kubo et al. [131].

7.5.4 Soft Sensors

Liquid metal-based soft wearable sensors have been widely adopted in various fields, such as healthcare and motion monitoring [54, 146], soft robots [147], and human implantable electronics [148], owing to the high stretchability and shape diversity. In contrast, the solid metallic-based sensor exhibit irreversible conductivity when recovering from large strains and display hysteresis in repeated cycles [117, 149–151]. Therefore, compared with solid metal not matching well with the natural soft tissue, liquid metal is more suitable option for the excellent compatibility with soft substrates [95]. Additionally, because of the great adaptation of liquid metal to the free movement of special parts of the human body (such as fingers), flexible sensors are also applied in soft robotics, motion tracking, and rehabilitation [146, 152–155].

In recent years, more and more researches have focused on liquid metal-based flexible sensors, because they can undergo larger strains without failure. When the sensor is deformed, the length or cross-sectional area of the liquid metal will change accordingly, so that it can be used to test stress and detect motion according to the change in resistance [156]. In addition, resistive strain sensors are one of the most widely used in flexible sensors based on liquid metal. Multiple liquid metal-based strain gauges are configured on a plane in a 2D orthogonal mesh format to become a tactile sensor [157]. Such tactile sensors can be fabricated even on an uneven surface. Yoon and coworkers reported a novel four degrees-of-freedom (4-DOF, XYZ and θ) direct writing of eutectic gallium indium (EGaIn) patterns on uneven surfaces [158].

The novel patterned direct writing technique could be achieved by virtue of 4-DOF (time delay in the feedback control) motion control of the pressurized dispenser of EGaIn and exact sensing of the distance between the dispenser tip and substrate. Since the time delay is involved in the feedback control, the writing system could realize the direct writing of different EGaIn patterns on uneven surfaces. Based on the patterning technique, a glove-type wearable device and a curved keypad can be successfully fabricated. Herein, taking the keyboard as an example, its model and the character configuration are shown in Figure 7.13a, b. The keypad was manufactured on a pre-prepared Ecoflex sheet on the saddle surface with a size of 50 mm × 50 mm. First, EGaIn is directly written to construct four 35-mm-long vertical lines with eight electrical wires configured, which are then molded with Ecoflex. Next, four 35-mm-long horizontal lines are also written by using EGaIn, connected with external eight electrical wires, and molded with Ecoflex. As a result, 16 intersecting points originating from the intersection of four horizontal lines and four vertical lines are formed and can finally achieve resistive tactual keypad of 4 × 4 2D array input keys (Figure 7.13c). When an input key (an intersecting point) is pressed, electrical resistances of relevant vertical and horizontal EGaIn lines are changed because EGaIn is partially squeezed out from the pressed area. The keypad application merely needs to figure out the number of keystrokes to switch between the assigned multiple characters, so the key should be pressed hard enough to surpass the single threshold. Then, the resistance changes of EGaIn lines are converted to voltage signals by utilizing simple voltage divider circuits (total eight same circuits were configured). When signals from both vertical and horizontal EGaIn lines surpass the threshold value, the corresponding key is determined to be one effective click. By virtue of the valid click times on the keys, particular characters could be presented in the liquid crystal display (LCD) (Figure 7.13d). Figure 7.13e lists the corresponding signals from four horizontal (X0, X1, X2, and X3) and four vertical (Y0, Y1, Y2, and Y3) lines of the keypad when the strings of "TACTILE KEYPAD!" are inputted and displayed in the LCD. Red dashed lines signify threshold values for each line. The success of the curved liquid metal keypad indicates that producing liquid metal circuits in multiple layers is possible to construct much complicated soft circuits by patterning and encapsulating repeatedly. Hence, it is possible to fabricate soft devices directly onto nonflat surfaces for the 4-DOF direct writing of EGaIn. In addition, the line width and the minimum distance between the two parallel lines can reach 70–80 μm and 200 μm, respectively, and the maximum local slope of getting reliable patterns is about 20° [158]. However, on a flat surface, the line width can reach ~2 μm [157]. More importantly, the measurable pressure range of this tactile sensor is 0–90 kPa. Obviously, its resolution is higher than the previously reported 5 kPa, which is the value required to identify a gentle touch on human skin [159]. Thus, it is very suitable for electronic skin applications.

In order to realize the practical application of the liquid metal-based sensor in the electronic skin, it can be prepared into an appropriate shape, due to inherently soft and deformable of the liquid metal and elastomer, according to the actual application. For instance, the sensor can be prepared as a wristband to sense the pulse of the human body. Lu and coworkers have realized a pulse oximeter sensor that can

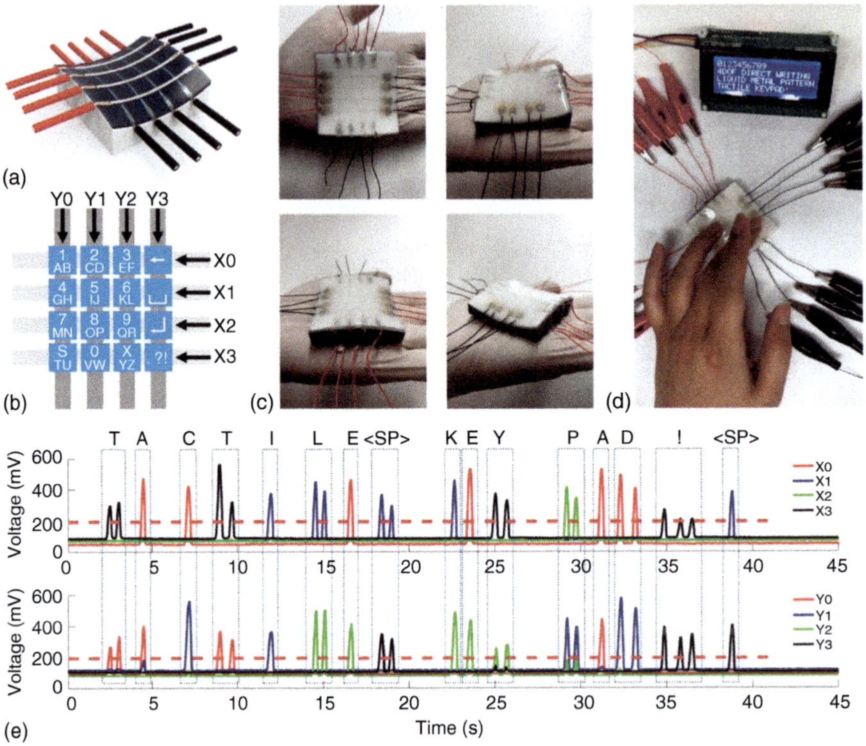

Figure 7.13 The liquid metal-based keypad and corresponding application, testing. (a) 3D schematic and (b) text configuration of the keypad. (c) Pictures of the keypad constructed on a saddle surface. (d) Operation of the keypad showing multiline text input and display. (e) Voltage signals from four horizontal (X0, X1, X2, and X3) and four vertical (Y0, Y1, Y2, and Y3) lines of the keypad when text strings of "TACTILE KEYPAD!" are sequentially input and displayed in the LCD. Dashed lines in red represent threshold values for detecting each line pressed [158]. Source: Yoon et al. [158].

be worn on the wrist (Figure 7.14d) [160]. First, they reported a unique soft-matter electronics consisting of core components (including solid-state technologies [SSTs] chip) and circuits of EGaIn traces, where the circuits were patterned with ultraviolet (UV) laser micromachining and embedded in elastomer. Then, several representative circuits together with patterned traces of EGaIn were fabricated through packaged SSTs (Figure 7.14). Figure 7.14a shows a circuit containing an array of LEDs and 331 Ω resistor chips. Figure 7.14b shows a soft-matter pulse oximeter circuit, which contains an integrated pulse oximetry unit (MAX30100; Maxim Integrated Products, Inc.) for noninvasive measurements of arterial blood oxygenation (SpO_2). Figure 7.14c shows photoplethysmogram (PPG) waveforms of the pulse oximeter at 200 Hz by using LED pulse width of 400 μs and the current of 20.8 mA. The results indicate that the sensor does have great potential in pulse measurement. In addition, Gao and coworkers also developed a pressure sensor prepared by injecting liquid gallium into the microchannel of the pre-prepared patterned PDMS and fabricated the PDMS wristband sensor [161]. When the

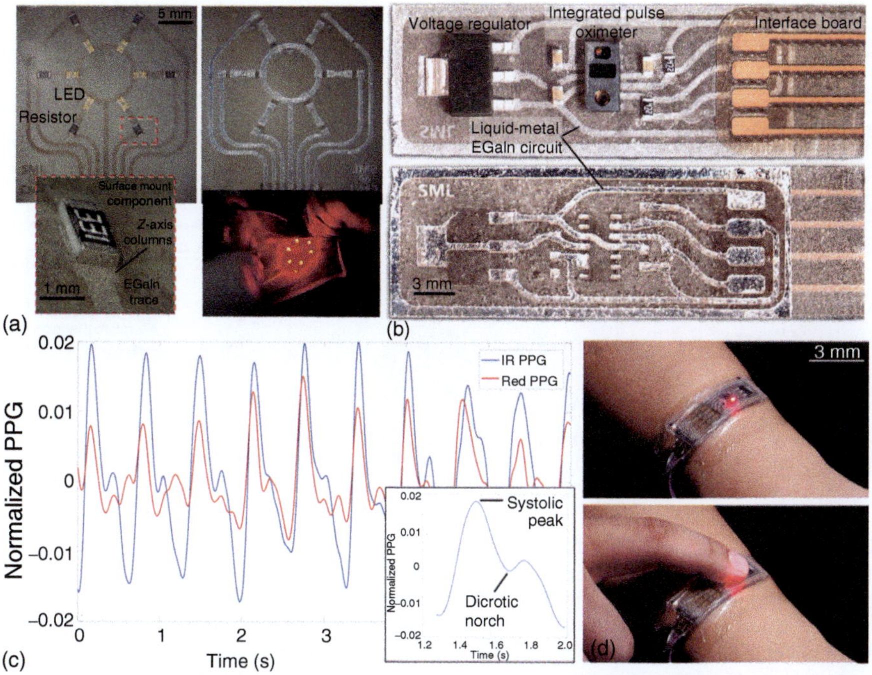

Figure 7.14 Soft-matter printed circuit board. (a) Top and bottom of a circuit containing LEDs in a circular array. Inset, bottom left: close up of interface between EGaIn trace and resistor; inset, bottom right: soft-matter LED array arising deformation during operation. (b) Top and bottom circuits of soft-matter pulse oximeter. (c) PPG waveform recorded from soft-matter pulse oximeter shown in (b); (c) inset: amplified view of single cardiac beat from the tested PPG waveform. (d) Possible effectuation of soft-matter pulse oximeter worn as wristband. PPG waveform can be obtained by placing finger on wristband while applying light pressure [160]. Source: Lu et al. [160].

embedded microfluidic diaphragm pressure sensor (Figure 7.15a) was connected to voltage input and output terminals, a similar Wheatstone bridge circuit can be formed (Figure 7.15b, c). Herein, this circuit makes full use of radial and tangential strain fields, allowing the sensor to have the advantages of high sensitivity, high resolution, low limit of detection, linearity, and temperature self-compensation. The potential working principle of the sensor is that the deformation of the elastomer will change the cross-sectional area and length of the microchannels, leading to resistance changes of the liquid metal in the microchannels. To better optimize the sensor, they calculated the optimal region on the diaphragm under different applying pressure. As shown in Figure 7.15d, e, when exerted pressure decreased on the optimal region (a 9-mm-diameter circle) marked for pressure detection, the ratio of the compression area to tension area can increase. Finally, the sensor exhibited excellent sensitivity of 0.0835 kPa^{-1} with high linear response. Additionally, the detection limit of the sensor was below 100 Pa with a sub-50 Pa resolution and a fast response time of 90 ms. Thus, it could be used for heart-rate monitoring. Subsequently, a PDMS wristband with an embedded microfluidic diaphragm

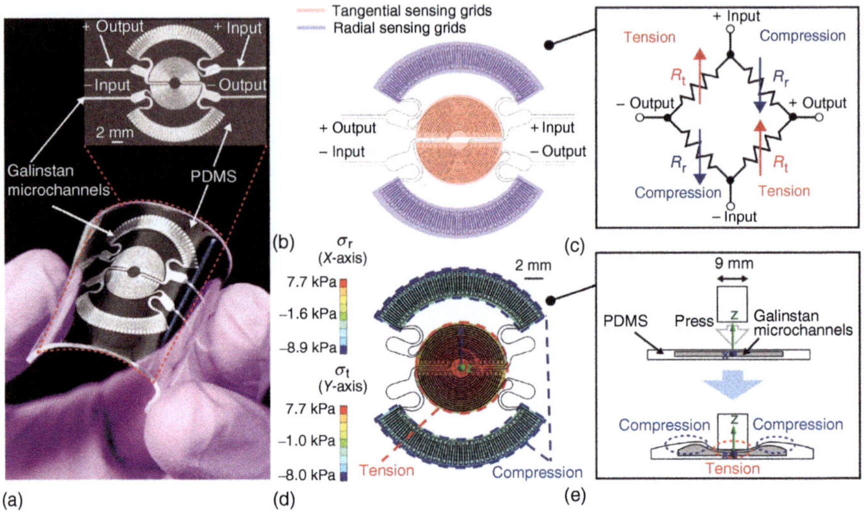

Figure 7.15 Microfluidic haptic diaphragm pressure sensor. (a) Picture of a completed microfluidic diaphragm sensor. Schematic layouts of (b) the diaphragm sensor and (c) the equivalent Wheatstone bridge circuit. (d) Simulation of the normal stress for the radial sensing grids (σ_r) (along X-axis) and the tangential sensing grids (σ_t) (along Y-axis) of the sensor under 1 kPa pressure applied over a 9 mm diameter, where positive and negative values refer to tension and compression, respectively. (e) Schematic of measurement conditions [161]. Source: Gao et al. [161].

pressure sensor (Figure 7.16a) was designed and constructed. The PDMS wristband was worn directly around the subject's (Subject A) wrist to measure dynamic pulse (Figure 7.16a–c), compared with a commercial TICKR heart-rate monitor (Wahoo fitness) during monitoring process. The pulse rate of Subject A in the three stages (initial, during exercise, and after exercise) was shown in Figure 7.16e, and the output data of both the sensors are very similar. Figure 7.16f, g shows the real-time output voltage measurements within 10 seconds periods in rest and at peak pulse rate, respectively. The differences in the magnitude of the output voltage change correspond to the differences of the strength of the heartbeats due to the exercise. All these indicated that the sensor (the PDMS wristband) can sense dynamic pulse well and give corresponding signal. In fact, it is possible to achieve the monitoring of more complex "motion" based on the sensor.

For resistive liquid metal-based soft sensors, besides bending stress, tensile stress can also change its resistance. Thus, the sensor may also adapt to extreme deformation of human joints [162]. Liu and coworkers fabricated human-friendly, tensile, and reusable electronics based on hydrogels, which can be used as a wearable force sensor [156]. The hydrogel pressure sensor, with vertically arranged helical arrangement of conductive liquid metal, exhibits excellent performances in detecting different pressure levels (i.e., 1, 10, and 100 kPa). Besides, it has been proven that the sensor has a stable response to a rapid finger press period (≈50 Hz). The strain-dependent resistance response is important for strain sensing capabilities. The sensor with high sensitivity (100 Pa) and fast response can promote

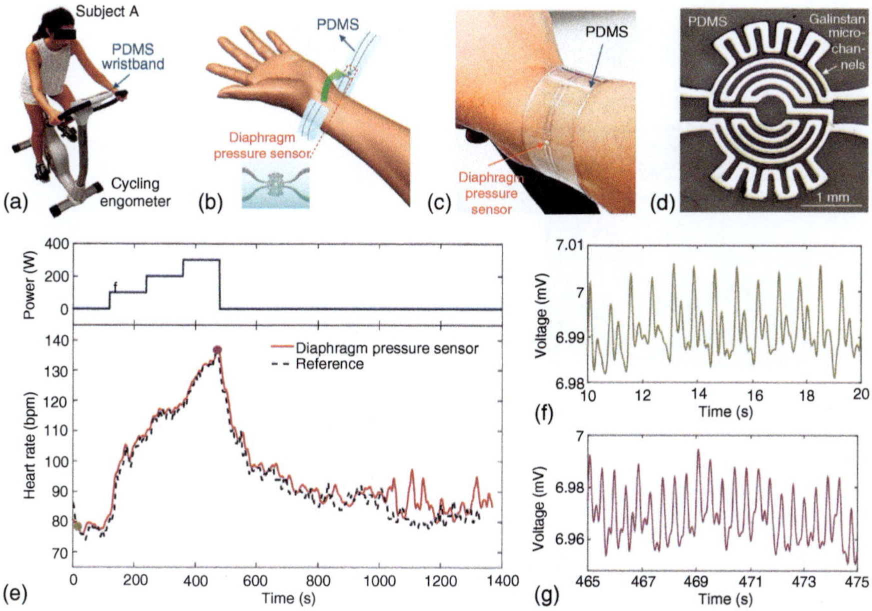

Figure 7.16 Heart-rate monitoring. (a–c) Schematic of how the PDMS sensor wristband is worn and tested. (d) Photograph of the 1.8 mm sensor used. (e) Plot of the pulse rate tested by the sensor and a commercial reference monitor during exercise. (f) Real-time testing data over a 10 seconds span taken during rest and (g) at peak pulse rate [161]. Source: Gao et al. [161].

the potential applications, such as wearable healthcare sensing systems for blood pressure, heartbeat, and pulse monitoring. For instance, a hydrogel sensor was attached to the finger joint for human motion monitoring (Figure 7.17a) [156]. The movement of the finger can be well-recognized with the help of an external-wired oscilloscope. The monitoring mechanism is quite simple, which is related to the resistance of the sensor. In detail, the sensor keeps unstretched when the finger is level, and the movement of the finger, for example, bending, will induce the stretch and the relevant resistance increase of the sensor, consequently leading to obvious waveform changes shown on the oscilloscope (Figure 7.17b). In addition, a double-finger movement monitoring system composed of two hydrogel sensors can be proposed to simultaneously track the movement of both index (the fresh hydrogel sensor) and middle fingers (the hydrogel sensor endured 12 drying–soaking cycles) (Figure 7.17c). The testing results indicated that the difference between sensing performances of the reusable and fresh hydrogel sensors can be ignored. Besides, it can also be concluded that both the fresh and reusable sensors can work collaboratively but independently for accurate and rapid detection of finger movement (with response time of 0.1 second) (Figure 7.17d) without any cross-effects. All in all, the hydrogel sensor is able to withstand large tensile strains and shows excellent electrical stability under cyclic mechanical stretch ascribed to the 3D helical layout, which allows the stress to be evenly distributed in helical structure inside hydrogel matrix even at large strain. Furthermore, on account of the swelling capacity

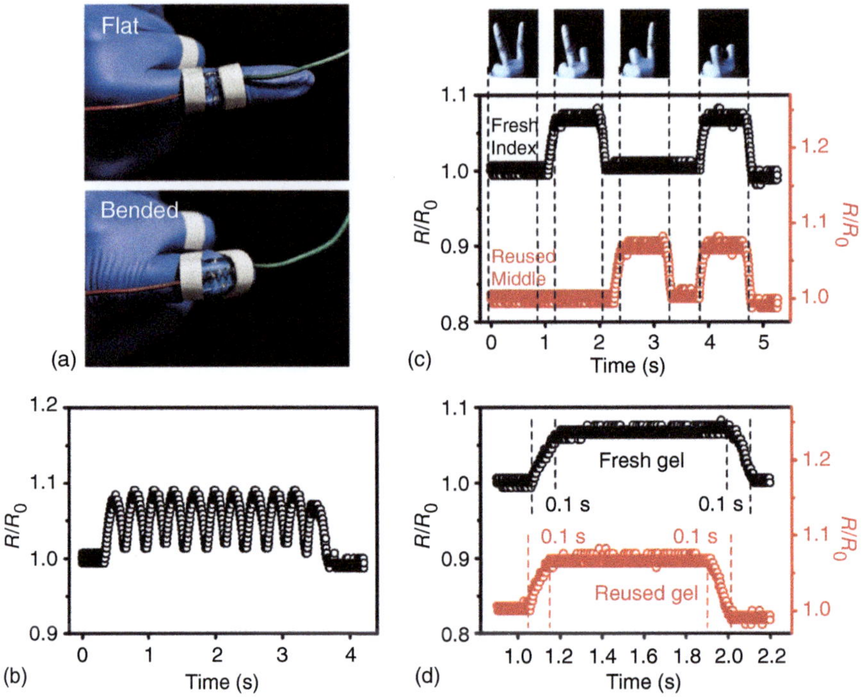

Figure 7.17 Application demonstration of a wearable sensor based on hydrogel electronics. (a) Photographs of the hydrogel strain sensor mounted to finger joint. (b) Electrical response of the hydrogel sensor when finger is stretched or bended for several times. (c) The dual sensing arise from a fresh hydrogel sensor (index finger, black) and a reusable (after 12 drying–soaking cycles) hydrogel sensor (middle finger, red) unveils that the as-prepared hydrogel electronics can simultaneously detect movements of multiple objects without cross-impact. (d) Quick response of the hydrogel sensor (≈0.1 second) [156]. Source: Liu et al. [156].

and excellent biological compatibility of hydrogel, the reusable sensor will be more popular in the market. In addition, Hirsch and coworkers designed biphasic solid–liquid thin metal films compatible with large areas and microfabrication, and applied it to stretchable electronic conductors. Such films are bilayer metallized films consisting of the sputtered alloying gold layer and thermally evaporated liquid gallium layer. In detail, the heterogeneous films contained uninterrupted networks and dispersive bulges formed by the solid intermetallic alloy $AuGa_2$ clusters and supercool liquid gallium (Figure 7.18a–c) [153]. Based on the films, several integrated circuits and devices with fine patterns on elastomeric substrates were designed and constructed (Figure 7.18d–i). Figure 7.18f–i displays the advanced applications (stretchable optoelectronic circuits, wearable sensors and heaters, and dielectric elastomer actuators) based on the solid–liquid metallization. Figure 7.18f illustrates characteristic I–V transfer curves of a soft LED microcircuit, and the result shows that the applied uniaxial tensile strain still has very small impact on the response of the diode circuit even though the strain reaches 0.8. As shown in

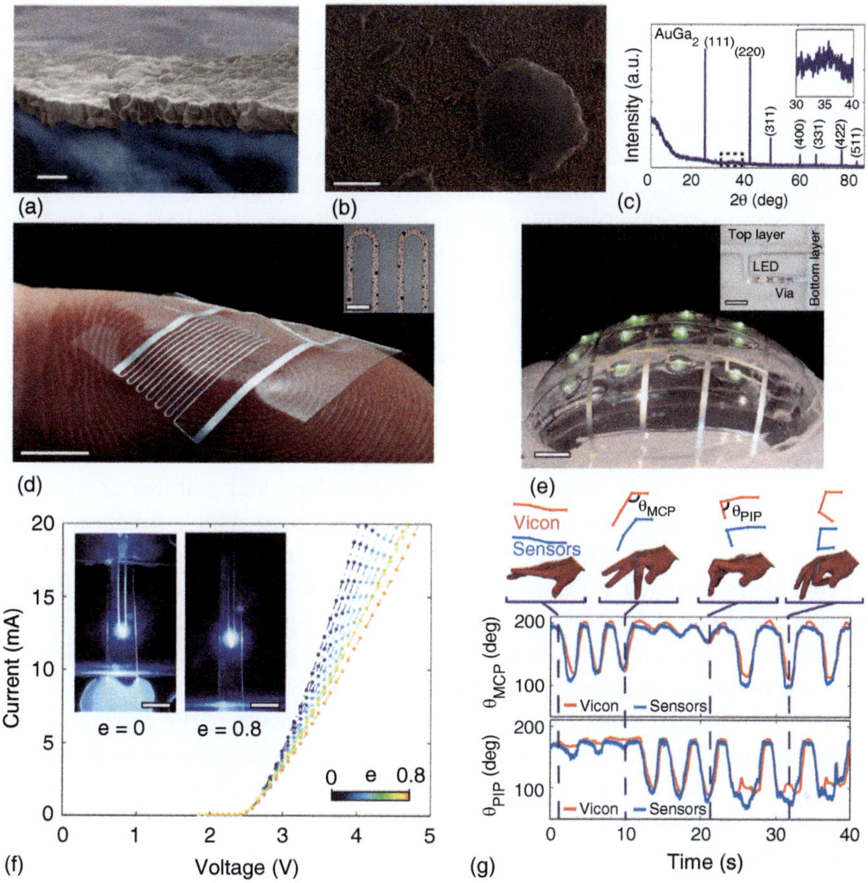

Figure 7.18 The biphasic gold–gallium thin films and the sensors based on the film. (a) False color SEM image of a slightly tilted cross section of the biphasic film on a PDMS substrate. The blue, yellow, and gray colors represent the PDMS substrate, the biphasic $AuGa_2$/Ga film, and a gallium bulge in the background, respectively. Scale bar: 500 nm. (b) False color SEM image of the surface of the biphasic film. The yellow and gray colors represent the $AuGa_2$/Ga film and the liquid gallium, respectively. Color mask was gained from backscattered electron detector (BSE) image. Scale bar: 5 μm. (c) X-ray diffraction pattern of the biphasic film deposited on a PDMS substrate clearly revealing the position of the peaks corresponding to $AuGa_2$ compound (ICDD PDF Card 01-072-5268). Inset: the increase in baseline signal around 35° is ascribed to liquid gallium. (d) Photograph of the biphasic film patterned by photolithography with critical size of 100 μm on a 40 μm thick PDMS elastomer membrane. Scale bar: 5 mm; Inset scale bar: 500 μm. (e) Stretchable multilayered matrix of green surface installed LEDs interconnected and powered through the biphasic conductors. Scale bar: 15 mm. Inset: the LEDs are interconnected with two biphasic conductor planes; scale bar: 2 mm. (f) *I*–*V* curves of an LED in PDMS interconnected with the biphasic conductors as a function of applied uniaxial strain. Scale bar: 2 mm. (g) Epidermal flexion sensor skin monitoring the position of the metacarpophalangeal (MCP) and proximal interphalangeal (PIP) joints of a finger. Joint angles as a function of time monitored by a soft biphasic gold–gallium sensor (blue) and Vicon motion capture system (orange) [153]. Source: Hirsch et al. [153].

Figure 7.18g, soft sensors attached to human hand can monitor the whole range of movement of the human finger, and the result in terms of kinematic reconstruction is almost identical to the benchmark motion tracking system. Moreover, the Joule heater prepared by using the biphasic metal film exhibited a thermal resistance coefficient of $1.43 \times 10^{-3}\ K^{-1}$ and high surface current densities (up to $200\ A\ m^{-1}$). The epidermal thermal actuators offered a well-controlled heat source on the skin, providing opportunities for minimally invasive or noninvasive biomedical mapping [163]. In summary, the feasibility of the earlier discussed devices is based on biphasic solid–liquid thin metal lines, which can be used as conductors to transmit electrical signals and can still do this even under multiaxial mechanical loading. Apart from the aforementioned metals, this method can apply to non-noble metals, as long as the as-obtained alloy films are not oxidized. In addition, the biphasic solid–liquid metal film together with arbitrarily patterning technology will provide a promising means to construct different stretchable devices.

Further, if multiple liquid metal-based strain sensors are integrated on each finger on the glove, it is possible to monitor complex hand gestures or motions [164]. By combining movement of each finger measured with a separate strain sensor, various hand gestures can be distinguished and thus detected [158]. Gao and coworkers prepared a PDMS "smart glove" with 17 embedded microfluidic diaphragm pressure sensors (7 on the palm and 10 on the fingers) can respond dynamically to various hand movements such as holding, grasping, gripping, lifting, squeezing, moving, or touching objects [161]. The PDMS "smart glove" was worn on a subject's (Subject B) hand, and then the real-time response of the corresponding thumb and index finger sensors was recorded by lightly grasping and releasing a grape (Figure 7.19c, d). The outputs voltage of both sensors increased simultaneously at the same time when Subject B touched the objects, and then returned to initial values after releasing. Figure 7.19e indicated Subject B gripping a bat, with the inset presenting a color intensity map of the corresponding voltage change ($\Delta V/V_0$) feedback from the sensors, exhibiting the distribution of contact pressure. Therefore, the pressure-sensitive glove may be used to provide tactile feedback for potential applications in augmented/virtual reality interactive environments, remote healthcare, and robotics. Additionally, compared with solid state, in case of liquid metals, the sensors can be directly printed and seamlessly integrated on gloves. Furthermore, to satisfy the demands for different applications, a patterning and extensible method needs to be developed.

Nonetheless, existing preparation technologies cannot manufacture the submicron-scale flexible sensors demanded for healthcare and medical applications involving single cells. Therefore, liquid metal patterning techniques with higher resolution need further development. Kim and coworkers proposed a nanofabrication strategy based on a hybrid method using soft lithography and electron beam lithography [165]. Additionally, the biphasic structure (a metallic adhesion layer coated with EGaIn) can ensure uniform distribution of EGaIn and avoid EGaIn aggregation or loss during separation from the carrier, thus ensuring no structural deformation of patterned EGaIn and resulting in the high resolution and integration density. Surprisingly, the method achieves patterned

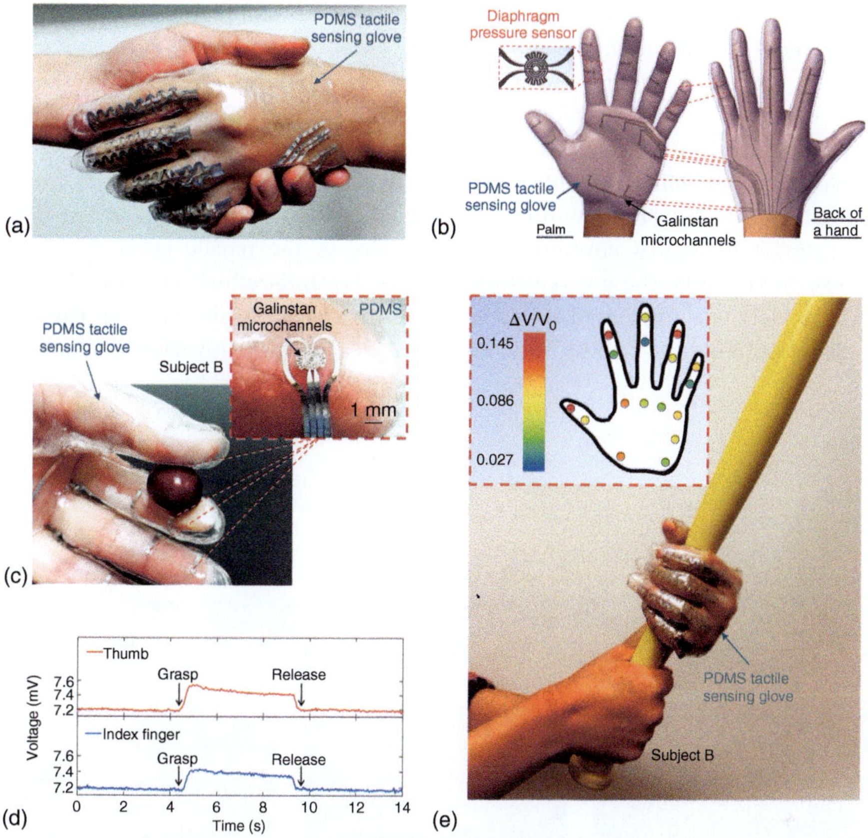

Figure 7.19 Tactile sensing glove. (a) Optical image of handshaking wearing the PDMS "smart glove." (b) Schematic of the PDMS "smart glove." (c) Photograph of the "smart glove" worn while grasping a grape. (d) Real-time response recorded from the relevant thumb and index finger sensors for lightly grasping and releasing the grape. (e) Photograph of the "smart glove" worn while gripping a bat and the relevant output voltage map across the sensors within the glove [161]. Source: Gao et al. [161].

submicron-scale EGaIn films with line width and line spacing as small as 180 nm and 1 μm, respectively. In brief, the method may furnish the currently unrivaled combination of resolution, electronic/wiring density, and electrical conductivity for soft sensors.

7.5.5 Others

Liquid metal can also be used in the field of wearable electronics to solve some problems. The primary difficulty in the practical application of such smart wearable devices is in need of a sustainable power source suitable for the biological environment. Inspired by the liquid metal-based self-powered motor [166–169] and triboelectric nanogenerator (TENG) [170–174], a flexible liquid metal-based triboelectric nanogenerator (LM-TENG) is expected to solve this problem. The development of

LM-TENG is an overwhelming trend. In this section, we will briefly introduce the LM-TENG.

Taking both conductivity and fluidity into consideration, liquid metal electrodes may provide a possible strategy for flexible power sources. For instance, Yang and coworkers proposed a LM-TENG by applying galinstan (EGaInSn) as the electrode and silicone rubber as the triboelectric and encapsulation layer (Figure 7.20a) [175]. The low Young's modulus of the liquid metal ensures that the electrode maintains continuously conductive at the tensile strain as large as 300%. Meanwhile, the surface oxide layer of EGaInSn effectively protects the EGaInSn electrode from further oxidation and permeation into silicone rubber, which can enhance the stability and lifetime of LM-TENG greatly. Due to coupling

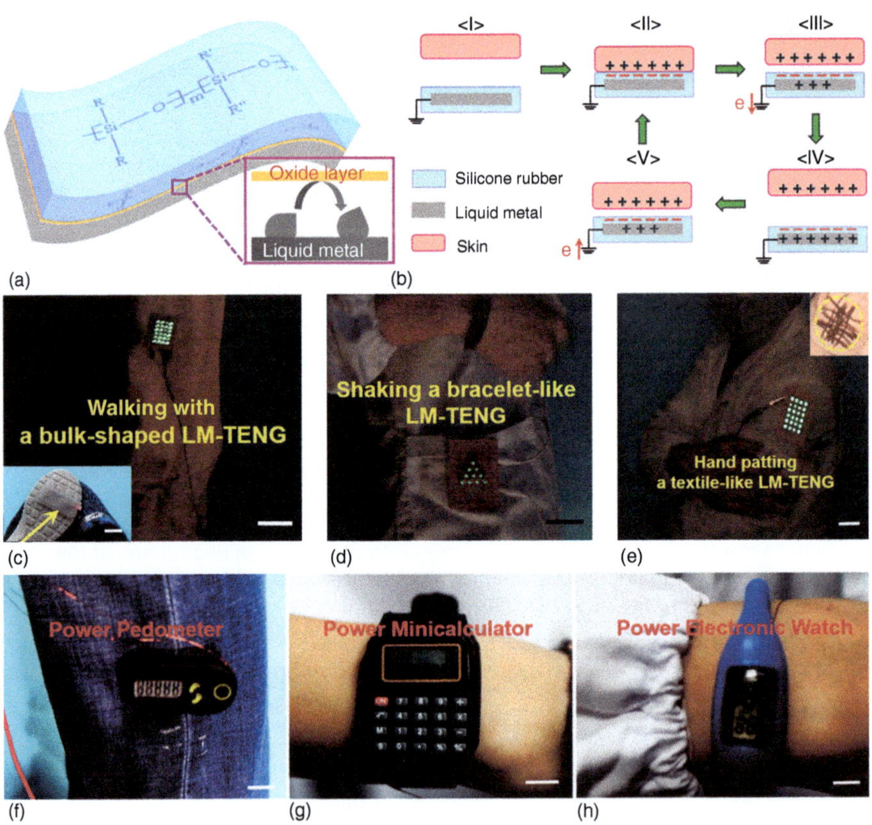

Figure 7.20 The mechanism and applications of LM-TENG (a) Structure of LM-TENG (inset indicates the protection of oxide layer). (b) Schematic of the electricity generating mechanism of the bulk-shaped LM-TENG. (c) Photograph of the LM-TENG driving LEDs by harvesting human walking mechanical energy (inset: bulk-shaped LM-TENG stuck to a shoe sole). (d) Photograph of the bracelet-like LM-TENG worn on the wrist and driving LEDs by harvesting mechanical energy from arm shaking. (e) Photographs of the LM-TENG powering LEDs by harvesting mechanical energy from hand patting (inset: textile-like LM-TENG fabricated on the clothes). Pictures of the LM-TENG sustainably powering (f) a pedometer, (g) a minicalculator, and (h) an electronic watch [175]. Source: Yang et al. [175].

effects of triboelectrification and electrostatic induction, LM-TENG will generate electricity after touching [176–178], and the underlying mechanism is illustrated in Figure 7.20b. On account of the difference in electronegativity between the skin and silicone rubber [179, 180], when they are in contact with each other, the charge transfer from the skin to silicone rubber will result in the positively charged skin and the negatively charged silicone rubber (state II). As the skin moves away, the liquid metal induced to take on positive charges can drive the free electrons to flow from the liquid metal to the ground through the external circuit, and then forming an electrical signal (state III). When the distance between the skin and silicone rubber reaches a maximum – the electrostatic balance – no signal would be detected (state IV). When the skin slowly approached the silicone rubber again, the positive charges on the liquid metal would reduce, and the free electrons would flow reversely, forming a reverse electrical signal (state V). By virtue of the circulation of the contact and separation between the LM-TENG and skin, AC can be generated to power the device. For instance, LM-TENGs in different forms (e.g., bulk-shaped, bracelet-like, and textile-like objects) can all harvest mechanical energy from walking, arm shaking, or hand patting to continuously drive wearable electronic devices (Figure 7.20c–e). To further expand the application of the LM-TENG, the self-charging power system was structured with several components including the LM-TENG, commercial capacitor, bridge rectifier, and wearable electronic device. A pedometer (Figure 7.20f), a minicalculator (Figure 7.20g), or an electronic watch (Figure 7.20h) can be powered by the wearable self-powered system harvesting enough mechanical energy from human motion. Considering the fluidity and great conductivity of liquid metal, Parida and coworkers combined liquid metal and silver flakes to develop a composite of thermoplastic elastomer to improve the performance of the soft LM-TENG. Thus, the LM-TENG (Figure 7.21a) based on the composite may also have these excellent characteristics [180]. The energy-harvesting performance of the LM-TENG was assessed by measuring the output voltage (V_{op}), short-circuit current density (I_{sc}), and charge-transfer density (σ_T) on the compressive mechanical impact. The LM-TENG generates a V_{op} of about 100 V (Figure 7.21b), σ_T of 12 nC cm^{-2} (Figure 7.21d), and I_{sc} of 4 μA cm^{-2} (Figure 7.21c). To simulate a practical working condition, they also measured the energy-harvesting performance of the LM-TENG under mechanical impact of diverse magnitudes. The result shows that the performance enhances with the increase of the magnitude of the mechanical force, which leads to the effective contact area between the latex and polyurethane acrylate (PUA) owing to the microscopic deformations at the interface. In addition, they also measured the output voltage of the LM-TENG with different triboelectric positive materials (such as latex, aluminum, polyethylene terephthalate, and PUA) (Figure 7.21e). The LM-TENG with latex displays the highest voltage, for the largest difference between the negative and positive triboelectric layers in the triboelectric series [177]. Furthermore, they also detected the performance of the LM-TENG with different load resistances from 1 kΩ to 1 GΩ. As shown in Figure 7.21f, the output voltage increases along with the increase of resistance, while the output current is opposite. Finally, they found that the performance of the device showed no obvious degradation after

50 000 consecutive cycles of mechanical impact, displaying mechanically robust and reliable performances for practical applications. Here, photos of the all-3D printable LM-TENG (consisting of PUA and (PUA + Ag + LM)) in different shapes are shown in Figure 7.21g–j. The above results indicate extremely stretchable, healable, and highly conductive properties of the generator. To demonstrate practical applications, the LM-TENG was utilized to power LEDs (Figure 7.21k–m). The results show that the LM-TENG can power LEDs even under extreme stretchability (2000%), thus confirming its ability to serve as a patternable power source for deformable electronics (Figure 7.21l). Additionally, the high conductivity of the LM-TENG at extreme stretchability (2500%) was proved by connecting it in series with a power supply and LEDs (Figure 7.21m). To sum up, the LM-TENG exhibited record-high stretchability of 2500%, fine conductivity (6250 $S\,cm^{-1}$), and healability. Additionally, after healing from an extreme mechanical damage, owing to the supramolecular hydrogen bonding of the elastomer, its energy-harvesting performance is recovered, and conductivity can be recovered to 96% of the original value. Thus, the all-3D printed LM-TENG can be an ideal energy harvester for portable self-powered electronics.

In summary, the LM-TENG can be designed into various shapes for stretching, folding, or twisting, but it can still maintain continuous conductivity because of the fluidity and deformability of liquid metals [181]. Therefore, the further optimized LM-TENG-based self-charging system could drive high-power electronic devices. Meanwhile, such LM-TENG may also provide an exciting platform for the field of intelligent soft robots in the near future.

7.6 Conclusion and Prospect

In summary, flexible liquid metal-based electronics have gradually played an important role in flexible applications. Excellent deformation-dependent conductivity and high compatibility with flexible substrates result in outstanding performances of liquid metal-based circuits and sensors, which are vital to practical instruments and biological healthcare. As typical conductors, liquid metals can be utilized to fill the as-designed microchannels to construct interconnecting wires and basic electrical elements including resistors, capacitors, inductors, and memristors, which are recognized as the cornerstones of flexible highly integrated circuits. However, compared to traditional flexible electronics, there are still some limitations existing in flexible liquid metal-based electronics. Flexible electronics based on traditional materials mainly rely on the function units of transistors and diodes, which often utilize semiconductors as channel materials. Traditional flexible electronics can easily achieve the signal amplification and rectification, as well as signal transformation. That is, traditional flexible electronics can provide diverse sensing properties for practical applications to deal with various signals. Actually, previously reported liquid metal-based electronics can only transform the mechanical motion to electrical signals, which means the limitation of functional integration. Nevertheless, flexible liquid metal-based electronics are beneficial

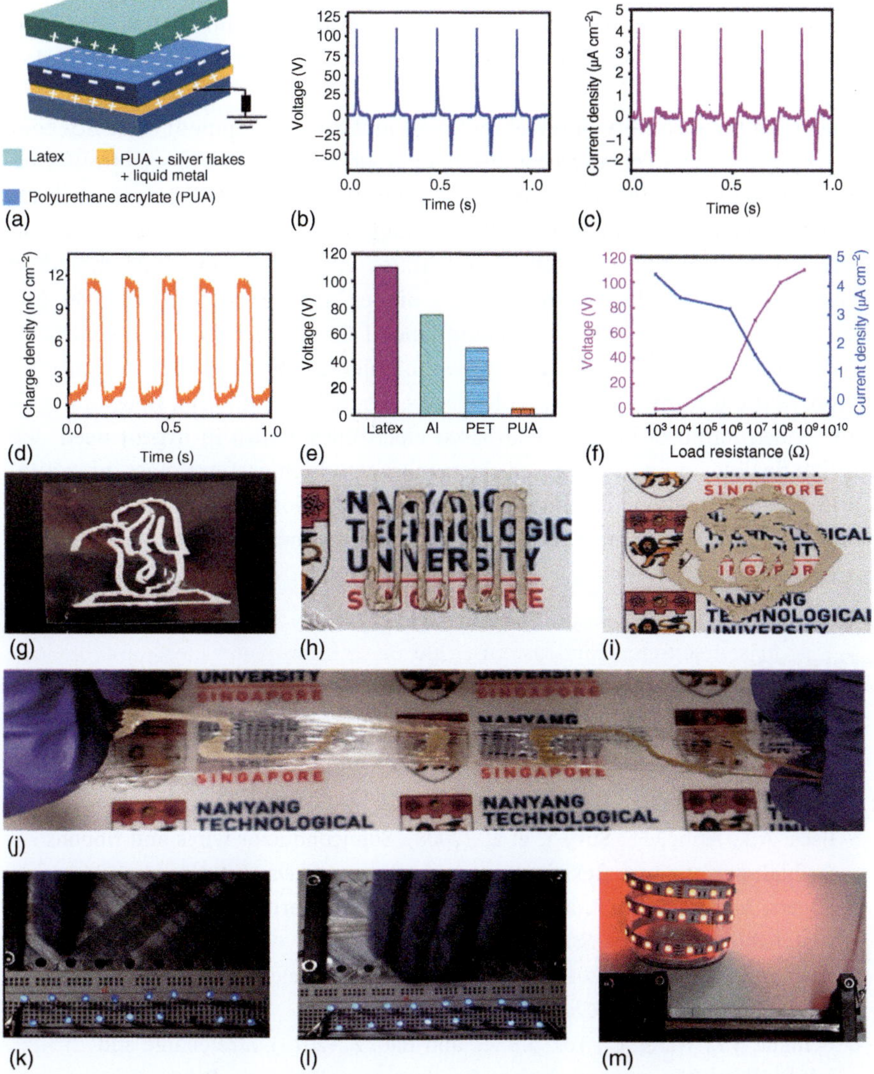

Figure 7.21 Schematic of performance and application of SH-LM-TENG. (a) Schematic of the complete stack of SH-LM-TENG, where latex and PUA are used as the positive and negative triboelectric layers, respectively, and (PUA+ silver flakes + liquid metal) is used as the current collector. (b) Output voltage (Vop), (c) current density (Is), (d) charge-transfer density (σT), and (e) output voltage of the SH-LM-TENG with PUA as the negative triboelectric material and various positive materials (latex, Al, polyethylene terephthalate (PET), and PUA). (f) Variations in output voltage, current, and power density as functions of external impedance. Photographs of the 3D printable PUA layer and elastic conductor in various shapes, (g) Merlion shape, (h) serpentine shape, (i) flower shape, (j) stretched serpentine pattern. (k) Digital photo of the SH-LM-TENG powering LEDs by finger tapping. (l) Photograph of the SH-LM-TENG under extreme deformation (2000% axial strain) and powering LEDs by palm tapping. (m) Photograph demonstrating the high conductivity of the SH-LM-TENG [180]. (a–d) Source: Parida et al. [180]. Licensed under CC BY 4.0; (g–m) Source: Parida et al. [180].

for practical applications because of the relatively simple fabrication process. In consideration of the construction strategies, substrate compatibility, excellent conductivity, and low cost, flexible liquid metal-based electronics have promising prospects for practical instruments.

Along with the discovery of novel materials and the development of device construction strategies, further explorations of high-performance flexible electronics are of urgent need for new devices, which can combine liquid metals with new semiconductors (for example, 2D materials). The combination can result in the integration of a diversity of elements on the flexible substrates with high compatibility including resistors, capacitors, inductors, memristors, diodes, and transistors. Moreover, it is necessary to improve the precision of liquid metal circuits. Printing with a micrometer resolution cannot meet the requirement of microelectronics or nanoelectronics to improve the integrating density of circuits. Therefore, exploring new strategies for the construction of liquid metal-based electronics is also in urgent need. We hope that flexible liquid metal-based electronics with high performances, high integrating density, excellent sensitivity, and various functions can be developed and utilized practically to promote the development of healthcare, communication, and electronic equipment.

References

1 Fjelstad, J. (1994). *Flexible Circuit Technology*. Seaside: *BR Publishing, Inc.*

2 Brunetti, C. and Curtis, R.W. (1948). Printed-circuit techniques. *Proc. IRE* 36: 121.

3 Baca, A.J., Ahn, J.H., Sun, Y. et al. (2008). Semiconductor wires and ribbons for high-performance flexible electronics. *Angew. Chem. Int. Ed.* 47: 5524.

4 Rogers, J.A., Someya, T., and Huang, Y. (2010). Materials and mechanics for stretchable electronics. *Science* 327: 1603.

5 Yang, Y. and Gao, W. (2019). Wearable and flexible electronics for continuous molecular monitoring. *Chem. Soc. Rev.* 48: 1465.

6 Benight, S.J., Wang, C., Tok, J.B.H., and Bao, Z. (2013). Stretchable and self-healing polymers and devices for electronic skin. *Prog. Polym. Sci.* 38: 1961.

7 Khang, D.-Y., Rogers, J.A., and Lee, H.H. (2009). Mechanical buckling: mechanics, metrology, and stretchable electronics. *Adv. Funct. Mater.* 19: 1526.

8 Park, S., Vosguerichian, M., and Bao, Z. (2013). A review of fabrication and applications of carbon nanotube film-based flexible electronics. *Nanoscale* 5: 1727.

9 Rogers, J.A., Bao, Z., Baldwin, K. et al. (2001). Paper-like electronic displays: large-area rubber-stamped plastic sheets of electronics and microencapsulated electrophoretic inks. *Proc. Natl. Acad. Sci. U S A* 98: 4835.

10 Uchikoga, S. (2011). Low-temperature polycrystalline silicon thin-film transist or technologies for system-on-glass displays. *MRS Bull.* 27: 881.

11 Duan, X., Niu, C., Sahi, V. et al. (2003). High-performance thin-film transistors using semiconductor nanowires and nanoribbons. *Nature* 425: 274.

12 McAlpine, M.C., Friedman, R.S., Jin, S. et al. (2003). High-performance nanowire electronics and photonics on glass and plastic substrates. *Nano Lett.* 3: 1531.

13 Takei, K., Takahashi, T., Ho, J.C. et al. (2010). Nanowire active-matrix circuitry for low-voltage macroscale artificial skin. *Nat. Mater.* 9: 821.

14 Sun, Y. and Rogers, J.A. (2007). Inorganic semiconductors for flexible electronics. *Adv. Mater.* 19: 1897.

15 Sun, Y., Kim, S., Adesida, I., and Rogers, J.A. (2005). Bendable GaAs metal-semiconductor field-effect transistors formed with printed GaAs wire arrays on plastic substrates. *Appl. Phys. Lett.* 87: 083501.

16 Sun, Y., Menard, E., Rogers, J.A. et al. (2006). Gigahertz operation in flexible transistors on plastic substrates. *Appl. Phys. Lett.* 88: 183509.

17 Ju, S., Facchetti, A., Xuan, Y. et al. (2007). Fabrication of fully transparent nanowire transistors for transparent and flexible electronics. *Nat. Nanotechnol.* 2: 378.

18 Wang, C., Chien, J.C., Takei, K. et al. (2012). Extremely bendable, high-performance integrated circuits using semiconducting carbon nanotube networks for digital, analog, and radio-frequency applications. *Nano Lett.* 12: 1527.

19 Kim, Y.H., Heo, J.S., Kim, T.H. et al. (2012). Flexible metal-oxide devices made by room-temperature photochemical activation of sol-gel films. *Nature* 489: 128.

20 Nomura, K., Ohta, H., Takagi, A. et al. (2004). Room-temperature fabrication of transparent flexible thin-film transistors using amorphous oxide semiconductors. *Nature* 432: 488.

21 Petti, L., Münzenrieder, N., Vogt, C. et al. (2016). Metal oxide semiconductor thin-film transistors for flexible electronics. *Appl. Phys. Rev.* 3: 021303.

22 Jeong, J.K. (2011). The status and perspectives of metal oxide thin-film transistors for active matrix flexible displays. *Semicond. Sci. Technol.* 26: 034008.

23 Zeng, M., Xiao, Y., Liu, J. et al. (2018). Exploring two-dimensional materials toward the next-generation circuits: from monomer design to assembly control. *Chem. Rev.* 118: 6236.

24 Akinwande, D., Petrone, N., and Hone, J. (2014). Two-dimensional flexible nanoelectronics. *Nat. Commun.* 5: 5678.

25 Zhao, J., Li, N., Yu, H. et al. (2017). Highly sensitive MoS_2 humidity sensors array for noncontact sensation. *Adv. Mater.* 29: 1702076.

26 Song, J., Xu, L., Li, J. et al. (2016). Monolayer and few-layer all-inorganic perovskites as a new family of two-dimensional semiconductors for printable optoelectronic devices. *Adv. Mater.* 28: 4861.

27 Das, S., Gulotty, R., Sumant, A.V., and Roelofs, A. (2014). All two-dimensional, flexible, transparent, and thinnest thin film transistor. *Nano Lett.* 14: 2861.

28 Kim, D.H., Xiao, J., Song, J. et al. (2010). Stretchable, curvilinear electronics based on inorganic materials. *Adv. Mater.* 22: 2108.

29 Kim, D.H., Song, J., Choi, W.M. et al. (2008). Materials and noncoplanar mesh designs for integrated circuits with linear elastic responses to extreme mechanical deformations. *Proc. Natl. Acad. Sci. U. S. A.* 105: 18675.

30 Gray, D.S., Tien, J., and Chen, C.S. (2004). High-conductivity elastomeric electronics. *Adv. Mater.* 16: 393.

31 Huang, K., Dinyari, R., Lanzara, G. et al. (2007). An approach to cost-effective, robust, large-area electronics using monolithic silicon. Paper presented at: 2007 IEEE International Electron Devices Meeting.

32 Qi, Y., Kim, J., Nguyen, T.D. et al. (2011). Enhanced piezoelectricity and stretchability in energy harvesting devices fabricated from buckled PZT ribbons. *Nano Lett.* 11: 1331.

33 Fan, Z., Ho, J.C., Takahashi, T. et al. (2009). Toward the development of printable nanowire electronics and sensors. *Adv. Mater.* 21: 3730.

34 Liu, Z., Xu, J., Chen, D., and Shen, G. (2015). Flexible electronics based on inorganic nanowires. *Chem. Soc. Rev.* 44: 161.

35 Long, Y.Z., Yu, M., Sun, B. et al. (2012). Recent advances in large-scale assembly of semiconducting inorganic nanowires and nanofibers for electronics, sensors and photovoltaics. *Chem. Soc. Rev.* 41: 4560.

36 Sirringhaus, H., Tessler, N., and Friend, R.H. (1998). Integrated optoelectronic devices based on conjugated polymers. *Science* 280: 1741.

37 Jones, B.A., Facchetti, A., Marks, T.J., and Wasielewski, M.R. (2007). Cyanonaphthalene diimide semiconductors for air-stable, flexible, and optically transparent n-channel field-effect transistors. *Chem. Mater.* 19: 2703.

38 Kelley, T.W., Baude, P.F., Gerlach, C. et al. (2004). Recent progress in organic electronics: materials, devices, and processes. *Chem. Mater.* 16: 4413.

39 Crone, B., Dodabalapur, A., Lin, Y.Y. et al. (2000). Large-scale complementary integrated circuits based on organic transistors. *Nature* 403: 521.

40 Sheraw, C.D., Zhou, L., Huang, J.R. et al. (2002). Organic thin-film transistor-driven polymer-dispersed liquid crystal displays on flexible polymeric substrates. *Appl. Phys. Lett.* 80: 1088.

41 Cho, J.H., Lee, J., Xia, Y. et al. (2008). Printable ion-gel gate dielectrics for low-voltage polymer thin-film transistors on plastic. *Nat. Mater.* 7: 900.

42 Briseno, A.L., Mannsfeld, S.C., Ling, M.M. et al. (2006). Patterning organic single-crystal transistor arrays. *Nature* 444: 913.

43 Usta, H., Facchetti, A., and Marks, T.J. (2011). n-Channel semiconductor materials design for organic complementary circuits. *Acc. Chem. Res.* 44: 501.

44 Beaujuge, P.M. and Frechet, J.M. (2011). Molecular design and ordering effects in pi-functional materials for transistor and solar cell applications. *J. Am. Chem. Soc.* 133: 20009.

45 Briseno, A.L., Mannsfeld, S.C., Lu, X. et al. (2007). Fabrication of field-effect transistors from hexathiapentacene single-crystal nanowires. *Nano Lett.* 7: 668.

46 Min, S.Y., Kim, T.S., Kim, B.J. et al. (2013). Large-scale organic nanowire lithography and electronics. *Nat. Commun.* 4: 1773.

47 Yin, D., Feng, J., Ma, R. et al. (2016). Efficient and mechanically robust stretchable organic light-emitting devices by a laser-programmable buckling process. *Nat. Commun.* 7: 11573.

48 Huang, J., Zhu, H., Chen, Y. et al. (2013). Highly transparent and flexible nanopaper transistors. *ACS Nano* 7: 2106.

49 Zhang, S., Hubis, E., Tomasello, G. et al. (2017). Patterning of stretchable organic electrochemical transistors. *Chem. Mater.* 29: 3126.

50 Yi, H.T., Payne, M.M., Anthony, J.E., and Podzorov, V. (2012). Ultra-flexible solution-processed organic field-effect transistors. *Nat. Commun.* 3: 1259.

51 Forrest, S.R. (2004). The path to ubiquitous and low-cost organic electronic appliances on plastic. *Nature* 428: 911.

52 Lewis, J. (2006). Material challenge for flexible organic devices. *Mater. Today* 9: 38.

53 Chen, D. and Pei, Q. (2017). Electronic muscles and skins: a review of soft sensors and actuators. *Chem. Rev.* 117: 11239.

54 Zang, Y., Zhang, F., Di, C.-a., and Zhu, D. (2015). Advances of flexible pressure sensors toward artificial intelligence and health care applications. *Mater. Horiz.* 2: 140.

55 Wang, L., Yoon, M.H., Lu, G. et al. (2006). High-performance transparent inorganic-organic hybrid thin-film n-type transistors. *Nat. Mater.* 5: 893.

56 Ju, S., Li, J., Liu, J. et al. (2008). Transparent active matrix organic light-emitting diode displays driven by nanowire transistor circuitry. *Nano Lett.* 8: 997.

57 Lau, P.H., Takei, K., Wang, C. et al. (2013). Fully printed, high performance carbon nanotube thin-film transistors on flexible substrates. *Nano Lett.* 13: 3864.

58 Wan, C., Gu, X., Dang, F. et al. (2015). Flexible n-type thermoelectric materials by organic intercalation of layered transition metal dichalcogenide TiS_2. *Nat. Mater.* 14: 622.

59 Guillén, C. and Herrero, J. (2011). TCO/Metal/TCO structures for energy and flexible electronics. *Thin Solid Films* 520 (1): 1–17.

60 Siegel, A.C., Phillips, S.T., Dickey, M.D. et al. (2010). Foldable printed circuit boards on paper substrates. *Adv. Funct. Mater.* 20: 28.

61 Kim, D.H., Ahn, J.H., Choi, W.M. et al. (2008). Stretchable and foldable silicon integrated circuits. *Science* 320: 507.

62 Kim, H.J., Sim, K., Thukral, A., and Yu, C. (2017). Rubbery electronics and sensors from intrinsically stretchable elastomeric composites of semiconductors and conductors. *Sci. Adv.* 3: e1701114.

63 White, M.S., Kaltenbrunner, M., Głowacki, E.D. et al. (2013). Ultrathin, highly flexible and stretchable PLEDs. *Nat. Photonics* 7: 811.

64 Kang, D., Pikhitsa, P.V., Choi, Y.W. et al. (2014). Ultrasensitive mechanical crack-based sensor inspired by the spider sensory system. *Nature* 516: 222.

65 Matsuhisa, N., Kaltenbrunner, M., Yokota, T. et al. (2015). Printable elastic conductors with a high conductivity for electronic textile applications. *Nat. Commun.* 6: 7461.

66 Gong, S., Zhang, B., Zhang, J. et al. (2020). Biocompatible poly(lactic acid)-based hybrid piezoelectric and electret nanogenerator for electronic skin applications. *Adv. Funct. Mater.* 30: 1908724.

67 Cao, Q. and Rogers, J.A. (2009). Ultrathin films of single-walled carbon nanotubes for electronics and sensors: a review of fundamental and applied aspects. *Adv. Mater.* 21: 29.

68 Da, Y., Liu, J., Zhou, L. et al. (2019). Engineering 2D architectures toward high-performance micro-supercapacitors. *Adv. Mater.* 31: 1802793.

69 Chen, Y., Liu, J., Liu, K. et al. (2019). GaN in different dimensionalities: properties, synthesis, and applications. *Mater. Sci. Eng. R* 138: 60.

70 Liu, J., Cao, H., Jiang, B. et al. (2016). Newborn 2D materials for flexible energy conversion and storage. *Sci. China Mater.* 59: 459.

71 Dickey, M.D. (2017). Stretchable and soft electronics using liquid metals. *Adv. Mater.* 29: 1606425.

72 Wagner, S. and Bauer, S. (2012). Materials for stretchable electronics. *MRS Bull.* 37: 207.

73 Rus, D. and Tolley, M.T. (2015). Design, fabrication and control of soft robots. *Nature* 521: 467.

74 Li, T., Huang, Z., Suo, Z. et al. (2004). Stretchability of thin metal films on elastomer substrates. *Appl. Phys. Lett.* 85: 3435.

75 Park, J., Wang, S., Li, M. et al. (2012). Three-dimensional nanonetworks for giant stretchability in dielectrics and conductors. *Nat. Commun.* 3: 916.

76 Zhu, S., So, J.-H., Mays, R. et al. (2013). Ultrastretchable fibers with metallic conductivity using a liquid metal alloy core. *Adv. Funct. Mater.* 23: 2308.

77 Markvicka, E.J., Bartlett, M.D., Huang, X., and Majidi, C. (2018). An autonomously electrically self-healing liquid metal-elastomer composite for robust soft-matter robotics and electronics. *Nat. Mater.* 17: 618.

78 Zhang, J., Zhu, X., Zeng, M., and Fu, L. (2020). Magnetically controlled on-demand switching of batteries. *Adv. Sci.* 7: 2000184.

79 Hirsch, A., Dejace, L., Michaud, H.O., and Lacour, S.P. (2019). Harnessing the rheological properties of liquid metals to shape soft electronic conductors for wearable applications. *Acc. Chem. Res.* 52: 534.

80 Palleau, E., Reece, S., Desai, S.C. et al. (2013). Self-healing stretchable wires for reconfigurable circuit wiring and 3D microfluidics. *Adv. Mater.* 25: 1589.

81 Yoon, J., Hong, S.Y., Lim, Y. et al. (2014). Design and fabrication of novel stretchable device arrays on a deformable polymer substrate with embedded liquid-metal interconnections. *Adv. Mater.* 26: 6580.

82 Kim, H.-J., Son, C., and Ziaie, B. (2008). A multiaxial stretchable interconnect using liquid-alloy-filled elastomeric microchannels. *Appl. Phys. Lett.* 92: 011904.

83 So, J.H. and Dickey, M.D. (2011). Inherently aligned microfluidic electrodes composed of liquid metal. *Lab Chip* 11: 905.

84 Jeong, S.H., Hagman, A., Hjort, K. et al. (2012). Liquid alloy printing of microfluidic stretchable electronics. *Lab Chip* 12: 4657.

85 Wang, Q., Yu, Y., Yang, J., and Liu, J. (2015). Fast fabrication of flexible functional circuits based on liquid metal dual-trans printing. *Adv. Mater.* 27: 7109.

86 Park, Y.G., An, H.S., Kim, J.Y., and Park, J.U. (2019). High-resolution, reconfigurable printing of liquid metals with three-dimensional structures. *Sci. Adv.* 5: eaaw2844.

87 Crabb, R.L. and Treble, F.C. (1967). Thin silicon solar cells for large flexible arrays. *Nature* 213: 1223.

88 Brody, T.P. (1984). The thin film transistor – a late flowering bloom. *IEEE Trans. Electron Devices* 31: 1614.

89 Drack, M., Graz, I., Sekitani, T. et al. (2015). An imperceptible plastic electronic wrap. *Adv. Mater.* 27: 34.

90 Nassar, J.M., Rojas, J.P., Hussain, A.M., and Hussain, M.M. (2016). From stretchable to reconfigurable inorganic electronics. *Extreme Mech. Lett.* 9: 245.

91 Kim, D.H., Lu, N., Ma, R. et al. (2011). Epidermal electronics. *Science* 333: 838.

92 Castano, L.M. and Flatau, A.B. (2014). Smart fabric sensors and e-textile technologies: a review. *Smart Mater. Struct.* 23: 053001.

93 Lu, N. and Kim, D.-H. (2013). Flexible and stretchable electronics paving the way for soft robotics. *Soft Rob.* 1: 53.

94 Majidi, C. (2013). Soft robotics: a perspective – current trends and prospects for the future. *Soft Rob.* 1: 5.

95 Hammock, M.L., Chortos, A., Tee, B.C. et al. (2013). 25th anniversary article: the evolution of electronic skin (e-skin): a brief history, design considerations, and recent progress. *Adv. Mater.* 25: 5997.

96 Sun, Y., Choi, W.M., Jiang, H. et al. (2006). Controlled buckling of semiconductor nanoribbons for stretchable electronics. *Nat. Nanotechnol.* 1: 201.

97 Lacour, S.P., Wagner, S., Huang, Z., and Suo, Z. (2003). Stretchable gold conductors on elastomeric substrates. *Appl. Phys. Lett.* 82: 2404.

98 Lacour, S.P., Jones, J., Wagner, S. et al. (2005). Stretchable interconnects for elastic electronic surfaces. *Proc. IEEE* 93: 1459.

99 Guo, C.F., Sun, T., Liu, Q. et al. (2014). Highly stretchable and transparent nanomesh electrodes made by grain boundary lithography. *Nat. Commun.* 5: 3121.

100 Blees, M.K., Barnard, A.W., Rose, P.A. et al. (2015). Graphene Kirigami. *Nature* 524: 204.

101 Fan, J.A., Yeo, W.-H., Su, Y. et al. (2014). Fractal design concepts for stretchable electronics. *Nat. Commun.* 5: 3266.

102 Shyu, T.C., Damasceno, P.F., Dodd, P.M. et al. (2015). A Kirigami approach to engineering elasticity in nanocomposites through patterned defects. *Nat. Mater.* 14: 785.

103 Lee, P., Lee, J., Lee, H. et al. (2012). Highly stretchable and highly conductive metal electrode by very long metal nanowire percolation network. *Adv. Mater.* 24: 3326.

104 Kim, Y., Zhu, J., Yeom, B. et al. (2013). Stretchable nanoparticle conductors with self-organized conductive pathways. *Nature* 500: 59.

105 Sekitani, T., Nakajima, H., Maeda, H. et al. (2009). Stretchable active-matrix organic light-emitting diode display using printable elastic conductors. *Nat. Mater.* 8: 494.

106 Park, M., Im, J., Shin, M. et al. (2012). Highly stretchable electric circuits from a composite material of silver nanoparticles and elastomeric fibres. *Nat. Nanotechnol.* 7: 803.

107 Sekitani, T., Noguchi, Y., Hata, K. et al. (2008). A rubberlike stretchable active matrix using elastic conductors. *Science* 321: 1468.

108 Le Bideau, J., Viau, L., and Vioux, A. (2011). Ionogels, ionic liquid based hybrid materials. *Chem. Soc. Rev.* 40: 907.

109 Whitney, R.J. (1953). The measurement of volume changes in human limbs. *J. Physiol.* 121: 1.

110 So, J.-H., Thelen, J., Qusba, A. et al. (2009). Reversibly deformable and mechanically tunable fluidic antennas. *Adv. Funct. Mater.* 19: 3632.

111 Li, G., Wu, X., and Lee, D.W. (2016). A Galinstan-based inkjet printing system for highly stretchable electronics with self-healing capability. *Lab Chip* 16: 1366.

112 Blaiszik, B.J., Kramer, S.L., Grady, M.E. et al. (2012). Autonomic restoration of electrical conductivity. *Adv. Mater.* 24: 398.

113 Feng, X., Yang, B.D., Liu, Y. et al. (2011). Stretchable ferroelectric nanoribbons with wavy configurations on elastomeric substrates. *ACS Nano* 5: 3326.

114 Xu, S., Zhang, Y., Cho, J. et al. (2013). Stretchable batteries with self-similar serpentine interconnects and integrated wireless recharging systems. *Nat. Commun.* 4: 1543.

115 Shin, G., Yoon, C.H., Bae, M.Y. et al. (2011). Stretchable field-effect-transistor array of suspended SnO_2 nanowires. *Small* 7: 1181.

116 Dickey, M.D., Chiechi, R.C., Larsen, R.J. et al. (2008). Eutectic gallium-indium (EGaIn): a liquid metal alloy for the formation of stable structures in microchannels at room temperature. *Adv. Funct. Mater.* 18: 1097.

117 Xu, F. and Zhu, Y. (2012). Highly conductive and stretchable silver nanowire conductors. *Adv. Mater.* 24: 5117.

118 Lazarus, N., Meyer, C.D., and Turner, W.J. (2015). A microfluidic wireless power system. *RSC Adv.* 5: 78695.

119 Ladd, C., So, J.H., Muth, J., and Dickey, M.D. (2013). 3D printing of free standing liquid metal microstructures. *Adv. Mater.* 25: 5081.

120 Lin, Y., Cooper, C., Wang, M. et al. (2015). Handwritten, soft circuit boards and antennas using liquid metal nanoparticles. *Small* 11: 6397.

121 Kim, M.-g., Alrowais, H., Pavlidis, S., and Brand, O. (2017). Size-scalable and high-density liquid-metal-based soft electronic passive components and circuits using soft lithography. *Adv. Funct. Mater.* 27: 1604466.

122 Koo, H.J., So, J.H., Dickey, M.D., and Velev, O.D. (2011). Towards all-soft matter circuits: prototypes of quasi-liquid devices with memristor characteristics. *Adv. Mater.* 23: 3559.

123 So, J.-H., Koo, H.-J., Dickey, M.D., and Velev, O.D. (2012). Ionic current rectification in soft-matter diodes with liquid-metal electrodes. *Adv. Funct. Mater.* 22: 625.

124 Liu, S., Sun, X., Hildreth, O.J., and Rykaczewski, K. (2015). Design and characterization of a single channel two-liquid capacitor and its application to hyperelastic strain sensing. *Lab Chip* 15: 1376.

125 Ota, H., Chen, K., Lin, Y. et al. (2014). Highly deformable liquid-state heterojunction sensors. *Nat. Commun.* 5: 5032.

126 Wu, W., DeConinck, A., and Lewis, J.A. (2011). Omnidirectional printing of 3D microvascular networks. *Adv. Mater.* 23: H178.

127 Lazarus, N., Meyer, C.D., Bedair, S.S. et al. (2014). Multilayer liquid metal stretchable inductors. *Smart Mater. Struct.* 23: 085036.

128 Fassler, A. and Majidi, C. (2013). Soft-matter capacitors and inductors for hyperelastic strain sensing and stretchable electronics. *Smart Mater. Struct.* 22: 055023.

129 Entesari, K. and Saghati, A.P. (2016). Fluidics in microwave components. *IEEE Microw. Mag.* 17: 50.

130 Cheng, S. and Wu, Z. (2012). Microfluidic electronics. *Lab Chip* 12: 2782.

131 Kubo, M., Li, X., Kim, C. et al. (2010). Stretchable microfluidic radiofrequency antennas. *Adv. Mater.* 22: 2749.

132 Cheng, S., Rydberg, A., Hjort, K., and Wu, Z. (2009). Liquid metal stretchable unbalanced loop antenna. *Appl. Phys. Lett.* 94: 144103.

133 Aïssa, B., Nedil, M., Habib, M.A. et al. (2013). Fluidic patch antenna based on liquid metal alloy/single-wall carbon-nanotubes operating at the S-band frequency. *Appl. Phys. Lett.* 103: 063101.

134 Qusba, A., RamRakhyani, A.K., So, J.-H. et al. (2014). On the design of microfluidic implant coil for flexible telemetry system. *IEEE Sensors J.* 14: 1074.

135 Hayes, G.J., Desai, S.C., Liu, Y. et al. (2014). Microfluidic coaxial transmission line and phase shifter. *Microwave Opt. Technol. Lett.* 56: 1459.

136 Morishita, A.M., Kitamura, C.K.Y., Shiroma, W.A., and Ohta, A.T. (2014). Two-octave tunable liquid-metal monopole antenna. *Electron. Lett.* 50: 19.

137 Cheng, S. and Wu, Z. (2010). Microfluidic stretchable RF electronics. *Lab Chip* 10: 3227.

138 Rashed Khan, M., Hayes, G.J., So, J.-H. et al. (2011). A frequency shifting liquid metal antenna with pressure responsiveness. *Appl. Phys. Lett.* 99: 013501.

139 Wang, M., Trlica, C., Khan, M.R. et al. (2015). A reconfigurable liquid metal antenna driven by electrochemically controlled capillarity. *J. Appl. Phys.* 117: 194901.

140 Khan, M.R., Hayes, G.J., Zhang, S. et al. (2012). A pressure responsive fluidic microstrip open stub resonator using a liquid metal alloy. *IEEE Microwave Wireless Compon. Lett.* 22: 577.

141 Hayes, G.J., Ju-Hee, S., Qusba, A. et al. (2012). Flexible liquid metal alloy (EGaIn) microstrip patch antenna. *IEEE Trans. Antennas Propag.* 60: 2151.

142 Huang, Y., Wang, Y., Xiao, L. et al. (2014). Microfluidic serpentine antennas with designed mechanical tunability. *Lab Chip* 14: 4205.

143 Mohammed, M.G. and Dickey, M.D. (2013). Strain-controlled diffraction of light from stretchable liquid metal micro-components. *Sens. Actuators, A* 193: 246.

144 Li, M., Yu, B., and Behdad, N. (2010). Liquid-tunable frequency selective surfaces. *IEEE Microwave Wireless Compon. Lett.* 20: 423.

145 Deng, Y. and Liu, J. (2015). Liquid metal based stretchable radiation-shielding film. *J. Med. Devices* 9: 014502.

146 Gerratt, A.P., Michaud, H.O., and Lacour, S.P. (2015). Elastomeric electronic skin for prosthetic tactile sensation. *Adv. Funct. Mater.* 25: 2287.

147 Hao, Y., Wang, T., Xie, Z. et al. (2018). A eutectic-alloy-infused soft actuator with sensing, tunable degrees of freedom, and stiffness properties. *J. Micromech. Microeng.* 28: 024004.

148 Pang, C., Lee, C., and Suh, K.-Y. (2013). Recent advances in flexible sensors for wearable and implantable devices. *J. Appl. Polym. Sci.* 130: 1429.

149 Rathmell, A.R., Bergin, S.M., Hua, Y.-L. et al. (2010). The growth mechanism of copper nanowires and their properties in flexible, transparent conducting films. *Adv. Mater.* 22: 3558.

150 Rathmell, A.R. and Wiley, B.J. (2011). The synthesis and coating of long, thin copper nanowires to make flexible, transparent conducting films on plastic substrates. *Adv. Mater.* 23: 4798.

151 Zeng, X.-Y., Zhang, Q.-K., Yu, R.-M., and Lu, C.-Z. (2010). A new transparent conductor: silver nanowire film buried at the surface of a transparent polymer. *Adv. Mater.* 22: 4484.

152 Segev-Bar, M. and Haick, H. (2013). Flexible sensors based on nanoparticles. *ACS Nano* 7: 8366.

153 Hirsch, A., Michaud, H.O., Gerratt, A.P. et al. (2016). Intrinsically stretchable biphasic (solid-liquid) thin metal films. *Adv. Mater.* 28: 4507.

154 Zhou, H. and Hu, H. (2008). Human motion tracking for rehabilitation – a survey. *Biomed. Signal Process. Control* 3: 1.

155 Johnson, A.N., Huo, X., Ghovanloo, M., and Shinohara, M. (2012). Dual-task motor performance with a tongue-operated assistive technology compared with hand operations. *J. NeuroEng. Rehabil.* 9 (1): 1–16.

156 Liu, H., Li, M., Ouyang, C. et al. (2018). Biofriendly, stretchable, and reusable hydrogel electronics as wearable force sensors. *Small* 14: 1801711.

157 Kim, D., Yoon, Y., Kauh, S.K., and Lee, J. (2018). Towards sub-microscale liquid metal patterns: cascade phase change mediated pick-n-place transfer of liquid metals printed and stretched over a flexible substrate. *Adv. Funct. Mater.* 28: 1800380.

158 Yoon, Y., Kim, S., Kim, D. et al. (2019). Four degrees-of-freedom direct writing of liquid metal patterns on uneven surfaces. *Adv. Mater. Technol.* 4: 1800379.

159 Dellon, A.L. and Keller, K.M. (1997). Computer-assisted quantitative sensorimotor testing in patients with carpal and cubital tunnel syndromes. *Ann. Plast. Surg.* 38: 493.

160 Lu, T., Markvicka, E.J., Jin, Y., and Majidi, C. (2017). Soft-matter printed circuit board with UV laser micropatterning. *ACS Appl. Mater. Interfaces* 9: 22055.

161 Gao, Y., Ota, H., Schaler, E.W. et al. (2017). Wearable microfluidic diaphragm pressure sensor for health and tactile touch monitoring. *Adv. Mater.* 29: 1701985.

162 Mengüç, Y., Park, Y.-L., Pei, H. et al. (2014). Wearable soft sensing suit for human gait measurement. *Int. J. Robot. Res.* 33: 1748.

163 Webb, R.C., Pielak, R.M., Bastien, P. et al. (2015). Thermal transport characteristics of human skin measured in vivo using ultrathin conformal arrays of thermal sensors and actuators. *PLoS One* 10: e0118131.

164 Kim, S., Oh, J., Jeong, D., and Bae, J. (2019). Direct wiring of eutectic gallium–indium to a metal electrode for soft sensor systems. *ACS Appl. Mater. Interfaces* 11: 20557.

165 Kim, M.G., Brown, D.K., and Brand, O. (2020). Nanofabrication for all-soft and high-density electronic devices based on liquid metal. *Nat. Commun.* 11: 1002.

166 Zhang, J., Yao, Y., Sheng, L., and Liu, J. (2015). Self-fueled biomimetic liquid metal mollusk. *Adv. Mater.* 27: 2648.

167 Tsai, J.T.H., Ho, C.-M., Wang, F.-C., and Liang, C.-T. (2009). Ultrahigh contrast light valve driven by electrocapillarity of liquid gallium. *Appl. Phys. Lett.* 95: 251110.

168 Xu, S., Yuan, B., Hou, Y. et al. (2019). Self-fueled liquid metal motors. *J. Phys. D. Appl. Phys.* 52: 353002.

169 Tan, S., Zhou, Y., Wang, L., and Liu, J. (2015). Electrically driven chip cooling device using hybrid coolants of liquid metal and aqueous solution. *Sci. China Technol. Sci.* 59: 301.

170 Niu, S., Liu, Y., Wang, S. et al. (2014). Theoretical investigation and structural optimization of single-electrode triboelectric nanogenerators. *Adv. Funct. Mater.* 24: 3332.

171 Wang, X., Yin, Y., Yi, F. et al. (2017). Bioinspired stretchable triboelectric nanogenerator as energy-harvesting skin for self-powered electronics. *Nano Energy* 39: 429.

172 Wen, Z., Chen, J., Yeh, M.-H. et al. (2015). Blow-driven triboelectric nanogenerator as an active alcohol breath analyzer. *Nano Energy* 16: 38.

173 Wen, Z., Yeh, M.-H., Guo, H. et al. (2016). Self-powered textile for wearable electronics by hybridizing fiber-shaped nanogenerators, solar cells, and supercapacitors. *Sci. Adv.* 2: e1600097.

174 Yi, F., Wang, X., Niu, S. et al. (2016). A highly shape-adaptive, stretchable design based on conductive liquid for energy harvesting and self-powered biomechanical monitoring. *Sci. Adv.* 2: e1501624.

175 Yang, Y., Sun, N., Wen, Z. et al. (2018). Liquid-metal-based super-stretchable and structure-designable triboelectric nanogenerator for wearable electronics. *ACS Nano* 12: 2027.

176 Deng, J., Kuang, X., Liu, R. et al. (2018). Vitrimer elastomer-based jigsaw puzzle-like healable triboelectric nanogenerator for self-powered wearable electronics. *Adv. Mater.* 30: 1705918.

177 Lai, Y.-C., Deng, J., Niu, S. et al. (2016). Electric eel-skin-inspired mechanically durable and super-stretchable nanogenerator for deformable power source and fully autonomous conformable electronic-skin applications. *Adv. Mater.* 28: 10024.

178 Wang, Z.L., Chen, J., and Lin, L. (2015). Progress in triboelectric nanogenerators as a new energy technology and self-powered sensors. *Energy Environ. Sci.* 8: 2250.

179 Wang, Z.L. (2013). Triboelectric nanogenerators as new energy technology for self-powered systems and as active mechanical and chemical sensors. *ACS Nano* 7: 9533.

180 Parida, K., Thangavel, G., Cai, G. et al. (2019). Extremely stretchable and self-healing conductor based on thermoplastic elastomer for all-three-dimensional printed triboelectric nanogenerator. *Nat. Commun.* 10: 2158.

181 Niu, S., Wang, S., Lin, L. et al. (2013). Theoretical study of contact-mode triboelectric nanogenerators as an effective power source. *Energy Environ. Sci.* 6: 3576.

8

Liquid Metal for Biomedicine

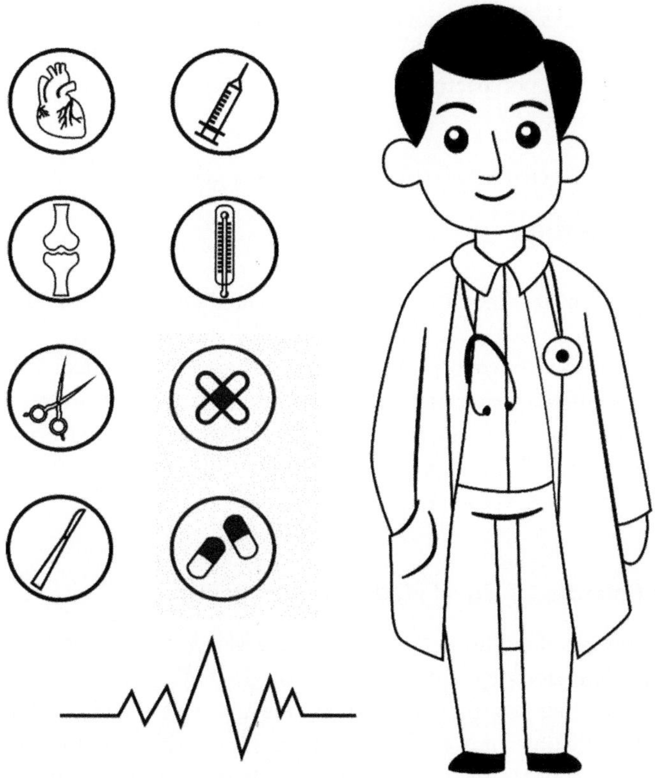

The sole purpose of science is to alleviate the hardships of human existence.
Source: Bertolt Brecht.

Liquid Metals: Properties, Mechanisms, and Applications, First Edition. Lei Fu and Mengqi Zeng.

8.1 Introduction

Liquid metal (LM) with fascinating physicochemical performances is emerging as a novel functional material. Recently, LMs, especially gallium-based LMs show great potential in biomedicine owing to their low toxicity and good biocompatibility. Because of the possession of both metallic and fluidic properties, LM-based functional materials at or near room temperature exhibit better versatility and adaptability to complex biological systems, which are superior to conventional biomaterials [1, 2].

This chapter provides a comprehensive summary on various biomedical applications of LMs-based functional materials in terms of therapeutics, bioimaging, biomimetics, and biodevices. In each part, the major modern biomedical scientific issues will be pointed out and the potential of LMs to resolve these issues will be discussed. As the previous chapters have detailed the basic properties of LMs, this chapter will mainly focus on the major advancements made in biomedicine and the perspective for future development in biomedicine based on LMs.

8.2 Therapeutics

Disease therapeutics is a momentous part of biomedicine. With advances in modern science and technology, various materials and strategies have been developed in biomedicine to solve current problems. To date, novel biomedical applications of LMs with extraordinary properties have aroused great research interests. Research on the therapeutic applications of LMs mainly focuses on drug delivery and therapy of tumors. This part will illustrate the potential of LMs applied in therapeutics and summarize the recent developments in this field.

8.2.1 The Potential of Liquid Metal for Therapeutics

As novel materials for tumor therapeutics, LMs possess promising potential with unique characteristics for therapeutics: (i) LMs with intrinsic flexibility and liquidity can fuse together or deform, offering a new strategy for delivering drugs to realize flexible and efficient therapy [3, 4]. (ii) Low toxicity and good biocompatibility of LMs are essential factors for therapy. More recently, both in vitro and in vivo studies have confirmed that LMs can be degraded as they can react with acid in cell microenvironment and subsequently be cleared [5]. (iii) LMs can be flexibly functionalized. Appropriate modifying or functionalizing the surface of LM is beneficial to enhance their biomedical performances, including their loading ability, targeting capacity, biocompatibility, and thermal stability. (iv) LMs are able to respond to external stimuli including pH, laser, magnetic field, electric field, radiofrequency (RF), and microwave, providing new opportunities for flexible and controllable therapeutics.

8.2.2 Drug Delivery

Chemotherapy is one of the indispensable approaches for tumor therapy. However, it has not reached the ideal state of treatment, there are still many problems including low bioavailability, strong side effect, drug resistance, and metastasis. Drug delivery system (DDS) is proposed to overcome the problems of free drugs, it is hopeful in enhancing the therapeutic efficacy of medications and mitigate the side effects [6, 7]. Drug carrier as the key of the DDS brings many advantages. It can protect the drug from premature interaction with biological environment and degradation, control the pharmacokinetics and drug distribution, as well as enhance absorption of drugs to selected tissues. It can also influence the duration of therapeutic effect, excretion, and toxicity [8, 9]. To achieve efficient drug delivery, a carrier should be rationally selected, designed, and modified to make it suitable for specific DDSs. Nowadays, a great amount of carriers have been explored for DDSs, such as polymers [10, 11], lipids [12], and inorganic particles [13]. However, there are still lots of limitations including the instability, poor biodegradability, systemic toxicity, and lack of specificity among various carriers.

Different from traditional carriers, LMs with the functional modifiability, intrinsic fluidity, and field-responsiveness have great potential for drug delivery. For instance, the surface of LMs functionalized with various ligands can serve as drug loading matrix or active targeting moiety for the target drug delivery. And they can also encapsulate drug molecules inside due to their good fluidity. More importantly, under external stimuli such as change of pH and temperature, LMs-based carriers can achieve morphological transformation, this process contributes to regulate the amount of chemotherapeutic drug release at the site of target tumor. Hence, the curative effect of chemotherapy can be enhanced, the side effect can be lowered as well. Subsequently, the carrier will be biodegraded in an acidic cellular environment, thus avoiding potential toxicity.

Lu and coworkers fabricated pH-responsive LM nanocarriers for drug delivery, enhancing the anticancer effect. The LM nanocarriers were prepared via assembling the thiolated ligands onto the surface of the eutectic alloy of gallium indium (EGaIn), which is comprised of 25% In and 75% Ga by weight [3]. The ligands consisted of the thiolated (2-hydroxypropyl)-*b*-cyclodextrin (CD) and thiolated hyaluronic acid (HA) (Figure 8.1a, b). The CD consisted a seven-membered sugar ring, which could provide reliable loading sites for anticancer drug doxorubicin (Dox) [14] (Figure 8.1e), and the HA supported active tumor targeting to receptors such as the CD44 receptor overexpressed on the surface of tumor cells. The schematic of pH-responsive delivery of Dox was illustrated in Figure 8.1c. The as-synthesized LM nanoparticles (designated LM-NP/L) with acid-triggered deformability was capable of fusing in mildly acidic-endosome/lysosome microenvironments, thus facilitating Dox release after cellular internalization [15] (Figure 8.1d). The intracellular delivery of Dox-loaded LMNP/L (LM-NP/Dox-L) toward the HeLa cell was studied using a confocal laser scanning microscopy. After incubating with LM-NP/Dox-L for one hour and four hours, HeLa cells showed clear fluorescence of Dox, confirming the internalization of LM-NP/Dox-L cells. Besides, ex vivo imaging proved that the

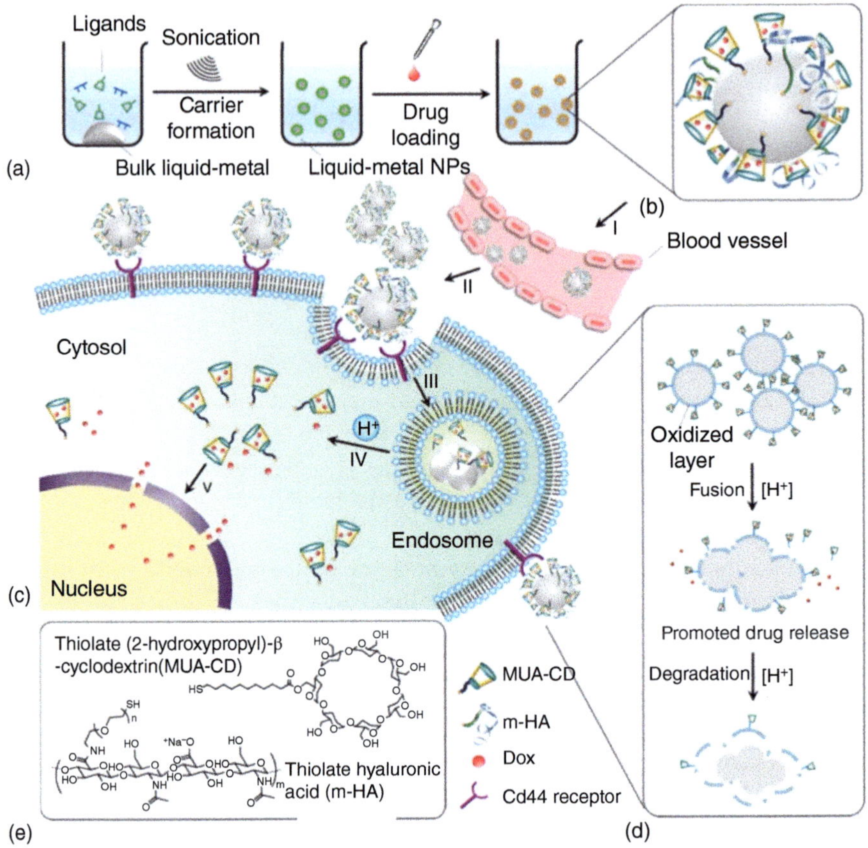

Figure 8.1 pH-responsive drug delivery based on LMs. (a) Preparation diagram of LM-NP/Dox-L. (b) Compositions and construction of LM-NP/Dox-L. (c) Schematic of pH-stimuli Dox delivery for the targeted tumor therapy. (d) Process of fusing and degrading of LM-NP/Dox-L under acid environment. (e) Structural formula of m-HA and MUA-CD [3]. Source: Lu et al. [3]. Licensed under CC BY 4.0.

LM-NP/Dox-L possessed the potent ability to target tumors. The tumor site showed higher fluorescence intensity than other tissues. Meanwhile, the tumor site treated with LM-NP/Dox-L exhibited stronger fluorescence signal than that treated with LM-NP/Dox. The tumor-targeting ability was due to the strengthened permeability and retention effect as well as the mechanisms of active targeting. The massive remission of cancer cell treated with LM-NP/Dox-L verifying the significant tumor suppression effects. A systematic study of the toxicology showed that the LM-NP/L was nontoxic at the therapy dose.

Considering the needs of controlling the drug release via external field, LMs possess the performance of photochemical and photothermal conversion. Therefore, they can be applied as light-responsive carriers for controllable drug release. Kim and coworkers [16] engineered LM particles with 1,2-distearoyl-*sn*-glycero-3-phosphocholin (DSPC)/Dox. The shape of LM/DSPC/Dox particles can deform to release the Dox (up to 50%) under laser or heat. The cancer cell viability was less than 50% in vitro experiment, which was more effective than free Dox, indicating better anticancer effects of particles. Besides, this LM-based carrier exhibited nontoxicity according to the 48 hour-cell viability test.

To enhance the responsive ability and drug release efficiency, the combination of stimulus response material with LM provides a good strategy. Lu and coworkers [4] developed light-responsive EGaIn NPs equipped with graphene quantum dots (GQDs), which serve as a photosensitizer. The preparation route of the transformable NPs (designated tNPs) was shown in Figure 8.2a and the Dox could be loaded onto GQDs through π–π interaction. The incorporation of GQDs enables the efficient photothermal conversion capability, not only local heat but also reactive oxygen species (ROS) generated with the near-infrared (NIR) light could promote

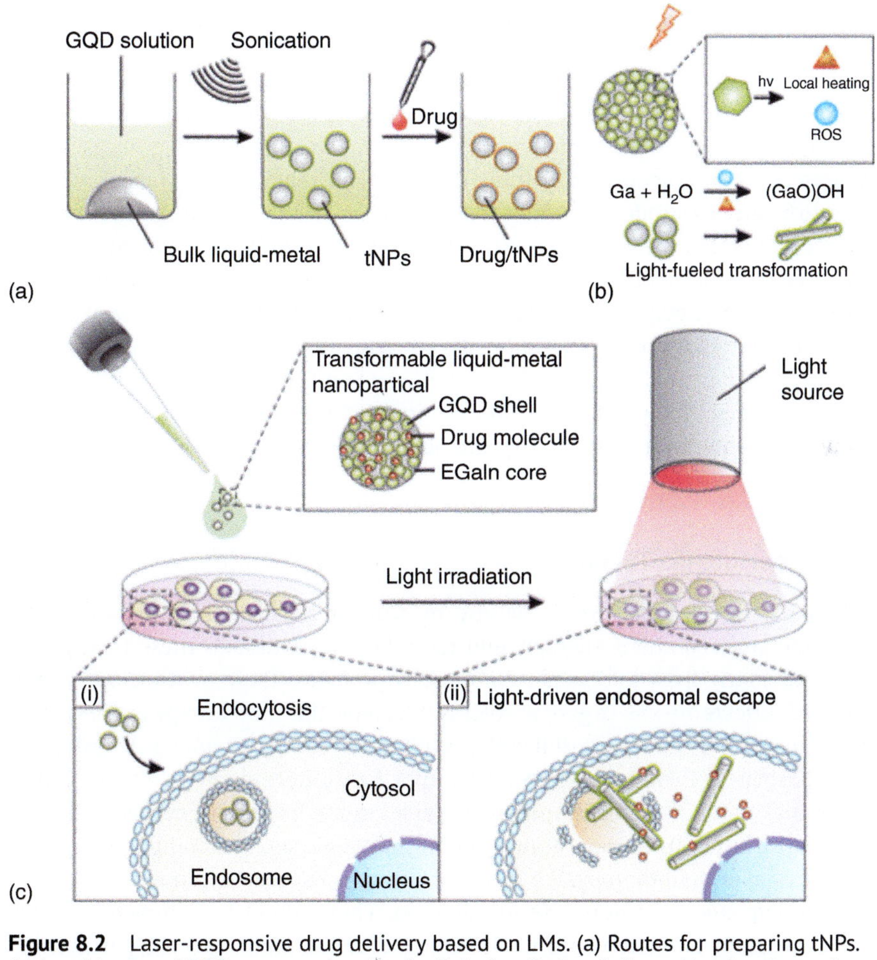

Figure 8.2 Laser-responsive drug delivery based on LMs. (a) Routes for preparing tNPs. (b) Local heat and ROS generated under the light irradiation, followed by the shape change from nanospheres into nanorods. (c) Light-responsive endosomal escape of GQD/tNPs. (i) Endocytosis of tNPs to enter a cell. (ii) Endosomal escape of tNPs under the light irradiation [4]. Source: Lu et al. [4].

the reaction of Ga with water, realizing the transition from Ga to (GaO)OH, this process was along with the phase separation of the alloy (Figure 8.2b). Simultaneously, due to the morphological change from nanospheres to nanorods, the endosomal membrane would be physically disrupted, thereby achieving efficient and accurate intracellular delivery of drugs (Figure 8.2c). Fluorescence quantitative analysis as well as fluorescence colocalization technology proved that the endosome escape could be enhanced in this way. The tumors were completely eliminated after 14-days treatment with Dox/tNPs under laser irradiation. In addition, LM can be considered to cooperate with the NIR laser-responsive polymeric hydrogels for improving the responsive capacity and controlling the drug release. Fan and coworkers [17] encapsulated LM (EGaIn) droplets to poly (NIPAm-coMBA) hydrogels (PNM), which could lock LM droplets and drugs tightly owing to their porous three-dimensional (3D) networks. EGaIn droplets could strengthen the overall photothermal conversion efficiency to promote the release of drugs encapsulated in 3D network under NIR laser irradiation.

At present, LMs as carriers for drug delivery are mainly based on the response to physical stimulus including pH and light, the former can make LM fuse together under endosomal/lysosomal microenvironments to promote the drug release with enhanced chemotherapeutic inhibition, the latter can generate local heat for shape change to facilitate the effective endosomal escape for controllable drug delivery. The LMs that cooperated with photosensitive materials also provide a promising strategy to enhance responsive efficiency and drug release efficiency for chemotherapeutic purpose. LMs modified with targeting moiety are endowed the capacity of target delivery. Therefore, rationally designing the structures and components of LM-based carriers will amplify the benefits of DDS for efficient and controlled therapy. Further, owing to the versatility of LMs, they are generally applied for combination therapy, and the combinations of drug delivery and other therapy are described in the following sections.

8.2.3 Hyperthermia Therapy

Traditional strategies of tumor therapy in clinical applications predominantly include the surgery, chemotherapy, and radiotherapy. Although these treatments have saved many lives and made continuous progress, their invasive ways and serious side effects are still urgent problems [8]. Hyperthermia therapy is a potential method for tumor therapy, as it utilizes physical energy to precisely generate the localized heat inside the cancerous tumor for tumor apoptosis. It can cause less thermal damage to surrounding normal tissues. Hyperthermia has always been a part of the multimodal tumor therapy strategy, such as combination with chemotherapy or radiotherapy. For instance, tumor cells located in the center of the tumor are hypoxic and insensitive to radiation. They cannot be completely killed after radiotherapy and often become the source of tumor recurrence. Hyperthermia is particularly effective for this type of tumor cells. As a result, hyperthermia can make up for the shortcomings of radiotherapy, and the combined use can improve the efficacy [18, 19].

Owing to the remarkable capacity of energy converting, LMs have also been exploited as energy-converting agents for hyperthermia therapy. External activation such as light, magnetic field, and RF field can induce LMs to absorb external energy and generate hyperthermia for cancer cell apoptosis/necrosis. For the high tissue penetration of these external stimuli, the therapeutic efficiency of tumors especially those in deep can be improved. This part summarizes the different hyperthermia therapies and the combined treatments.

8.2.3.1 Magnetic Hyperthermia

Magnetic hyperthermia has been a renaissance of tumor treatment modality, since it can remotely and precisely treat tumors especially those that are situated deep. This treatment is achieved by the heat generated from biological materials with the alternating magnetic field (AMF) [20, 21]. Although researchers have made progress in exploring the suitable magnetothermal agent for magnetic hyperthermia, there are still some critical drawbacks among these studies. For instance, high doses of injected magnetic NPs may induce biosafety issues related to metabolism and circulation. Particles in aqueous conditions tend to aggregate when they are exposed to the magnetic field, leading to the instability of this system. Besides, superparamagnetic particles show the low thermal efficiency, which constrains their applications [22, 23].

LMs can generate the inductive heat under AMF, exhibiting favorable magnetothermal effect. Therefore, they can be magnetothermal agents for magnetic hyperthermia. Meanwhile, magnetic hyperthermia based on LMs always combines with chemotherapy to improve the therapeutic efficiency.

Wang and coworkers [24] developed an oxidized gallium–indium (O-GaIn) for AMF-based tumor therapy (Figure 8.3a), this O-GaIn could serve as e-skin because of its excellent adhesion to the surface of skins. The generated heat from the eddy current on e-skin could ablate the tumor tissues below the e-skin. When the as-printed O-GaIn above the tumor site of the mouse was exposed to an AMF, the induction heat made the temperature to increase obviously, and a scar appeared in the area covered with O-GaIn after treatment. In addition, it is worth mentioning that combining chemotherapy and magnetic hyperthermia could further enhance the therapeutic effect. Based on this, a temperature-triggered chitosan (CS) hydrogel loaded with Dox was injected and released, the Dox via the AMF induced increase of temperature. The group using LM with the AMF exposure possessed remarkable tumor growth inhibition. As for treatment with only LM under the AMF, owing to the lack of the chemotherapeutic drug, higher temperature was required to thoroughly kill the tumor, which could reduce the survival rates of the mice. Furthermore, simultaneous multisite tumor treatment via multisite LM electrodes was performed with high oncotherapeutic efficacy. During the process of dual-site heating, there were temperature variations of tumors in different sites (Figure 8.3b). Therefore, satisfactory therapeutic effect would be achieved by controlling the temperature via tuning the space between the tumor site and the electromagnetic coil. Moreover, both in vitro and in vivo tests of O-GaIn verified the negligible toxicity and benign biosafety.

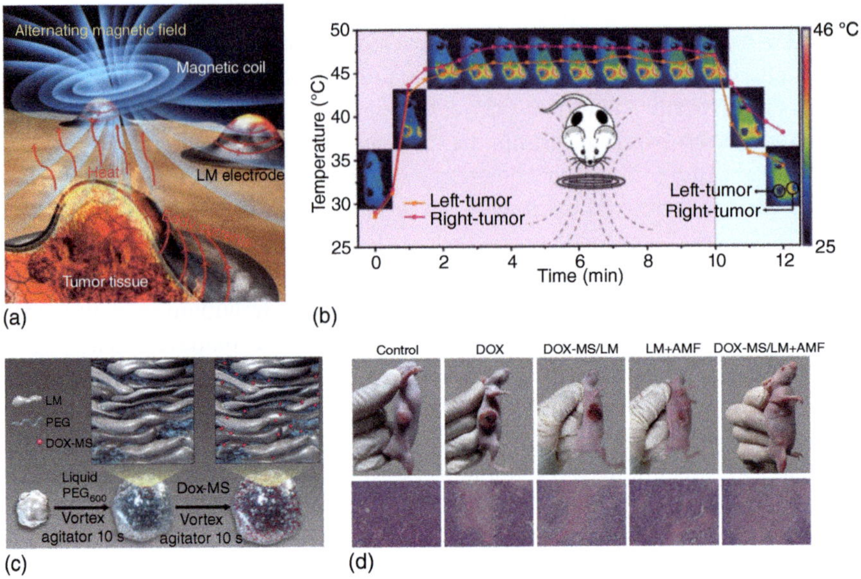

Figure 8.3 Magnetic hyperthermia based on LMs. (a) Schematic illustration of LM electrodes on skin with dual-site tumor therapy under an AMF. (b) Temperature curves of the mouse with dual sites treatment under an AMF and the images of IR thermography [24]. (a, b) Source: Wang et al. [24]. (c) Schematic illustration of combined therapy for breast tumors based on LM. (d) Representative images of tumor-bearing mice of different groups (up) and tumor sections of different groups stained with H&E (down) [25]. Source: Wang et al. [25].

During the LM-based magnetothermal treatment, LMs can also double as drug carriers to achieve synergistic effect of in vivo multiple therapeutic modalities. Wang and coworkers incorporated mesoporous silica loaded with Dox (Dox-MS) into the PEGylated LM, this LM hybrid owned the capability of pH/AFM dual response for combined therapy between drug delivery and magnetic hyperthermia [25] (Figure 8.3c). The pH-stimulus-responsive drug release was attributed to the abundant protonated silanols on surfaces of silica at low pH condition, promoting the separation of drugs from silica cores. The thermal induction of EGaIn induced by an AMF was proved to facilitate the drug release. The representative photographs of mice indicated that the tumors were thoroughly ablated with the treatment of LMs after the AMF exposure (Figure 8.3d). Except the control group, each group of tumors were necrotic and the inflammatory infiltration were observed, confirming the superior therapeutic efficacy of a combination therapy.

The alternative magnetic field enable the multisite and noninvasive tumor therapy, and LMs with its exposure can work in both internal and external environments, showing spatiotemporal controllability and operational flexibility in magnetic hyperthermia. More importantly, LMs-based magnetic hyperthermia united with other therapeutic modality will contribute to improving efficacy of comprehensive treatment.

8.2.3.2 Photothermal Therapy

Photothermal therapy (PTT) is a spatiotemporally controllable method of tumor thermal ablation, based on the photothermal effect of photothermal conversion agents [26] (PTAs). It has attracted wide attention for its target ability, minimal invasion, and obvious tumor ablation [27], since the heat energy can be deposited specifically in tumors to minimize damage of nearby normal tissue. Considering that the NIR region is a transparent window for biological tissues [28], the ideal photothermal agent should possess strong absorption capacity and effective conversion ability for the NIR light. Besides, the agents for PTT are required to be nontoxic and biocompatible, guaranteeing the therapeutic efficacy without causing toxic side effects. So far, a variety of NIR-sensitive agents have been developed, such as organic [29–32] and inorganic nanomaterials [33, 34]. Although these PTAs exhibit various strengths in PTT, there are still deficiencies among the PTAs. For instance, as for organic NPs, they have low toxicity and good biocompatibility [35]. However, as for organic small molecules, their photothermal conversion efficiency and stability are low, which hinder their clinical use. Additionally, gold nanomaterials as the common inorganic PTC materials have been extensively researched, their surface plasma resonance peak (SPR) can be adjusted in NIR region. They exhibit high photothermal stability even if they are exposed to repeated irradiation, and they are not vulnerable to photobleaching. But the rigidity of gold NPs leads to the lack of flexibility and adaptability, biodegradation and price are also hurdles for their applications [36]. Besides, the potential long-term toxicity of metal-based nanomaterials remains to be explored.

LMs with high NIR PTC efficiency and good thermal conductivity represent the new platform for PTT. The photothermal effect of LMs is based on optical resonance. As a newly emerging photothermal conversion agent, LM with the property of easy modification can be regulated to enhance the photothermal performance, stability under long period irradiation and achieve the multifunctional therapy.

Wang and coworkers prepared rodlike LM Ga via pressure-filter template method [37]. Propelled by acoustic radiation force from the external ultrasound field, the LMs could actively target tumor cells and drill into cells. Owing to their pH stimulus responsibility, they subsequently transformed from rod to droplet and fused together in the acidic tumor environment. The strong plasmon resonance of LMs under illumination of NIR light induced photothermal effect for tumor thermal ablation (Figure 8.4a). The two internalized LMs exhibited the morphology transition from rod to sphere, and started to fuse together after 3 hours (Figure 8.4b). The optical image in Figure 8.4c showed the membrane integrity of the cells with internalized LMs before the NIR laser exposure, and a region of the cell membrane was broken after the NIR laser exposure for 5 seconds. The fluorescence image also confirmed that the HeLa cell was necrotic after treatment. Meanwhile, the degradation of LMs in acidic conditions indicated the biodegradability for PTT. With the advancement of functional materials, hybrid functional biomaterials based on LMs fabricated by being mixed with a variety of metal or nonmetal materials can well adapt to practical conditions with desired performances. Wang and coworkers [38] developed a laser heating Mg–GaIn material for skin tumor therapy by mixing EGaIn

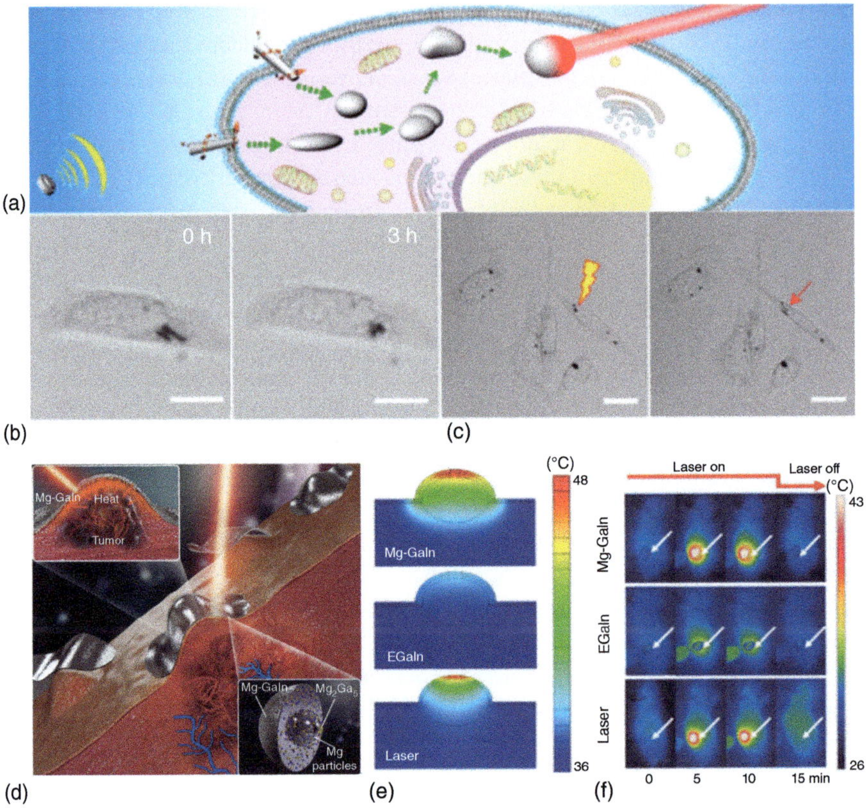

Figure 8.4 Photothermal therapy based on LMs. (a) Scheme of the PTT of Ga. (b) Intracellular deformation and fusion of two internalized Ga nanorods. Scale bars, 10 μm. (c) Optical images of HeLa cells with the fused Ga before and after the irradiation of NIR laser. Scale bars, 20 μm [37]. (a–c) Source: Wang et al. [37]. (d) Schematic illustration of Mg–GaIn in vivo tumor PTT. (e) Temperature distributions of tumor based on Fluent simulation under the laser irradiation. (f) IR thermography images of C8161 tumor-bearing mice with different treatments under the 12 minutes-laser exposure (2.0 W cm^{-2}) [38]. (d–f) Source: Wang et al. [38].

with Mg particles (Figure 8.4d). This soft Mg–GaIn could adhere to the tumor surface and cause the thermal ablation of tumor via laser heating. Fluent simulation showed the temperature distributions of tumor tissues under the irradiation of laser (Figure 8.4e), demonstrating that Mg–GaIn caused the deeper and wider heating range. And the IR thermography images of mice in Figure 8.4f also verified the larger thermal treatment region of Mg–GaIn, indicating good photothermal effects of Mg–GaIn on antitumor application. Herein, Mg particles were able to increase the diffuse reflection of LM surface, and the as-generated Mg_2Ga_5 was intermetallic phase that could benefit the light absorbance, which explained the reason of high photothermal efficacy of Mg–GaIn. The remarkably high tumor growth retardation was displayed in PTT with Mg–GaIn. In vitro cell and skin surface cytotoxicity studies proved the low toxicity of Mg–GaIn.

The shape and composition exert a significant influence on the photothermal performances of LMs, providing more possibilities to manipulate their photothermal efficacy in tumor ablation treatment. Yan and coworkers realized the controllable-morphology synthesis of LM droplets and investigated the PTT effect of different shapes [39]. The ultrasmall melanin nanoparticles (MNPs) owned strong metal chelation with the Ga oxide layer, leading to the enhanced solubility and stability of EGaIn nanodroplets. Various shapes of EGaIn including nanorice, nanosphere, and nanorod were obtained by tuning a series of reaction parameters such as the sonication time and the category of LMs (Figure 8.5a). EGaIn nanorices showed the highest absorbance among these three shapes, which may be due to its large size, and its temperature increased by 32.2 °C under laser irradiation (Figure 8.5b). Besides, as for the EGaIn nanorice, its calculated photothermal conversion efficiency based on Roper's method was 36.7%, which is the best one compared with the other two. The temperature in the irradiated region treated with EGaIn nanorice raised to about 50.3 °C for 5 minutes, which was enough to

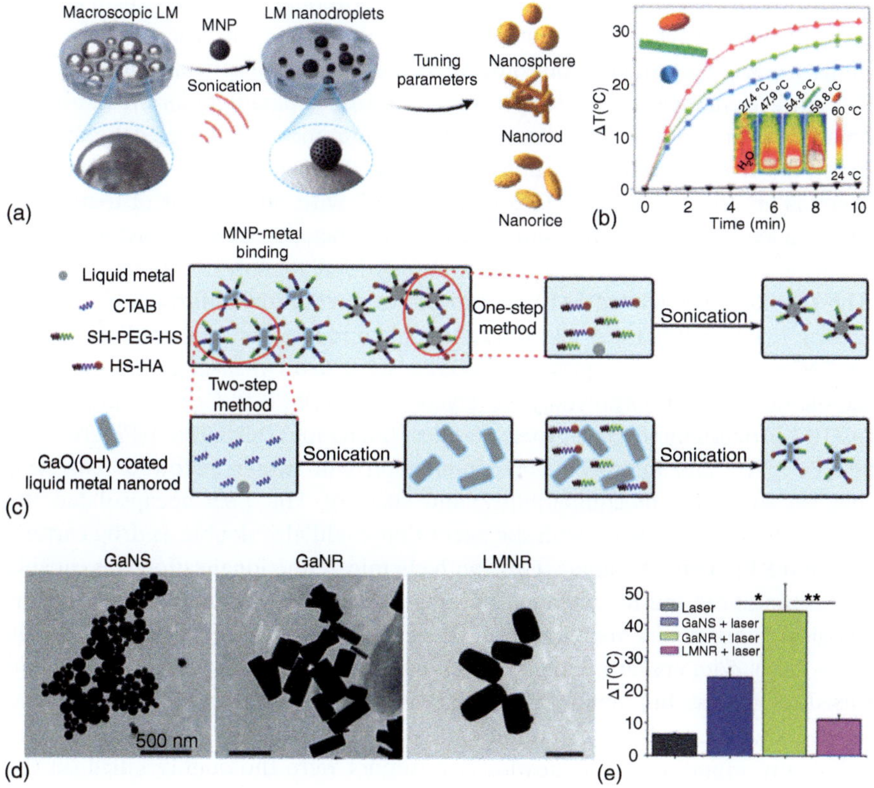

Figure 8.5 The shape control of LMs for PTT. (a) Schematic of LM nanodroplets synthesis. (b) Temperature variation of EGaIn in different shapes under 10 minutes-NIR laser exposure ($2.0\,W\,cm^{-2}$) [39]. Source: Yan et al. [39]. (c) Schematic of synthesizing LM nanoparticles (d) TEM images of different LM nanoparticles. (e) Average temperature variation of tumors with different treatments [40]. Source: Sun et al. [40].

effectively ablate the tumor. Moreover, good biocompatibility of EGaIn nanorices was confirmed by testing the body weight of treated mice.

Sun and coworkers [40] fabricated morphology tunable liquid metal nanoparticles including Ga nanosphere (GaNS), Ga nanorod (GaNR), and liquid metal nanorod (LMNR) consisting of Ga–In alloy. The nanospheres were fabricated through mixing LM droplets with thiol–polyethylene glycol–thiol (SH–PEG–HS) and HS–HA via a one-step sonication, while the synthesis of nanorods was in a method of the two-step sonication. First, LM droplets were coated in cetyltrimethylammonium bromide (CTAB), thus forming GaO (OH) on the LM surface via sonication, and then they were mixed with HS–HA and SH–PEG–HS by sonication (Figure 8.5c). The TEM images of as-made GaNS, GaNR, and LMNR were shown in Figure 8.5d, indicating that the little larger sizes of GaNR and LMNR than that of GaNS. The formation of different shapes was ascribed to the interactions between the different chemical molecules with LMs. The rod-shaped LM exhibited better photothermal performance than the sphere-shaped ones (Figure 8.5e). However, the addition of indium could weaken the photothermal performance of the nanorod, due to the decreased homogeneity of the nanorods of LM. Although the mechanism of how the morphology and composition affect PTT performance is not clear, it is concluded that the size and the composition could affect the ability to absorb and convert light.

The surface engineering of LMs provides a controlled paradigm for the LM-based biomaterials construction. In order to meet the requirements of expected therapeutic efficacy, the design and modification of LM surface to realize the synergistic tumor therapy is an effective way [41]. Combining PTT with other kinds of treatments provides a feasible strategy for better therapeutic efficiency. The complementary or synergistic effects can be elicited by the combinational therapy of tumors.

The NIR laser-stimulated LM nanocapsule was synthesized through photopolymerizing EGaIn with functional phospholipids under ultraviolet irradiation [42]. This LM nanocapsule comprised of EGaIn, 1,2-distearoyl-*sn*-glycero-3-phosphoethanolamine-*N*-[amino(polyethylene glycol)-2000] (DSPE-PEG_{2000}-Amine) and 1, 2-bis(10,12-tricosadiynoyl)-*sn*-glycero-3-phosphocholine (DC(8,9)PC) (Figure 8.6a). The PEG parts and hydrogen bond interaction with water endowed LMs with enhanced solubility, biocompatibility, and stability. The LMs encapsulated the carmofur, which is an antineoplastic agent that could also double as drug carriers for chemo-PTT. This LM nanocapsule with chemical functionalization was capable of generating heat and ROS while absorbing NIR light, resulting in dramatic morphological changes of LMs and the disintegration of nanocapsule to controllably release drugs. The epidermal growth factor receptor antibody (Anti-EGFR) could be used for further functionalizing LMs to achieve the targeting therapy. In vivo studies indicated that laser-induced antibody-modified LM nanocapsule could significantly suppress tumor growth, and tumors were thoroughly killed on the third day, showing the rapid therapeutic effects.

As for the photothermal agents, the photothermal stability should be of concern, which determines the long-term efficiency. Bare LMs with intrinsic low oxidative resistance may suffer from the photothermal performance deterioration under long-time laser irradiation. The modification of LMs is regarded as an effective

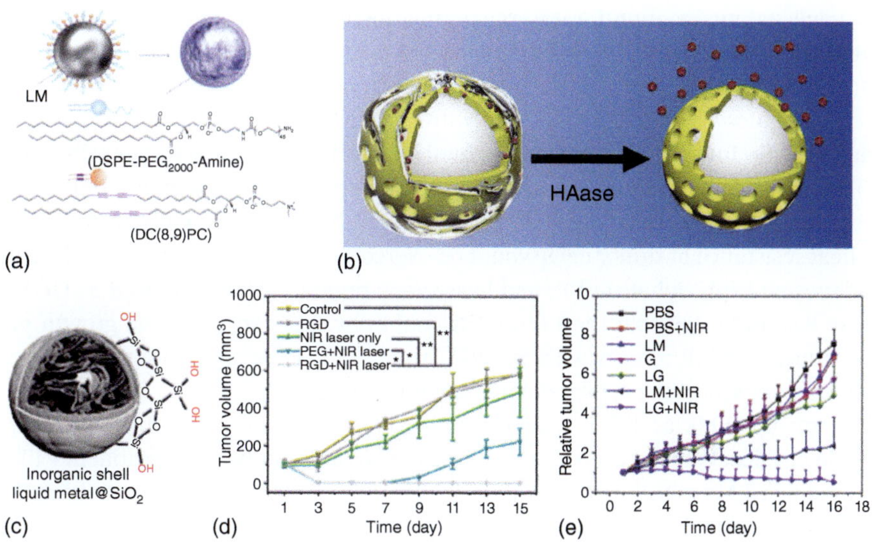

Figure 8.6 The surface engineering of LMs for combined therapy. (a) Schematic of the LM nanocapsule [42]. Source: Chechetka et al. [42]. Licensed under CC BY 4.0. (b) Schematic illustration of the drug release of the LM@MSN/Dox@HA [43]. Source: Hu et al. [43]. (c) Schematic illustration of the LM@SiO_2 nanoparticle. (d) Tumor growth curves of mice with different treatments. ($^{*}p < 0.05$, $^{**}p < 0.01$, $^{***}p < 0.001$) [44]. (c, d) Source: Zhu et al. [44]. (e) Relative tumor volume of mice bearing murine mammary carcinoma (4T1) with different treatments [45]. Source: Hu et al. [45].

approach to make up for this shortage of LMs. Besides the encapsulation of LMs via organic molecules [42, 46–48], a surface coating strategy with inorganic materials can also be proposed to solve the various instability and the subsequent issues of LMs. LM nanoparticles could be coated with multifunctional mesoporous silica shells (LM@MSN) to avoid outside damage [43]. The as-prepared LM@MSN integrated the performances of easy surface functionalizing, good drug loading efficiency [49], and the good photothermal performance, showing the potential for combinational therapy. The Dox was encapsulated into the pores of MSN, then HA was modified on LM@MSN (LM@MSN/Dox@HA) for combination therapy (Figure 8.6b). In vitro and in vivo experiments indicated the high combined-therapy efficiency of LM@MSN/Dox@HA with the NIR irradiation. In addition, the surface engineering of LMs with inorganic silica nanoshell can prevent further oxidation of LMs and avoid heat loss to prominently enhance their photostability for better PTT [44]. Further surface functionalizing could be realized owing to the anchoring points offered by rich silanol groups of SiO_2-coating (Figure 8.6c). The as-protected LMs with RGD peptides showed obvious inhibition of tumor growth, enhanced absorption, and high photothermal hyperthermia efficacy in comparison to other groups (Figure 8.6d).

In addition, the efficacy of tumor therapy can further be enhanced through combinational starvation/PTT. LMs with the photothermal conversion capability can also offer energy for certain reactions, in this case, it is possible to be used in the enzymic catalytic reaction for synergistic treatments. In this study, LM NPs were coated with glucose oxidase (GOX) via electrostatic interaction [45], designated as LM@GOX. The GOX served as the enzyme for starvation therapy and could convert glucose to gluconic acid and cytotoxic hydrogen peroxide, which could inhibit the tumor growth. Integrating it with LM offered an effective way to kill tumor cells. The PTT process resulted in the increase of local temperature, along with the improvement of the enzyme activity of GOX for starvation therapy. Meanwhile, as for PTT, the heat-resistance of tumor cells would be overcome owing to the decreased levels of adenosine triphosphate (ATP) and heat shock proteins (HSPs) induced by GOX. Figure 8.6e indicated that the LM@GOX could obviously suppress the growth of solid tumor and kill tumors with the NIR irradiation, achieving more remarkably antitumor activity of the synergetic therapeutic modes.

To sum up, LMs as PTAs with high photothermal conversion efficiency and tunable physicochemical properties have demonstrated significant potential in PTT, the modified LMs with functional materials promise a platform for enhanced PTT involving combinational tumor therapy.

8.2.3.3 Radio-frequency Hyperthermia

RF hyperthermia is a significant physical approach for minimally invasive treatment of tumors [50]. RF miniature electrodes guided by the imaging system are inserted into tumor center. Thus, high-frequency oscillation currents from the tiny electrode flow through the tissue, and ions present in the tumor tissue around electrodes are significantly agitated. This kind of ion agitation can generate frictional heating within the tissue to cause the tumor destruction, and the frictional heating can be strictly controlled via modulating the amount of deposited RF energy [51, 52]. In the clinical practice, treating the tumors that are irregular or located in a complex area increases the difficulty of operation using common rigid plate electrodes. This problem induces the low efficiency of hyperthermia and hampers the application of the RF therapy.

The soft nature and good conductivity endow LMs with the potential in conformable RF hyperthermia. LM bath-electrodes were proposed for radio-frequency ablation (RFA), which was achieved by directly making the tumor tissue sites immersed in the LM bath [53] (Figure. 8.7a). Compared with the conventional RFA based on rigid electrode, LM-based RF owned the better, achieving the rapid heating to the temperature for ablating neoplasm, longer heating, and larger necrosis areas, especially for the tissue with a complicated shape. The ablation zone and temperature distribution could be tuned through arranging the LM bath-electrodes (Figure 8.7b), indicating that the LM-based RFhyperthermia can be dynamically adjustable to satisfy the need for therapy.

In summary, LMs applied in RF ablation show the potential in dealing with realistic and complex irregular tumors. This conceptual treatment with good adaptability provides a possibility to achieve flexible operation and induce the heat injury, resulting in high efficiency of thermal ablation.

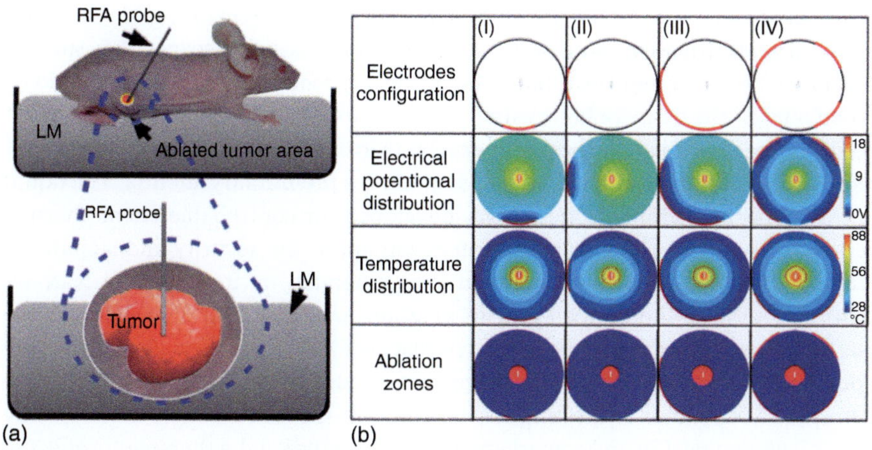

Figure 8.7 Radiofrequency hyperthermia based on LMs. (a) Schematic for LM bath electrode-based radiofrequency hyperthermia. (b) Various arrangements of the LM bath electrode and the corresponding influence on RF performance via numerical simulation [53]. Source: Sun et al. [53].

8.2.3.4 Chemothermal Therapy

Chemothermal treatment is a new field of hyperthermia therapy, which is based on heat generated by exothermic chemical reactions of tumor tissue, involving oxidation-reduction chemistry and acid-base neutralization chemistry [54]. This approach provides a possible modality for local hyperthermia in a minimally invasive way. As for chemothermal therapy, the appropriate reactants are expected to release enough heat for thermally ablating the tumor tissue during exothermic chemical reaction and minimize the systemic toxicity [55–59]. To meet these requirements, liquid alkali metals with injectable performance are attractive as reagents for chemothermal therapy. This part presents the concept and application of alkali metal-enabled chemothermal therapy [55–57, 60].

At present, liquid alkali metals used in chemothermal treatment refer to sodium–potassium (NaK) alloy, which is liquid at room temperature (m.p. is ~12.8 °C)[61], making it convenient for direct and noninvasive injection into target tumor sites using a syringe. Meanwhile, only an extremely small amount is sufficient to make the tumor temperature rise significantly or even burn at the target site, resulting in a strong thermal ablation effect. The high rate of heat generation is attributed to the severe exothermic reaction of the NaK fluid and the water that intrinsically existed in target tissues, and involved reaction equations are as follows:

$$\mathrm{Na} + \mathrm{H_2O} = \mathrm{NaOH} + \frac{1}{2}\mathrm{H_2} \uparrow + 140.886\ \mathrm{kJ\ mol^{-1}} \tag{8.1}$$

$$\mathrm{K} + \mathrm{H_2O} = \mathrm{KOH} + \frac{1}{2}\mathrm{H_2} \uparrow + 140.008\ \mathrm{kJ\ mol^{-1}} \tag{8.2}$$

According to the equations, alkali metals react violently with water. Therefore, appropriate packaging alkali metals should be taken into consideration. Besides, NaOH and KOH as the reaction products can also contribute to the further heating if they dissolve in the water of the targeted tissue. The remaining Na^+ and K^+ as products are friendly and able to be absorbed by tissues, because they are the basic elements of physiological fluid. As for the as-produced hydrogen, it should be guided outside the tissue or body using a hollow pipe. Hence, appropriate use of the alkalis for thermal ablation therapy is crucial to avoid significant side effects.

Research on chemothermal therapy using LM is a preliminary attempt. The liquid NaK alloy packed with silicone oil was developed for controllable chemothermal therapy in vivo. The oil-packaging strategy enhanced the security and stability of NaK alloy [62]. Significantly high heat was generated by injecting the NaK–oil mixture to tissues, the temperature in ablation area raised to 35.90 °C, and the instantaneous temperature of the center reached 75 °C. Additionally, using only about 10 μl pure NaK with oil package was enough to result in cell death and necrosis of VX2 tumor, indicating the potential in the treatment of tumors using liquid alkali metals. Liu and coworkers [63] further studied the therapeutic effect of NaK alloy at different growth stages of tumors in vivo. The initial body temperature could remain above 42 °C for ~50 seconds during injection. Based on the analysis of histopathology sections, tumors at an early stage received nearly thorough ablation while tumors at the developing stage were incompletely ablated owing to their large sizes and irregularly shapes. The tumor suppression rates of tumor treated at the early and developing stages were 88.5% and 67.6%, respectively.

In conclusion, the LM-based chemothermal therapy provides a possibility for effective oncotherapy. However, this new method is at the incubation stage, and it is more of a proof of clinical concept. It remains many challenges, such as the difficulty of accurately controlling the local temperature. Therefore, extensive research is needed before possible clinical implementation, the safety, and efficacy based on LM require more tests. Meanwhile, thermal dose studies and the corresponding medical equipment are indispensable as well.

8.2.4 Microwave Dynamic Therapy

Since there are interesting biological effects of microwave (MW) exposure, the MW as an electromagnetic spectrum, has sparked extensive research interest in tumor therapy. For its strong penetrability and effective heating capacity, MW has been applied in microwave thermal therapy (MWTT) [64–66]. Nevertheless, the low energy (3–10 eV) of the MW make it hard to split chemical bonds of materials to generate cytotoxic ROS, which limits the development of microwave dynamic therapy (MDT). Hence, exploring suitable MW sensitizers could open up opportunities for MDT.

Nanoscale LMs are able to enhance cytotoxic ROS triggered via the MW irradiation, because of the mechanism of electron transfer in hot spots [67]. For the difference of redox states, tumor cells tend to be preferentially killed using ROS compared with normal cells. As a result, LMs can be utilized as MW sensitizers for MDT.

Wu and coworkers made the EGaIn alloy and ionic liquid (IL) load into mesoporous ZrO_2 NPs to fabricate IL–LM–ZrO_2 nanoparticles (IL–LM–ZrO_2 NPs), realizing the combined therapy between MDT and MWTT [67] (Figure 8.8a). LMs played a

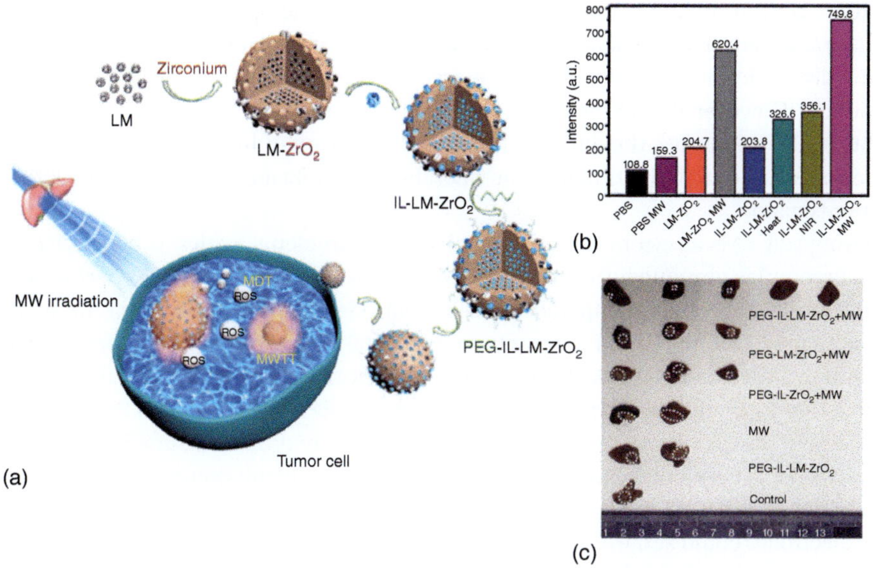

Figure 8.8 Microwave dynamic therapy based on LMs. (a) Diagram of MDT and the preparation of a PEG–IL–LM-ZrO_2 NP. (b) Intensity of fluorescence related to ROS amount under different treatments. (c) Digital photos of the livers from mice, which are under various treatments for 14 days; white circle refers to tumor area [67]. Source: Wu et al. [67].

crucial part in generating ROS through the irradiation of MW, inducing the apoptosis of tumors. The thermal effect of MW sensitive IL was concurrent with the ROS generation. The intensity of fluorescence reflected the concentration of generated ROS. Figure 8.8b indicated that the ROS generated from IL–LM–ZrO_2 NPs was associated to the MW irradiation. IL–LM–ZrO_2 NPs could produce abundant ROS via the MW. The generation of ROS could be influenced by the heat from MW heating effect of IL. Tumor photos in Figure 8.8c showed the inhibitory effect of PEG–IL–LM–ZrO_2 was relatively high with the MW irradiation, manifesting the combination between MDT with MWTT could achieve the significant therapeutic effect. Furthermore, the PEG modification endowed the ideal biocompatibility this LM-based material, as was confirmed by a toxicity test.

In conclusion, the application of LMs in the field of MW therapy is still in its infancy. LMs as MW sensitizers are able to be activated via MW for dynamic therapy, and they can be integrated with other biomaterials for combination therapy, paving a new route for dynamic treatment of tumors.

8.2.5 Electrochemical Therapy

As a minimally invasive, effective, and low-cost method, electrochemical therapy has been used for tumor therapy [68]. Generally, electrochemical therapy can be defined as imposing direct electric current or voltage through the insertion of electrodes on tumor areas. In this process, a variety of physiological and biochemical responses are induced, for instance, it can cause the change of pH and potentials of tumors, generate toxic products, induce the apoptotic, and stimulate immune system. The efficiency of the ablation of tumor depends on electrochemical reduction–oxidation reactions and the resulting drastic change of pH and toxic products. However, due to the unusual structure of the tumor and the nondeformability of the electrodes, the effective treatment area is often limited unless multiple electrodes are used, which makes the operations complicated and strengthens the patients' pain.

Using soft LMs as electrodes greatly solve these problems, which is beneficial to satisfy complex physiological conditions. In a recent study, LM electrodes performed better antitumor effect compared with Pt electrodes in vitro experiment, demonstrating the feasibility of LM to treat tumor cells [69] (Figure 8.9a). Electrochemical treatment based on EGaIn electrodes generated more toxic products to kill the tumor cell in the electrochemical process, with the remarkable enhancement of therapeutic efficacy. In the meantime, with high electrical conductivity and outstanding flexibility, LMs could completely attach to various electrical transmission-related surfaces, while it was difficult for the conventional rigid electrodes to make it. Moreover, the LM electrodes could achieve significant tumor growth regression and prolong life in the body, which indicated the development prospect of the present strategy. Overall, electrochemical therapy as a conceptual treatment with injectable LM electrodes is expected to promote individualized medical treatment in tumor and better relieve of patients' pain in the future.

8.2.6 Embolotherapy

Embolotherapy is a kind of endovascular therapy, this therapy aims to cut the supply for tumors by introducing embolic materials to embolize the arteries and capillaries. For its little harm and easy operation, it is widely applied for palliative treatment of tumors. Commonly, the proliferation and differentiation of tumors are triggered by various growth factors and sufficient provision of oxygen and nutrients by blood vessels [71]. Considering that tumor growth is dependent on the supply of blood vessels, using embolotherapy to starve tumors is an effective way. At the same time, research in this field has also received widespread attention [72]. At present, various embolization agents with the ability of blocking blood vessels have been developed such as small molecular materials. Meanwhile, several physical embolization methods have also appeared, such as coils and balloons. However, the conventional materials used commonly still have many shortcomings, restricting the application of embolization technology [73] For instance, despite the high mechanical strength of the metal microcoil, its filling rate is poor because of the poor packing ability in the blood vessel; the space at the interface can cause the permeation of nutrients, which would weaken the efficacy. In addition, its rigidity inevitably causes the damage to the vascular wall [74, 75]. Hence, proposing an alternative material is essential for vascular embolization therapy.

As a soft material with high fluidity, LMs can be easily injected and fully filled into blood vessels. At the same time, due to their plasticity, LMs can fit tightly to the blood vessels, thus filling the blood vessels sufficiently, preventing blood flow and

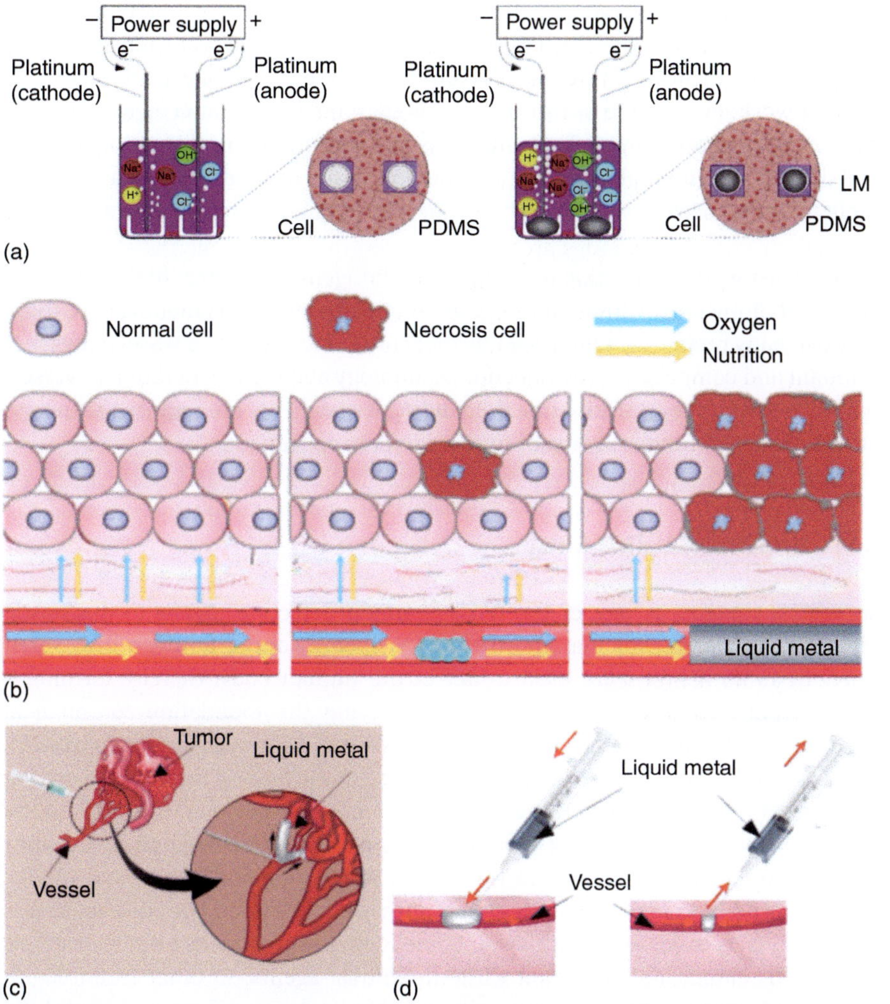

Figure 8.9 LMs for electrochemical therapy and embolotherapy. (a) Schematic of electrochemical therapy in vitro using platinum (Pt) electrodes (left) or LM electrodes (right) [69]. Source: Sun et al. [69]. (b) The principle illustration of LM-based tumor vascular embolotherapy. (c) Physical obstruction of blood supply for tumors. (d) Injection or suck of LMs on demand [70]. (b–d) Source: Wang et al. [70].

other related physiological influences. Based on this situation, Wang and coworkers demonstrated the potential of pure Ga and $Ga_{75.5}In_{24.5}$ alloy as the kind of unique embolization agents to starve the target tumor cells [70]. Figure 8.9b illustrated the mechanism of embolization therapy based on LMs. It is known that tissues acquire oxygen and nutrition transported by surrounding blood vessels to live, and when embolic agents partially occlude blood vessels, oxygen and nutrients can still reach necrotic cells through the blood vessels. Therefore, partly reducing the blood flow did not ensure a sufficient effect on starving tumor cells, and the unstable embolic agents can be carried away by the blood. With the ability to stay in liquid form at body temperature, LMs could easily plasticize along the shape of the blood vessel and completely occlude the blood vessel, resulting in a physical occlusion and tumor regressions (Figure 8.9c). Moreover, LMs could also be easily removed after completing their work, making the operation process easy and convenient (Figure 8.9d). In addition, the location and track of embolic agents were indispensable during the surgical operation. With high density, LMs could be easily tracked by X-ray and clearly display vascular occlusion, which offered an effective diagnostic tool to study tumor vascular.

In conclusion, the utilization of LMs as embolic agents is a conceptual strategy for future embolotherapy. Although LMs exhibit great potential in embolization, there still exist many problems, which need further consideration. As the vessel network is abundant and complicated, the injection technology and the regeneration of vessels should be further developed. More demanding assessments are required to ensure the safety of in vivo experiments or future clinical applications.

8.3 Biomedical Imaging

Biomedical imaging is one of the most common detection means to diagnostic diseases in modern medicine. Shortly after Wilhelm Konrad Rontgen discovered X-rays in 1985, X-rays were used to detect the human body, thereby forming a new discipline in diagnostic radiology and laying the foundation for medical imaging. Diverse imaging devices, consist of X-rays, computed tomography (CT), ultrasonography, and the like and are widely used in the field of biomedical imaging technology [1–6, 76–80]. According to different working principles, the information carriers of biomedical imaging include X-ray, gamma rays, magnetic resonance signal, reflected wave, transmitted wave, microwave radiation signal, and others. Since many tissues of the human body are not visible under X-rays, contrast agents are proposed to make them displayed. Some X-ray inspection techniques such as angiography cannot be performed without contrast agents. In other technologies such as CT, using contrast agents will enhance imaging effects and make the diagnostic image clearer, which helps physicians make a more reliable diagnosis. Nowadays, metal compounds and iodide have been widely used as contrast agents, such as gadolinium propellate glucamine solution (Gd-DTPA), mangafodipir trisodium (Mn-DPDP), and iron oxides for MRI, $BaSO_4$, and iohexol for CT and X-rays. However, there still exists the problems of large doses, low tumor targeting, high acute toxicity and side effects caused by metal compounds and iodides. LMs with high density, good biocompatibility, and low toxicity provide a new strategy to solve these problems [25, 42, 81, 82]. In this part, we will introduce the application of LMs in X-ray imaging and photoacoustic (PA) imaging.

8.3.1 X-ray Imaging

For a long time, X-ray diagnostic technology has been the fastest and most effective means applied in the noninvasive visceral inspection technology. As X-rays pass through the body structures, they will be absorbed to varying degrees based on the differences in density and thickness of human tissue. Then, images with different black and white contrast are formed on the screen or X-ray film. The effect of X-ray imaging is greatly improved after adding contrast agents. However, distinguishing the finer structures and tissues with high spatial resolution and high contrast remains an obstacle. There are two approaches to improve the fineness of X-ray imaging, one is to choose a higher energy X-ray source, and the other is to find an alternative high-density contrast agent. Alternatively, LMs were developed as a kind of X-ray sources due to higher energy, which could heighten the ability of high-resolution X-ray imaging in small animals [83]. The LMs used in this method were two alloys composed of GaInSn and GaIn, which can greatly improve the resolution of skeleton tomography and soft-tissue tumor imaging. Compared to the existing X-ray sources, LMs possesses higher penetrating power through thick tissue and shorter exposure times. As contrast agent, LMs are mainly used in angiography and CT scans, which would be discussed in detail below.

8.3.1.1 Angiography

X-ray imaging is commonly used in angiography. It is known that blood vessels throughout biological body continuously transport blood to provide necessary oxygen and nutrients for tissues and organs, while discharging metabolic waste. The health of blood vessels is vital for the body's physiological function. Through angiography, the distribution and structure of blood vessels can be clearly monitored, and the vascular conditions of the lesion can be found in time, including the existence, degree, location, and severity of the lesion. For X-ray angiography, the clarity of the image depends on the density of the contrast agent used, which can distinguish them from the surrounding tissues. After developing and studying many contrast agents, there are still insufficient in the resolution of angiography, especially the smallest capillary vessel networks. For example, when using X-rays with higher intensity to distinguish tiny capillaries, the effect of traditional contrast agent iodine and iodinated agents would decrease. Moreover, most contrast agents are solutions containing high concentrations of solutes, and the properties similar to blood vessels significantly reduce the possibility of distinguishing them in X-ray imaging. Considering the advantage of higher density and metal characteristics, LMs can be used as a novel contrast agent to visualize blood vessels, even the smallest [81]. The low viscosity and good fluidity of liquid Ga facilitated its direct injection into the targeted vessels in vitro, which included heart and kidney of pigs. The blood vessel distribution of the hearts was distinguished after infusing with liquid Ga. It was obvious that Ga provided superior contrast effect and a highly clearer coronary network than that with iohexol. The contrast between Ga-filled blood vessels and surrounding tissue increased significantly with X-ray intensity. Correspondingly, injecting liquid Ga into the pig's kidney had similar imaging effect as that of heart. Meanwhile, the veins of tiny blood vessels were very clear, and 0.1 mm vessel was even legibly seen. From the development mentioned earlier, the application of LMs angiography opens a brand new way for exploring finer structures under the X-ray scan. With strong

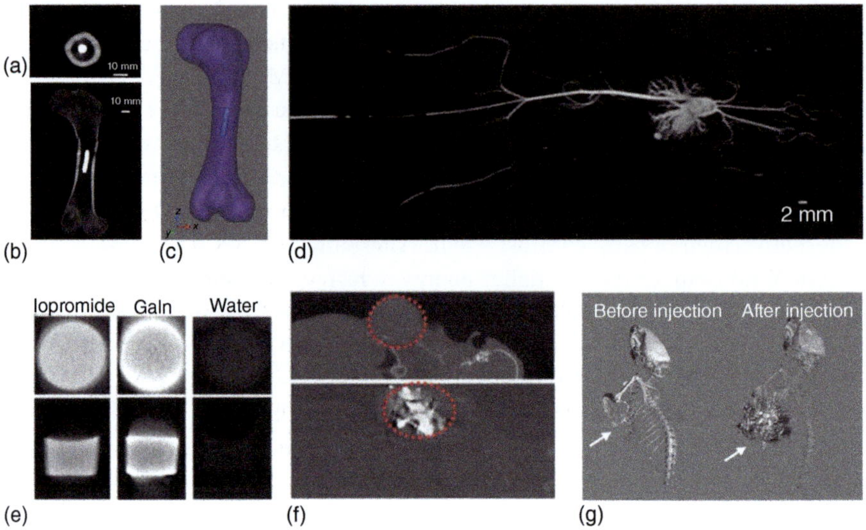

Figure 8.10 CT scans of various tissues and organs filled with LMs. (a, b) The transverse and coronal sections CT slices of bone with LMs. (c) The geometrical model of bone with LMs reconstructed by the CT slices [84]. Source: Yi et al. [84]. (d) The 3D reconstruction image of CT scan for the whole body's blood vessels of mouse injected with LM [70]. Source: Reproduced with permission of arXiv. (e) CT contrast of LM, iopromide, and water. (f) The comparison CT slices of the tumor site after injection of LMs. (g) 3D reconstruction comparison image of the mouse body after injection of LM [25]. Source: Wang et al. [25].

contrast and high resolution, LMs angiography helps people understand more about visualizing biological anatomy and vascular disease.

8.3.1.2 Computed Tomography

Since X-ray film is difficult to detect the lesions of those overlapping tissues, scientists start looking for alternative technology to make up for the lack of X-rays imaging. As another powerful technique, CT has been developed for imaging of soft tissue systems, which mainly irradiates the human body through the X-ray rotation of a single axis. Then, due to the different absorption capacity of X-rays in different tissues, the geometrical processing technology of the computer can be used to reconstruct the image of the slice to obtain the overall structure. As contrast agent, the role of LMs in CT is to enhance the effect of imaging and real-time monitor when using them to treat tumors and diseased tissues. For instance, when LM was used as a phase transition bone cement, it can also be used as an imaging contrast agent for further monitoring and evaluation of the cement in porcine femurs [84]. In Figure 8.10a, b, LM injected into the bone was clearly seen from the transverse and coronal sections of CT slices, indicating the excellent imaging ability and high contrast of LM. To understand the whole structure of the LM filling quality, the geometrical model of bone was reconstructed subsequently (Figure 8.10c). The obtained information about the shape of the filling metal could offer high contrast reference for medical measurement.

CT imaging-based LMs is often used as an observation mean for treatments. As for tumor therapy, the introduction of LMs as contrast agents can enhance the effect of CT scan [70]. In the study of starving diseased tumors, LMs were injected into blood vessels as embolization agents to block blood. As shown in Figure 8.10d, after the whole body's blood vessels of mouse injected with LM, the vessels were obviously distinguished in the reconstruction image of CT scan. In this image, the aorta of the entire mouse was clearly visible with LM, and was distinguished from the limb and tail vessels. Furthermore, the tumor vessels were also clearly shown because of ultra-high contrast of LM. Besides, there were some irregular vasculature of tumor vessels in their 3D reconstruction image, which clearly reflected the changes in shape caused by different growth processes of diseased blood vessels.

In cancer treatment, the radiopaque characteristic of LM under X-ray endows potential of LM compound to CT imaging [25]. The drug-loaded compound based on LM was evaluated with CT scan, by both in vitro and in vivo experiment. Figure 8.10e exhibited the CT contrast image of LM, iopromide, and water in vitro, in which LM demonstrated a better X-ray contrast intensity than of iopromide and water. The changes before and after the injection of LM at tumor site were shown in Figure 8.10f. After injection of LM, an enhanced signal was obviously observed, which was further demonstrated by the reconstructed 3D images of CT scans (Figure 8.10g).

8.3.2 Photoacoustic Imaging

In recent years, PA imaging as a new biomedical imaging method has been developed. When the laser light irradiates biological tissue, the light energy absorbed by contrast agents would convert to thermal energy. Then, biological tissue produces broadband acoustic waves at megahertz frequencies on account of the thermal energy, which is called a PA signal. PA signals generated by biological tissues can carry light absorption characteristics information of the tissue. An image of the light absorption distribution in the tissue can be reconstructed after receiving these PA signals [85]. PA imaging possesses not only high selection characteristics of optical imaging, but also deep penetration characteristics of ultrasonic imaging. Hence, tissue images with deeper imaging depth and higher spatial resolution can be obtained via this imaging method [86–88]. Virtually, the mechanical and elastic properties of tissues have not affected the contrast of PA images, but its optical properties, especially optical absorption, determine the contrast. Hence, developing optically contrast agents is expected to promote PA imaging technologies through the combination of optical effects and multiple imaging.

Recently, LMs displayed a PA effect under the irradiation of NIR laser. Specifically, multifunctional light-driven LM nanocapsules were constructed for the theranostics and elimination of cancer cell in living animals [42]. As seen in Figure 8.11a, the synthesized particles consisted of LM core and polymer shell, demonstrating the formation of capsule-like structures. It was found that LM nanocapsules would undergo morphological changes and thermal Brownian motion after laser irradiation. The PA signal of synthesized LM nanocapsules was measured in the wavelength range of 670–970 nm, in which the strongest signal appeared at 680 nm (Figure 8.11b). The ex vivo PA imaging of phantom samples indicated that the enhancement of the PA intensity followed the concentration of LM nanocapsules (Figure 8.11c). Then, in vivo experiments were carried out in mice by the injection of LM nanocapsules and distilled water. PA signal did not appear when only water was injected; however, a strong PA signal was generated in the mouse with LM nanocapsules, as shown in Figure 8.11d. Moreover, compared to the mouse injected with LM nanocapsules (PA signal is ~9–45%), the mouse injected with epidermal growth factor receptor (EGFR) antibody-functionalized LM nanocapsules displayed an augmented PA signal (~22–102%) (Figure 8.11e). 3D imaging in Figure 8.11f also verified the excellent imaging effect of antibody-functionalized LM nanocapsules in the tumors of the mouse. Therefore, the emerging LM nanocapsules are useful contrast agents in the PA imaging, providing a robust tool for biological imaging.

In summary, LMs can be easily injected into the tissues and organs and visualize them with high resolution under the X-ray and laser light. The application of LMs in biomedical imaging provides a mega contrast quality for further scientific and medical researches. After more security and stability studies, LM-based biomedical imaging has the capability to be used in clinical application. In addition, the research helps to deepen the understanding of body's organs and structures, which is meaningful for both pathology evaluation and forensic medicine. The in vivo study of LMs on imaging is expected to have a positive impact on future research and clinical application.

8.4 Biomimetics

Biomimetic materials generally refer to artificial materials that are designed and manufactured to imitate the operating mode of living things and the structure of biological tissues. Understanding the relationship between the microstructure of natural materials and specific properties is of great significance in fabricating biomimetic materials. With the continuous improvement of current medical standards, the provision of safe and effective biomimetic materials for tissue replacement and transplantation for patients has become a focused research, covering the fields of biology, medicine, and materials science. In particular, it remains challenging to design and synthesize biomimetic materials for biological tissues and organs with complex functions, such as artificial bone and artificial nerve. In order to prepare better performance bionic materials, it is necessary to combine the features of both engineering material and medical material. Generally, biomedical materials require safety, nontoxicity, biocompatibility, certain mechanical strength, and the like. Although there has been a lot of progress in this area, biomimetic materials developed in the past only meet part of the demand and are not capable of meeting challenges. Among them, LMs as the emerging materials possess some properties similar to biological materials, owing to their mechanical strength and softness based on the phase transition process. Hence, LMs have great potential in replacing living

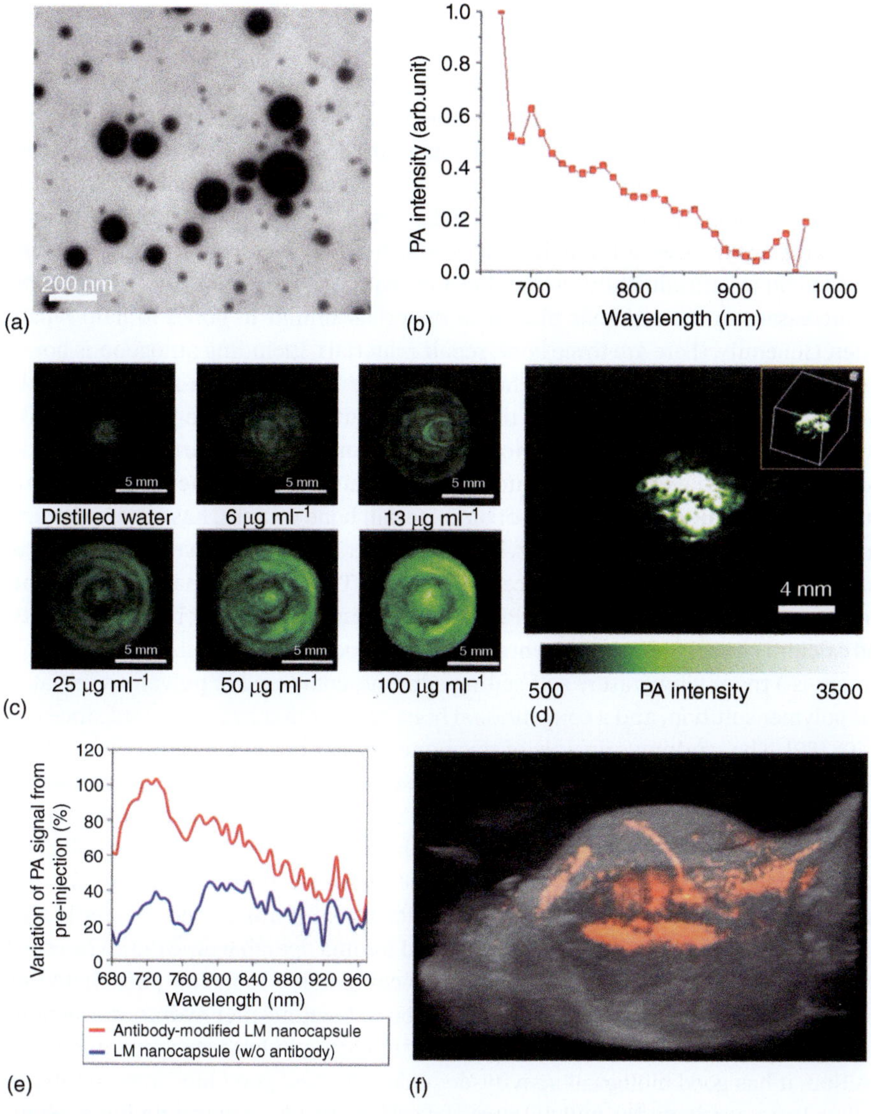

Figure 8.11 PA imaging based on LM nanocapsules. (a) TEM image of LM nanocapsules. Source: Chechetka et al. [42]. (b) PA intensity of LM nanocapsules at different wavelength. The concentration of LM nanocapsules is 100 mg ml^{-1}. Source: Chechetka et al. [42]. Licensed under CC BY 4.0. (c) Ex vivo PA imaging of phantom samples. (d) PA imaging of the mouse with LM nanocapsules under the skin. (c, d) Source: Chechetka et al. [42]. (e) PA intensity of tumor injected with LM nanocapsules with and without antibody in wide wavelength range under laser irradiation. Source: Chechetka et al. [42]. Licensed under CC BY 4.0. (f) 3D PA imaging of the tumor sites of the mouse injected with LM nanocapsules containing antibody [42]. Source: Chechetka et al. [42].

tissues and organs or restoring and maintaining their functions. This part mainly introduces the research progress of LMs for biomimetic bone and nerve connector.

8.4.1 Reversible Bone Cement

As an important part of the human body, bones play an indispensable role in protecting internal organs, forming a skeleton to maintain body posture, as well as the storage of minerals, hematopoiesis, and motor. However, diseases in bones may result in structural deterioration and functional disorders, such as osteoporosis, arthritis, and malignancy. Despite the ability of bones to repair themselves, some severely damaged bones are clinically irreparable and require surgery to recover. Therefore, it is necessary to use nontoxic biological materials similar to bones to help repair bones. Generally, there are three bone-repair materials, including autogenous bone, allogeneic bone, and artificial synthetic bone. The first two bone-repair materials have their own shortcomings, and the source of autogenous bone is limited while the allograft bone will bring immunological rejection. Therefore, artificial synthetic bone becomes the favorable candidate for bone-repair surgery. In the development of artificial bone material, a new type of biomaterial, bone cement, has attracted enormous interest. As the name suggests, bone cement is a kind of fixative agent, which is mainly used for artificial joint replacement surgery. There are two main types of bone cement: polymethyl-methacrylate (PMMA) bone cement with poor biocompatibility and calcium phosphate bone cement (CPC) with good biocompatibility. PMMA bone cement is a room temperature self-setting adhesive composed of pulverous initiator and polymer solution, and a conventional bone cement that has been used since the 1960s [89]. The working principle of this bone cement is a polymerization reaction of methylmethacrylate that occurs during mixing. When the reaction is completed, the PMMA bone cement becomes a solid with high mechanical strength, making the implant load distribution uniform. However, the heat generated during the polymerization of PMMA bone cement would cause the necrosis of the surrounding cells and tissues [90–92]. In addition, the clinical application of PMMA bone cement is also limited by the cytotoxicity of the monomer and limited operable time [93]. The other essential bone cements, CPC, consists of two components, calcium phosphate salt powder and liquid blending agent, which can be self-setting after hydration reaction. Since CPC is similar to human bone tissue after curing in both structure and composition, it has good biological activity, degradability, and good biocompatibility, as well as bone conductivity, which promotes self-repair of surrounding bone. However, CPC has low mechanical strength and high brittleness and cannot be used in load-bearing applications, such as leg bone-repair application. Hence, it is necessary to explore new materials for bone cement. In general, appropriate bone cement should possess the characteristic of proper injectability, biocompatibility, adequate mechanical strength, low curing temperature, and radiopacity.

The low-melting-point, excellent fluidity, outstanding mechanical strength, and good biocompatibility of LMs satisfy these requirements. As a result, LM can serve as an appropriate candidate of bone cement. Meanwhile, low-melting-point LMs behave like PMMA bone cement in many aspects, including the similar phase

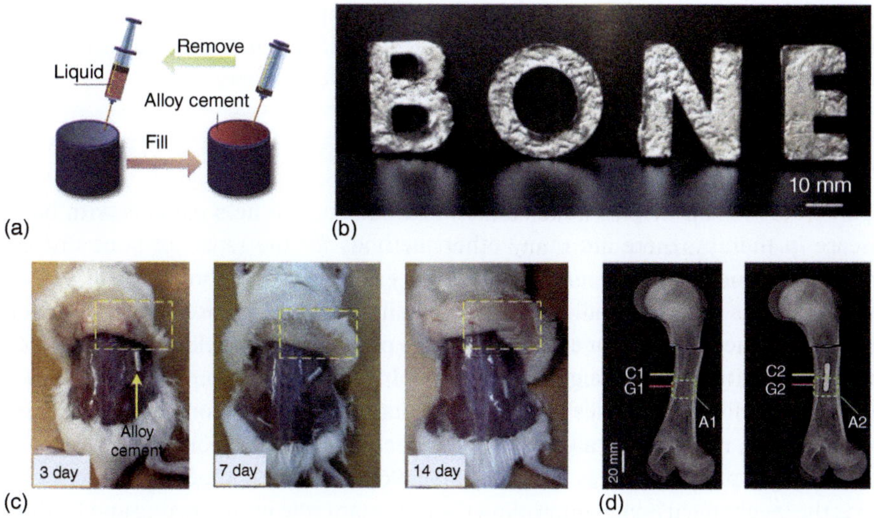

Figure 8.12 LMs bone cement. (a) Schematic of filling and removal process of LM bone cement. (b) A shaped model of bone cement based on LMs into "BONE." (c) The photos of mice implanted with LM bone cement after 3, 7, and 14 days. (d) X-ray imaging of bone with part marrow removed and the bone filled with LM bone cement [84]. Source: Yi et al. [84].

transition process and the ability to join two materials together. Remarkably, there have been precedents in the industry for using liquid eutectic alloys as binders. With the progress of research in LMs recently, $Bi_{35}In_{48.6}Sn_{16}Zn_{0.4}$ alloy was prepared as the candidate for novel bone cement, whose melting temperature was 58.3 °C [84]. The working principle of the alloy was the liquid–solid phase transition at low temperature. Due to the low-temperature fluidity, the melting LMs were easily injected to the target position by an ordinary syringe and then solidify in situ (Figure 8.12a). Based on the phase transition process of LMs bone cement, the solidified cement could also be removed by heating to liquid in the same way as injection (Figure 8.12a). This characteristic made LM bone cement only need a simple operation to complete the injection and removal, which is very convenient in clinical practice. Unlike the chemical reaction of the two substances of traditional materials, LMs bone cement only underwent a physical reaction of liquid–solid phase transition, which was more simple and controllable. Meanwhile, it was found that LMs bone cement could be shaped into different models at low temperature (Figure 8.12b), which guaranteed the completely filling the shaped bone cavity and randomly shaping according to the bone defect. The moderate operating temperature and a small amount of heat release in phase transition procedure of LMs bone cement also ensured that surrounding tissues were not damaged during operation procedure. Moreover, the good biocompatibility of such LMs bone cement was demonstrated by in vitro and in vivo experiments. The wounds caused by needle puncturing healed well after 3 days (inside the dotted frame in Figure 8.12c), and was mostly invisible after 7 days and 14 days. As mentioned in the previous section about the advantage of LMs-based bioimaging, the state of formed cement in the body can be evaluated by X-ray and CT scan. The shape of the alloy cement injected into the femur of the pig was clearly seen in Figure 8.12d. This outstanding imaging facilitated the monitoring of internal bone cement after surgery.

8.4.2 Flexible Exoskeleton

Apart from the injection of bone cement into the body to help patients with bone disease in therapy, there are many other methods for the repair of bone disease and performance enhancement of bone. The exoskeleton is representative among them, which is used for medical auxiliary healthcare. Exoskeleton is mainly made of materials such as metal or carbon fiber. The purpose of exoskeleton is to enhance human mobility, reduce weight load, and help people with complicated diseases overcome mobility difficulties [94]. In the military, the exoskeleton helps the soldier to fight better, because he can get better protection and bear more equipment and have stronger physical fitness than usual [95, 96]. The core component of exoskeleton is the mechanical joint, which plays the important role in supporting and loading [97]. Generally, conventional mechanical joints consist of bearings and a hydraulic mechanism, achieving flexibility and load bearing. However, because of the need for pumping machinery, the conventional mechanical joints have to cost much and is not stable and reliable.

The basic of preparing flexible mechanical joints was choosing materials with phase-change effect that can conveniently switch between liquid phases and solid phases. In other words, the flexible mechanical joint can be in a flexible state or rigid states respectively, as long as the heating or cooling process is performed, to perform bending and weight-bearing movements. LMs can not only meet the above requirements, but also overcome the shortcomings of the traditional mechanical joint. Appropriately, a flexible mechanical joint based on the LMs was proposed for human exoskeleton [98]. The melting point of the eutectic alloy $Bi_{32.5}In_{51}Sn_{16.5}$ was 60 °C and was a great phase-change material. In practical applications, the front arm and the rear arm of the joint were connected to the upper limb of human arm. This part of thermoelectric device was for heating to melt and cooling to solidify LMs. The metal changed from solid to liquid when the temperature rising to 60 °C. Compared with the traditional phase-change material paraffin, thermal conductivity of LMs was higher and heat capacity was smaller, indicating the faster heat transfer. As a consequence, the mechanical joint could respond very quickly and consume little power. Meanwhile, the mechanical strength of LMs and conventional paraffin were compared, and the results showed that low-melting-point alloy had larger loading capacity. The application of LMs in exoskeleton provides more options for powerful, convenient, and effective exoskeletons in the future.

Apart from being an important part of in flexible mechanical joint, LMs can also participate in the soft wearable robotic exoskeleton device, which helps the disabled to walk. For instance, a wearable robotic exoskeleton device was developed in ankle rehabilitation, which was driven by artificial muscle actuators [94]. Such an exoskeleton device simulated the musculoskeletal behavior of human's foot and lowered leg, reproducing an artificial muscle structure with similar morphology and functionality. LM was employed as one of the key components in the robotic device for the purpose of being flexible. EGaIn alloy was embedded in the microchannels made by a silicone elastomer to form the strain sensors, which was used to measure the angles of ankle joint [99]. When received axial strain, the length of entire channel increased and the cross-sectional areas decreased accordingly, resulting in the enhancement of electrical resistance. Comparing to traditional rigid sensors, the strain sensor filled with LM was highly flexible and stretchable, making medical device wearable and adaptable to complicated body shapes. With the development of technology, the flexible structure is becoming a new demand for biomedical devices. Thus, the application of LMs will facilitate the development of more flexible and intelligent medical devices.

8.4.3 Nerve Connector

Presently, peripheral nerve injury (PNI) as one of the important causes of movement disorders often caused by accidents, disease, and long-term overload [100]. As a result, the motor function will suffer from the partial or complete loss. With the advancement of microsurgical technology and anatomy, the structure and pathophysiology of peripheral nervous system have been deeply studied and understood. But the treatment of PNI is still a huge challenge in clinical and basic research. Depending on the location and extent of the damage, the repair and regeneration of PNI are mainly divided into nerve suture, cannula, graft, and transfer. The method of direct suturing of the cut ends (nerve stumps) is normally applied when the gap is less than 20 mm [101]. Conversely, if there is a significant nerve gap, nerve suture is incapable to repair the gap, and other methods need to be introduced including autograft and allograft. As the current gold standard, nerve autografting becomes the most common way to repair peripheral nerve in most clinical cases [102]. But the promotion of autografting is limited by scarce graft resource, significant donor site morbidity, and required dimensions for matching [103]. Different from nerve autografting, nerve allografting enables to bridge larger nerve gap, but the systemic immunosuppression lasts 18 months and may cause opportunistic infections and tumor formation during this time [104]. To find an alternative approach, various natural and artificial biomaterials have been investigated as nerve conduits and connectors in experimental and clinical conditions to promote the regeneration of peripheral nerve defects and functional recovery [105, 106]. After a large number of animal experiments, electrical stimulation has been proved the potential to promote the regeneration of injured nerves [107]. Thence, conductive materials have attracted a lot of attention in neural tissue engineering [108–110]. To satisfy the requirements of nerve conduits, the conductive materials also need to possess the mechanical stability, biocompatibility, and biodegradability. Conductive polymers have been widely studied and shown excellent performance, such as polypyrrole, polyaniline, and others [111]. However, such problems as the discharge of acidic degradation products, poor process capability, and premature failure of mechanical properties during degradation hindered the application of conductive polymers.

Researchers are committed to developing a novel material to meet the biological requirements and electrical conductivity.

As mentioned earlier, LM exhibits excellent electrical conductivity and flexibility, which may be beneficial for conducting the exciting signal of nerve. Meanwhile, LM has been proven to have good biocompatibility, which is possible to achieve the in vivo applications. Hence, such material is introduced to be peripheral nerve functional recovery channel in the regeneration of transected peripheral nerve. $Ga_{67}In_{20.5}Sn_{12.5}$ alloy was prepared for the first time to treat peripheral neurotmesis, which had aroused great attention from all over the world [112]. The blue frog as the experimental object, two broken sciatic nerves were reconnected by $Ga_{67}In_{20.5}Sn_{12.5}$ alloy (Figure 8.13a). Due to its liquid state at low temperatures, $Ga_{67}In_{20.5}Sn_{12.5}$ alloy was convenient in fabrication, encapsulation, and surgical operation. For comparison, the conventionally used Riger's solution was also filled into the two broken sciatic nerves, similarly setting on a glass slide (Figure 8.13a). With high conductivity, LMs and Riger's solution not only connected broken sciatic nerves, but also conducted the excited signal of nerve during the regeneration process in vivo. Figure 8.13b showed the measured waveform of electroneurographic signal conducted by intact nerve and reconnected nerve, which the connectors were $Ga_{67}In_{20.5}Sn_{12.5}$ alloy and Riger's solution. The electroneurographic signal of alloy was basically synchronized with the electric-stimuli signal. It was confirmed that LM could efficiently connect the broken nerve and simultaneously conduct the excited signal of nerve in vitro. Later, the contrast experiment revealed that LM was superior to Riger's solution in performance as the peripheral nerve functional recovery channel. This was because $Ga_{67}In_{20.5}Sn_{12.5}$ alloy had lower impedance than Riger's solution, resulting in better conduction of the weak electroneurographic

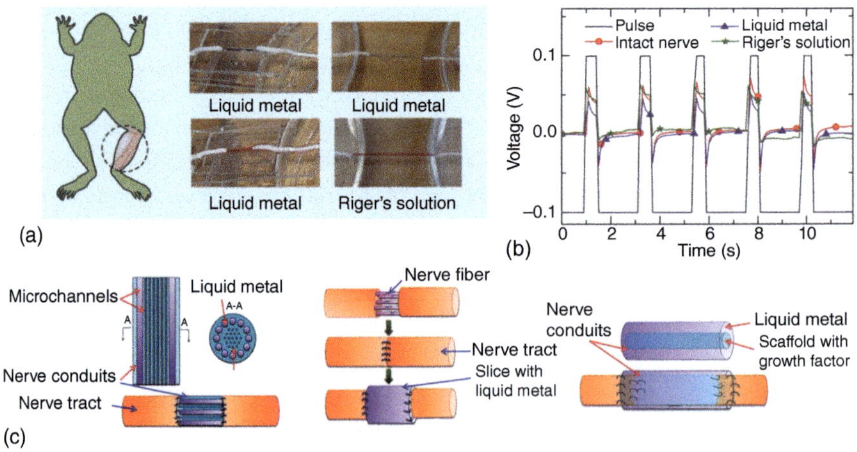

Figure 8.13 Repair of PNI based on LMs. (a) Schematic of the broken sciatic nerve reconnected by LM (upper) and Riger's solution (under). (b) The pulse signal and the conducted signal by intact nerve and reconnected nerve, and the connectors were LM and Riger's solution. (c) Three available configurations based on LM for the repair of nerve conduits [112]. Source: Zhang et al. [112].

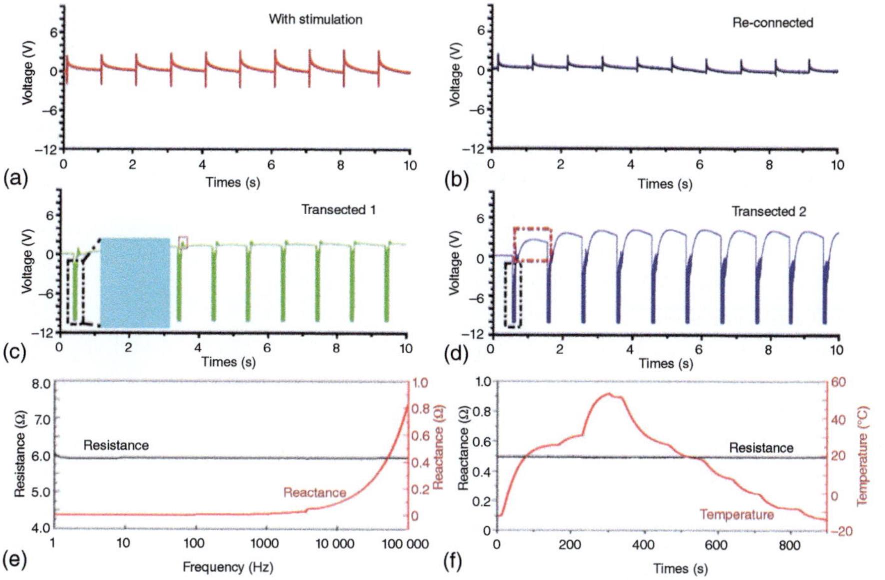

Figure 8.14 In vivo neurological signal detection based on the Ga-reconnected sciatic nerve. (a) The intact nerve with stimulation. (b) The sciatic nerve after reconnection operation. (c, d) Broken nerve with two electrodes attached, one of which is that both electrodes are onto the proximal nerve stump in (c), the other is one on the proximal stump and one on the distal stump in (d). The insert with dotted line in (c) was the negative bursting firing signal. (e, f) The impedance of Ga with different frequencies and temperatures [113]. Source: Liu et al. [113].

signal. Furthermore, several available configurations for the repair of nerve conduits by LM have been proposed, including microchannels, biocompatible slice coating as well as concentric tubes (Figure 8.13c). LMs are essential in the enhancement of the conduction of electroneurographic signals in these nerve conduits.

Subsequently, in vivo experiments were carried out to study the effect of LM employed as the connector [113]. Liquid Ga was employed to reconnect the broken sciatic nerve of mouse. As can be seen in Figure 8.14a, b, the nerve signals detected with the reconnection of Ga were comparable to that from the intact nerve with stimulation. There was a negative bursting firing signal when the nerve was transected (Figure 8.14c, d), which disappeared after nerve-reconnection (Figure 8.14b). This further confirmed restoration of the transmission of neural signal by the reconnection surgery. The study of the physical properties of Ga revealed that the impedance was almost unaffected when the frequency and temperature changed (Figure 8.14e, f), which indicated the stability of Ga in future clinical applications. Additionally, the pathological results confirmed that nerve reconnection by Ga could distinctly postpone the muscle atrophy.

After successfully reconnecting the nerves, researchers are trying to investigate the application of LMs in recovering animal locomotion functions. Electrical stimulation of nervous tissue is an effective means in the rapid restoration of the nerve function [114]. Since commonly stimulating electrodes are rigid conductive

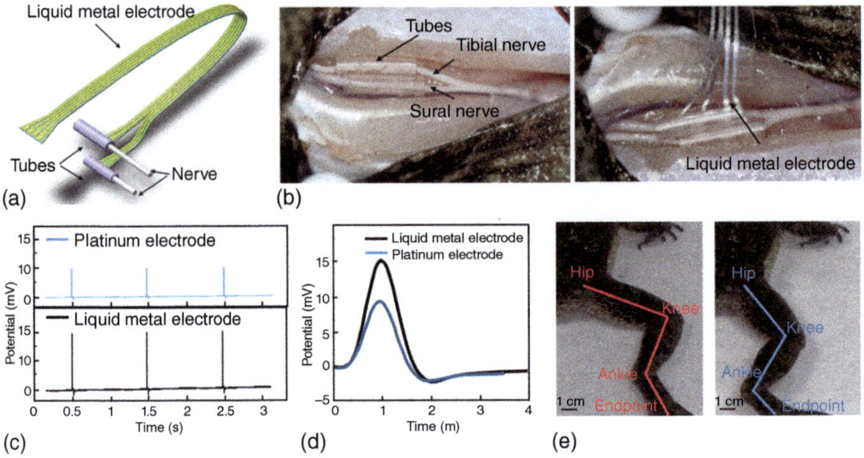

Figure 8.15 The restoration of the movement of bullfrog limb with LMs as connectors. (a) Schematic of the structure of the LM-based nerve electrodes. (b) Images of the implanted LM electrode between the sural nerve and tibial nerve of bullfrog. (c, d) The comparison of action potential and its amplitudes in sciatic nerve under the stimulus of Pt electrode and LM electrode. (e) The lower limb movement of the dead bullfrog under the electrical stimulations. [116]. Source: Guo and Liu [116].

materials, there are wear and injury between soft tissue and rigid electrodes and the corresponding electrical stimulation would reduce after long-term. Therefore, it is required to take advantage of flexible materials with similar properties to biological tissues as stimulation electrodes. Recently, LMs have been found to have great flexibility and deformability, which was the favored material for preparing implantable flexible electrode [115]. GaIn alloy was adopted as electrode material and encapsulated inside flexible PDMS polymer to fabricate an implantable electrode [116]. The structure of the nerve electrodes was shown in Figure 8.15a, and the LM electrode was connected to the nerve through a silicone tube. LM electrode was implanted into the blue frog body to connect the nerve and perform electrical stimulation experiment, while the Pt electrode was used as the comparison (Figure 8.15b). Under the stimulation of two electrodes, the action potential was detected and shown in Figure 8.15c, at the same time characteristics and inconsistent amplitude characteristics. The amplitudes of nerve action potentials stimulated by Pt electrode and LM electrode were exhibited in Figure 8.15d, which confirmed the better conductive performance of LM due to the higher amplitude of action potential. Then, LM electrode was used to restore the motor function of dead frogs by stimulating its nerves. As shown in Figure 8.15e, the lower limbs of the dead bullfrog could rhythmically lift up under the electrical stimulations from the electrodes, which demonstrated the ability of LM electrode to quickly recover biological nerve functions. These results indicated that LMs can not only repair nerve damage very well, but also help to restore motor function.

With the rapid advancement of technology, the bionic design of biomedical materials and the development of a new generation of biomedical materials are the inevitable trend in the development of materials science. Until now, LMs have been expected to be emerging biomimetic material due to their high mechanical strength, good electrical conductivity, biocompatibility, low toxicity, and low phase transition temperature. Thus, the medical diagnosis and healthcare based on LMs will promote precise and intelligent development in biomedicine. Although bone cement and nerve connectors composed of LMs are still in the preliminary research stage, they make a significant impact on future research.

8.5 Biodevices

Biodevices play an important role in biomedicine. With the exploration of biodevices with distinct characteristics and functions, it has an enormous application prospects in health care monitoring, drug delivery, and tissues/organs stimulation. Compared with traditional biomaterials, LMs as new functional materials demonstrate good biocompatibility and excellent deformability, which endow biomedical devices with soft, stretchable, and flexible ability. This part mainly illustrates the research progress of biosensor, implantable electrode, and microfluidic pump based on LMs.

8.5.1 Biosensor

Real-time monitoring of vital signs is generally considered as an important means of modern medical diagnosis. Changes in human physiological signals can be precisely detected and recorded by a variety of biomedical instruments. Currently, the development of biosensors for medical systems has received widespread attention [117–119]. Biosensors are integrated with microelectronic components attached to human skin for health assessment and collection of sensitive tactile information. It is worth noting that the emerging LMs are suitable for the production of high-performance biosensors, which is attribute to their desirable deformability, high conductivity, low viscosity as well as good safety in an ambient environment. Furthermore, the physical and chemical properties of LMs can be improved through size adjustment and surface functionalization [120, 121], which make LMs possess considerable development potential in biosensors.

A new type of stretchable noninvasive pulse sensor based on LM was fabricated through a selective LM plating technology [122]. The flexible pulse sensor could be conveniently adhered to human skin for personal health monitoring. In the process of sensor preparation, the formation of various complex LM patterns largely depended on the selective wetting behavior of hydrochloric acid reduced Ga droplets on different substrate surfaces. The LM patterns possessed S-shaped uniform surface with a width of about 2 μm. A transparent light-emitting diode (LED) with programmed circuits was used to display the working status of the stretchable pulse sensor. When a certain working voltage was applied to make the green LED light on, the sensor worked very well in any state of bending, twisting and stretching

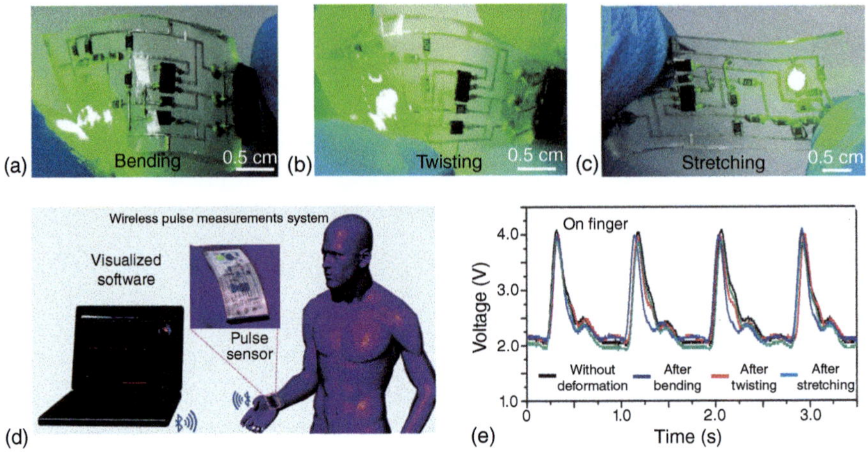

Figure 8.16 Movement monitoring of biosensors based on LMs. Top views of the transparent flowing LED lamp operating under (a) bending, (b) twisting, and (c) stretching. (d) Schematic of the wireless pulse sensor device. (e) The measurement results of human heartbeat under different movement states of finger [123]. Source: Li and Lee [123].

(Figure 8.16a–c). Moreover, owing to the liquid nature of electrical wires, the flexible circuit was able to self-healing when it was destroyed. Combining a retractable pulse sensor with a wireless device achieved noninvasive heart rate measurement (Figure 8.16d). When the pulse sensor was glued to the human skin, the collected electrical signal was transformed into digital signal via an Arduino board and wirelessly transported to a laptop computer, and then the monitoring data were presented on the displayer. The experimental results indicated that the soft pulse sensor still maintained superior electrical performance and stable mechanical features at various external stimulations, including deformation, bending, twisting, and stretching (Figure 8.16e). Thus, such portable LM pulse sensor exhibits enormous potential in monitoring the physiological health of the human body.

LM-based biosensor can not only realize noninvasive heart rate measurement, but also be able to accurately monitor the complex movement of fingers. A gel-like LM integrated system for wireless human motion sensing was proposed [124]. All active ingredients in the sensor were composed of Ga, which demonstrated unique skin-friendly and stretchable characteristics. Each deformation of the sensor system corresponded to a definite movement, and achieved simultaneous sensing of multisite. For example, the system was capable of attaching independent sensors to the joints of separate fingers to record complex movements at multiple points when monitoring finger flexion. The wireless detection ability provided by the sensor system did not cause performance degradation even in case of large strain (30%) deformation. This powerful sensor platform had received tremendous research interest in human–robotic interfaces for prosthetic control and other wearable areas. LMs-based sensor systems utilized active LMs to achieve good performance, but usually lack precise and suitable methods for assembling active liquids into functional devices. Using a digital fabrication platform and a full

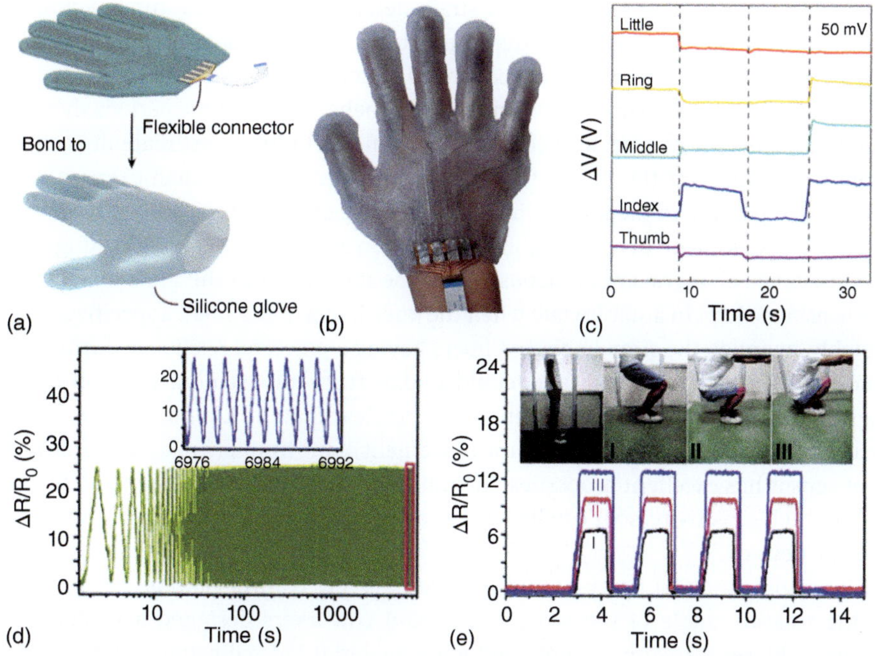

Figure 8.17 (a) Schematic of smart glove. (b) The image of a smart glove composed of liquid strain sensor, liquid interconnect, and flexible connector. (c) Real-time response of the strain sensor on a single finger to display various gestures [125]. (a–c) Source: Wang et al. [125]. (d) Cycle durability and stability of the LM fiber sensor under 70% strain deformation. (e) Exercise monitoring of squatting postures with different amplitudes [126]. Source: Wu et al. [126].

range of three-dimensional printing technology, a highly elastic three-dimensional integrated sensor system was produced [125]. The special flowing property of the elastomeric ink formulation made the printing technology exhibited high fidelity nature. Various elastomeric sensors with outstanding sensing properties can be manufactured via directly injecting LM galinstan into the microfluidic channels. A smart silicone-based glove with a fully integrated sensing system (Figure 8.17a) was manufactured by this attractive fabricating method. The skin-like electronic glove exhibited a soft three-dimensional structure, which was formed by interconnecting electronic devices and LM components (Figure 8.17b). Portable smart glove could capture a variety of gestures through monitoring the potential signal output during finger movement. When the user made various gestures continuously, the real-time response data record of the smart glove (Figure 8.17c) indicated the bending deformation induced by an individual finger.

In addition, a high-performance LM fiber sensor was used for exercise monitoring of the knee joints. As a kind of highly flexible inorganic conductor, LM was infused into the tubular channel forming a unique superelastic biocompatible fiber sensor [126]. The good combination of soft elastomer and inorganic conductor offered a facile method for manufacturing excellent biosensors. Figure 8.17d showed the features of the LM fiber sensor at 70% strain deformation. The real-time recording results of 3500 periods displayed that the output resistance data remain basically unchanged, achieving a favorable mechanical match properties of the LM and fiber channel. Moreover, the contraction and recovery behavior under the periods showed that the fiber sensor possesses good durability and stability. To estimate the actual monitoring ability of the highly flexible LM fiber sensors in human exercise, the sensor was adhered to the knee joints. The response of the knee-mounted sensor to continuous squat with diverse amplitude was shown in Figure 8.17e, indicating that the resistance changes simultaneously with the movement of the knee joints, and the resistance kept in a stable state when the knee joints maintained a specific squat. Notably, owing to the stretchable LM fibers having a powerful resilience during the deformation process, the resistance could quickly recover to its initial position while the knee joints were straightened rapidly. It was clear that the LM fiber sensor could detect the skin stretching caused by the movement of the human knee joint. This LM fiber sensor has excellent properties, including hysteresis-free, excellent stability as well as high sensitivity, so it can be used as an accurate and quantitative intelligent sensing device.

Biosensors can also be employed to quantitatively control motion states, such as the rotation angle of the wrist. Sheng and coworkers designed a stretchable capacitor biosensor composed of conductive material LM galinstan and dielectric material VHB tape [127]. The LM flexible capacitor sensor was used for frequency response measurement (Figure 8.18a), the output voltage corresponding to the frequency response signal indicated that the capacitor sensor possessed an extremely fast response speed. To study the practical application of the capacitor sensor in physiological signals detection, the movement of the wrist joint was tested by a capacitor sensor (diameter 30 mm) (Figure 8.18b). Quantitative analysis of the functional relationship between the rotation angle of wrist joint and capacitance data indicated that there was a great linear correlation (Figure 8.18c). The LM capacitor sensor has excellent performance, such as good stretchability, high sensitivity, and stable electrical features, making it a promising biomedical device that can be used for motion control, electronic skin as well as accident alarm.

Biosensors provide a simple and effective method for blood pressure measurement in clinical applications. The capacitive fluid pressure microsensor based on soft LM was composed of a detection microchannel and a couple of LM electrodes, which can detect the fluid pressure at any position in the target microchannel [122]. Pressure fluctuations caused interface movement between different fluids in the monitoring channel, leading to the changes in capacitance between LM electrodes. Thus, pressure can be acquired via recording changes in capacitance. By recording the change curve of the measured capacitance and pressure with time in real time (Figure 8.18d), it can be known that the changing trend of capacitance and pressure was linear correlation. In addition, a classic carotid blood pressure detection experiment was performed in rabbits using the pressure sensor (Figure 8.18e). Experimental data showed that the LM-based pressure sensor had a high sensitivity to the changes of blood pressure in animals, such as rabbits. The pressure sensor

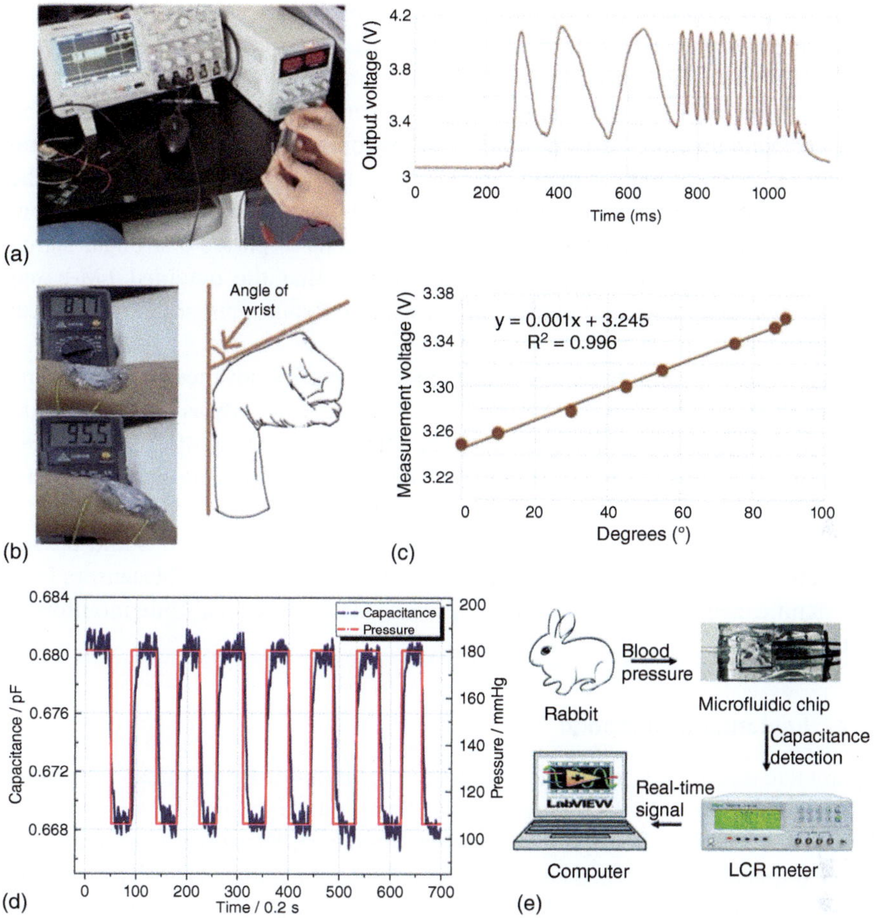

Figure 8.18 Biological signals analysis of biosensors based on LMs. (a) The output voltage of LM capacitor sensor at different time showed the corresponding frequency change. (b) Dynamic analysis of wrist joint movement, the left pictures showed the capacitance change of the sensor, and the right picture showed the rotation angle of the wrist. (c) Use the linear regression formula to analyze the relationship between the measured voltage and the wrist angle [127]. (a–c) Source: Sheng et al. [127]. (d) The relationship between the tested capacitance of the pressure sensor and pulse fluid pressure. (e) Experimental device diagram of microfluidic blood pressure sensor [122]. (d, e) Source: Zhou et al. [122].

shows good flexibility and stability within a certain pressure range, laying a good application foundation for biomedical monitoring.

High sensitivity monitoring of exosomes is extremely crucial for the diagnosis of diseases, such as tumors. LMs were used in intelligent respond to tissues or cells in organisms due to their excellent flexibility, high conductivity, and low toxicity. A high-performance electrogenerated chemiluminescence (ECL) biosensor for the detection of exosomes and their surface proteins was prepared using LM galinstan [128]. LM nanoparticles played the role of nanoprobes and had excellent properties to promote electron transfer during electrochemical reduction process.

The fabricated biosensor system was utilized to detect the concentration response of exosomes from HeLa cells, indicating that the ECL signal intensity gradually enhanced with the increase of exosome concentration. What's more, ECL signal was associated with the expression level of exosome surface proteins. Act as the cell correspondents, the surface proteins of exosomes had the ability to regulate cell biology. Such biosensor could be applied to analyze the ECL signals of the surface proteins derived exosomes from various cell lines at the same concentration. The analysis indicated that exosomes from HeLa cells displayed a stronger signal than other cell lines. These consequences proved that the obtained LM-based biosensor possesses a wide application prospect in exosomes research and medical diagnosis.

In conclusion, the emerging LMs have made impressive advances in biosensors. LMs can provide good skin-friendly and excellent deformability as an active liquid ingredient for the fabrication of biosensors. High conductivity of LMs enhances the electrical properties of biosensors in biomedical monitoring and diagnosis. LMs-based biosensors are used for real-time detection of a variety of biological signals, including heart rate and blood pressure measurements, joints and tissues deformation movements. The development of LMs in the field of biosensors is of great significance, which provides an important platform for real-time monitoring of vital signs.

8.5.2 Implantable Electrode

Implantable electrode is an efficient approach for monitoring and treating serious physiological diseases in the biological body. Traditional implantable electrode usually requires complicated surgery and later maintenance. These processes often lead to huge physical trauma and a heavy financial burden on patients [129]. As a bridge for transmitting signals between biological system and external device, bioelectrodes need to exhibit good conductivity and adaptability to target biological body. Due to the excellent properties of LMs, including high conductivity, good wettability, great fluidity and more, making it an increasingly popular material for preparing implantable electrodes. LM-based implantable electrodes display superior flexibility and adaptability in the biological body, which promotes their better compatibility and integration with biological tissues.

Jin and coworkers developed a novel method to directly fabricate three-dimensional compatible LM electrodes in the biological body [115]. LMs with high conductivity and controlled wettability, providing an economical and convenient material for the preparation of implanted electrodes. The implantable electrode was constructed by sequentially injecting conductive LM and corresponding packaging material into the target biological tissues. Conventional implantable surgery included incision, electrode implantation as well as suture, which inevitably caused excessive emotional stress and severe physical pain to patients (Figure 8.19a). Unlike conventional surgery, this approach used a friendly minimally invasive injection technique that significantly reduced the patient's pain. The packaged electrode with LM ink as the main component was implanted into the porcine

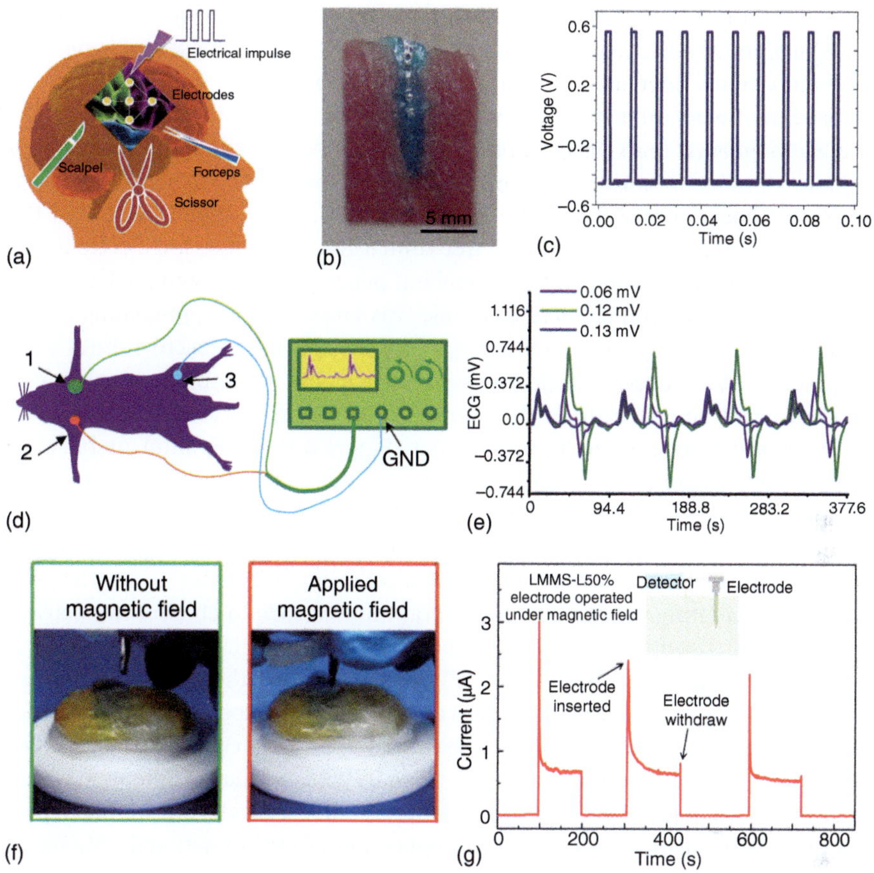

Figure 8.19 Simulation and in vivo experiments of implantable electrodes based on LMs. (a) Schematic of the traditional surgery for implanting electrodes into the brain. (b) Cross-sectional image of the injectable electrode in isolated porcine tissues. (c) Electronic measurement results of the formed injectable electrode. (d) Schematic of an experimental device for measuring signals in the mouse body with an injection electrode. (e) The electrocardiogram signal recorded in vivo experimental mouse under different electrical stimulations [115]. (a–e) Source: Jin et al. [115]. (f) The penetration ability of magnetic LM electrode to the artificial biofilm. (g) The electrochemical current response of magnetic LM electrode in the process of implantation or removal of electrolyte [130]. (f, g) Source: Ren et al. [130].

tissue, and its cross-sectional photograph as shown in Figure 8.19b. Electrical tests showed that the output signal displayed similar characteristics to the input signal, such as amplitude, frequency, and so on (Figure 8.19c). Therefore, it was feasible to the implantable electrode acted as an excellent medium to transport ideal stimulation into the biological body. To test the working characteristics of the injected electrode, a mouse (weight 20 g) was selected as the subject to conduct the in vivo experiment. Figure 8.19d presented the schematic of an experimental device for testing the electrocardiograph of mouse. When the stimulation signal was 0.6 mV, a significant change in the electrocardiogram of the mouse was observed (Figure 8.19e). Electrical stimulation caused an obvious increase in heart rate. A similar physiological response phenomenon was also observed when the stimulation signal increased to 1.2 mV. Whereas, when the stimulation signal increased to 1.3 mV, the physiological response phenomenon recorded on the electrocardiogram almost disappeared. It was explained that the enhancement of electrical stimulation caused muscle spasm in mouse, so the stimulation signal cannot be transmitted to the body. These characterization results illustrated the great biomedical practicability of the three-dimensional LM injection electrode.

The implantable electrode requires excellent penetration ability to the biofilm. A penetrable LM-based implantable electrode was fabricated by spin-coating a thin layer of polydimethylsiloxane on the needle-shaped LM magnetoactive slurry sample [130]. Without the application of a magnetic field, the soft implantable electrode could not penetrate the artificial biofilm and thus could not achieve the contact with the bioelectrolyte of the internal tissues (Figure 8.19f). The implantable electrode promptly became hard enough to penetrate the membrane when a magnetic field was applied. It was known that the implantable electrode could be reversibly switched between the flexible and rigid modes by applying or removing the magnet. The implantable electrode was connected to a collection device of the electrical signal, and a stimulating voltage was applied on the electrode to obtain the electrical response of the system. When the electrode was implanted into the organism, an obvious current signal can be detected (Figure 8.19g). After the stimulated current was stabilized, the electrode was removed and the current response signal vanished. Different locations were selected to repeat the implantation and removal experiments, the results showed that the current signals possess small differences, which was caused by changes in the bioelectrolyte around the electrode. This research provides new ideas for the fabrication and application of multifunctional LMs-based bioelectrodes with stimuli-responsive.

Overall, the introduction of LMs makes the implantable electrode show high flexibility, which is conducive to the minimally invasive surgery of the implantation electrode in biological tissues. Moreover, LM-based implantable electrode exhibits good adaptability and rapid stimulus response to organisms, offering a favorable guarantee for its application in biomedicine.

8.5.3 Microfluidic Pump

Microfluidics has become an essential biological analysis approach in modern clinical diagnosis and medical research. As the "heart" of the microfluidic system, the microfluidic pump is not only the power source of microfluidic transportation, but also an important symbol of the development of microfluidic technology. A microfluidic pump is capable of delivering small volumes of liquid accurately [131, 132]. LMs are considered as a promising material for the preparation of microfluidic pumps owing to their high conductivity, good deformability, and low volatility in microchannels [133, 134]. LMs-based microfluidic pump can continuously transport fluids into organisms, such as drugs and blood, and it will not cause damage to biological tissues during the entire conveying process. In

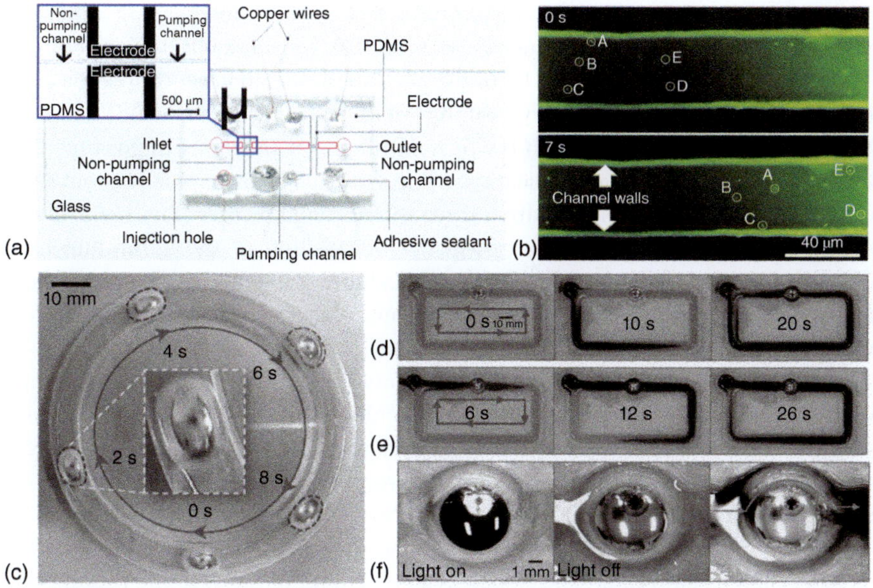

Figure 8.20 Characterization of fluid motion in microfluidic pump based on LMs. (a) Optical photograph of the microfluidic chip, the enlarged illustration showed the LM electrodes of the EOF pump. (b) Optical photograph of fluorescent particles at different times in the flow channel of the EOF pump. A, B, C, D, and E marked with circles represented five different tracking fluorescent particles [135]. (a, b) Source: Gao and Gui [135]. (c) The movement of LM droplet in a circular channel containing NaOH solution, the inset was an enlarged image of the squeezed LM droplet. (d, e) Continuous images of the solution flowing (d) counterclockwise and (e) clockwise in a closed-loop channel. (f) Continuous digital microscope picture of the LM droplet flowing in the direction of the arrow [136]. (c–f) Source: Zhang et al. [136].

addition, LMs-based microfluidic pump can accurately monitor and control the flow of fluids, which shows great application prospects in the field of biomedicine.

A simple electroosmotic flow (EOF) pump integrated with room temperature LM was developed by Gao and Gui [135] Room temperature LM was used in EOF pump due to its excellent properties, namely, low melting point and good deformability. This noncontact EOF pump was fabricated by filling LM into a polydimethylsiloxane microchannel. Figure 8.20a showed the optical image of the microfluidic chip. The LM alloy was infused into the electrode channels to constitute the pump electrode. By measuring the continuous movement state of the fluorescent particles in the EOF pump (Figure 8.20b), the average velocity of the EOF in the pump channel was obtained. The injection of room temperature LM into the microchannel provided a promising method for the integration of the EOF pump. LM-based microfluidic pump has enormous application prospects in the field of biomedicine, as microdevice that drives the movement of various fluids, such as cell suspensions and drug reagents.

In addition, a self-propelled motor device based on LM exhibited good self-deformation ability, and had strong adaptability to complex tasks in special environments [136]. Such motor displayed a fast swimming speed in different structural channels containing aqueous solution. The movement of the LM engine in a circular channel as shown in Figure 8.20c. Due to the good fluidity of the LM, the droplet displayed extremely low rolling friction, so the motor could maintain continuous and stable operation. It is worth noting that self-driving motor provides possibilities for the practical applications. The self-driven LM motor was put into the NaOH solution and printed with a three-dimensional printer (Figure 8.20d–f). Its flow direction was mainly controlled by the position of the aluminum and the local rotation of the LM. This LM motor exhibited unusual performance and longevity, and can be used as a self-driven biomimetic mollusk, paving the way for the manufacture of soft machines in a biomedical field.

The working principle of the human heart is very similar to the microfluidic pump. LM with unique behavior can be utilized to achieve the pump function that controls the flow of liquid like an artificial heart. A new method using electrically tuned LM droplet to realize an artificial heart was proposed by Li and coworkers [137]. An electric field excitation equipment was fabricated to redistribute the charges of LM droplets in the NaOH solution, thereby inducing the LM formed the desired structure, such as ellipsoidal and puddle shape. Dropping LM into the round copper electrode could cause diffuse and wetting on the copper surface. When the size

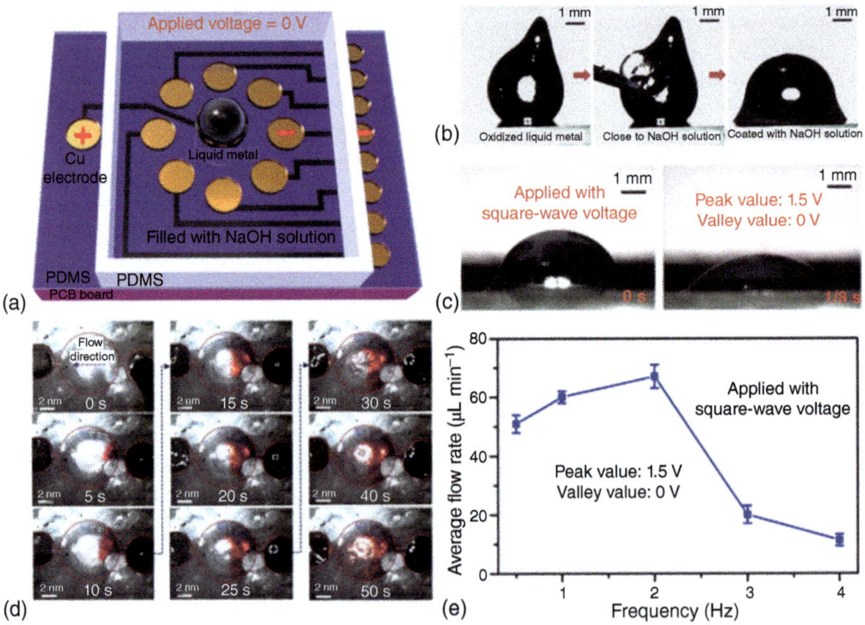

Figure 8.21 The artificial heart based on LMs. (a) Schematic of an experimental device for deformation of LM droplet filled in NaOH solution. (b) A solution film was formed by immersing NaOH solution into the interface of LM droplet. (c) The optical images of LM droplet after applying a driving voltage. (d) The top view optical images of a set of the LM artificial heart flow (red ink). (e) Control the average flow rate of artificial blood by adjusting the driving voltage [137]. Source: Li et al. [137].

of the copper electrode was smaller than the LM droplet, a spherical shape with a large contact angle was formed (Figures 8.21a). The main component of the LM was Ga, and the LM droplets attached to the copper electrode reacted slowly with NaOH solution. Owing to the difference in surface tension, LM droplet was put on the substrate, and the presence of a thin gallium oxide layer made the droplet appeared peach-shaped (Figure 8.21b). When NaOH solution wet the surface of LM droplet, the oxide layer was dissolved to form a solution film. Figure 8.21c showed that the LM droplet encased in the solution produced a heartbeat motion. The artificial heart LM pump operated persistently under the applied voltage (Figure 8.21d). With the increased of time, the artificial heart LM pump was gradually filled with red ink, showing that it had a good working performance. The average velocity of artificial blood depends on the frequency and amplitude of the driving voltage. Hence, the artificial heart LM pump able to accurately regulate blood flow by controlling the frequency and amplitude.

In summary, LMs with good fluidity can be used as a desired fluid for microfluidic pump. The soft and stretchable LM fluid maintains excellent dynamic continuity and self-recovery ability during the deformation movement in the microfluidic channel. Thus, the unique properties and functions of LMs have promoted the development of high-performance microfluidic pump systems, paving the way for the exploration of self-driven artificial biological tissues or organs.

8.6 Conclusion and Prospect

In the whole chapter, we have systematically demonstrated the great potential of LMs applied in biomedicine. Compared with traditional biomaterials, LMs as the new generation of multifunctional biomaterials exhibit superior advantages, involving good adaptability, excellent thermal and electrical conductivities, controllable behavior, convenience for operation, as well as good biocompatibility. More significantly, these characteristics enable LMs to be utilized in various scopes of bioapplications including therapeutics, imaging, biomimetic materials, and biomedical device. It is believed that the development of LM-based biomaterials can promote the basic medical research as well as inspire more technological innovations to treat biomedical challenges.

Despite the great progress in biomedicine, many problems still remain to be solved. The long-term toxicity as well as the toxicity of LM-based materials in different scales require more comprehensive and systematic research. Meanwhile, the stability of LMs is also one of the concerns, which will affect the efficacy in complex biological systems. In order to satisfy the technical requirements, a complete standard system for a series of operations on LMs should be taken into consideration. The safe injection and surgery, accurate package and positioning, and convenient molding and removal are crucial to possible clinical applications of LMs. In addition, some bioapplications of LMs are in their infancy stage and still considered as novel concepts, which means that the proposals for biomedicine of LMs require more realistic assessments. More efforts are desired to further promote the development of intelligent biomedical materials and bring about technological innovation to solve the current medical challenges. With continuous technical progresses, LM-based biomaterials tend to be smarter, more flexible, and more precisely controllable for complex biosystems.

References

1 Yan, J., Lu, Y., Chen, G. et al. (2018). Advances in liquid metals for biomedical applications. *Chem. Soc. Rev.* 47: 2518.

2 Liu, J. (2018). *Liquid Metal Biomaterials~Principles and Applications.* Berlin: Springer.

3 Lu, Y., Hu, Q., Lin, Y. et al. (2015). Transformable liquid-metal nanomedicine. *Nat. Commun.* 6: 10066.

4 Lu, Y., Lin, Y., Chen, Z. et al. (2017). Enhanced endosomal escape by light-fueled liquid-metal transformer. *Nano Lett.* 17: 2138.

5 Chandler, J.E., Messer, H.H., and Ellender, G. (1994). Cytotoxicity of gallium and indium ions compared with mercuric ion. *J. Dent. Res.* 73: 1554.

6 Mitragotri, S. and Lahann, J. (2012). Materials for drug delivery: innovative solutions to address complex biological hurdles. *Adv. Mater.* 24: 3717.

7 Tibbitt, M.W., Dahlman, J.E., and Langer, R. (2016). Emerging frontiers in drug delivery. *J. Am. Chem. Soc.* 138: 704.

8 Peer, D., Karp, J.M., Hong, S. et al. (2007). Nanocarriers as an emerging platform for cancer therapy. *Nat. Nanotechnol.* 2: 751.

9 van der Meel, R., Sulheim, E., Shi, Y. et al. (2019). Smart cancer nanomedicine. *Nat. Nanotechnol.* 14: 1007.

10 Hinde, E., Thammasiraphop, K., Duong, H.T. et al. (2017). Pair correlation microscopy reveals the role of nanoparticle shape in intracellular transport and site of drug release. *Nat. Nanotechnol.* 12: 81.

11 Xu, F., Huang, X., Wang, Y., and Zhou, S. (2020). A size-changeable collagenase-modified nanoscavenger for increasing penetration and retention of nanomedicine in deep tumor tissue. *Adv. Mater.* 32: 1906745.

12 Yu, M., Song, W., Tian, F. et al. (2019). Temperature- and rigidity-mediated rapid transport of lipid nanovesicles in hydrogels. *Proc. Natl. Acad. Sci. USA* 116: 5362.

13 Haume, K., Rosa, S., Grellet, S. et al. (2016). Gold nanoparticles for cancer radiotherapy: a review. *Cancer Nanotechnol.* 7: 8.

14 Davis, M.E. and Brewster, M.E. (2004). Cyclodextrin-based pharmaceutics: past, present and future. *Nat. Rev. Drug Discov.* 3: 1023.

15 Ganta, S., Devalapally, H., Shahiwala, A., and Amiji, M. (2008). A review of stimuli-responsive nanocarriers for drug and gene delivery. *J. Control. Release* 126: 187.

16 Kim, D., Hwang, J., Choi, Y. et al. (2019). Effective delivery of anti-cancer drug molecules with shape transforming liquid metal particles. *Cancer* 11: 1666.

17 Fan, L., Sun, X., Wang, X. et al. (2019). NIR laser-responsive liquid metal-loaded polymeric hydrogels for controlled release of doxorubicin. *RSC Adv.* 9: 13026.

18 Beik, J., Abed, Z., Ghoreishi, F.S. et al. (2016). Nanotechnology in hyperthermia cancer therapy: from fundamental principles to advanced applications. *J. Control. Release* 235: 205.

19 Otte, J. (1988). Hyperthermia in cancer therapy. *Eur. J. Pediatr.* 147: 560.

20 Dutz, S. and Hergt, R. (2014). Magnetic particle hyperthermia—a promising tumour therapy? *Nanotechnology* 25: 452001.

21 Jang, J.T., Nah, H., Lee, J.H. et al. (2009). Critical enhancements of MRI contrast and hyperthermic effects by dopant-controlled magnetic nanoparticles. *Angew. Chem. Int. Ed. Engl.* 48: 1234.

22 Noh, S.-h., Moon, S.H., Shin, T.-H. et al. (2017). Recent advances of magneto-thermal capabilities of nanoparticles: from design principles to biomedical applications. *Nano Today* 13: 61.

23 Pankhurst, Q.A., Thanh, N.T.K., Jones, S.K., and Dobson, J. (2009). Progress in applications of magnetic nanoparticles in biomedicine. *J. Phys. D. Appl. Phys.* 42: 224001.

24 Wang, X., Fan, L., Zhang, J. et al. (2019). Printed conformable liquid metal e-skin-enabled spatiotemporally controlled bioelectromagnetics for wireless multisite tumor therapy. *Adv. Funct. Mater.* 29: 1907063.

25 Wang, D., Xie, W., Gao, Q. et al. (2019). Non-magnetic injectable implant for magnetic field-driven thermochemotherapy and dual stimuli-responsive drug delivery: transformable liquid metal hybrid platform for cancer theranostics. *Small* 15: 1900511.

26 Cheng, L., Wang, C., Feng, L. et al. (2014). Functional nanomaterials for phototherapies of cancer. *Chem. Rev.* 114: 10869.

27 Zhang, Z., Wang, L., Wang, J. et al. (2012). Mesoporous silica-coated gold nanorods as a light-mediated multifunctional theranostic platform for cancer treatment. *Adv. Mater.* 24: 1418.

28 Weissleder, R. (2001). A clearer vision for in vivo imaging. *Nat. Biotechnol.* 19: 316.

29 Xia, Y., Li, W., Cobley, C.M. et al. (2011). Gold nanocages: from synthesis to theranostic applications. *Acc. Chem. Res.* 44: 914.

30 Yang, K., Feng, L., Shi, X., and Liu, Z. (2013). Nano-graphene in biomedicine: theranostic applications. *Chem. Soc. Rev.* 42: 530.

31 Huang, X., Tang, S., Mu, X. et al. (2011). Freestanding palladium nanosheets with plasmonic and catalytic properties. *Nat. Nanotechnol.* 6: 28.

32 Zhou, M., Zhang, R., Huang, M. et al. (2010). A chelator-free multifunctional [64Cu]CuS nanoparticle platform for simultaneous micro-PET/CT imaging and photothermal ablation therapy. *J. Am. Chem. Soc.* 132: 15351.

33 Tian, Q., Tang, M., Sun, Y. et al. (2011). Hydrophilic flower-like CuS superstructures as an efficient 980 nm laser-driven photothermal agent for ablation of cancer cells. *Adv. Mater.* 23: 3542.

34 Chou, S.S., Kaehr, B., Kim, J. et al. (2013). Chemically exfoliated MoS_2 as near-infrared photothermal agents. *Angew. Chem. Int. Ed. Engl.* 52: 4160.

35 Zheng, X., Zhou, F., Wu, B. et al. (2012). Enhanced tumor treatment using biofunctional indocyanine green-containing nanostructure by intratumoral or intravenous injection. *Mol. Pharm.* 9: 514.

36 Zha, Z., Yue, X., Ren, Q., and Dai, Z. (2013). Uniform polypyrrole nanoparticles with high photothermal conversion efficiency for photothermal ablation of cancer cells. *Adv. Mater.* 25: 777.

37 Wang, D., Gao, C., Wang, W. et al. (2018). Shape-transformable, fusible rodlike swimming liquid metal nanomachine. *ACS Nano* 12: 10212.

38 Wang, X., Yao, W., Guo, R. et al. (2018). Soft and moldable Mg-doped liquid metal for conformable skin tumor photothermal therapy. *Adv. Healthcare Mater.* 7: 1800318.

39 Yan, J., Zhang, X., Liu, Y. et al. (2019). Shape-controlled synthesis of liquid metal nanodroplets for photothermal therapy. *Nano Res.* 12: 1313.

40 Sun, X., Sun, M., Liu, M. et al. (2019). Shape tunable gallium nanorods mediated tumor enhanced ablation through near-infrared photothermal therapy. *Nanoscale* 11: 2655.

41 Xia, N., Li, N., Rao, W. et al. (2019). Multifunctional and flexible ZrO_2-coated EGaIn nanoparticles for photothermal therapy. *Nanoscale* 11: 10183.

42 Chechetka, S.A., Yu, Y., Zhen, X. et al. (2017). Light-driven liquid metal nanotransformers for biomedical theranostics. *Nat. Commun.* 8: 15432.

43 Hu, J.J., Liu, M.D., Chen, Y. et al. (2019). Immobilized liquid metal nanoparticles with improved stability and photothermal performance for combinational therapy of tumor. *Biomaterials* 207: 76.

44 Zhu, P., Gao, S., Lin, H. et al. (2019). Inorganic nanoshell-stabilized liquid metal for targeted photonanomedicine in NIR-II biowindow. *Nano Lett.* 19: 2128.

45 Hu, J.J., Liu, M.D., Gao, F. et al. (2019). Photo-controlled liquid metal nanoparticle-enzyme for starvation/photothermal therapy of tumor by win-win cooperation. *Biomaterials* 217: 119303.

46 Ren, L., Zhuang, J., Casillas, G. et al. (2016). Nanodroplets for stretchable superconducting circuits. *Adv. Funct. Mater.* 26: 8111.

47 Lin, Y., Cooper, C., Wang, M. et al. (2015). Handwritten, soft circuit boards and antennas using liquid metal nanoparticles. *Small* 11: 6397.

48 Hohman, J.N., Kim, M., Wadsworth, G.A. et al. (2011). Directing substrate morphology via self-assembly: ligand-mediated scission of gallium–indium microspheres to the nanoscale. *Nano Lett.* 11: 5104.

49 Hu, J.J., Xiao, D., and Zhang, X.Z. (2016). Advances in peptide functionalization on mesoporous silica nanoparticles for controlled drug release. *Small* 12: 3344.

50 Alhafid, H.T. (2006). *Critical Reviews™ in Biomedical Engineering*. Danbury, CT: Begell House.

51 Gillams, A.R. (2005). The use of radiofrequency in cancer. *Br. J. Cancer* 92: 1825.

52 Chu, K.F. and Dupuy, D.E. (2014). Thermal ablation of tumours: biological mechanisms and advances in therapy. *Nat. Rev. Cancer* 14: 199.

53 Sun, X., He, Z.Z., Deng, Z.S. et al. (2018). Liquid metal bath as conformable soft electrodes for target tissue ablation in radio-frequency ablation therapy. *Minim. Invasive Ther. Allied Technol.* 27: 233.

54 Rao, W. and Liu, J. (2009). Injectable liquid alkali alloy based-tumor thermal ablation therapy. *Minim. Invasive Ther. Allied Technol.* 18: 30.

55 Deng, Z.S. and Liu, J. (2007). Minimally invasive thermotherapy method for tumor treatment based on an exothermic chemical reaction. *Minim. Invasive Ther. Allied Technol.* 16: 341.

56 Rao, W. and Liu, J. (2008). Tumor thermal ablation therapy using alkali metals as powerful self-heating seeds. *Minim. Invasive Ther. Allied Technol.* 17: 43.

57 Rao, W., Liu, J., Zhou, Y.X. et al. (2008). Anti-tumor effect of sodium-induced thermochemical ablation therapy. *Int. J. Hyperth.* 24: 675.

58 Freeman, L.A., Anwer, B., Brady, R.P. et al. (2010). In vitro thermal profile suitability assessment of acids and bases for thermochemical ablation: underlying principles. *J. Vasc. Interv. Radiol.* 21: 381.

59 Misselt, A.J., Edelman, T.L., Choi, J.H. et al. (2009). A hydrophobic gel phantom for study of thermochemical ablation: initial results using a weak acid and weak base. *J. Vasc. Interv. Radiol.* 20: 1352.

60 Deng, Z.-S. and Liu, J. (2013). Chemothermal therapy for localized heating and ablation of tumor. *J. Healthcare Eng.* 4: 409.

61 MacDonald, D.K.C., Pearson, W.B., and Towle, L.T. (1956). An investigation of the sodium-potassium equilibrium diagram. *Can. J. Phys.* 34: 389.

62 Guo, Z., Zhang, Q., Li, X., and Jing, Z. (2015). Thermochemical ablation therapy of VX_2 tumor using a permeable oil-packed liquid alkali metal. *PLoS One* 10: 0123196.

63 Wang, W., Liu, Y., Liu, H. et al. (2018). NaK alloy-induced in vivo tumor ablation therapy. *Minim. Invasive Ther. Allied Technol.* 27: 90.

64 Vogl, T.J., Naguib, N.N., Gruber-Rouh, T. et al. (2011). Microwave ablation therapy: clinical utility in treatment of pulmonary metastases. *Radiology* 261: 643.

65 Long, D., Liu, T., Tan, L. et al. (2016). Multisynergistic platform for tumor therapy by mild microwave irradiation-activated chemotherapy and enhanced ablation. *ACS Nano* 10: 9516.

66 Long, D., Mao, J., Liu, T. et al. (2016). Highly stable microwave susceptible agents via encapsulation of Ti-mineral superfine powders in urea-formaldehyde resin microcapsules for tumor hyperthermia therapy. *Nanoscale* 8: 11044.

67 Wu, Q., Xia, N., Long, D. et al. (2019). Dual-functional supernanoparticles with microwave dynamic therapy and microwave thermal therapy. *Nano Lett.* 19: 5277.

68 Nilsson, E., von Euler, H., Berendson, J. et al. (2000). Electrochemical treatment of tumours. *Bioelectrochemistry* 51 (1).

69 Sun, X., Yuan, B., Rao, W., and Liu, J. (2017). Amorphous liquid metal electrodes enabled conformable electrochemical therapy of tumors. *Biomaterials* 146: 156.

70 Wang, Q., Yu, Y., Liu, J. (2014). Delivery of liquid metal to the target vessels as vascular embolic agent to starve diseased tissues or tumors to death. arXiv preprint. arXiv:1408.0989 0989.

71 Witsch, E., Sela, M., and Yarden, Y. (2010). Roles for growth factors in cancer progression. *Physiology (Bethesda)* 25: 85.

72 Thorpe, P.E. (2004). Vascular targeting agents as cancer therapeutics. *Clin. Cancer Res.* 10: 415.

73 Poursaid, A., Jensen, M.M., Huo, E., and Ghandehari, H. (2016). Polymeric materials for embolic and chemoembolic applications. *J. Control. Release* 240: 414.

74 Zhang, Y., Gao, H., Wang, H. et al. (2018). Radiopaque highly stiff and tough shape memory hydrogel microcoils for permanent embolization of arteries. *Adv. Funct. Mater.* 28: 1705962.

75 Romeike, B.F. and Feiden, W. (1998). Demonstration of platinum microcoils in embolized blood vessels in-situ using a modified methyl methacrylate embedding method, and a special cutting and grinding technique. *Biotech. Histochem.* 73: 198.

76 Puylaert, J.B., Rutgers, P.H., Lalisang, R.I. et al. (1987). A prospective study of ultrasonography in the diagnosis of appendicitis. *N. Engl. J. Med.* 317: 666.

77 Shapiro, D., Thibault, P., Beetz, T. et al. (2005). Biological imaging by soft x-ray diffraction microscopy. *Proc. Natl Acad. Sci. USA* 102: 15343.

78 Scarfe, W.C., Farman, A.G., and Sukovic, P.J.J.-C.D.A. (2006). Clinical applications of cone-beam computed tomography in dental practice. *J. Can. Dent. Assoc.* 72: 75.

79 Momose, A., Takeda, T., Itai, Y., and Hirano, K.J.N.m. (1996). Phase–contrast X–ray computed tomography for observing biological soft tissues. *Nat. Med.* 2: 473.

80 Agatston, A.S., Janowitz, W.R., Hildner, F.J. et al. (1990). Quantification of coronary artery calcium using ultrafast computed tomography. *J. Am. Coll. Cardiol.* 15: 827.

81 Wang, Q., Yu, Y., Pan, K., and Liu, J. (2014). Liquid metal angiography for mega contrast X-ray visualization of vascular network in reconstructing in-vitro organ anatomy. *IEEE Trans. Biomed. Eng.* 61: 2161.

82 Fan, L., Duan, M., Xie, Z. et al. (2020). Injectable and radiopaque liquid metal/calcium alginate hydrogels for endovascular embolization and tumor embolotherapy. *Small* 16: 1903421.

83 Larsson, D.H., Lundstrom, U., Westermark, U.K. et al. (2013). First application of liquid-metal-jet sources for small-animal imaging: high-resolution CT and phase-contrast tumor demarcation. *Med. Phys.* 40: 021909.

84 Yi, L., Jin, C., Wang, L., and Liu, J. (2014). Liquid-solid phase transition alloy as reversible and rapid molding bone cement. *Biomaterials* 35: 9789.

85 Kruger, R.A., Liu, P.Y., Fang, Y.R., and Appledorn, C.R. (1995). Photoacoustic ultrasound (PAUS)—reconstruction tomography. *Med. Phys.* 22: 1605.

86 Weber, J., Beard, P.C., and Bohndiek, S.E. (2016). Contrast agents for molecular photoacoustic imaging. *Nat. Methods* 13: 639.

87 Taruttis, A., van Dam, G.M., and Ntziachristos, V. (2015). Mesoscopic and macroscopic optoacoustic imaging of cancer. *Cancer Res.* 75: 1548.

88 Beard, P. (2011). Biomedical photoacoustic imaging. *Interface Focus* 1: 602.

89 Charnley, J. (1960). Anchorage of the femoral head prosthesis to the shaft of the femur. *J. Bone Joint Surg.* 42B: 28.

90 Ormsby, R., McNally, T., Mitchell, C., and Dunne, N. (2010). Influence of multiwall carbon nanotube functionality and loading on mechanical properties of PMMA/MWCNT bone cements. *J. Mater. Sci. Mater. Med.* 21: 2287.

91 Yang, J.-M., Lu, C.-S., Hsu, Y.-G., and Shih, C.-H. (1997). Mechanical properties of acrylic bone cement containing PMMA-SiO_2 hybrid sol-gel material. *J. Biomed. Mater. Res.* 38: 143.

92 Stanczyk, M. and van Rietbergen, B. (2004). Thermal analysis of bone cement polymerisation at the cement-bone interface. *J. Biomech.* 37: 1803.

93 Cristofolini, L., Minari, C., and Viceconti, M. (2000). A methodology and criterion for acrylic bone cement fatigue tests. *Fatigue Fract. Eng. Mater. Struct.* 23: 953.

94 Park, Y.L., Chen, B.R., Perez-Arancibia, N.O. et al. (2014). Design and control of a bio-inspired soft wearable robotic device for ankle-foot rehabilitation. *Bioinspir. Biomim.* 9: 016007.

95 Ferris, D.P., Sawicki, G.S., and Daley, M.A. (2007). A physiologist's perspective on robotic exoskeletons for human locomotion. *Int. J. Hum. Robot.* 4: 507.

96 Jarrasse, N. and Morel, G. (2012). Connecting a human limb to an exoskeleton. *IEEE T. Robot.* 28: 697.

97 Berring, J., Kianfar, K., Lira, C. et al. (2010). A smart hydraulic joint for future implementation in robotic structures. *Robotica* 28: 1045.

98 Deng, Y. and Liu, J. (2014). Flexible mechanical joint as human exoskeleton using low-melting-point alloy. *J. Med. Dev.* 8: 044506.

99 Park, Y.-L., Majidi, C., Kramer, R. et al. (2010). Hyperelastic pressure sensing with a liquid-embedded elastomer. *J. Micromech. Microeng.* 20: 125029.

100 Lee, S.K. and Wolfe, S.W. (2000). Peripheral nerve injury and repair. *J. Am. Acad. Orthop. Sur.* 8: 243.

101 Yannas, I.V., Zhang, M., and Spilker, M.H. (2007). Standardized criterion to analyze and directly compare various materials and models for peripheral nerve regeneration. *J. Biomater. Sci. Polym. Ed.* 18: 943.

102 Evans, G.R. (2001). Peripheral nerve injury: a review and approach to tissue engineered constructs. *Anat. Rec.* 263: 396.

103 Gordon, T., Sulaiman, O., and Boyd, J.G. (2003). Experimental strategies to promote functional recovery after peripheral nerve injuries. *J. Peripher. Nerv. Syst.* 8: 236.

104 Pabari, A., Yang, S.Y., Seifalian, A.M., and Mosahebi, A. (2010). Modern surgical management of peripheral nerve gap. *J. Plast. Reconstr. Aesthet. Surg.* 63: 1941.

105 Xu, H., Holzwarth, J.M., Yan, Y. et al. (2014). Conductive PPY/PDLLA conduit for peripheral nerve regeneration. *Biomaterials* 35: 225.

106 A Moss, R. and V Martin, J. (2014). Debating new theory in neuroscience. *AIMS Neuroscience* 1: 1.

107 Wilson, D.H. and Jagadeesh, P. (1976). Experimental regeneration in peripheral nerves and the spinal cord in laboratory animals exposed to a pulsed electromagnetic field. *Paraplegia* 14: 12.

108 Balint, R., Cassidy, N.J., and Cartmell, S.H. (2014). Conductive polymers: towards a smart biomaterial for tissue engineering. *Acta Biomater.* 10: 2341.

109 Zhang, J., Qiu, K., Sun, B. et al. (2014). The aligned core-sheath nanofibers with electrical conductivity for neural tissue engineering. *J. Mater. Chem. B* 2: 7945.

110 Hardy, J.G., Lee, J.Y., and Schmidt, C.E. (2013). Biomimetic conducting polymer-based tissue scaffolds. *Curr. Opin. Biotechnol.* 24: 847.

111 Abidian, M.R., Daneshvar, E.D., Egeland, B.M. et al. (2012). Hybrid conducting polymer-hydrogel conduits for axonal growth and neural tissue engineering. *Adv. Healthcare Mater.* 1: 762.

112 Zhang, J., Sheng, L., Jin, C. et al. (2014). Liquid metal as connecting or functional recovery channel for the transected sciatic nerve. arXiv preprint. arXiv: 1404.5931.

113 Liu, F., Yu, Y., Yi, L., and Liu, J. (2016). Liquid metal as reconnection agent for peripheral nerve injury. *Sci. Bull.* 61: 939.

114 Scheib, J. and Hoke, A. (2013). Advances in peripheral nerve regeneration. *Nat. Rev. Neurol.* 9: 668.

115 Jin, C., Zhang, J., Li, X. et al. (2013). Injectable 3D fabrication of medical electronics at the target biological tissues. *Sci. Rep.* 3: 3442.

116 Guo, R. and Liu, J. (2017). Implantable liquid metal-based flexible neural microelectrode array and its application in recovering animal locomotion functions. *J. Micromech. Microeng.* 27: 104002.

117 Wu, Y.H., Liu, H.Z., Chen, S. et al. (2017). Channel crack-designed gold@PU sponge for highly elastic piezoresistive sensor with excellent detectability. *ACS Appl. Mater. Interfaces* 9: 20098.

118 Wang, Y., Wang, L., Yang, T. et al. (2014). Wearable and highly sensitive graphene strain sensors for human motion monitoring. *Adv. Funct. Mater.* 24: 4666.

119 Ye, X., Yuan, Z., Tai, H. et al. (2017). A wearable and highly sensitive strain sensor based on a polyethylenimine-rGO layered nanocomposite thin film. *J. Mater. Chem. C* 5: 7746.

120 Ilyas, N., Cook, A., and Tabor, C.E. (2017). Designing liquid metal interfaces to enable next generation flexible and reconfigurable electronics. *Adv. Mater. Interf.* 4: 1700141.

121 Gan, T., Shang, W., Handschuh-Wang, S., and Zhou, X. (2019). Light-induced shape morphing of liquid metal nanodroplets enabled by polydopamine coating. *Small* 15: 1804838.

122 Zhou, X., Zhang, R., Li, L. et al. (2019). A liquid metal based capacitive soft pressure microsensor. *Lab Chip* 19: 807.

123 Li, G. and Lee, D.W. (2017). An advanced selective liquid-metal plating technique for stretchable biosensor applications. *Lab Chip* 17: 3415.

124 Jeong, Y.R., Kim, J., Xie, Z. et al. (2017). A skin-attachable, stretchable integrated system based on liquid GaInSn for wireless human motion monitoring with multi-site sensing capabilities. *NPG Asia Mater.* 9: 443.

125 Wang, J., Yang, S., Ding, P. et al. (2019). Omnidirectional printing of soft elastomer for liquid-state stretchable electronics. *ACS Appl. Mater. Interf.* 11: 18590.

126 Wu, Y.-h., Zhen, R.-m., Liu, H.-z. et al. (2017). Liquid metal fiber composed of a tubular channel as a high-performance strain sensor. *J. Mater. Chem.* 5: 12483.

127 Sheng, L., Teo, S., and Liu, J. (2016). Liquid-metal-painted stretchable capacitor sensors for wearable healthcare electronics. *J. Med. Biol. Eng.* 36: 265.

128 Zhang, Y., Wang, F., Zhang, H. et al. (2019). Multivalency interface and g-C_3N_4 coated liquid metal nanoprobe signal amplification for sensitive electrogenerated chemiluminescence detection of exosomes and their surface proteins. *Anal. Chem.* 91: 12100.

129 Hlatky, M.A. and Mark, D.B. (2007). The high cost of implantable defibrillators. *Eur. Heart J.* 28: 388.

130 Ren, L., Sun, S., Casillas-Garcia, G. et al. (2018). A liquid-metal-based magnetoactive slurry for stimuli-responsive mechanically adaptive electrodes. *Adv. Mater.* 30: 1802595.

131 Debesset, S., Hayden, C.J., Dalton, C. et al. (2004). An AC electroosmotic micropump for circular chromatographic applications. *Lab Chip* 4: 396.

132 Harnett, C.K., Templeton, J., Dunphy-Guzman, K.A. et al. (2008). Model based design of a microfluidic mixer driven by induced charge electroosmosis. *Lab Chip* 8: 565.

133 Tang, S.Y., Khoshmanesh, K., Sivan, V. et al. (2014). Liquid metal enabled pump. *Proc. Natl. Acad. Sci. USA* 111: 3304.

134 Liu, T., Sen, P., and Kim, C.-J. (2012). Characterization of nontoxic liquid-metal alloy galinstan for applications in microdevices. *J. Microelectromech. Syst.* 21: 443.

135 Gao, M. and Gui, L. (2014). A handy liquid metal based electroosmotic flow pump. *Lab Chip* 14: 1866.

136 Zhang, J., Yao, Y., Sheng, L., and Liu, J. (2015). Self-fueled biomimetic liquid metal mollusk. *Adv. Mater.* 27: 2648.

137 Li, G., Du, J., Zhang, A., and Lee, D.w. (2019). Artificial heart based on electrically controlled non-toxic liquid metal pump. *Adv. Eng. Mater.* 21: 1900381.

9

Liquid Metal for Energy

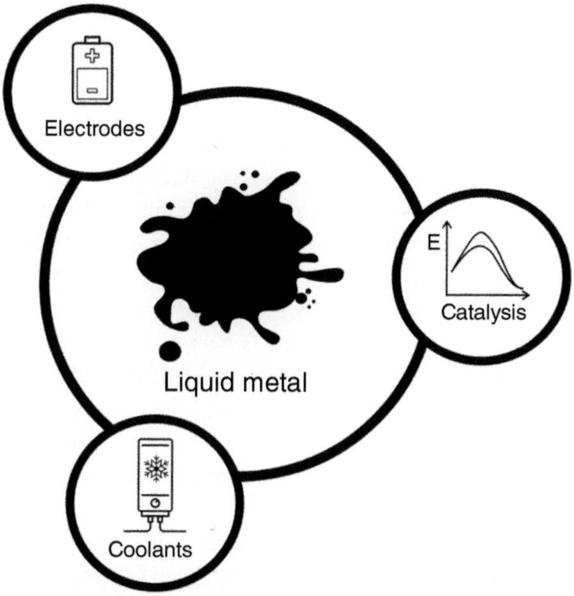

The progress of scientific research and its increasingly expanding fields will arouse our hope.
Source: Alfred Nobel. Public Domain.

9.1 Introduction

Energy is essential for human life and social development. The process of energy generation, transformation, and storage is of great importance to the end use of energy. Over the years, people have been exploring and using different materials and methods to make a big step forward in energy development. Liquid metal has become more and more important material due to its special properties such as low

Liquid Metals: Properties, Mechanisms, and Applications, First Edition. Lei Fu and Mengqi Zeng.

melting point and fluidity. In recent years, it has also been widely used in the field of energy. Herein, liquid metal electrode, liquid metal coolant, and liquid metal catalyst are introduced, which may include the basic and typical application of liquid metal in energy.

Batteries are very important energy storage systems, and electrode as an important component of battery greatly affects the performance of a battery. In recent years, with the rapid development of liquid metal in the energy storage, liquid metal electrodes have also attracted widespread attention [1, 2]. Liquid metal has unique advantages as an electrode due to its fluidity. In this chapter, the situation when liquid metal is used as electrode is described in detail. The explorations include design principle, high-temperature liquid metal electrodes, and room-temperature liquid metal electrodes. The two kinds of electrode systems are mentioned as follows: (i) high temperature liquid metal system: although the temperature of some metals in this part is little higher than that of liquid metals described in other parts of the book, It is still described in this chapter due to liquid state during the cycling process, and it is also an indispensable part of the battery system, such as Na–S electrode [3]; (ii) Room temperature liquid metal electrodes have gradually attracted attention because they do not require additional functions. It is described here including Ga-based, Na–K and semi-fluid electrode, and the like [1].

In addition to battery energy storage, liquid metal coolant will be introduced. Here, liquid metal coolant can be considered as part of the energy conversion process. Nuclear energy system is developed as an important driving force for fast reactors [4]. The function of coolant is to timely export the huge heat generated in the reaction process. In addition, thermal management is also required in the advanced electronic device due to possession of related properties [5]. The low melting point liquid metal may have better cooling capability than the conventional coolant when used for cooling. Therefore, this chapter mainly introduces the application of liquid metal coolant in two aspects, which includes application as traditional nuclear energy and new electronic devices [6].

Finally, the liquid metal serves as catalysts for energy conversion. Its applications include electrocatalytic system [7] and photocatalytic system [8], as well as some common organic reactions. It is an important medium for energy conversion [9]. However, investigations on liquid metal catalysts are rare at present, particularly to that of catalysis of single elements [10]. The reaction mechanism is usually the change of metal density state and the activation of site isolation. Liquid metal catalysts often show excellent catalytic performance and unique advantages, which provides opportunities for improving the products of many reactions. Therefore, although the research on liquid metal as catalyst is still in the initial stage, it is worth mentioning here.

9.2 Liquid Metal as Electrodes

As one of the components of the energy storage systems, rechargeable batteries are capable of facing the rapidly developing requirements due to small footprint and flexibility in siting [11–14]. Hereinto, lithium-ion batteries (LIBs) are considered to be the hopeful energy storage devices in modern society, which are widely used in our daily life [15–17]. Recently, LIBs are gradually approaching its limited capability and then Li metal electrode is deemed as an alternative power source, thanks to high theoretical specific capacity [18]. However, the practical applications of Li metal are suffered from safety hazards, such as Li dendrites and volume expansion. Compared with conventional solid-state metal electrodes, liquid metal as electrodes comes into view due to fluidity, self-healing effect, and high conductivity. Moreover, less restriction on the mass transport and the contact interface also can endow rechargeable batteries with unexpected electrochemical performance. In the following, we will introduce the liquid metal electrodes from design principle, high-temperature liquid metal electrodes, and room-temperature liquid metal electrodes and discuss their potential applications.

9.2.1 Design Principle

9.2.1.1 Advantages of Liquid Metal Electrodes

Li source batteries typically consist of two solid-state electrodes and liquid electrolyte. Currently, the high performance of LIBs are mainly impeded by capacity decay during the long-term cycling, which results from the pulverization of electrode materials [12, 19, 20]. In this regard, as a class of emerging electrode materials, liquid metal possesses high fluidity, high surface tension, and self-healing effect, which is a potential solution for the pulverization of electrode materials. Meanwhile, the liquid-state surface of liquid metal electrode is possible to intrinsically avoid uneven dendrites issues. These properties have propelled liquid metal into researchers' sights, hoping to fulfill the requirements of energy storages.

In detail, liquid metal can be transported to the damaged sites because of fluidity, leading to the self-healing of mechanical and electrical properties [21]. As shown in Figure 9.1a, Blaiszik and coworkers chose liquid gallium–indium (Ga–In) alloy as healing agent to autonomously restore the broken circuit [22]. Under the critical bending load, the cracks appeared on device were likely to cut-off the conductive pathways. Nonetheless, these Ga–In capsules also could be broken by propagating cracks, and then liquid Ga–In alloy released to heal the damaged sites. According to the high conductivity of Ga–In alloy ($3.40 \times 10^4\ \mathrm{S\,cm^{-1}}$), these cracks can be recovered over 99% conductance. The autonomously restorable system employing liquid metal can effectively promote fault-tolerance and increase devices' reliability. In addition, the high surface tension is also beneficial to maintain anode self-healing. Wu and coworkers reported a novel anode consisting of liquid gallium–tin (Ga–Sn) nanoparticles embedded in a carbon skeleton [23]. The Ga–Sn nanoparticles presented a spherical shape, meanwhile, the size of Ga–Sn nanoparticles ranged from 50 to 500 nm. Owing to the high surface energy and fluidity of liquid metal, the expanded Ga–Sn nanoparticles tended to recover to spherical shape after full delithiation process. As exhibited in Figure 9.1b, the size and morphology of Ga–Sn nanoparticles have no obvious changes after cycling, confirming the self-healing

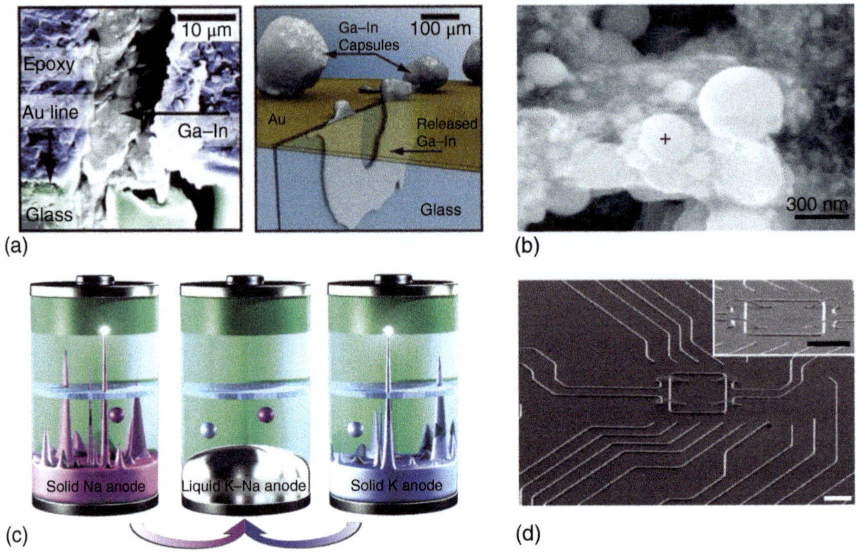

Figure 9.1 (a) SEM image and schematic of the damaged site healed by released liquid metal [22]. Source: Blaiszik et al. [22]. (b) SEM imaging of Ga–Sn nanoparticles have no obvious changes after 4000 cycles [23]. Source: Wu et al. [23]. (c) The dendrite-free behavior of liquid metal anode compared with solid-state metal anodes [24]. Source: Xue et al. [24]. (d) A high-resolution 3D electrode by directly printing on liquid Ga–In alloy. Scale bar, 100 μm. The inset exhibits the magnifying liquid metal electrode. Scale bar, 100 μm [25]. Source: Park et al. [25].

property. The self-healing anode can greatly prolong cycling life and has no obvious capacity decay after 4000 cycles.

As for the serious safety hazards derived from dendrite issue, using liquid metal as anode can perfectly avoid the uncontrolled deposition. Xue and coworkers presented a dendrite-free battery by utilizing room-temperature liquid sodium–potassium (Na–K) alloy as anode [24]. In Figure 9.1c, the surface of solid Na or K metal had abundant dendrites growth during the cycling process, while the surface of liquid Na–K alloy maintained a dendrite-free state. The liquid metal can provide a liquid–liquid interface to realize uniform deposition, and the self-healing property can remain as a smooth surface after discharging. More than the intrinsic liquid properties, liquid metal still has good plasticity. For example, Park and coworkers successfully designed a high-resolution three-dimensional (3D) electrode by directly printing liquid Ga–In alloy (Figure 9.1d) [25]. The thin gallium oxide of Ga–In alloy was formed once exposed to air, which could maintain various shapes and reconfigure 3D structure. Furthermore, liquid metal electrodes with different sizes would be printed by changing the various nozzles. The direct printing method can give full play to the plasticity of liquid metal and facilitate the preparation of complex electrodes. Hence, based on the earlier discussed advantages, liquid metal is a good candidate to serve as an alternate electrode for various concerns.

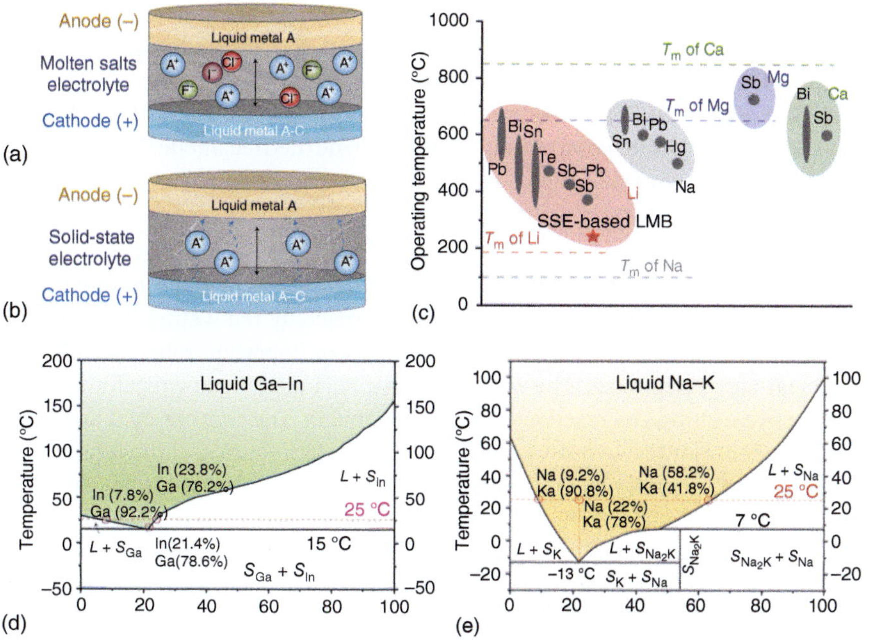

Figure 9.2 (a) Conventional configuration of liquid metal consists of two liquid metal electrodes and molten salts electrolyte. (b) Liquid metal battery based on the solid-state electrolyte. (c) Statistics of operating temperature in the various high-temperature liquid metal batteries [26]. Source: Li [26]. (d, e) Phase diagrams of liquid Na–K alloy and liquid Ga–In alloy, respectively [1]. Source: Guo et al. [1].

9.2.1.2 Potential Liquid Metal Candidates

Liquid metal electrodes can be divided into high-temperature electrodes and room-temperature electrodes. Traditional high-temperature liquid metal batteries are composed of molten salt electrolytes and molten metal electrodes (Figure 9.2a) [26]. Some solid-state electrolytes also can be used between molten anode and molten cathode (Figure 9.2b). Since alloying reaction occurs at high-temperature system, alloying-based elements and less electronegative metals are generally suitable for the anodes, such as Li and Na (Figure 9.2c). Meanwhile, some materials with more electronegativity are served as the cathode to match cell voltage, including palladium (Pd), bismuth (Bi), Sn, and antimony (Sb). The reasonable electrode matching is in favor of keeping the high-temperature liquid metal batteries running in a high-voltage window.

As for room-temperature liquid metal electrodes, Ga (melting point, 29 °C) first stands out because of nontoxic property and low evaporation pressure [27]. Subsequently, Ga-based alloys develop several rich systems to lower melting point, including Ga–In, Ga–Sn, or Ga–In–Sn [1]. As shown in Figure 9.2d, the Ga–In alloy with the Ga content of 78.6% displayed a melting point of 15 °C, which could keep liquid under the operating room-temperature. In addition, Na and K metals are inclined to form room-temperature eutectics alloy. Both alkali metals have extremely low reduction potentials and high theoretical capacity during cycling. It's worth noticing that the composition of Na–K alloy must be adjusted in a suitable region ranging from 9.2 to 58.2 wt% Na [1]. Only in this region, the Na–K alloy can maintain a dendrite-free liquid surface at room temperature. As exhibited in Figure 9.2e, the Na–K liquid alloy with 22 wt% Na had an extremely low eutectic point [1, 28, 29], which was an appropriate composition to design dendrite-free anode. In summary, there are many elements that can be used as liquid metal electrodes.

9.2.1.3 Wetting Behavior

In order to promote the practical application of liquid metal, some fundamental chemistry issues deserve more research efforts, especially the wetting behavior of liquid metal. In practice, the wetting behavior plays an important role in stabilizing liquid metal-based electrodes. The wetting behavior of liquid metal typically consists of nonreactive wetting and reactive wetting phenomena. The conventional nonreactive wetting behavior has two key parameters: the surface tension of the liquid (σ_{lv}) and the contact angle of liquid metal on the substrate (θ) [1]. The nonreactive wetting can be defined by classical wetting theory, as follows:

$$\cos\theta = \frac{\sigma_{sv} - \sigma_{sl}}{\sigma_{lv}} \tag{9.1}$$

$$w_\alpha = \sigma_{lv} \times (1 + \cos\theta) \tag{9.2}$$

where σ_{sv}, σ_{sl}, and σ_{lv} present the surface energy of liquid metal at different interfaces. w_α is the adhesion work. As for reactive wetting behavior, the adhesion work is better for wetting when the formed Gibbs free energy is more negative.

To get an objective understanding about the wettability of liquid metal, we choose the most explored liquid alkali metals as examples. The influence factors of nonreactive wetting mainly involve atomic composition, temperature, substrate, surface topography, and so on [30]. For example, Lu and coworkers markedly improved the wettability of pure Na metal via changing atomic composition [31]. As shown in Figure 9.3a, molten pure Na appeared as a nonwetting behavior on β"-Al_2O_3 solid electrolyte (BASE), and no difference was observed even at 250 °C. Oppositely, the sodium–cesium (Na–Cs) alloy showed a contact angle of 0° on the same BASE at 250 °C. That indicated that the addition of Cs thoroughly influenced the wettability of liquid Na, leading to the liquid metal to full wetting. In this case, authors utilized a simulation to further analyze the changes of wettability. Figure 9.3b could confirm that the liquid Na–Cs alloy gradually spread with the content of Cs increasing. By virtue of classical wetting theory, it could infer that liquid Cs improved the adhesion work and formed a better wettability than liquid pure Na. Besides the liquid metal itself, the researchers also focus on adjusting substrates to enhance the wettability of liquid metal. Wang and coworkers successfully designed a setup to explore the influences of substrate temperature, substrate type, and surface topography on the wetting behavior of molten Li (Figure 9.3c) [30]. As shown in Figure 9.3d, all kinds of substrates had a better wettability as the increase of temperature. This was attributed to the high entropy deriving from the fact that high temperature could decrease the viscosity and surface tension of molten Li. Moreover, different surface topography

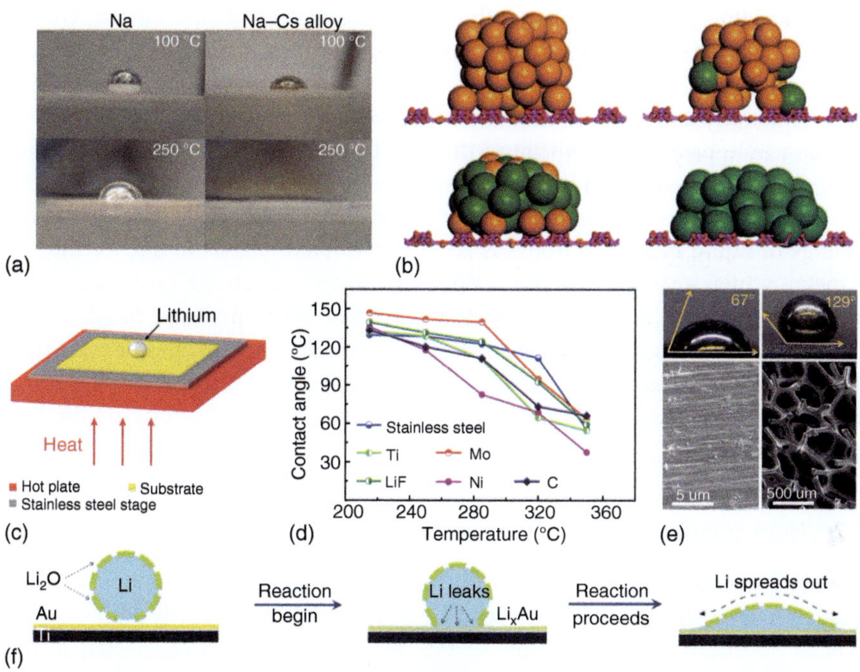

Figure 9.3 The influence factors of wetting behavior of liquid metal. (a) Contact angle of pure Na and Na–Cs alloy on the BASE. Source: Lu et al. [31]. (b) Simulated structures of liquid Na–Cs alloy with the increase of Cs content [31]. Source: Lu et al. [31]. (c) Experimental setup to explore the influences on the wettability of liquid Li. (d) Wettability of liquid Li on various substrates as the increase of temperature. (e) Contact angle of liquid Li on the Ni with different surface topographies. (f) Mechanism of improving wettability by the Li-reactive coating [30]. (c, d, f) Source: Wang et al. [30].

also had different wetting behavior. Nickel (Ni) foil had a contact angle of 67°, while Ni foam had a contact angle of 129° at the same temperature (Figure 9.3e), suggesting that the smooth and continuous surface facilitated molten Li to spread.

Other than the nonreactive wetting, covering up substrates with Li-reactive materials is a conventional approach for the reactive wetting behavior. Wang and coworkers compared with various coating materials and proposed that gold (Au) coating could easily turn lithiophobic titanium (Ti) foil into lithiophilic one [30]. The reaction between molten Li and Au coating released a negative Gibbs free energy to drive the spread of molten Li (Figure 9.3f). Generally speaking, a more negative Gibbs free energy offered a stronger driving force, therefore the Au coating had a better wettability. On the other hand, the authors found that the inert oxide layer (Li_2O) of Li surface played a hindering effect on spreading Li. Hence, keeping a clean liquid surface was important to promote the wetting behavior. Recently, Ga-based alloys systems also sparked a research boom into deeper understanding wetting behavior [32]. Kramer and coworkers studied the wettability of Ga-based alloys on various substrates with different surface topographies [33]. The wettability of Ga-based alloys was consistent with the molten alkali metals, where both reactive component and smooth morphology had a positive influence on the wetting of liquid metal. On the contrary, the oxide of Ga played a negative role on forming a reactive layer to some extent.

In conclusion, many elements can be formed as liquid metal electrodes in the field of rechargeable batteries. In virtue of the novel liquid properties including fluidity and self-healing effect, liquid metal electrodes enable thorough avoidance of the growth of Li dendrites and ensure excellent electrochemical cyclicity. Herein, the key challenge of liquid metal electrodes is seeking a suitable matrix to anchor the liquid metal, which ensured that the obtained composite electrode possesses stable interface chemistry. Currently, many strategies have been employed to effectively facilitate the wetting behavior of liquid metal on conductive substrates or solid electrolytes, which are useful in the practical application of liquid metal electrodes. To date, liquid metal battery is promising to fulfill the requirements of long-life and high-power device.

9.2.2 High-temperature Liquid Metal Electrodes

Large-scale energy storage plays a vital part in enhancing the stability, reliability, and security of grid-scale devices [12, 13]. And it can greatly improve the efficiency of intermittent renewable energy technologies when connected to the grid. High-temperature liquid metal electrodes have been considered as strong candidate solutions for the large-scale energy storage because of long cycling life, high energy density, low cost, and simple scalability. Early high-temperature liquid metal electrodes mainly referred to full liquid batteries, consisting of two molten metal electrodes and molten salt electrolyte. Recently, sodium–sulfur (Na–S) and Na–metal halide high-temperature batteries have emerged thanks to their high theoretical capacity and high energy density [13, 15]. Hereinto, we will discuss the high-temperature liquid metal electrodes from metal–metal and metal–nonmetal design.

9.2.2.1 Metal–metal Design

Metal–metal design refers to molten metals such as the anode and cathode. Traditional metal–metal design is composed of three liquid layers. By reason of mutual immiscibility and different densities, the metal–metal electrodes are separated by a layer of molten salt. For instance, Ning and coworkers developed a self-healing Li–Bi system, where the formed intermetallic compound could lead to a much deeply discharge [34]. As shown in Figure 9.4a, Li^+ ions were reduced and alloyed to form liquid Li–Bi alloy at the positive electrode during the discharge process. Subsequently, the formation of solid compound (Li_3Bi) further deepened the discharge process, indicating that the Li–Bi system was able to discharge deeply. More importantly, the dealloying of solid Li_3Bi was fully reversible, meanwhile the system fully became liquid state again when charged. The self-healing property greatly facilitated Li–Bi battery cycling stability, in which no obvious capacity fade was observed at 3C rate after 1000 cycles. Another metal–metal system worthy of attention is lithium–antimony–lead (Li||Sb–Pb) battery. On account of high cell

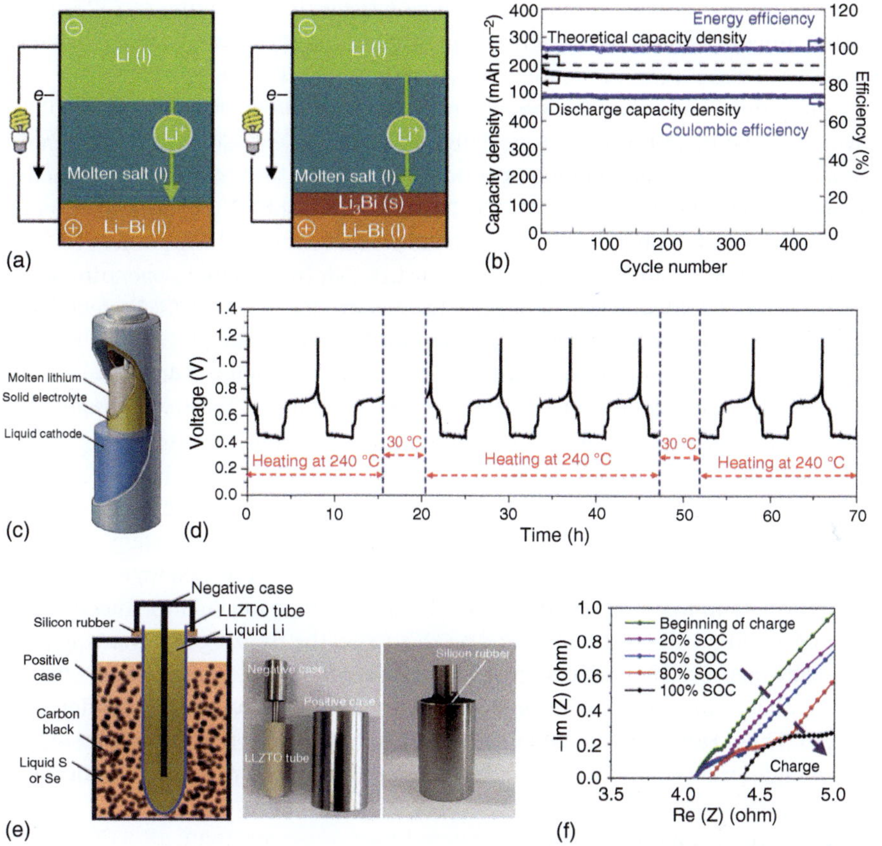

Figure 9.4 (a) Mechanism of self-healing Li–Bi liquid metal system during discharging process, forming a solid compound (Li_3Bi) [34]. Source: Ning et al. [34]. (b) Capacity retention of liquid Li||Sb–Pb battery over 450 cycles at 275 mA cm^{-2} [11]. Source: Wang et al. [11]. (c) Structure graphing of liquid Li||Sb–Pb battery. (d) Voltage curve of Li||Sb–Pb system during the freezing and thawing process [35]. (c, d) Source: Jin et al. [35]. (e) Schematic and optical image of Li-S and Li-Se high-temperature liquid batteries. (f) Electrochemical impedance of liquid Li-S batteries during the charging process [36]. (e, f) Source: Jin et al. [36].

voltage and low cost, pure Sb is a class of desirable positive material. In this case, by introducing low-cost Pb, the obtained Sb–Pb alloy served as positive electrode and can almost retain the pristine voltage of Sb, meanwhile further reducing the operating temperature [37]. Wang and coworkers proposed a high-energy-density Li||Sb–Pb (Sb–Pb 30 : 70 mol%) liquid metal battery with ultralong cycling life [11]. The introduction of Sb–Pb liquid alloy reduced the operating temperature to 450 °C. In Figure 9.4b, the battery showed a capacity fade of 0.004% per cycle ranging from 100th to 450th cycles even at 275 mA cm^{-2}. By inferring that the Li||Sb–Pb liquid system still had 85% capacity retention during the 10 years of daily operation, it was ascribed to the avoidable failure mechanisms of liquid metal electrodes.

The operating temperature of Li||Sb–Pb liquid system is almost over 450 °C, which possibly brought about serious corrosion and high cost [11, 34]. Moreover, Li has an inevitable dissolution in the molten salt, causing a self-discharged behavior and low Coulombic efficiency [2]. Consequently, lowering the melting point of Sb–Pb alloy and adjusting the type of electrolyte are needed to provide a safety guarantee for high-performance liquid metal batteries. Recently, Jin and coworkers designed a molten Li||Sb–Pb battery in a Sb–Pb 75 : 25 mol% cathode [35]. The introduction of 25 mol% Pb made the prepared Sb–Pb cathode possess a low melting point of 183 °C. In this work, the liquid Li||Sb–Pb battery ensured an intermediate operating temperature of 240 °C. Meanwhile, a tubular $Li_{6.4}La_3Zr_{1.4}Ta_{0.6}O_{12}$ (LLZTO) electrolyte was located between the molten Li anode and the molten Sn–Pb cathode to protect the molten Li anode from dissolving in electrolyte (Figure 9.4c). In virtue of liquid properties of liquid metal electrodes, the solid–liquid interface remained a good contact and reduced interfacial resistance to 0.4 Ω cm^2, laying a foundation for high-power cycling of the battery. The liquid Li||Sb–Pb battery had an average Coulombic efficiency of 99.98% during the long-term cycling, where the electrochemical property was superior to that of the traditional Li||Sb–Pb system. Furthermore, as for high-temperature liquid metal electrode, the volume changes resulting from freezing and thawing process were a huge challenge to avoid the mechanical failure of liquid metal electrode. Based on this issue, the freezing and thawing processes were measured in both charging and discharging states. Figure 9.4d displayed that the voltage curve of Li||Sb–Pb system kept stable and no evident fluctuation was observed. These results verified that the reliability of reducing the operating temperature of high-temperature battery, and further confirmed the practical significance.

9.2.2.2 Metal–nonmetal Design

Currently, metal–chalcogen battery and metal–metal halide battery are two typical metal–nonmetal designs in the field of high-temperature liquid batteries. Considering the high melting point and expensive cost of cathode metals, metal–nonmetal design mostly has potential to lower operating temperature and cost. For instance, Jin and coworkers assembled Li–S and Li–selenium (Se) liquid batteries with an operating temperature of 240 °C, focusing on eliminating dendritic issue and shuttling effect [36]. Figure 9.4e exhibited the schematic of Li–S battery, in which molten Li and molten S adding conductive carbon black were separated by a tube-based LLZTO. Compared with the solid-state electrodes of conventional Li–S, the high-temperature Li–S battery at an operating temperature of 240 °C fundamentally enabled to avoid the dendritic growth of Li metal thanks to liquid properties. Then the evenly dispersed conductive carbon black effectively improved the conductivity of molten S. This further helped Li_2S component to recover the molten electrode after full charging, eliminating the abscission of S electrode caused by volume expansion. As seen in Figure 9.4f, the impedance of charging process gradually increased and eventually remained the same as the original value, proving the stability and reversibility of high-temperature Li–S system. Moreover, the shuttling behavior of polysulfide was also prevented because of the packaging of LLZTO tube. In this liquid system, the unstable cycling and low

Coulombic efficiency of conventional Li–S battery were solved simultaneously. Additionally, different assembly methods also affect the operating temperature of high-temperature liquid batteries. Li and coworkers assembled a planar Na–nickel chloride ($NiCl_2$) battery [38]. Compared with the conventional tubular Na–$NiCl_2$, the planar assembly reduced the operating temperature to 190 °C. Owing to the lower operating temperature, the Na–$NiCl_2$ battery possessed a more stable capacity retention and Coulombic efficiency.

In summary, high-temperature liquid metal electrode has been considered as one of the strongest chooses for large-scale energy storage thank to the high energy density and theoretical capacity. The unique properties derived from liquid metal electrode can fundamentally overcome many concerned issues, such as volume expansion, dendritic growth, and structure pulverization, ensuring the great cycling stability and long survivability life. However, high temperature accompanied by high cost, easy corrosion, and safe problem still hinders the practical steps of high-temperature liquid electrodes. Consequently, reducing the operating temperature has attracted much attention in the field of high-temperature liquid batteries. There are many effective strategies to obtain low operating temperature, such as changed alloying composition and different assembly methods. These advances accomplished many breakthroughs toward making high-temperature liquid metal electrodes more competitive for large-scale energy storage.

9.2.3 Room-temperature Liquid Metal Electrodes

Room-temperature liquid metal is capable of remaining liquid-state at room temperature. On account of the rigorous thermal management and battery packaging of high-temperature liquid metal electrodes, recently room-temperature liquid metal electrodes have been widely considered as promising anode materials [1, 39]. Compared with high-temperature liquid metal, room-temperature liquid metal can keep the same advantageous features and need no external energy input. So far, two typical room-temperature liquid metal including Ga-based alloys and Na–K alkali metal have received a lot of attention for energy storage. Moreover, the development of some unexpected properties also cannot be neglected. Thus, we will describe the room-temperature liquid metal electrodes from three sections: Ga-based metal anodes, Na–K alkali metal anodes, and others.

9.2.3.1 Ga-based Metal Anodes

Ga-based liquid metals, such as pure Ga and Ga–In alloy have been widely researched as promising anode materials in alkali ion batteries in virtue of their high capacity base alkali ions storage [40, 41]. For instance, Ga can be discharged with Li^+ ions to form Li_2Ga alloy, accompanied by delivering a highest capacity of 768.8 mA h g^{-1}. Analogously, Ga also can react with Na^+ ions to release a capacity of 216.8 mA h g^{-1}, which correspond to $Na_{22}Ga_{39}$ alloy. It's worth noting that the phase transition accompanied by lithiation or sodiation process is inclined to introduce volume expansion and further leads to electrode pulverization. Fortunately, their intrinsic features, such as fluidity and self-healing effect, ensure that it can return

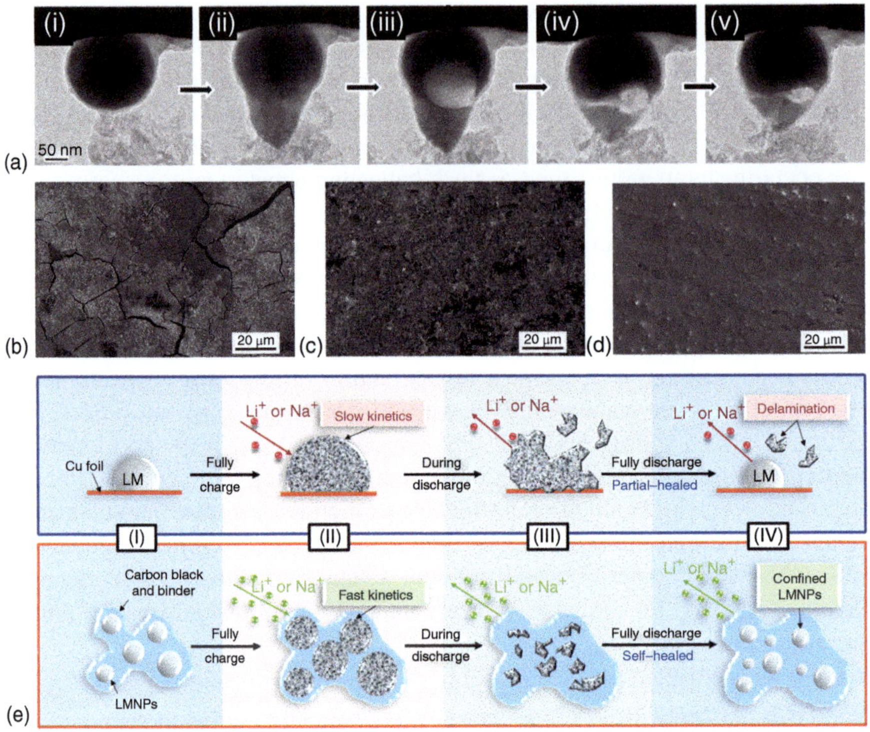

Figure 9.5 (a) TEM revealing the real-time morphological evolution of Ga droplet during the electrochemical process [42]. Source: Liang et al. [42]. (b, c) Morphologic changes of pure Ga electrode during the delithiation process. (d) Morphologic image of pure Ga electrode before cycling [43]. (b, c, d) Source: Deshpande et al. [43]. (e) Operating mechanisms of bulk liquid metal electrode and LMNP electrode in a cycling [44]. Source: Guo et al. [44].

to the liquid state after delithiation or desodiation process. Hence, Ga-based liquid metals have great potential for the self-healing electrodes. To verify the self-healing ability of Ga electrode, Liang and coworkers developed an in situ transmission electron microscopy (TEM) technology to explore the real-time morphological evolution of Ga droplet during the electrochemical process [42]. The pristine Ga droplet showed a spherical shape due to the isotropic surface energy of liquid Ga (Figure 9.5a-i). After full lithiation, the shape of Ga droplet turned into a highly distorted morphology under the uneven Li flow and had a 160% volume expansion, reflecting a clear liquid-solid phase transition (Figure 9.5a-ii). Subsequently, the solid-state Li_xGa alloy could convert back to liquid Ga after full delithiation, which was confirmed by a nanovoid originating from the outward flow of Li. As exhibited in Figure 9.5a-iii–v, the nanovoid inside the Ga electrode annihilated gradually because of the fluidity, suggesting the self-healing ability of liquid Ga. When directly integrated liquid Ga as anodes in LIBs, the self-healing ability will effectively alleviate the fracture of electrodes. Deshpande and coworkers reported pure Ga anode with no binder or carbon black and successfully explored the morphologic changes by an ex situ scanning electron microscopy (SEM) [3, 4]. As shown in Figure 9.5b, owing to the volume expansion and diffusion stress after full lithiation, many cracks formed in the Ga electrode during the delithiation process. Whereafter, SEM image displayed that these cracks disappeared with the help of solid–liquid transformation of Ga (Figure 9.5c). The surface of cyclic Ga electrode was similar to the pristine Ga surface (Figure 9.5d), indicating that the self-healing effect of liquid Ga electrode was able to eliminate mechanical degradation during the phase transformation and ensured the durability of electrode. However, the direct Ga electrode delivered a depressed capacity of less than 500 mA h g^{-1}, which was far below the theoretical capacity of Li–Ga system. The result may impute the weak diffused kinetics of liquid metal surface; these partial-healed cracks can lead to the delamination of electrodes and further generate irreversible capacity loss. Currently, preparing dispersive liquid metal nanoparticles (LMNPs) as electrodes is a useful strategy to achieve full self-healing ability. For instance, Guo group used dispersive LMNPs as electrode to improve the capacity and cycling lifetime [44]. As for the bulk liquid metal electrode, the concentration gradient of electrolyte caused a faster volume diffusion of alkali ions than that of the surface as depicted in Figure 9.5e. Therefore, the lithiated bulk electrode with serious volume expansion was likely to delaminate and irreversibly lose capacity during the delithiation. On the contrary, the LMNP electrodes eliminated volume expansion and promoted the diffused kinetics of ions, which were beneficial to fully heal cracks in the electrode during cycling. Based on above advantages, LMNPs electrode delivered a high capacity of 706.0 mAh g^{-1} at 0.1C and kept a remarkable stability over 500 cycles. Based on the full self-healing ability, the Ga-based liquid metals possessing high theoretical capacity are expected to pave the way toward high energy and long lifetime energy storage.

9.2.3.2 Na–K Alkali Metal Anodes

Both Na and K metals have extremely low reduction potentials, therefore the emerging Na–K alloy is expected to run at a high voltage [45]. The Na–K alloy is considered as room-temperature liquid metal in a wide region of composition. The intrinsic liquid properties are capable of addressing not only the dendritic issues of solid metal anodes, but also the interfacial contact resistance between the metal electrode and solid electrolyte. Besides, the Na–K alloy possesses a high surface tension. Based on this feature, Na–K alloy is not easy to penetrate a membrane containing liquid electrolyte, providing a possible application for energy storage. However, the surface tension also causes a considerable difficulty, where the eutectic Na–K alloy is hardly possible to wet the conductive matrix at room temperature and pressure. Currently, seeking suitable conductive matrix is the key factor in the practical application of Na–K alkali metal anodes. Guo and coworkers successfully absorbed Na–K alloy (33.7 wt% of Na) into a porous carbon paper by heating up to 420 °C, obtaining a sheet electrode [1]. As displayed in Figure 9.6a, when a carbon paper was immersed in Na–K alloy at 420 °C, 5 g Na–K alloy was gradually absorbed and immobilized in 1 g porous skeleton. Subsequently, a symmetric battery assembled by the sheet electrode containing Na–K alloy remained an excellent cycling stability without obvious voltage fluctuations over 2800 hours, which was far more than a solid

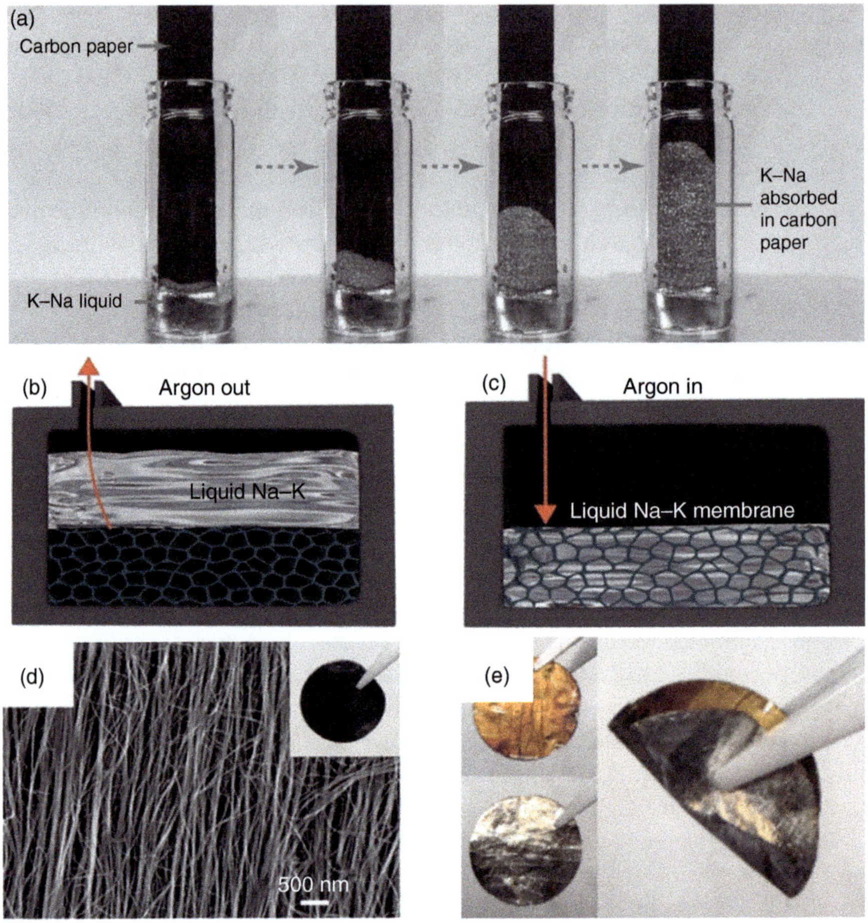

Figure 9.6 (a) Adsorption of Na–K alloy in a porous carbon paper by heating up to 420 °C [24]. Source: Xue et al. [24]. (b, c) Adsorption of Na–K alloy in various metal foams by a vacuum infiltration method [46]. (b, c) Source: Xue et al. [46]. (d) SEM image of nanotube membrane possessed a super-aligned structure. (e) Optical images of nanotube membrane after adsorbing Na–K alloy [47]. (d, e) Source: Qin et al. [47].

Na||Na battery. The difference was attributed to a dendrite-free deposition behavior of Na–K alloy. Moreover, the absorbed behavior of Na–K alloy was a recycled process. The liquid Na–K alloy could be eluted by dropping tetramethylene glycol dimethyl ether on the sheet electrode without affecting the next use. Besides porous carbon paper, other porous materials also could be used to absorb and form electrodes. Recently, Xue group fused Na–K alloy into various metal foams by a vacuum infiltration method at room temperature [46]. In Figure 9.6b, c, the argon was emitted from the metal foam at the bottom by virtue of a vacuum environment, and then liquid Na–K alloy was pushed into the metal foam by releasing argon. The Na–K alloy was evenly distributed within the metal foam and did not flow out even if the foam was bent. Additionally, a full battery consisting of Na–K alloy anode and

$Na_{2/3}Ni_{1/3}Mn_{2/3}O_2$ cathode exhibited outstanding capacity retention over 500 cycles, confirming the high stability and reliability of the Na–K electrode. More importantly, the method could be applied to many porous skeletons, including copper, nickel, aluminum foam, and even carbon paper. Comparing with high-temperature treatment, the vacuum infiltration method at room temperature was more economical, more convenient, and safer [1]. Except for the driving forces derived from temperature and pressure, the strong affinity of matrix is also beneficial to liquid metal to wet the conductive matrix. For example, Qin and coworkers reported that the capillary force of nanotube membrane could realize the adsorption of liquid Na–K alloy without extra energy input [47]. In Figure 9.6d, SEM image found that the nanotube membrane possessed a super-aligned structure. When contacted with liquid Na–K alloy, the nanotube membrane delivered a strong affinity to the liquid Na–K alloy due to the capillary force of structure. As shown in Figure 9.6e, the color changes from black to silver on the surface of nanotube membrane were evidence of Na–K alloy penetration. Oppositely, nanotube membrane with irregular structure had poor wettability, illustrating the key role of the arrangement structure in adsorbing liquid Na–K alloy. Furthermore, in this work, the nanotube membrane adsorbed a mass of Na–K alloys and meanwhile remained the intrinsic flexibility of the membrane, which reflected the possibility of wearable devices.

Recently, many researches have focused on the in situ preparation of Na–K alloy composite electrodes to obtain homogeneous testing electrodes. Zhang and coworkers used carbon fiber cloth (CFC) to anchor Na and K metal, respectively, to form Na–K alloy electrode [48]. The formation of Na–K alloy electrode was an in situ alloying at room temperature. As exhibited in Figure 9.7a, CFC/Na and CFC/K electrodes were prepared by pressing the Na or K into the CFC matrix, respectively. And then the authors physically stacked the two electrodes and motivated the in situ alloy reaction at room temperature. Inside the as-obtained CFC/Na–K anode, the Na–K alloy was filled with the CFC, demonstrating a great compatibility. It was worth mentioning that the in situ alloy process formed a solid K_2O layer (KOL) because of tiny oxygen and thermal activation on the surface of Na–K alloy (Figure 9.7b). On the one hand, the KOL could avoid the Na–K alloy reacting with organic electrolyte, decreasing the consumption of active material. Furthermore, the KOL was able to provide a better wetting behavior compared with bare Na–K alloy. Figure 9.7b reflected that the CFC/Na–K had a contact angle of 50° rather than that of 150° for bare Na–K alloy, where the better wettability reinforced the practical application of CFC/Na–K electrode. Another recent work by Zhang and coworkers utilized graphite intercalation compound (GIC) to anchor Na–K alloy via an in situ synthesis [49]. The preparation of CGC/Na–K electrode was based on the faster diffusion of K than that of Na. As described in Figure 9.7c, K rapidly diffused along with the carbon fiber and turned the carbon skeleton into a uniform GIC skeleton (KC_8) once the carbon fiber was immersed in Na–K alloy. Subsequently, the obtained GIC skeleton enhanced the wettability of Na–K alloy and attracted the infiltration of Na–K alloy, ultimately constituting a homogeneous CGC/Na–K anode. Owing to the superior conductivity of the GIC skeleton, the in-situ synthesis effectively combined the self-healing ability of liquid metal with the fast kinetics of CGC. Based on this in situ preparation, a

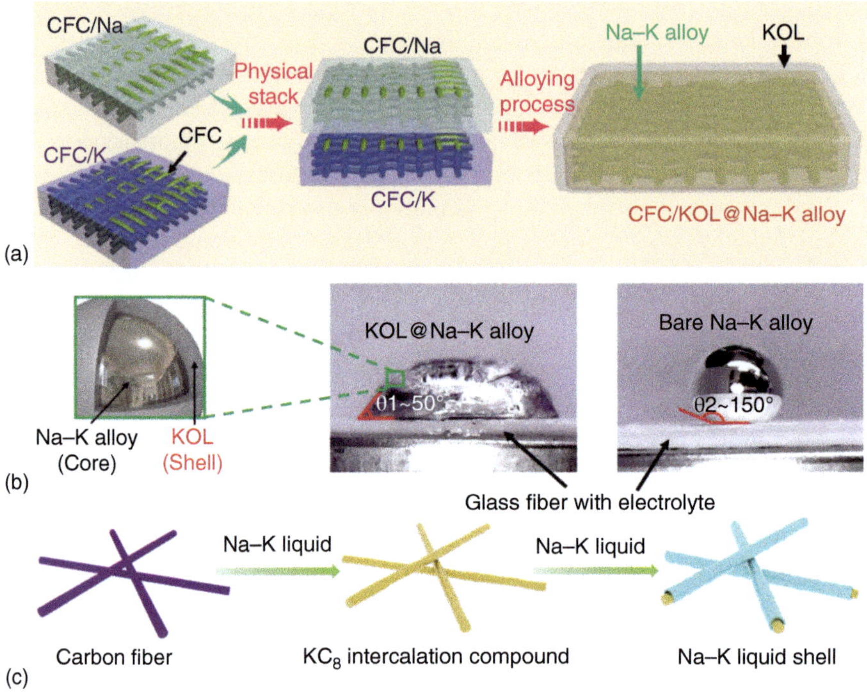

Figure 9.7 (a) Preparation process of CFC/Na–K electrode at room temperature. Source: Zhang et al. [48]. (b) Wetting behaviors of CFC/Na–K alloy and pure Na–K alloy at the surface of glass fiber. Source: Zhang et al. [48]. (c) Schematic of preparation of CGC/Na–K electrode [48].

symmetric battery consisting of CGC/Na–K electrodes showed an excellent cycling stability over 5000 hours at 20 mA cm^{-2}. Moreover, at a super-high current density of 80 mA cm^{-2}, the battery could still circulate for more than 1200 hours. In a word, the preeminent electrodeposition behavior was ascribed to both the dendrite-free growth of liquid Na–K electrode and the fast kinetics of conductive skeleton.

To sum up, the emerging Na–K alloy is a class of promising anode materials, which not only own the unique liquid properties to eliminate the structural degradation, but also have a high theoretical capacity. Currently, the conductive matrix of Na–K alloy mainly focuses on carbon materials because of fast mass transport and cost considerations. Moreover, some metal foams are also utilized to absorb Na–K alloy. To deliver great electrochemical performance, many strategies are explored to improve electrode uniformity by enhancing the wettability of Na–K alloy, such as high temperature, super-aligned structure, and in situ alloy. Hereinto, the infiltration of Na–K alloy at room temperature is a hot tendency, which is more economical, convenient, and safer.

9.2.3.3 Others

Except for serving as electrodes directly, liquid metal as an additional agent to endow electrodes with unexpected properties has attracted wide attention.

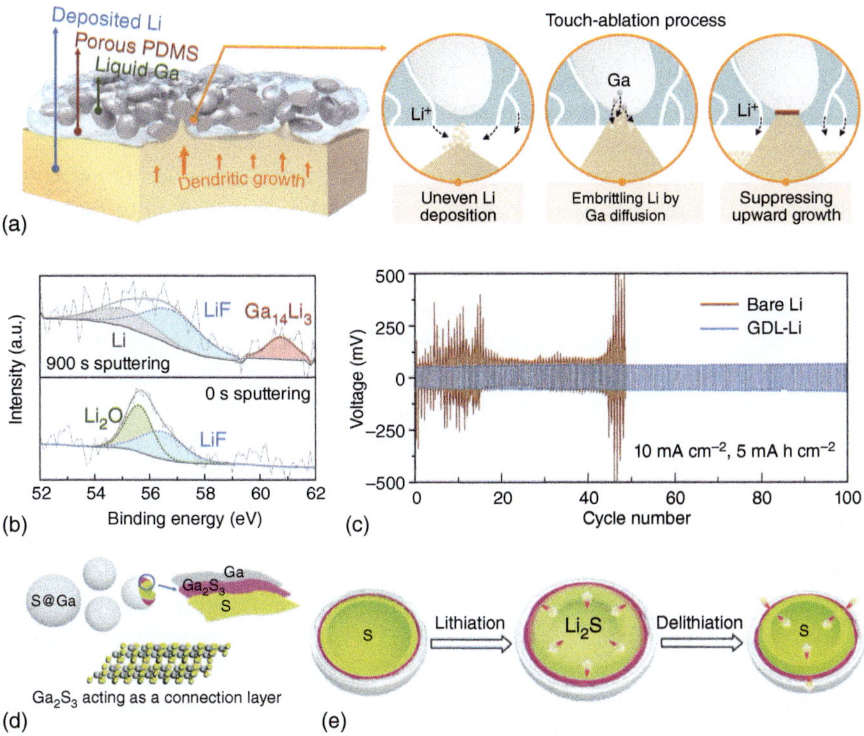

Figure 9.8 (a) Mechanism of touch-ablation process when Li plated. (b) XPS depth profiling of different sputtering time for Ga-based defense layer after 50 cycles. (c) Voltage profiles of symmetric Li||Li batteries with/without the Ga-based defense layer protection at high current density [50]. (a–c) Source: Wang et al. [50]. (d) Schematic of the formation of S@Ga composite. (e) Self-adaptive characteristic to solve the volume expansion issue [51]. Source: Zhu et al. [51].

Utilizing the embrittlement phenomenon of liquid Ga, Wang and coworkers designed a defense layer to suppress the uneven dendrites of Li metal electrode by touch-ablation (Figure 9.8a) [50]. In this work, the Ga-based defense layer was coated on Li metal and allowed Li^+ ions to deposit between the electrode and the Ga-based layer. When the uneven dendrites touched liquid Ga droplets, Ga diffused along the grain boundary of Li dendrites due to the weakly liquid Ga–Ga bonding interaction and then ablated dendrites to form dendritic-free deposition ultimately. Moreover, the touch-ablation process was accompanied by the alloying reaction between Ga and Li. As seen in Figure 9.8b, X-ray photoelectron spectroscopy (XPS) depth profiling demonstrated the generation of $Ga_{14}Li_3$ during the cycling process. Based on the nonbonding state of $Ga_{14}Li_3$ component, the reversible alloying and dealloying behavior of Ga-based layer ensured the continuous inhibition of Li dendrites in cycling. Additionally, because of high deformability and elasticity of liquid Ga, the Ga-based defense layer also played a buffer role in relieving upward stress of Li dendrites, further guaranteeing the reliability and stability of Ga-based defense layer during the cycling process. As illustrated in Figure 9.8c, a symmetric Li||Li battery with Ga-based layer protection kept a stable cycling even at strict conditions (10 mA cm^{-2} and 5 mA h cm^{-2}), which verified the excellent performance of touch-ablation. Owing to the unique liquid characters of Ga, the design of Ga-based defense layer provided a novel idea to eliminate Li dendrites by reacting with existing dendrites.

Another example of unexpected properties of liquid metal in the additional agent is self-adaptive characteristic. Looking at Li–S batteries, they have ultrahigh theoretical capacity but their applications are plagued by the volume expansion and shuttling behavior of S electrode. Liquid Ga is regarded as an ideal candidate to perfectly adapt the changes of volume expansion. For instance, Zhu and coworkers prepared a S@Ga electrode with a core–shell structure by a mixing way [51]. During the mixing process, the surface energy of S and Ga could be effectively balanced by in situ formed Ga_2S_3 layer, promoting liquid Ga to infiltrate S and prepare uniform core–shell S@Ga composite (Figure 9.8d). In this case, the high conductivity of liquid Ga shell was capable of increasing the electronic/ion transport, which facilitated the rate performance of Li–S battery. Meanwhile, accompanied by the volume changes of S electrode, the uniform liquid shell always coated the S electrode and revealed no fracture (Figure 9.8e). The uniform liquid shell not only utilized self-adaptive characteristic to solve the volume expansion issue, but also acted as a protective layer to inhibit the polysulfide shuttle. These superiorities of S@Ga composite helped the Li–S battery to display a stable cycling without evident capacity decrease at 1C even after 1000 cycles. In fact, as an additional agent, there are many unexpected properties of liquid metal, such as self-healing protective layer [52] and magnetic field-controlled electrode [53]. More explorations are worth exploring, which will enable liquid metal to have more wide applications.

9.3 Liquid Metal as Coolants

Looking at excellent thermal conductivity, low viscosities, and other intrinsic properties, liquid metals show potential of being the most attractive and promising candidates for effective heat transport fluids. Progresses in the corresponding research and novel technologies are emerging and increasing. The development of liquid metal coolants has advanced considerably in the past decades, offering the open sesame to broader applications. Especially, several areas requiring high heat loads are very pertinent for national welfare and people's livelihood, including nuclear applications and electronic cooling. Focusing on two typical applications, this chapter involves in unique physical properties, optimal designs, and pros and cons of liquid metal coolants.

9.3.1 Traditional Nuclear Application

Liquid metal coolant employed in the first fast neutron reactor as early as 1946 in the United States, which used mercury (Hg) due to its weak moderation of neutrons. Up to now, liquid-metal coolants have been widely used in traditional nuclear applications, such as cooling the core of accelerators and transferring heat to turbines.

Aimed at different nuclear applications, scientists have investigated a series of liquid metals to make further development, and meanwhile they have overcome many technological restrictions. Considering their features, we divide these liquid metals into three categories, namely, liquid alkali metals, liquid heavy metals, and fusible metals. Notably, the first two are most representative coolants used in nuclear applications, but the last one mainly refers to Ga or Ga-based alloys with relatively few studies. Hence, the next two sections are devoted to the motivations for alkali metals and heavy metals serving as reactor coolants, as well as their respective challenges encountered in practical applications.

9.3.1.1 Liquid Alkali Metal

In nuclear applications, long-lived fission fuel transmutes and nuclear reactions take place accompanied by the generation of activated materials. Without considering breeding, liquid metal, water, helium, and even carbon dioxide (CO_2) can act as fast reactor coolants in principle. Nevertheless, liquid-metal coolants are a more efficient and feasible choice for cooling fast reactors, in particular, which are equipped with tremendous volumetric power density and metal-fueled cores. Among liquid-metal coolants, liquid alkali metals mostly include Li, Na, and Na-K alloys, all of which are relatively light and possess high thermal conductivities and heat capacities.

As for liquid Li, it has the lowest density in the alkali metals, ensuring light in weight of coolants applied to space systems [54]. Due to the lowest vapor pressure, liquid Li performed in the nuclear power system contributes to reducing components workload at high temperature. It possesses the highest specific heat capacity among liquid metals, which guarantees high cooling efficiency and low heating of the core. Additionally, prominent tritium breeding and neutron moderation render good performance of liquid Li when cooling mixed breeder blankets. Equally as important, liquid Li has substantial shortcomings, particularly a high melting point (nearly 180 °C) and severe corrosion activity. Although the corresponding technology has already been progressed and refractory metals have been designed as circulation pumps, the practical applications of liquid Li still lack sufficient research [55].

Among alkali metals, liquid Na coolants are the most common for fast reactors in large nuclear power stations. This is because pure Na presents superior thermophysical properties, simple technological operations, relatively low cost, and good compatibility with structural materials [56]. These characteristics make Na keep stable coolant quality during the cooling process, enriching operational practical experience than other liquid metals. For example, Na-cooled fast reactor with BN-350 enjoyed a good reputation because it operated satisfactorily and had few accidents. Meanwhile, it has demonstrated that the intermediate loop can overcome tough difficulties caused by water leakage [55]. Experience has been gained in aerospace applications with using nuclear power systems. As the undisputed leader of this area, liquid Na coolant can greatly simplify boot-up of reactors in orbit, and apply to a compact power system. However, there are some shortcomings in liquid Na. Compared to Li, Na possesses higher chemical activity when being exposed to water or oxidative mediums. Such intense interaction tends to produce hydrogen gas, resulting in safety issues. At the same time, ^{22}Na and ^{24}Na will be induced radioactivity, which have half-lives 2.6 years and 15 hours, respectively. Thus, liquid Na coolant requires two loops in order to hinder the products of the Na–water reaction from getting into the core.

As for Na–K alloy, one outstanding feature is that it can maintain its liquid form at room temperature, due to its low melting temperature (even down to −12.3 °C). This long-standing liquid state is conducive to the simplicity of construction and operation [55]. It acts as a good candidate for experimental rigs or small power stations. Notably, the neutronic, thermal, and mechanical properties of Na–K alloy offer advantages of safety as facing operating failure and large-scale postulated accidents [57]. Unfortunately, thermophysical performance of this alloy is far less than that of Na, so it is difficult for it to fulfill the needs of large power stations. Besides, eutectic components have high vapor pressure relative to Na. In spite of enhanced safety to some extent, the chemical activity of Na–K alloy is still too high. Moist air can lead to fatal results of alloy, such as spontaneous inflammation or even generating explosive peroxide compounds of potassium.

In conclusion, liquid alkali metals meet basic requirements of low moderation of neutrons or charged particles and small absorption cross section, ensuring efficient utilization. Furthermore, in view of excellent thermophysical and neutronic properties, liquid-metal coolant is a reliable, simple, and efficient heat-transfer fluid, allowing for operating at low pressure and moderate temperature. Additionally, liquid alkali metals have better compatibility with structural materials compared to liquid heavy metals. The most challenging issue is facing chemical properties of liquid alkali metals, since they can lead to a violent reaction with water [58]. To address the challenges, people have devoted a great deal of research to suppress fires of alkali metals. For instance, adding a nonvolatile component into Na coolants can extinguish Na fires. The nonvolatility of the impurity creates a barrier to prevent Na from diffusing to the melt surface. Meanwhile, it can decrease the chemical activity via clusters of alloying.

9.3.1.2 Liquid Heavy Metal

Since the 1950s, the exploration of liquid heavy metals has begun in cooling nuclear applications, keeping development in parallel to liquid alkali metals. The primary members of liquid heavy metals are Hg, lead–bismuth eutectic (LBE) alloy, and Pb. All of these members possess similar characters and overall performance in cooling.

Liquid Hg was used for fast breeder reactors at the earliest time. However, its high toxicity, high vapor pressure, and finite reserves greatly limit its further applications. Despite wide usage as an experimental model fluid in the 1960s, it has ruled out all possibilities of Hg for cooling fast reactors.

Among liquid heavy metals, LBE alloy servers as a good candidate for nuclear-powered submarines, owing to its appropriate melting point (~125 °C), low activity toward the surrounding environment, and high boiling point reaching 1638 °C [55]. Especially, when compared to alkali metals, LBE alloy has a slow oxidation with water, and its oxides are insoluble in water, controlling risk caused by water leakage. As a result, a nuclear-powered submarine equipped with LBE alloy coolant was successfully launched in 1963 [59]. The LBE alloy coolant has obvious drawbacks in low specific heat capability as well as thermal conductivity, reducing its cooling performance. To make it worse, high viscosity and severe corrosivity greatly reduce the compatibility with the structural materials.

As for nuclear power stations with high capacity, liquid Pb as the fast reactor coolant has aroused great interest among developers. In contrast to alkali metals, liquid Pb weakly interacts with air or water. Additionally, the boiling point of liquid Pb reaches up to 1745 °C. Another advantage is the extremely low accumulation of polonium when Pb exposes to the irradiation.

For liquid heavy metals, high and special requirements of structural materials always restrict their widespread applications, deriving from strong corrosion and complex handling. It is necessary to carry out research on the anticorrosion of protective coatings, for example by developing passivated steel.

Aside from liquid alkali metals and heavy metals, fusible metals have promising applications in cooling fast nuclear reactors. Sawada and coworkers reported to mix Ga coolant with liquid metallic fuels at high temperature, indicating the possibility of a compact nuclear reactor [60]. Nevertheless, Ga and its alloys have low melting temperatures, as well as corrosion effect and embrittlement phenomenon at room temperature. These disadvantages of the fluid will cause huge safety concerns when under the extreme condition in nuclear applications. Therefore, fusible metals in nuclear applications have little research and operation experience.

9.3.2 Advanced Electronic Cooling

With the rapid development of electronic devices, a growing number of social demands focus on data processing, instant messaging, and online entertainment. As a result, surface heat flux of electronic devices on working is increasing dramatically along with enhancing power densities. For example, to enhance user experience, realizing various functionalities requires a greatly high processing speed, suggesting a central processing unit (CPU) with high performance. Unfortunately, it is not easy to dissipate heat produced by the CPU, owing to low conduction of supporting substrate and limited space. The accumulated heat will shorten service life of the CPU and even make it out-of-order. Therefore, thermal management of electronic devices encounters an extremely tough challenge of releasing ever-larger heat when social demands of high-heat-flux electronic devices become broader.

So far, to address this issue, various cooling systems have been presented for thermal management, such as liquid cooling, air cooling, heat pipe cooling, phase change cooling, thermoelectric cooling, and vapor chambers [61–63]. Among them, air cooling with fans has advantages of simple construction, light weight, low cost, and high reliability. Nevertheless, this cooling system is not adequate for electronic devices with a heat flux density more than 10 W cm^2, owing to the low specific heat capacity of air. In addition, when facing a heat flux density less than

$100\,W\,cm^2$, the heat pipe cooling is a mainstream approach. If the heat flux density exceeds $100\,W\,cm^2$, liquid cooling will be a more suitable and efficient method, compared to other cooling systems. Thereinto, water is widely used in liquid cooling. Its convective heat transfer efficiency is nearly 10 times more than that of the air cooling. However, the use of water is still incapable in offering sufficient cooling to emerging sophisticated electronic devices. It confronts some problems including low thermal conductivity, limited temperature area, and drive mode. Water has thermal conductivity as low as $0.6\,W\,m^{-1}\,K^{-1}$ (at 20 °C), greatly reducing cooling efficiency. Water as heat transfer fluid requires mechanical pumps to drive, leading to noise, large room, and vibration. Additionally, it is difficult to apply the water to dissipate heat from heat source over 100 °C, owing to its low evaporation point. Thus, it is urgently necessary to explore novel liquid coolants for efficient electronic cooling.

9.3.2.1 Proposition of Liquid-metal Coolants

In searching for efficient liquid coolants, high thermal conductivity is the primary consideration. As mentioned in Section 9.3.1, liquid metal and its alloys have long been applied as coolants for nuclear reactors, by virtue of their excellent thermo-physical and nuclear–physical properties. Both heavy metal and alkali metal possess high thermal conductivity, which is dozens of times more than that of water. Therefore, liquid metal has immense potential to stand out as alternatives for electronic coolants.

In fact, though liquid-metal coolants have been prominent in traditional nuclear applications, they started really late in nonmilitary fields. As for electronic cooling, traditional liquid-metal coolants do contain some problems. For instance, liquid Na may cause severe safety concerns especially for portable electronic devices, resulting from strong chemical activity and serious corrosion. In addition, high toxicity and cost of widely studied heavy metals are unavoidable, such as Hg and Pb–Bi alloys. Nonetheless, some advances and new discoveries in traditional nuclear applications finally offer guidance to research for cooling high-heat-flux electronic devices. Liu and coworkers pioneered the concept of room-temperature liquid-metal coolants, which sprout and arose in 2002 [64]. Herein, room-temperature liquid metal mainly refers to liquid Ga and Ga-based alloys. With extremely low melting points, these liquid metals are able to remain in liquid state around room temperature. In addition, they are noncorrosive and nontoxic, suggesting the possibility of practical application. This concept aroused widespread concern, setting off a new wave of research.

To demonstrate the merits of room-temperature liquid metals, people have conducted plenty of studies to compare their cooling performance to that of water coolant. Among high-heat-flux devices, light emitting diodes (LEDs) are of significance in the area of high brightness lighting, because of their low power consumption and long service life. When lacking rapid and effective heat dissipation, junction in LEDs will exceed a safe temperature, worsening optical performance. Deng and coworkers proposed a cooling system using liquid metal for LED with high power [65]. As illustrated in Figure 9.9a, the experimental setup was mainly composed of two parts, namely, liquid cooling system and LED heat source.

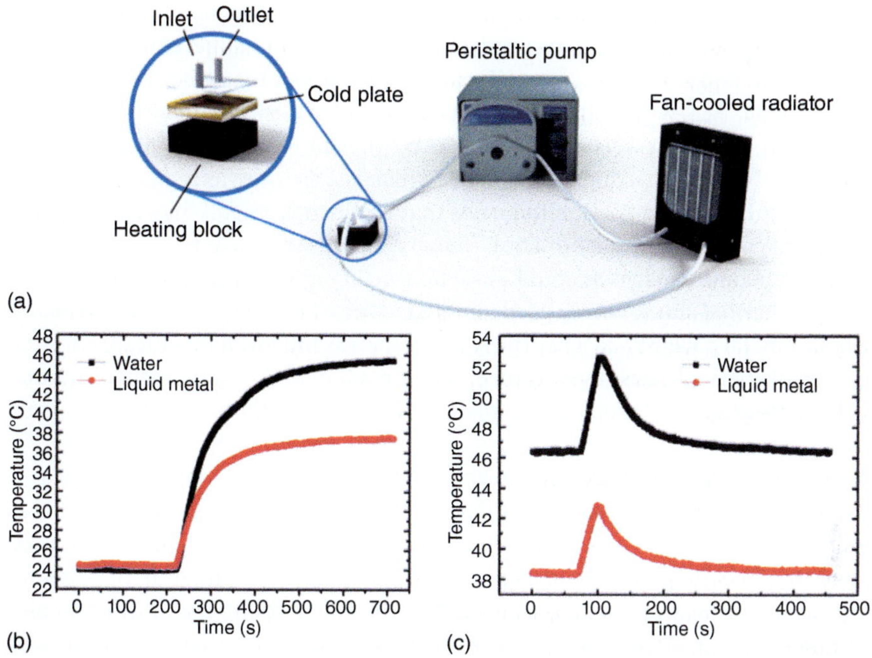

Figure 9.9 Appearance of a prototype device of optical near-field lithography. Advantages of Ga-based coolants in high-heat-flux electronic devices. (a) Schematic representation of experimental platform. (b) Temperature curves on same heat load of 100 W. (c) Comparison of temperature response curves using liquid metal and water [65]. Source: Deng and Liu [65].

A heating block simulated the LED, which generated heat and transferred to the neighboring cold plate. The opposite side of the cold plate was in contact with liquid coolant. The cold plate, a peristaltic pump, and a fan-cooled radiator constituted the liquid cooling system. Notably, when using liquid-metal coolant, the cooling system adopted a mechanical pump than an electromagnetic pump, in order to facilitate comparison with water. During the circulation of liquid coolants, thermocouples recorded temperature changes of cold plate, environment, and heating block. At first, heat dissipation performance of two coolants was illustrated in Figure 9.9b, which were under a same heating power of 100 W. Evidently, when using water, the time–temperature curve of the LED substrate exhibited a rise of 22 °C. Meanwhile, a reduced temperature rise about 13 °C was observed when using $GaIn_{20}$ (fusion point 16 °C), proving that liquid-metal coolant has a better cooling effect. Based on temperature curves, we can calculate the system resistance (R_{sys}) and the convective heat transfer coefficient (h) easily as follows:

$$R_{sys} = \frac{T_{substrate} - T_{ambient}}{Q} \tag{9.3}$$

$$Q = hA(T_{cold\ plate} - \overline{T_f}) \tag{9.4}$$

where, T_{ambient}, $T_{\text{substrate}}$, and $T_{\text{cold plate}}$ are temperature of ambient environment, LED substrate, and cold plate, respectively; T_f is the mean temperature of coolant; Q is heat flux; and A is the heat transfer area. Thus, under the same flow condition and heat load, R_{sys} with water coolant is about 0.22 °C W^{-1}, while the R_{sys} with liquid-metal coolant is very small (0.13 °C W^{-1}). The calculated h reaches up to 9343 W m^{-2} °C^{-1} when using liquid metal, which is 2.54 times as high as that of the water coolant. This comparison indicates that highly increased h played a key role in excellent cooling capability of liquid-metal coolant. As shown in Figure 9.9c, the system temperature response could evaluate stability of the cooling system. During the test, a thermal shock was applied and then it lasted up to 30 seconds. Obviously, the liquid-metal coolant had a relatively fast response and low temperature rise, due to its small thermal resistance. Hence, when coping with instantaneous thermal shock, cooling system using liquid metal has a better ability than that of using water.

Except for high brightness LEDs, good Ga-based liquid coolants can act as a broad cooling technology in other high-heat-flux applications. Li and coworkers performed a conceptual experiment to evaluate the feasibility of liquid metal used in cooling CPUs [66]. They found that liquid Ga exhibited a much more powerful performance of heat dissipation than that of the water cooling, and in particular when high heat generation or low flow rate was required. Furthermore, Deng and Liu designed practical prototypes for cooling CPUs with high powers and drew a similar conclusion [67]. This work studied heat dissipation process of a typical cooling system. Coolants transferred the heat produced by CPUs to the removed radiator by the liquid convection, and then they got recycled. According to basic heat transfer theory, the R_{sys} in this work was comprised of radiator thermal resistance (R_{radiator}), heat capacity thermal resistance (R_c), convective thermal resistance in the cold plate (R_{conv}), and contact thermal resistance (R_{TIM}). Through comparing the foregoing resistances, it was easy to know how to optimize the cooling system. Taking into account the practical applications, the authors presented a tower structure to save the dosage of liquid metal. Additionally, they also constructed two prototypes with different coolant amount, geometry size, and matching fan. $Ga_{66}In_{20.5}Sn_{13.5}$ used in those prototypes has a low melting point of 10 °C. What's more, they compared the thermal resistance distributions of liquid-metal prototypes with a commercial device of water cooling under a same flow condition. The comparison result with water cooling showed that liquid-metal cooling system had significantly smaller R_{conv} than that of a water-cooling system, relying heavily on the large convective coefficient of liquid metal. The prototype with a flat structure had the lowest thermal resistance as its structure is in favor of great airflow and heat dissipation area. From the above, the presented liquid-metal cooling device after further optimization can match or even exceed the latest cooling system of CPUs in the market, revealing tremendous potential in applications.

So far, Ga-based liquid cooling has emerged as a good effective technology for high-heat-flux electronic devices. Considering from excellent cooling performance of liquid metal, it mainly relies on superior thermophysical properties. On one hand, characteristics such as high thermal conductivity, vapor pressure, and specific heat per unit volume provide liquid metals with core competitiveness as liquid coolants. In this view, these merits ensure liquid metal fulfill efficient heat transfer, as well as meeting the requirements of high-heat-flux applications. On the other hand, characteristics such as high boiling point, surface tension, electrical conductivity, and low dynamic viscosity guarantee liquid-metal cooling system to adapt to market demand. From safety aspect, stable thermophysical properties and the boiling point reaching up to 2000 °C make liquid metal perfectly adapted to extreme conditions and keep single-phase flow. Therefore, liquid-metal system can effectively eliminate explosion risks caused by coolants depleted in the heat pipes. Because of high surface tension, liquid metal highly raises toleration for heat pipes with small cracks or poor seals, thereby avoiding fluid leakage. From energy-saving aspect, excellent electrical conductivity makes it possible to introduce electromagnetic pumps into liquid cooling systems. Electromagnetic pumps do not contain mechanical moving components, which reduces flow resistance and prevent mechanical efficiency loss. Compared to peristaltic pumps using water cooling systems, electromagnetic pumps are more energy-efficient, and meanwhile this unique driven mode can help realize the silent operation within electronic devices. Low dynamic viscosity enables liquid metal to flow through small pipes, indicating easy miniaturization. In order to account for the significance of Ga-based coolants, we take several typical liquid metals and water as examples, listing basic features in Table 9.1. The comparison result suggests an overwhelming victory of liquid metals, especially for thermophysical properties. Notably, mercury is too toxic so that it is not suitable for electronic cooling. In spite of thermophysical properties comparable to Ga-based alloys, $Na_{23.3}K_{76.7}$ alloy has high activity to water and air, which limits its practical applications. In short, liquid Ga or its alloys ought to be an ideal candidate for heat transfer fluid in the field of electronic cooling. Recently, the burgeoning liquid-metal coolants have gradually annihilated the public stereotypes of unsafe liquid metals, which demonstrated to be more efficient and more reliable than conventional electronic cooling systems.

9.3.2.2 Design of Hybrid Coolants

Though the liquid-metal coolants have excellent cooling efficiency, Ga or its alloys have a relatively high cost. Reducing dosage of liquid metal turns out to be the most critical challenge to further large-scale utilization. Meanwhile, the liquid cooling system faces challenges as the miniaturization of electronic devices are progressing. Typically, heat sinks, external pumps, and connecting pipes require adequate room in the liquid cooling system. To solve these problems, scientists presented hybrid coolants to lower the amount of liquid metal needed and even simplify the complete cooling system.

In the infancy of hybrid coolant studies, scientists focused on replacing partial liquid metal with the cheapest coolants, which means combining liquid-metal cooling with water cooling. Deng and coworkers presented a hybrid liquid metal–water-cooling system, realizing similar cooling performance and only a half cost of pure liquid-metal cooling [69]. The key here was using liquid metal as "heat spreader" to contact with a water-cooling module, directly dissipating heat from liquid metal to water. During this process, the heat source generated high heat flux,

Table 9.1 Properties of water and typical liquid metals [68].

Properties	Liquid metals				Water
	Mercury	Gallium	Galinstan	$Na_{23.3}K_{76.7}$	
Melting point (°C)	−38.87	29.8	−19	−12.6	0
Vapor pressure (mmHg)	1.68×10^{-3} a)	10^{-12}	$<10^{-8}$ d)	2.5×10^{-8} c)	17.54
Mass density ($kg\,m^{-3}$)	13 546 a)	5907 b)	6440 a)	855 c)	988 b)
Viscosity (mPa s)		1.2 e)			1.002
Thermal conductivity ($W\,m^{-1}\,°C^{-1}$)	8.34 a)	29.4 b)	16.5	23	0.6
Specific heat ($kJ\,kg^{-1}\,K^{-1}$)	0.139 a)	0.37 b)		0.9538 b)	4.182
Surface tension ($N\,m^{-1}$)	0.455 a)	0.707 b)	0.718 a)	0.11 f)	0.072

a) At 25 °C.
b) At 50 °C.
c) At 100 °C.
d) At 500 °C.
e) At 77 °C.
f) At melting point.
Source: Wang and Liu [68].

and then liquid metal efficiently carried away heat flux. Subsequently, the heated liquid metal flowed into a heat exchanger to dissipate the heat to the water. Finally, water moved to the remote radiator to complete removal of the heat. The area of the heat exchanger was large enough so that "heat spreader" could run well. By virtue of the "heat spreader," the high heat flux expanded from a finite heat source to the heat dissipation surface with a much larger area. Therefore, this function greatly reduced heat transfer area and the total convective resistance.

Considering the demand of miniaturization, the hybrid liquid metal–water system integrates the strengths of two coolants but still lacks a simple and efficient design. External field tuning for Ga-based alloys has attracted intensive research attention (details seen in Chapter 4). It was reported that a low voltage could drive and control locomotion of a large-sized liquid metal immersed in water [70]. Inspired by this, Tan and coworkers developed a novel cooling device consisting of hybrid coolants, which employed a flow driving method by electric fields instead of pumps [71]. Notably, a $GaIn_{20}$ liquid droplet was soaked in 0.5 mol l^{-1} NaOH solution as a whole, serving as a hybrid coolant. The working prototype was comprised of an annular open-top channel with the injected coolant. The prototype was equipped with two pairs of electrodes, and the corresponding circuit design enabled continuous circular motion of the droplet. Once applying a voltage to the loop, the circuits could automatically actuate the droplet to keep counterclockwise movement to transfer heat. According to this novel cooling device, the authors confirmed the feasibility of hybrid coolants based on a liquid metal droplet and highlighted the purely electrical driving mode. However, it should be pointed out that this design does not make the best of liquid metal and requires further optimization. Actually, the major heat transfer in this

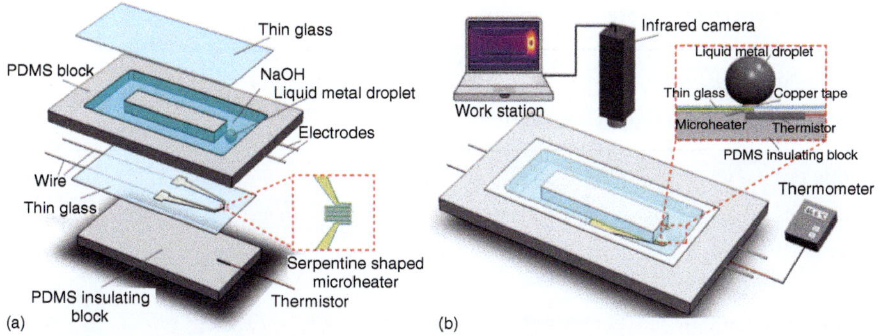

Figure 9.10 An integrated cooling system with hybrid coolants. (a) Exploded schematic of experimental setup. Inset, top-view of the microheater. (b) Schematic of the whole assembly. Inset, close-up diagram of the droplet placed over the heat source [72]. Source: Zhu et al. [72].

design depends on the solution flow caused by the movement of the droplet. Because liquid metal was not in direct contact with the heat source, the droplet only took away tiny heat. Low thermal conductivity of the solution limited heat dissipation. The authors observed that the heat dissipation performance did not improve while increasing voltage. This phenomenon was due to that the solution flow rate did not change clearly.

The same year, Zhu and coworkers developed an ingenious integrated liquid cooling system based on a small Galinstan droplet, achieving a high flow rate and quick cooling of hot spots [72]. As illustrated in Figure 9.10a, the circular cooling channel was formed by grooving polydimethylsiloxane, which was sealed with two thin glass slides to avoid leakage. On one side of the glass slide, a serpentine-shaped microheater was adhered to act as a localized hot spot. A square wave signal (200 Hz, 5 V_{p-p}, 2.5 V dc offset and 50% duty cycle) was set between two electrodes with a spacing distance of 10 mm, which were stainless steel and anti-corrosive in the aqueous solution. Notably, liquid metal kept immobilized in a copper tape, connecting with the hot spot (see an inset in Figure 9.10). To monitor the temperature change, the system was equipped with a thermistor and an infrared camera.

Figure 9.11a recorded the rapid cooling process of the channel, exhibiting the temperature changes of the hybrid coolant over time. At the beginning ($t = 0$ s), the hot spot accumulated heat, leading to a hot region with temperature up to 36.5 °C. Meanwhile, without an electric stimulus, the motionless Galinstan droplet reflected the infrared radiation and then formed a cavity-like image. Obviously, once a square wave signal activates the droplet, the surrounding solution around the droplet will start to flow. At this point, the heat transfer process would start and lead to quick diffusion of heat in the whole channel. After short electrical activation of Galinstan droplet ($t = 5$ s), the distribution of temperature was well-uniform and lowered to 29.5 °C, confirming rapid migration and heat dissipation of this integrated cooling system.

Such a forced convection based on the droplet is induced by the Marangoni flow. The corresponding mechanism is based on the surface tension gradient formed on

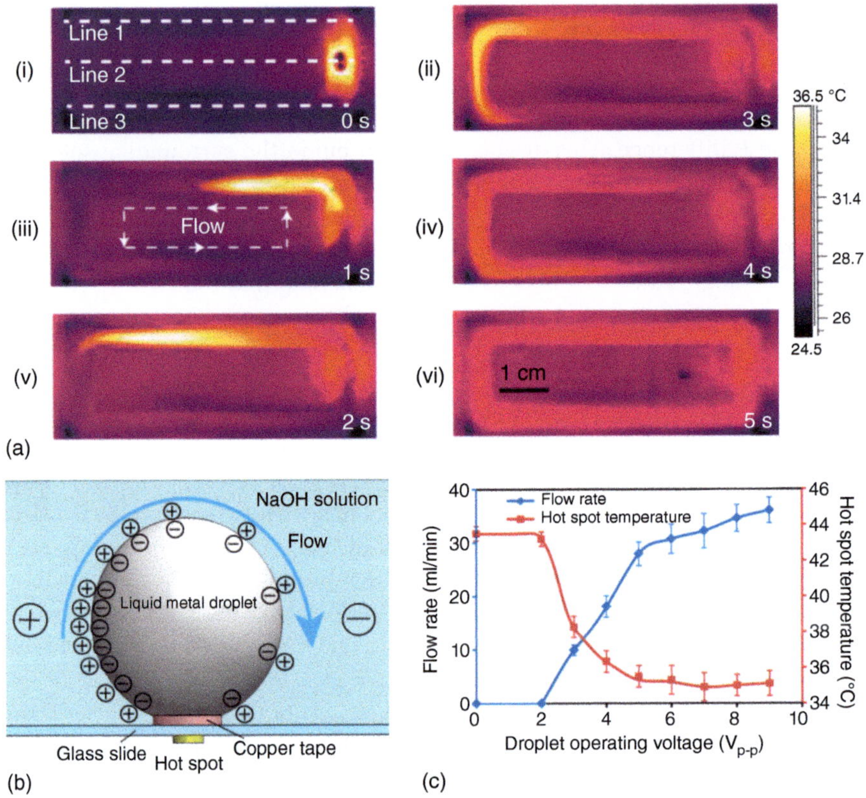

Figure 9.11 Heat dissipation process of the integrated cooling device. (a) A series of infrared images were sequentially snapshotted, exhibiting temperature variation over time. Source: Zhu et al. [72]. (b) Surface charge distribution of Galinstan under electric field. The Marangoni flow was continuously generated at the interface. (c) Variation of hot spot temperature and flow rate with respect to applied voltages, when setting a square wave signal as 200 Hz [72]. Source: Zhu et al. [72].

the droplet. A chemical reaction will take place if the surface of Ga touches alkali solution [73]. The products contain $[Ga(OH)_4]^-$ anions make the interface between solution and droplet electrically charged. Hence, the interface tends to attract positively charged ions, resulting in the evolution of an electrical double layer (EDL). Without external potential, the potential of the EDL remains constant across the droplet. As a voltage is applied, the interfacial tension of the droplet will alter according to Lippman's equation:

$$\gamma = \gamma_0 - \frac{1}{2}C(V - V_0)^2 \tag{9.5}$$

where, the surface tension is γ, the maximum value is γ_0, C is the capacitance of the EDL per unit area, the applied voltage on the interface is referred to V, V_0 is expressed by dividing initial charge into C. As shown in Figure 9.11b, the balance of charge distribution throws off, which is caused by external potential. The leftward γ is lower than that on the right. The relationship between γ and pressure (P) on the interface can be described as follows through the Young–Laplace equation:

$$P = \gamma\left(\frac{2}{r}\right) \tag{9.6}$$

where r is the radius of curvature. Taking into account this equation, the asymmetric γ causes the P difference. This driving force can pump the surrounding solution through the channel, which was named Marangoni flow. In other words, liquid metal herein serves as a pump. Unlike mechanical or electromagnetic pumps, this integrated cooling system is capable of no moving parts, no connecting tubes, instantaneous high flow rates, and no risk of degaussing under high temperature. To exploring the influence factors of cooling performance, authors assessed the flow rate curve and hot spot temperature variation under different operating voltages (Figure 9.11c). When the operating voltage was in the range from 2 V_{p-p} to 5 V_{p-p}, the flow rate kept linear growth, decreasing the hot spot temperature of 43.5 to 35.5 °C. Operating voltages exceeding 5 V_{p-p} could bring about electrolysis of electrodes to produce hydrogen bubbles, which in turn lowered flow rates. This work provides a novel technology of hybrid coolants with simple fabrication and short response time, which is suitable for electronic devices against an unexpected thermal shock and limited space.

Except for designing novel cooling systems, optimizing coolant constituents is one of the most significant research on hybrid coolants. In particular, nanofluids can greatly enhance thermal conductivity of liquid coolants and their relative research has undergone an explosive growth. To prepare the nanofluids, nanoparticles (size less than 100 nm) are added into the heat transfer fluids to form suspension, which is first presented in 1995 [74]. In early researches about nanofluids, researchers have attempted that dispersing highly conductive nanoparticles in water, which could effectively improve the convective heat transfer of water [75]. Surprisingly, a small amount of nanoparticles significantly improves performance of the original coolants. For instance, the inherent thermal conductivity of ethylene glycol has been reported to be increased by nearly 40% after mixing with 0.3 vol% copper nanoparticles [76]. Nonetheless, the base fluids actually restricts further improvement, because of their conductive characteristics, surface tension, and liquid density. Among widely used base fluids, oil, water, and ethylene glycol suffer from poor thermal conductivity. Due to their small surface tension, the volume fraction of nanoparticles is very confined, which is less than 10 vol%. As nanoparticles overload, additives will depart from the bulk of suspension and then deposit at the bottom, leading to degeneration of nanofluid quality or even blocking pipes.

To address those problems, Ma and coworkers, for the first time, proposed nanofluid using liquid metal as the base fluid [77]. The surface tension of Ga was 7–10 times larger than that of traditional fluids. Therefore, the nanofluid based on Ga showed a much high load of nanoparticles up to 20 vol%, which was conducive to maximizing thermal conductivity. In Figure 9.12a, the mixed results remained good since the volume fraction of carbon nanotubes (CNTs) increased to 20 vol%. The enhancement tendency of nanofluid with several typical nanoparticles was predicted by simulation based on Bruggeman model (Figure 9.12b). When using CNTs, the thermal conductivity had the largest improvement compared to other metal

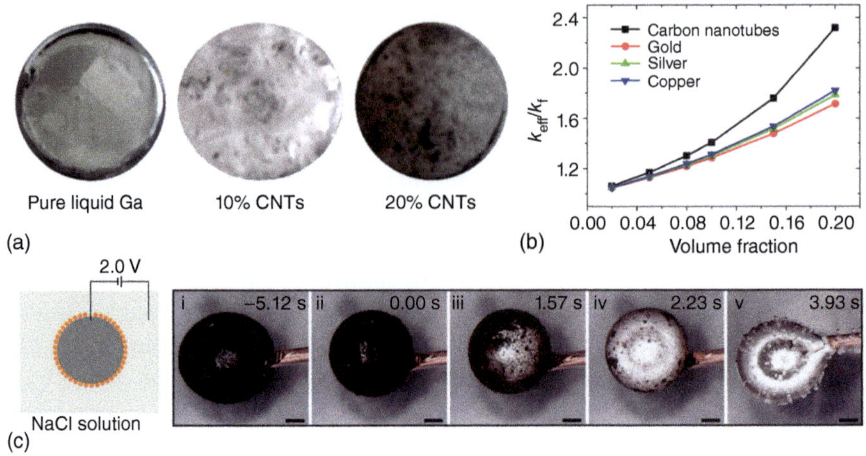

Figure 9.12 Nanofluids based on liquid metals. (a) Photographs of pure liquid Ga and nanofluids with different volume fractions of CNTs. Source: Ma and Liu [77]. (b) Thermal conductivity enhancement ratio (expressed as k_{eff}/k_f) changes of nanofluids containing typical nanoparticles with varying volume fraction [77]. Source: Ma and Liu [77]. (c) Electrical polarization facilitates liquid metal's phagocytosis in neutral solution. Scale bars: 200 μm [78]. Source: Tang et al. [78].

particles, suggesting a very bright prospect of liquid-metal nanofluids for electronic cooling. In fact, uniform dispersion of nanoparticles plays a vital role in nanofluids. Especially for liquid metal, evenly dispersing "naked" metal into the base fluid is challenging, due to its passivity, surface tension, and other aspects [79]. Fortunately, Tang and coworkers reported particle internalization of Ga-based liquid metal in solutions, providing an efficient strategy for constructing nanofluids with high quality [78]. In Figure 9.12c, the liquid metal was coated with copper nanoparticles in a solution. After applying a cathode polarization voltage to the droplet, liquid metal totally engulfed these nanoparticles, accompanied with dense bubbles. Considering contact angle tests and the theoretical study, they disclosed that the reactive wetting between two metallic phases dominated this process. This funding shows a unique wet-processing method, which provides a new perspective to prepare nanofluids. Other research has demonstrated that optimizing size effect, particle morphology, concentration, and dispersants could greatly affect the thermal properties of nanofluids [80], creating an immense potential of preparing ultimate coolants.

9.3.2.3 Optimization of Cooling Systems

According to the detailed discussion in the last section, researches about designing hybrid coolants mostly focus on pursuit of excellent cooling efficiency as well as reduced cost. Apart from these two critical factors, the cooling system is supposed to achieve low energy consumption, small occupation of room, and no noise, which is vital to support further industrialization. Therefore, optimizing cooling system has aroused a great deal of attention. Specifically, some novel cooling systems, such as self-supporting modules and mini-channel heat sinks, make full use of liquid metal's merits.

As is well known, high-heat-flux electronic devices always produce a lot of heat, which is passed to the environment through cooling systems. If the waste heat is harvested to convert to electricity, it will make enormous contributions to energy sustainability. Aiming at this target, Li and coworkers developed a self-driven cooling system, which utilized waste heat to realize flow of the liquid-metal coolant [81]. Amazingly, the liquid cooling system is self-powered and has no pumps, which is completely different from other cooling systems. For existing liquid cooling systems, extra power supply was always required to drive pumps to circulate fluids, regardless of mechanical pumps or electromagnetic pumps. In this work, the authors utilized thermosiphon effect of liquid metal to provide the driving force. This phenomenon refers to that temperature differences of fluid that can generate density gradients so that buoyancy is generated to drive the fluid flow. Unfortunately, conventional coolants barely realize a good flow driven by this effect, because the generated buoyancy is not enough to provide strong driving force. However, liquid metals with excellent thermal conductivity makes it come true. As displayed in Figure 9.13a, an experimental platform consisted of tube, heat source, cooler, heater, and heat-to-electricity generator. Liquid metal was injected into the tube, and then absorbed heat produced by the heat source. Then, commercial thermoelectric (TE) modules served as generators to utilize heat transferred by liquid metal. Due to the thermosiphon effect, liquid metal was subjected to the force of buoyancy. When the force was large enough to get rid of the viscous drag, liquid metal automatically began to flow in the closed loop. Subsequently, the heat could be passed to TE modules, and partial heat would be converted to electricity. In Figure 9.13b, the liquid-metal coolant revealed an obvious growth of the heat transfer coefficient as increasing heating power. At the same time, the value of velocity increased from 0.60 to 1.46 cm s^{-1}. This result indicated that enhanced heat load could enlarge the temperature difference, and thereby liquid metal driven by powerful force led to rapid heat dissipation. Moreover, authors also assessed the energy conversion efficiency (Figure 9.13c). Clearly, a large temperature difference facilitates the energy conversion efficiency. Despite relatively low efficiency, the energy conversion capability of the system can be boosted after further modifications, such as increasing TE modules and reducing heat leakage. As a result, this strategy based on the thermosiphon effect successfully harvests the waste heat, keeping cooling system in operation without pumps. This cooling system exhibits a promisingly wide application in electronic devices with high power, covering smartphones, LEDs, computers, and so on.

As for liquid cooling systems, the optimization of heat sinks is also of importance. As integrated circuits and miniaturization continue to flourish, laser diode array with ultrahigh power proposes great demands on thermal management. To remove large heat, Tuckerman's group brought micro/minichannel heat sink in the water cooling system [61]. By means of this compact heat sink, the cooling system exhibited reduced thermal resistance and good cooling performance even tested at a power density of 790 W cm^{-2}. However, pumping devices suffer from limited cooling capability because of intrinsic thermal conductivity [82]. As for liquid metal, some

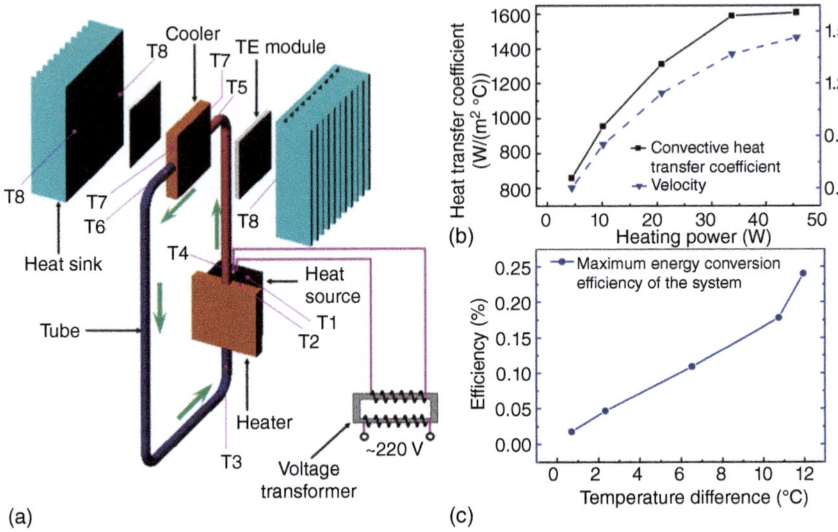

Figure 9.13 Self-driven cooling system via thermosiphon effect. (a) The corresponding prototype schematic. The heat source was connected to the heater by silicone grease, which was wrapped with insulated plastic foam and cotton to eliminate heat loss. Tube was made of silicone. T_m ($m = 1, 2, 3, 4, 5, 6, 7, 8$, and 9) referred to temperatures at different positions. (b) The convective heat transfer coefficient versus heating power, and the velocity versus heating power. (c) The energy conversion efficiency yielded by the system as varying temperature difference [81]. Source: Li and Liu [81].

fundamental strengths are high thermal conductivity and high electrical conductivity, which seem to be more appropriate for using the advanced heat sink. Currently, Zhang and coworkers presented a vascularized heat sink using liquid-metal coolants for laser diodes [83]. In the cooling system, the vascularized structure was composed of 70 channels with a diameter of 1.8 mm. This system achieved heat dissipation reaching 2000 W and its heat sink surface kept a low temperature of 54 °C. Such excellent heat removal performance demonstrates that the optimization of heat sinks offers an effective way for advanced electronic cooling.

9.4 Liquid Metal as Catalysis

Due to the possession of special properties, such as liquid surface, composition tunability by opting different metals, and free electrons and ions on the surface, liquid metal has great advantages in catalysis [84]. Now, there is little research on liquid metal catalysis, especially single-element liquid metal. The study of liquid metal alloy is relatively more than that of single-element liquid metal, but limitation still exists. Although there are few researches on liquid metal catalysis, the existing research shows that liquid metal has a good catalytic prospect. Here, the catalyst mechanism and catalytic advantages of liquid metals and their alloys will be described.

For liquid metal with single element as catalyst, some organic reactions are mainly reported at present [10]. However, the catalytic performance of single-element liquid metal is not excellent. It is reported that transition metals have better catalytic performance for the narrow d-band structure. The main group metals mainly use the delocalized s/p orbit as the host-orbital, and "weak chemsorption" broadens the adsorbate state [85]. The interaction between adsorbate and adsorbate is either too strong to cause catalytic poisoning or too weak, exhibiting poor catalytic activity finally. Liquid metals are mainly main group elements, such as Ga, In, Sn, Pb, Bi, Sb and the like. It was reported earlier that liquid metal Ga was used to catalyze hydrogen production. The results show that single-element liquid metal does have low catalytic activity [86]. In addition, in the reaction of coal liquefaction, liquid metal Sn, In, Bi, and Ga were applied as catalysis [10]. The result shows that the catalytic activity of liquid element is not particularly well either. However, different from single-element liquid metals, their alloy showed excellent catalytic performance, as discussed in the following section.

Since liquid metal alloy has its special properties, it is increasingly used in catalytic research. Regarding liquid metal alloy as catalyst, Ga-enriched Pd–Ga alloy was used as a catalyst for alkane dehydrogenation [87]. The catalytic mechanism of the Pd–Ga mixture was explained as follows. Previous studies had suggested that active sites composed of isolated monoatomic can significantly improve the catalytic properties [88]. The mechanism here can be considered as an extension of the concept. Because Ga has a lower melting point, the ability to dissolve other transition metals is easy. By dissolving catalytically active metals into a liquid environment, a completely different electronic and spatial properties can be provided from solid metal catalysts. In addition, in order to prove that the liquid alloy worked in a liquid state and the Pd content on the alloy surface, the state and surface composition of the catalyst in the reaction was characterized in detail. The phase state of Ga–Pd alloy was shown in Figure 9.14a. It can be seen from the SEM image that the alloy exists in liquid form. In theory, according to the result of normalization, the integrated Pd signal should be the same at 0° and 80°. The signal at 80° was obviously smaller than the signal at 0°, which indicated that Pd in the top layer is rare (Figure 9.14b). Figure 9.14c and inset graph were vertical density profiles of Ga–Pb and pure Ga, respectively. They showed that the structure of Ga has not changed basically, but the density of Pd has changed. Whether in Figure 9.14b or c, it indicated that Pb exists in the area below the surface because its content on the surface is very small. In the liquid phase, atoms have higher mobility, and Pd atoms enriched in the area below the surface can contact with reactants close to the surface, and finally changed the atom distribution of the liquid catalyst surface. This may explain that the observed catalytic activity is related to a relatively small number of isolated Pb atoms on the liquid surface. Therefore, the metal Ga environment created a very unusual space and electronic environment for a small number of surface Pd atoms and determines their catalytic performance.

Besides, the catalyst has high catalytic activity and stability, as shown in Figure 9.14d. In order to manifest the excellent catalytic performance of liquid metal clearly, it was compared with commercial catalysts Cr_2O_3/Al_2O_3 and

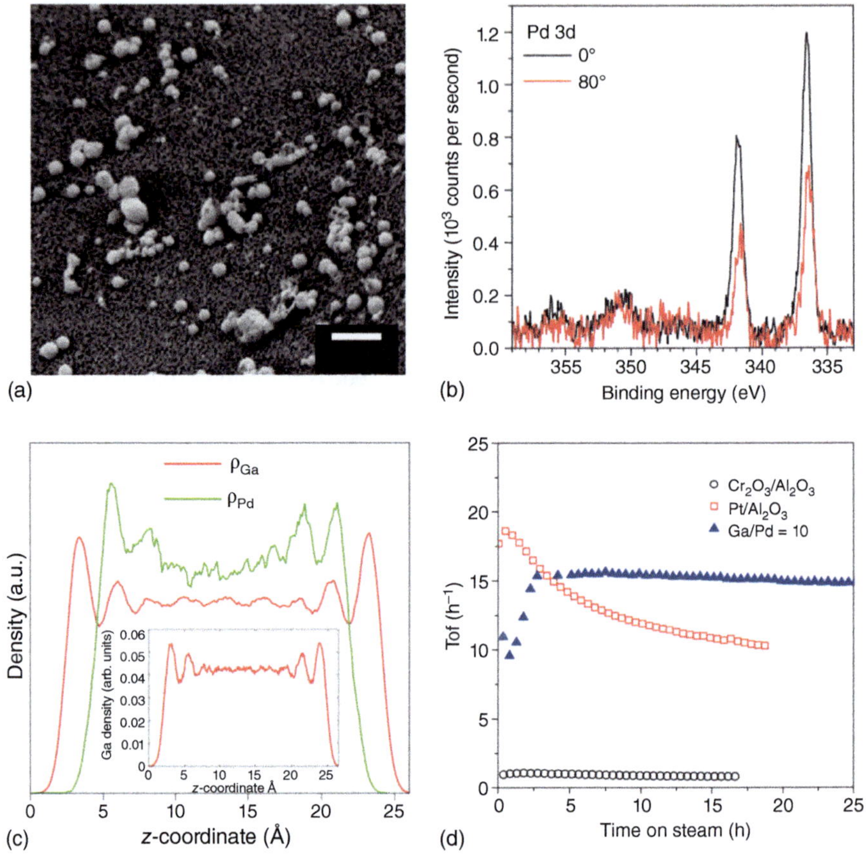

Figure 9.14 (a) SEM image of Pd–Ga alloy. (b) The AR-XPS of Pd 3d in the mixture of Pd–Ga 477 °C. Source: Taccardi et al. [87]. (c) Density distribution of Pd–Ga alloy and pure Ga (inset) in z direction. (d) Comparison of catalytic activity between Pd–Ga alloy and commercial catalyst [87]. Source: Taccardi et al. [87].

Pt/Al_2O_3. It showed the catalytic activity of liquid metal catalyst is much higher than that of Cr_2O_3/Al_2O_3. Although it was lower than Pt/Al_2O_3 at first, the activity of Pt/Al_2O_3 decreased rapidly over time. The activity of liquid metal catalyst nearly showed no decreases, which indicated the excellent stability of liquid metal catalyst. The result can be attributed to the fluidity of liquid metal, thereby coke formation can be effectively avoided, thus preventing the reduction of activity. At the same time, isolation of active sites also can be contributed to the improvement of activity. In addition, the reactions mentioned earlier are all carried out above 200 °C, which is another advantage of liquid metal catalysis. Due to limited thermal stability, the reaction temperature of traditional catalysts is often limited below 200 °C, which greatly hinders the progress of many catalytic reactions. However, the catalytic temperature of liquid metal catalyst can exceed this value, which is related to the insolubility between liquid metal and organic matter.

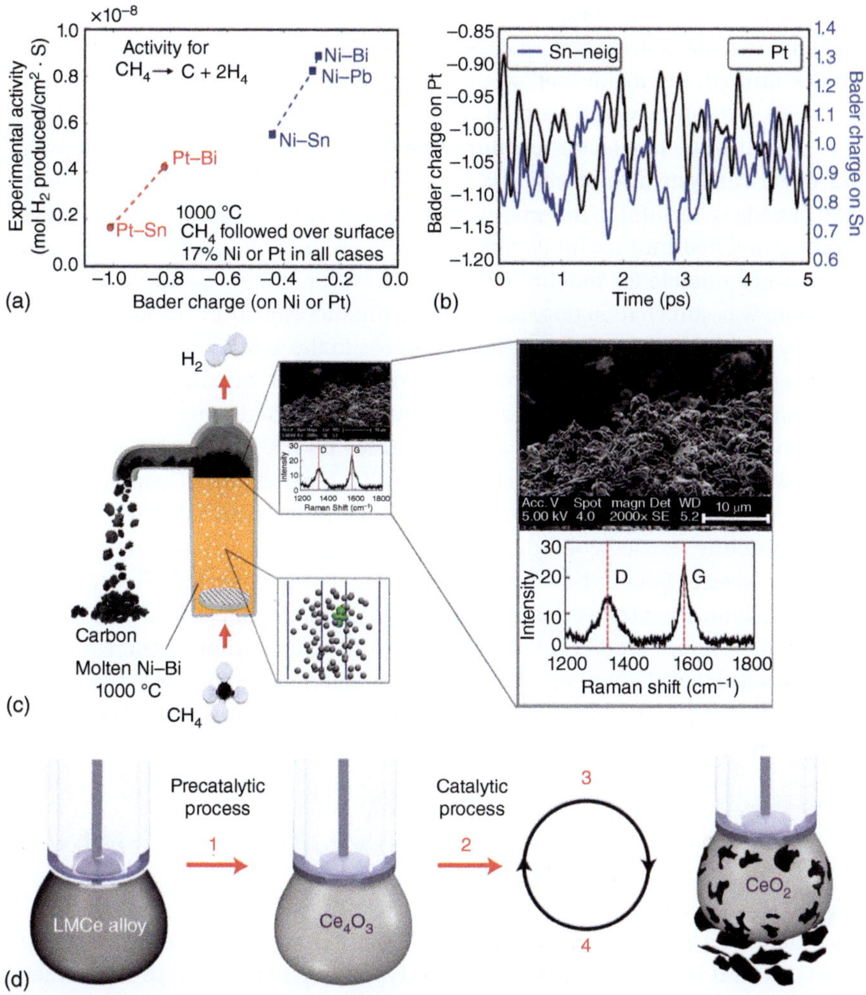

Figure 9.15 Mechanism of Ga–Ce alloy catalyst in catalytic conversion of CO_2 to solid C species. Source: Esrafilzadeh et al. [90].

In addition, it has also been reported that molten metal catalyst can be applied to catalyze methane to hydrogen [89]. In the experiment, after forming the alloy, the activity of the catalyst and the reaction rate have changed. Changes in the electronic properties of the alloy are used to explain its catalytic mechanism. It reported that the charge of active atoms is related to the catalytic activity. The less negative charge, the better the catalytic activity. For example, adjacent Sn atoms provided electronic charge to Ni, and the positive charge in each Sn atom was slightly more than the situation when they existed alone in the melt because the existed of Ni in Ni–Sn Alloy. At the same time, the stable phase of the catalyst has been greatly improved after forming the molten alloy catalyst, because the catalyst can remove coke, which will be introduced below.

In the process of studying liquid metal catalysis, it reported that liquid metal also has a function in preventing carbon deposition. In the process of reaction, if a carbon layer was formed, the metal melt would dissolve, and the carbon would continuously float to the surface and then be removed. Carbon deposition on the catalyst surface was also the key factor leading to catalyst deactivation. The emergence of liquid metal catalysts also played a key role in solving this problem. In addition, carbon dioxide is harmful to the environment as a greenhouse gas, and it had potential to be converted into useful chemicals in the production process. Solid carbon products were suitable for indefinite storage. However, reducing carbon dioxide to a solid state was still challenging, because any product can cover the surface of the catalyst by van der Waals adhesion, hinder access to the active sites of the catalyst, and damage the catalyst in a process called coking. In the experiment of reducing CO_2 to solid carbon at room temperature, liquid metal containing Ce was used as catalyst, in which Ce was solid. Ce catalyst formed on the liquid metal/electrolyte interface was used to promote the reduction of CO_2 together with Ce nanoparticles [90]. Because van der Waals bonding at the liquid interface was inhibited, the electrode prominently resisted coking caused by solid carbonaceous materials. Finally, solid carbon species fall off the electrode surface and were collected (Figure 9.15).

In a word, liquid metal can be used as catalyst in various applications. It shows that the research on liquid metal catalytic system is still in embryonic form. No matter whether it is a single element or an alloy, there is still much room for development. Because of the unique properties of liquid metal, the research on liquid metal catalyst is still worthy of further study.

9.5 Conclusion and Prospect

As mentioned earlier, this chapter introduces liquid metal electrodes from high temperature and low temperature, respectively. For high-temperature liquid metal electrodes, due to their high theoretical capacity and energy density, they are considered as the most favorable choice for extensive energy storage systems. For room-temperature liquid electrodes, their greatest advantage is that it does not need additional energy supplies. Besides, the liquid metal fluidity can lead to self-healing, which obviously improves the service life of the battery. Meanwhile, it can also solve the dendrite problem that has plagued battery development for many years. Therefore, liquid metal electrodes are expected to have a widespread application in the battery system in the future. However, there are still some challenges in the popularization of liquid metal electrodes. First, Na–K based metal has higher activity, the exposure of which to air is very dangerous. Therefore, it is necessary to find suitable components to inhibit this property while maintaining high catalytic activity. Second, for Ga-based liquid metal electrode, its cost is a problem to be considered, and better metallurgical technology is needed to reduce the cost. Finally, if liquid metal electrodes are going to popularize, it is necessary to further improve their volumetric density and capacity.

Traditional nuclear energy and the new electronic field show the progress and development of science and technology. They are finking the common problem in the process of use, which is heat generation. This chapter describes the Na-based and Pb–Bi coolants for conventional fast reactors and Ga-based metal in the cooling of new electronic chips. With the deepening of the research on liquid metal coolant and its excellent properties, it is expected to be widely used in industry in the future. However, there are still exist some problems that need to be solved before that. The first problem is cost, liquid metal is expensive now and that means more economical metallurgical technology is needed. The second problem is corrosion. As far as liquid gallium is concerned, it can dissolve most metals, causing damage to equipment during use, such as motherboard, CPU, and graphics card. Therefore, in order to make use of this material better, more research must be done.

According to the current research situation, we can conclude that the advantages of liquid metal in the field of catalysis are obvious, which shows liquid metal can occupy a significant state in the field of catalysis in the future. However, the liquid metal-based catalysis is still in the primary stage compared with other aspects of liquid metal research. Few mature systems have been studied, especially reports on single element systems, such as Ga. Alloys may provide important opportunities for the development of liquid metal catalysts due to optional adjustment of components, the formation of interfacial compounds as thin as atoms and its fluidity, and the like. Moreover, compared with traditional catalysts, liquid metal catalysis has a wider temperature range. In addition, except for Ga and its alloys, other metals with low melting points such as In, Sn, Bi, and corresponding alloys have great development potential. Therefore, in the future research, liquid metal may bring great opportunities to the catalytic field.

In short, liquid metal has played a certain role in the energy field. Especially in the field of battery energy storage, there have been a wide range of researches and have shown unique advantages. Although the research on using it as coolant and catalyst is rare, it still shows obvious value from the current research. Consequently, it is believed that in-depth research on liquid metal will definitely have bright prospects in the field of energy in the future.

References

1 Guo, X., Zhang, L., Ding, Y. et al. (2019). Room-temperature liquid metal and alloy systems for energy storage applications. *Energy Environ. Sci.* 12: 2605.
2 Li, H., Yin, H., Wang, K. et al. (2016). Liquid metal electrodes for energy storage batteries. *Adv. Energy Mater.* 6: 1600483.
3 Lin, D.C., Liu, Y.Y., and Cui, Y. (2017). Reviving the lithium metal anode for high-energy batteries. *Nat. Nanotechnol.* 12: 194.
4 Wydler, P. (2005). Liquid metal cooled reactors. *Chimia* 59: 970.
5 Ma, K. and Liu, J. (2007). Liquid metal cooling in thermal management of computer chips. *Front. Energy* 1: 384.

6 Miner, A. and Ghoshal, U. (2004). Cooling of high-power-density microdevices using liquid metal coolants. *Appl. Phys. Lett.* 85: 506.

7 Datta, R.S., Haque, F., Mohiuddin, M. et al. (2017). Highly active two dimensional α-MoO_{3-x} for electrocatalytic hydrogen evolution reaction. *J. Mater. Chem. A* 5: 24223.

8 Zhang, W., Naidu, B.S., Ou, J.Z. et al. (2015). Liquid metal/metal oxide frameworks with incorporated Ga_2O_3 for photocatalysis. *ACS Appl. Mater. Interfaces* 7: 1943.

9 Ralphs, M.I., Kemme, N., Vartak, P.B. et al. (2018). In situ alloying of thermally conductive polymer composites by combining liquid and solid metal microadditives. *ACS Appl. Mater. Interfaces* 10: 2083.

10 Matsuura, M., Matunaga, S.-I., Oazwa, S., and Ogin, Y. (1981). Volcano-shaped relation between heats of oxide formation of metals and coal-liquefaction activities of liquid-metal catalysts. *J. Chem. Soc., Chem. Comm.* 508: 721.

11 Wang, K., Jiang, K., Chung, B. et al. (2014). Lithium-antimony-lead liquid metal battery for grid-level energy storage. *Nature* 514: 348.

12 Dunn, B., Kamath, H., and Tarascon, J.M. (2011). Electrical energy storage for the grid: a battery of choices. *Science* 334: 928.

13 Yang, Z., Zhang, J., Kintner-Meyer, M.C. et al. (2011). Electrochemical energy storage for green grid. *Chem. Rev.* 111: 3577.

14 Barnhart, C.J. and Benson, S.M. (2013). On the importance of reducing the energetic and material demands of electrical energy storage. *Energy Environ. Sci.* 6: 1083.

15 Sun, Y., Wang, C., Xue, Y. et al. (2016). Coral-inspired nanoengineering design for long-cycle and flexible lithium-ion battery anode. *ACS Appl. Mater. Interfaces* 8: 9185.

16 Xue, Y.H., Deng, J.W., Wang, C. et al. (2016). A pinecone-inspired nanostructure design for longcycle and high performance Si anodes. *J. Mater. Chem. A* 4: 5395.

17 Zhang, Q., Tan, S., Mendes, R.G. et al. (2016). Extremely weak Van Der Waals coupling in vertical Res_2 nanowalls for high-current-density lithium-ion batteries. *Adv. Mater.* 28: 2616.

18 Zhang, W.D., Zhuang, H., L, L. et al. (2018). A "cation-anion regulation" synergistic anode host for dendrite-free lithium metal batteries. *Sci. Adv.* 4: eaar4410.

19 Huang, X., Cui, S., Chang, J. et al. (2015). A hierarchical tin/carbon composite as an anode for lithium-ion batteries with a long cycle life. *Angew. Chem. Int. Ed.* 54: 1490.

20 Chan, C.K., Peng, H., Liu, G. et al. (2008). High-performance lithium battery anodes using silicon nanowires. *Nat. Nanotechnol.* 3: 31.

21 Chu, K., Song, B.G., Yang, H.-I. et al. (2018). Smart passivation materials with a liquid metal microcapsule as self-healing conductors for sustainable and flexible perovskite solar cells. *Adv. Funct. Mater.* 28: 1800110.

22 Blaiszik, B.J., Kramer, S.L., Grady, M.E. et al. (2012). Autonomic restoration of electrical conductivity. *Adv. Mater.* 24: 398.

23 Wu, Y.P., Huang, L., Huang, X.K. et al. (2017). A room-temperature liquid metal-based self-healing anode for Lithium-Ion batteries with an ultra-long cycle life. *Energy Environ. Sci.* 10: 1854.

24 Xue, L., Gao, H., Zhou, W. et al. (2016). Liquid K–Na alloy anode enables dendrite-free potassium batteries. *Adv. Mater.* 28: 9608.

25 Park, Y.-G., An, S.K., Kim, J.Y., and Park, J.-U. (2019). High-resolution, reconfigurable printing of liquid metals with three-dimensional structures. *Sci. Adv.* 5: eaaw2844.

26 Li, G. (2018). Turning cooler. *Nat. Energy* 3: 714.

27 Chen, Y., Zhou, T., Li, Y. et al. (2018). Robust fabrication of nonstick, noncorrosive, conductive graphene-coated liquid metal droplets for droplet-based, floating electrodes. *Adv. Funct. Mater.* 28: 1706277.

28 Webster, D. (1987). The effect of low melting point impurities on the properties of aluminum-lithium alloys. *Mater. Trans.* 18: 1987.

29 Shpil'rain, E.E., Savchenko, V.A., Mozgovoi, V.A.G., and Skovorod'ko, S.N. (2003). Experimental investigation of the activity of sodium in a liquid sodium–potassium alloy in a wide temperature range. *High Temp.* 41: 23.

30 Wang, J., Wang, H., Xie, J. et al. (2018). Fundamental study on the wetting property of liquid lithium. *Energy Storage Mater.* 14: 345.

31 Lu, X., Li, G., Kim, J.Y. et al. (2014). Liquid-metal electrode to enable ultra-low temperature sodium-beta alumina batteries for renewable energy storage. *Nat. Commun.* 5: 4578.

32 Naidich, J.V. and Chuvashov, J.N. (1983). Wettability and contact interaction of gallium-containing melts with non-metallic solids. *J. Mater. Sci.* 18: 2071.

33 Kramer, R.K., Boley, J.W., Stone, H.A. et al. (2014). Effect of microtextured surface topography on the wetting behavior of eutectic Gallium-Indium alloys. *Langmuir* 30: 533.

34 Ning, X.H., Phadke, S., Chung, B. et al. (2015). Self-healing Li–Bi liquid metal battery for grid-scale energy storage. *J. Power Sources* 275: 370.

35 Jin, Y., Liu, K., Lang, J. et al. (2018). An intermediate temperature garnet-type solid electrolyte-based Molten Lithium battery for grid energy storage. *Nat. Energy* 3: 732.

36 Jin, Y., Liu, K., Lang, J. et al. (2020). High-energy-density solid-electrolyte-based liquid Li–S and Li–Se batteries. *Joule* 4: 262.

37 Weppner, W. and Huggins, R.A. (1978). Thermodynamic properties of the intermetallic systems lithium–antimony and lithium–bismuth. *J. Electrochem. Soc.* 125: 7.

38 Li, G., Lu, X., Kim, J.Y. et al. (2016). Advanced intermediate temperature sodium–nickel chloride batteries with ultra-high energy density. *Nat. Commun.* 7: 10683.

39 Kim, H., Boysen, D.A., Newhouse, J.M. et al. (2013). Liquid metal batteries: past, present, and future. *Chem. Rev.* 113: 2075.

40 Lee, K.T., Jung, Y.S., Kim, T. et al. (2008). Liquid gallium electrode confined in porous carbon matrix as anode for lithium secondary batteries. *Electrochem. Solid-State Lett.* 11: A21.

41 Zhu, J., Wu, Y., Huang, X. et al. (2019). Self-healing liquid metal nanoparticles encapsulated in hollow carbon fibers as a free-standing anode for lithium–ion batteries. *Nano Energy* 62: 883.

42 Liang, W., Hong, L., Yang, H. et al. (2013). Nanovoid formation and annihilation in gallium nanodroplets under lithiation-delithiation cycling. *Nano Lett.* 13: 5212.

43 Deshpande, R.D., Li, J., Cheng, Y.-T., and Verbrugge, M.W. (2011). Liquid metal alloys as self-healing negative electrodes for lithium ion batteries. *J. Electrochem. Soc.* 158: A845.

44 Guo, X., Ding, Y., Xue, L. et al. (2018). A self-healing room-temperature liquid-metal anode for alkali–ion batteries. *Adv. Funct. Mater.* 28: 1804649.

45 Bard, A.J. and Faulkner, L.R. (2001). *Electrochemical Methods: Fundamentals and Applications*, 2e. New York: Wiley.

46 Xue, L.G., Zhou, W.D., Xin, S. et al. (2018). Room-temperature liquid Na–K anode membranes. *Angew. Chem. Int. Ed.* 57: 14184.

47 Qin, L., Yang, W., Lv, W. et al. (2018). Room-temperature liquid metal-based anodes for high-energy Potassium-based electrochemical devices. *Chem. Commun.* 54: 8032.

48 Zhang, L., Xia, X., Zhong, Y. et al. (2018). Exploring self-healing liquid Na–K alloy for dendrite-free electrochemical energy storage. *Adv. Mater.* 30: 1804011.

49 Zhang, L.Y., Peng, S.S., Ding, Y. et al. (2019). A graphite intercalation compound associated with liquid Na–K towards ultra-stable and high-capacity alkali metal anodes. *Energy Environ. Sci.* 12: 1989.

50 Wang, W.J., Zhu, X.H., and Fu, L. (2020). Touch ablation of Lithium dendrites via liquid metal for high-rate and long-lived batteries. *CCS Chem.* 2: 686.

51 Zhu, M., Li, S., Li, B., and Yang, S. (2019). A liquid metal-based self-adaptive sulfur–gallium composite for long-cycling lithium–sulfur batteries. *Nanoscale* 11: 412.

52 Zhang, G., Deng, H., Tao, R. et al. (2020). Constructing a liquid-metal based self-healing artificial solid electrolyte interface layer for Li metal anode protection in Lithium metal battery. *Mater. Lett.* 262: 127194.

53 Li, W., Liang, Z., Lu, Z. et al. (2015). Magnetic field-controlled lithium polysulfide semiliquid battery with ferrofluidic properties. *Nano Lett.* 15: 7394.

54 Subbotin, V.I., Arnol'dov, M.N., Ivanovskii, M.N. et al. (1999). *Lithium*. Moscow: IzdAT.

55 Subbotin, V.I., Arnol'dov, M.N., Kozlov, F.A., and Shimkevich, A.L. (2002). Liquid-metal coolants for nuclear power. *Atom. Energy.* 92: 29.

56 Efanov, A.D. and Kozlov, F.A. (eds.) (2001). *Liquid Metals from the First Thermophysical Stand to Large-Scale Nuclear Power*. Obninsk: GNTS RF–FÉI.

57 Sofu, T. (2015). A review of inherent safety characteristics of metal alloy sodium-cooled fast reactor fuel against postulated accidents. *Nucl. Eng. Technol.* 47: 227.

58 Addison, C. (1984). *The Chemistry of the Liquid Alkali Metals*. Chichester: Wiley.

59 Zrodnikov, A.V., Efanov, A.D., Orlov, Y.I. et al. (2004). Heavy liquid metal coolant–lead–bismuth and lead–technology. *Atom. Energy* 97: 534.

60 Sawada, T., Netchaev, A., Ninokata, H., and Endo, H. (2000). Gallium-cooled liquid metallic-fueled fast reactor. *Prog. Nucl. Energy* 37: 313.

61 Tuckerman, D.B. and Pease, R.F.W. (1981). High-performance heat sinking for VLSI. *IEEE Electron Dev. Lett.* 2: 126.

62 Ge, H. and Liu, J. (2013). Keeping smartphones cool with Gallium phase change material. *J. Heat Transf.* 135: 054503.

63 Walsh, E. and Grimes, R. (2007). Low profile fan and heat sink thermal management solution for portable applications. *Int. J. Therm. Sci.* 46: 1182.

64 Liu, J. and Zhou, Y.X. (2002). Heat sink for chip cooling with low melt-ing point metal or its alloy as working fluid: CN1489020 [P].

65 Deng, Y. and Liu, J. (2010). A liquid metal cooling system for the thermal management of high power leds. *Int. Commun. Heat Mass* 37: 788.

66 Li, T., Lv, Y.-G., Liu, J., and Zhou, Y.-X. (2006). A powerful way of cooling computer chip using liquid metal with low melting point as the cooling fluid. *Forschung. Ingenieurwes.* 70: 243.

67 Deng, Y. and Liu, J. (2010). Design of practical liquid metal cooling device for heat dissipation of high performance CPUs. *J. Electron. Packag.* 132: 031009-1.

68 Wang, L. and Liu, J. (2013). Liquid metal material genome: initiation of a new research track towards discovery of advanced energy materials. *Front. Energy* 7: 317.

69 Deng, Y. and Liu, J. (2010). Hybrid liquid metal–water cooling system for heat dissipation of high power density microdevices. *Heat Mass Transf.* 46: 1327.

70 Sheng, L., Zhang, J., and Liu, J. (2014). Diverse transformations of liquid metals between different morphologies. *Adv. Mater.* 26: 6036.

71 Tan, S., Zhou, Y., Wang, L., and Liu, J. (2015). Electrically driven chip cooling device using hybrid coolants of liquid metal and aqueous solution. *Sci. China. Technol. Sci.* 59: 301.

72 Zhu, J.Y., Tang, S.Y., Khoshmanesh, K., and Ghorbani, K. (2016). An integrated liquid cooling system based on Galinstan liquid metal droplets. *ACS Appl. Mater. Interf.* 8: 2173.

73 Tanga, S.-Y., Khoshmanesha, K., Sivana, V. et al. (2014). Liquid metal enabled pump. *Proc. Natl Acad. Sci. USA* 111: 3304.

74 Choi, S.U.S. (1995). *Enhancing Thermal Conductivity of Fluids with Nanoparticles*, 99–105. San Francisco: International Mechanical Engineering Congress and Exhibition.

75 Buongiorno, J. (2006). Convective transport in nanofluids. *J. Heat Transf.* 128: 240.

76 Eastman, J.A., Choi, S.U.S., Li, S. et al. (2001). Anomalously increased effective thermal conductivities of ethylene glycol-based nanofluids containing copper nanoparticles. *Appl. Phys. Lett.* 78: 718.

77 Ma, K.-Q. and Liu, J. (2007). Nano liquid-metal fluid as ultimate coolant. *Phys. Lett. A* 361: 252.

78 Tang, J., Zhao, X., Li, J. et al. (2017). Liquid metal phagocytosis: intermetallic wetting induced particle internalization. *Adv. Sci.* 4: 1700024.

79 Xiong, M., Gao, Y., and Liu, J. (2014). Fabrication of magnetic nano liquid metal fluid through loading of Ni nanoparticles into gallium or its alloy. *J. Mang. Mang. Mater.* 354: 279.
80 Hong, T.-K., Yang, H.-S., and Choi, C.J. (2005). Study of the enhanced thermal conductivity of Fe nanofluids. *J. Appl. Phys.* 97: 064311.
81 Li, P. and Liu, J. (2011). Harvesting low grade heat to generate electricity with thermosyphon effect of room temperature liquid metal. *Appl. Phys. Lett.* 99: 094106.
82 Adham, M.A., Mohd-Ghazali, N., and Ahmad, R. (2013). Thermal and hydrodynamic analysis of microchannel heat sinks: a review. *Renew. Sust. Energ. Rev.* 21: 614.
83 Zhang, X.-D., Li, X.-P., Zhou, Y.-X. et al. (2019). Vascularized liquid metal cooling for thermal management of Kw high power laser diode array. *Appl. Therm. Eng.* 162: 114212.
84 Allioux, F.M., Merhebi, S., Tang, J. et al. (2019). Catalytic metal foam by chemical melting and sintering of liquid metal nanoparticles. *Adv. Funct. Mater.* 30: 1907879.
85 Liu, S., Li, Z., Wang, C. et al. (2020). Turning main-group element Magnesium into a highly active electrocatalyst for oxygen reduction reaction. *Nat. Commun.* 11: 938.
86 Butler, J. and Meehan, M. (1966). Hydrogen evolution on gallium, indium-gallium and mercury-gallium electrodes. *Trans. Faraday Soc.* 62: 3524.
87 Taccardi, N., Grabau, M., Debuschewitz, J. et al. (2017). Gallium-rich Pd–Ga phases as supported liquid metal catalysts. *Nat. Chem.* 9: 862.
88 Liu, J. (2016). Catalysis by supported single metal atoms. *ACS Catal.* 7: 34.
89 Upham, D.C., Agarwal, V., Khechfe, A. et al. (2017). Catalytic molten metals for the direct conversion of methane to hydrogen and separable carbon. *Science* 358: 917.
90 Esrafilzadeh, D., Zavabeti, A., Jalili, R. et al. (2019). Room temperature CO_2 reduction to solid Carbon species on liquid metals featuring atomically thin ceria interfaces. *Nat. Commun.* 10: 865.

Index

Liquid Metals: Properties, Mechanisms, and Applications, First Edition. Lei Fu and Mengqi Zeng.

d

e

l